333 .79 G77w 2186731
Grathwohl, Manfred.
World energy supply

Grathwohl, World Energy Supply

Manfred Grathwohl

World Energy Supply

Resources · Technologies · Perspectives

Walter de Gruyter · Berlin · New York · 1982

Author

Dipl.-Phys. Dr. rer. nat. Manfred Grathwohl
Scientific Director at the Führungsakademie der Bundeswehr
(German Armed Forces Command and General Staff College), Hamburg

CIP-Kurztitelaufnahme der Deutschen Bibliothek

Grathwohl, Manfred:
World energy supply : resources, technologies, perspectives / Manfred Grathwohl. – Berlin ; New York : de Gruyter, 1982.
ISBN 3-11-008153-9

Library of Congress Cataloging in Publication Data

Grathwohl, Manfred.
World energy supply.

Bibliography: p.
Includes index.
1. Power resources. 2. Energy consumption.
3. Energy development. 4. Energy policy. I. Title.
HD9502.A2G694 1982 333.79 82-9292
ISBN 3-11-008153-9 AACR2

Typesetting and Printing: Sulzberg-Druck GmbH, Sulzberg im Allgäu. – Binding: Lüderitz & Bauer Buchgewerbe GmbH, Berlin. – Cover design: K. Lothar Hildebrand, Berlin. **Printed in Germany**

Preface

Energy is one of the foundations on which our civilization rests, and one for which there is no substitute. Until some time ago, an adequate supply of energy was usually taken for granted. This has changed. In view of the complexity of the tasks associated with it, guaranteeing energy supplies for the future has become one of the greatest problems facing humanity. Many call it the task of the century.

Energy problems are both interdisciplinary and international. Their solution will require strenuous scientific, technical, economical and political efforts. Many aspects of energy provision in the future will require international cooperation. For example, the geographical distribution of individual primary energy sources is very uneven. An immense expenditure will be required to develop new energy technologies and supply systems. The problems of environmental pollution and safety as nuclear installations spread across the world are global problems. The many unsolved problems and disputes should be reason enough to take up the question of energy.

The publication of yet another book on energy requires some justification. Other publications have addressed essentially individual fields within the total area of energy problems, e.g. energy economics, techniques for energy conversion, or environmental aspects. This is clearly needed for analysis of a complex problem. However, the question of the energy supply of the future has often been reduced to the problems of a single source of energy, as can be seen in some publications on nuclear or solar energy. It would appear that this approach has sometimes slighted important relationships.

It is the purpose of this book to treat the problems of world energy supply for the future as a whole, and to deal with interrelationships within this field. The author intends primarily to inform the reader by providing reliable data, but also to illuminate the connections and interactions between different aspects of the total energy problem.

This book is intended for scientists and technicians active in energy-related fields of teaching, research and development, and also for those responsible for insuring the supply of energy. Some interesting things had to be left out for reasons of space, but where a question could only be mentioned briefly, literature references are given for the reader who wishes to study the matter in more depth. The book is also intended for students whose field of study includes the matter of energy supply, and for everyone who wishes to become better informed about it.

I would like to express my thanks to everyone who has supported me in this work, and above all, to my wife.

The manuscript was written in German, my native language, and translated into English by Dr. Mary E. Brewer, in Portland, Oregon, USA. The translation was done rapidly and carefully, so that it could be sent to press without delay, and I wish to express my thanks to her for this.

Hamburg, January 1982 Manfred Grathwohl

Contents

1. Introduction

A secure supply of energy is unquestionably a matter of life and death, not only for an individual country, but also for all of humanity. It is therefore understandable that for years, the question of future energy supply has been a central issue of public debate in all parts of the world. In spite of many efforts, it is very difficult, as experience shows, for a country to develop an appropriate plan for a future-oriented energy supply system. Although efforts are being made all over the world, it appears only partly possible for this civilization to switch from petroleum as the primary energy source to other forms of energy (1, 2).

Thorough studies indicate that the total world energy demand will continue to rise, even if great efforts are made to save energy and to use it in more rational fashion. Also, the fossil fuels petroleum, natural gas and coal will continue to be the basis of the world's energy supply in the foreseeable future. Nevertheless, our civilization must not lose site of the long-term goal of becoming independent of fossil fuels.

After the introduction (1.), the important questions for the future energy supply are treated in the following chapters: Primary energy sources and world economics; The world's energy potential; Energy supply systems; Environmental impact and safety problems; Conclusions. The extent and complexity of the energy problem prevent a complete treatment – if such is possible – within the framework of this book.

The connection between the consumption of primary energy carriers and economic development is unmistakable (2.), and an adequate supply of energy is probably a requirement for the successful handling of global human problems like hunger and poverty, and thus, in the end, for the achievement of world peace. Since the amount and kinds of energy available at various times were undoubtedly milestones in human history, the discussion begins with a short history of energy (2.1). This is followed by a few physical and technical principles (2.2), insofar as they are relevant to an understanding of energy problems: the concepts of work, energy and power (2.21), the first and second laws of thermodynamics (2.22) and a few basic facts of energy conversions (2.23).

The development of primary energy consumption in the world (2.3) has been due partly to the increasing world population (2.31). Also, the apparent (within certain limits) connection between the per capita primary energy consumption and the gross national product (2.32) means that the rising per capita national products, especially in developing countries, will increase the world demand for primary energy. This will come in spite of the possibility of saving energy and using it more rationally, especially in industrialized countries (2.33). (Some techniques for more rational use of energy are discussed in the corresponding sections of chapter 4, e.g. 4.61 and 4.62). It must be emphasized at this point that it is extraordinarily difficult to predict future energy demands, either for individual countries or for the entire world, because of the difficulty of predicting important factors like the availability of primary energy carriers, their prices, the energy policies of individual countries or groups of countries, developments in technology, consumer behavior, and so on. Most predictions of the future energy demand, both for single countries and for the entire world, have been revised downward several times in the last few years. It is the opinion of the author that this trend will continue, because, especially in the industrialized countries, the possibilities for saving energy have not been exhausted.

Because of the fundamental importance of the energy supply to a national economy, energy economics are vitally important in every country (2.4). It is especially important to remember that in the future, the percentage of every gross national product spent on energy will be considerably higher than in the past. After a short discussion of the development of the energy economy (2.41), the predicted investment needs of this branch of the economy will be treated. (The economy of individual energy supply systems or conversion technologies is discussed, if not especially emphasized, in the corresponding sections of chapters 3 and especially 4.)

The energy potential of the world is analysed in the third chapter. The distribution of the available energy is discussed first (3.1), then the world reserves and lifetimes of primary energy carriers (3.2).

The security of the primary energy supply of a country, region or economic-political block of countries depends not only on the global reserves of a given primary energy carrier, but also very critically on the geographical distribution of that energy carrier. The fact that there are adequate reserves, in the world as a whole, does not guarantee that a given country will always have access to them. The security of the energy supply depends critically on the geographical distribution of primary energy carriers, and this distribution may well determine the choice of carrier or carriers on which a country builds its economy. In addition to the rather uneven distribution of individual primary energy carriers, there is a discrepancy in many countries and regions between the amount of production and the size of the reserves, so that the lifetimes of the reserves vary; also, there is a discrepancy between production and consumption, resulting in various types of political dependence where the consumption exceeds production. Finally, in addition to the geographical distribution of production and consumption centers, the particulars of the technology of prospecting, tapping, production, storage, refining and transporting an energy carrier play important parts. Therefore, these aspects of the use of coal (3.31), petroleum (3.32), natural gas (3.33) and oil shales, oil sands and heavy oils (3.34) are discussed in detail. Because it might be possible to use underground nuclear explosions to release hydrocarbons from oil shales and oil sands, some of the physical and political aspects of this prospect are discussed here.

The geographic distribution and centers of production and consumption of nuclear fuels for fission reactors are discussed in 3.35.

Although there are still difficult physical and technical problems to solve, there is growing confidence that controlled nuclear fusion will eventually be possible. Everything we know about it suggests that this will make possible a practically unlimited supply of energy (3.36). Therefore the geographical distribution of fuels for nuclear fusion and, as far as presently possible, the probable costs of this fuel, are discussed.

In contrast to controlled nuclear fusion, the technology for utilization of solar energy exists. The enormous potential for solar energy is hardly being tapped in significant amounts in any country. Because there are serious problems associated with other sources of energy, and because there are certain advantages to solar energy with respect to the security of national (regional) supplies, the resources required for its utilization and the relatively low environmental impact (compared to other energy sources), solar energy unquestionably should be given more serious consideration internationally. Therefore the basic considerations determining applicability of solar energy, even in temperate latitudes, are discussed at length (3.37).

The tidal energy potential is very small, compared to other sources of energy (3.38). Although geothermal energy represents a large potential (3.39), there are many difficulties in its utilization, so that on a global scale, the contribution of this source of energy will probably remain small, even in the future.

The 4th chapter deals with energy supply systems. The role of secondary energy carriers is explained (4.1), then various methods of producing secondary energy carriers are discussed. At the present level of technology, it seems that of the "new" sources of energy, nuclear fission and solar energy have the greatest chances of making significant contributions to the world energy supply in future. Therefore the technologies appropriate to these sources are discussed first.

The production of secondary energy from nuclear fuels (4.2) is subdivided, corresponding to the two basic methods, into nuclear fission (4.21) (the principles of reactor physics applicable to the most important reactor types are discussed) and nuclear fusion (4.22), although this has not yet been realized.

There are various reasons for a greater utilization of solar energy. For the purpose of comparing the chances of different methods of converting solar energy to secondary energy (4.3), they have been classified as direct methods (4.31) (solar thermal conversion, solar thermal power plants, photoelectric conversion) or indirect methods (4.32) (water power, wave and wind energy, ocean heat and currents, utilization of stored solar heat with heat pumps, photochemical conversion).

Because tidal and geothermal energy can certainly be of local importance, concrete examples are used to illustrate the production of secondary energy from the tides (4.4) or geothermal sources (4.5).

Secondary energy carriers will probably become even more important in future as efforts are made to use primary energy more efficiently. Therefore promising secondary energy carriers (4.6) are discussed. Electricity (4.61) can be expected to continue to play an important role in future. District heating (4.62), especially in population centers, may assume an important function in a more rational utilization of primary energy, as it can supply the demand for low temperature heat with heat otherwise wasted in the production of electricity. Refined petroleum products will still dominate, especially in transportation, for a long time to come (4.63). Because of the existence of enormous coal reserves, refined coal products (4.64) have potential for the future. The system of long distance energy (4.65) could become important, especially for the transport of energy over longer distances. In particular, hydrogen (4.66) could, in the long run, find many applications. Because of the heavy dependence of the transportation sector on petroleum products, possible alternative drive systems for mobile consumers (4.67) deserve serious attention.

The fifth chapter deals with the environmental impact of the use of energy carriers. Possible safety problems are included in the discussion. After an introduction to the problems which are created by the release of energy (5.1), the direct anthropogenic heat release caused by energy conversion processes is treated in 5.2. Then the environmental problems arising specifically from the use of fossil fuels (5.3) are discussed: 5.31 deals with the emission of pollutants, and 5.32 with the global problem of carbon dioxide release. Everything we know suggests that there is a danger of climatic changes (5.4) if the release of energy is thoughtlessly increased. To obtain a point of reference, we first consider climatic variation in the past (5.41), and thereafter, possible effects of carbon dioxide (5.42) and other anthropogenic emissions (5.43) on the climate are discussed. The environmental impacts of solar (5.5), tidal (5.6) and geothermal energy (5.7) are considered.

The environmental impact and safety problems specific to nuclear fission are treated in 5.8. After the introduction (5.81), the questions related to the fuel cycle are discussed (5.82). A special problem of political safety arises from the fact that nuclear fuels like uranium and plutonium can be used not only for producing energy, but also for producing nuclear weapons. Therefore problems of the peaceful utilization of nuclear energy are related to the problem of non-proliferation of nuclear weapons (5.83). Questions related to the safety of nuclear installations under certain conditions are discussed in 5.84. The specific environmental impact and safety problems associated with nuclear fusion (5.9) make up the conclusion of this chapter.

The complexity of many of the questions discussed, and the unsolved or not yet satisfactorily solved problems make it difficult to draw unequivocal conclusions. In spite of the problems, some conclusions are drawn, especially at the end. These may be worthy of more careful consideration (6). I hope that the book will provide worthwhile ideas.

2. Primary energy sources and world economics

2.1 *A short history of energy*

Every form of plant and animal life depends on some sort of energy supply, be it radiation, conduction, convection, or chemically bound energy in the form of nutrients.

While amount and kinds of energy available to mankind have been hallmarks in its development, the history of energy has been accorded little importance. In the approximately 1 000 000 years of his history, man had to rely on muscle power exclusively for the first 600 000 years. Fire was discovered by man about 400 000 years ago, and was used even in prehistoric times to improve living conditions. He heated his dwellings, whenever he needed it, and with its light he was no longer limited to daylight. The use of fire to prepare food also widened his nutritional repertoire.

In the third millenium B.C. the utilization of fire was greatly expanded. Ores were melted to extract metals, and man entered a new phase of development. The stone age gave way to the bronze age; about 700 B.C., this in turn yielded to the iron age.

Up to the bronze age and a few millenia thereafter, mechanical energy was supplied by human and animal muscle power. The Romans, particularly, were known to subject their slaves to hard physical labor. Domestic animals, originally kept as a meat supply, were already used by the Sumerians as draft animals, and the inhabitants of the Asian steppe introduced horse-back riding about the 8th century B.C. One of the prerequisites for the high civilizations which have arisen since the year 3000 B.C. was improved working of the soil, so that a small segment of the population did not have to toil for its daily food supply and thus was freed for other cultural pursuits.

Successive development of energy technology led to an increase in the kinds of energy utilized and led eventually to the first industrial revolution. The water wheel was known to Byzantium as early as 200 B.C. The first purely mechanical energy source was used in 200 A.C. to drive a grist mill in Arles, France, which could produce nearly 28 tons of flour in 24 hours. Water power was much used in the middle ages.

The wind mill is reported to have been known to the Arabs in the 9th century. With the crusades they were introduced into Europe in the 12th century. Wind power had been used in Europe until then only to propel sailing ships. Now wind mills became useful not only in the grinding of grain, but in the pumping of water and in mining technology. The disadvantage of the intermittent operation of wind mills at unfavorable changes in the wind was solved in part by Leonardo da Vinci, who designed a wind mill with a rotatable roof.

Light technology remained unchanged over a long time. Man burned animal and plant fats in lamps equipped with wicks, with a light yield of the order of magnitude of 0.1 lm/W. Torches and resinous wooden kindling gave a somewhat greater light yield but, owing to obnoxious smells, had no significant advantages.

The need for mechanical energy grew continously toward the end of the middle ages, in part due to increasing population. People also began to search for energy sources independent of place and time, in order to avoid the disadvantages of wind and water power. Many age-old experiments were re-examined in the beginning of the 17th century. Heron of Alexandria is said to have developed power from hot air and steam as early as 100 B.C. The decisive experiments leading to realization of the steam engine came from England. There was a need for a new power source to pump water out of mine shafts. The first pistonless engines of Thomas Savery (1698) were replaced by the first useful steam engines of James Watt (1736–1819). His earliest model could produce 7.5 kW.

The steam engine, in fact, as the first driving engine, started the industrial revolution. There were new demands for manufacturing techniques, economics, industry and traffic, and with them the life style of man underwent tremendous changes. Within two centuries, a new environment based on technical progress has been created.

In addition to steam engines, the 19th century saw the development of hot-air engines and internal combustion engines, which, thanks to the four-stroke-cycle invented by Nikolaus August Otto, were far superior to hot-air engines.

Turbine engineering can be viewed as a logical extension of water wheels and wind mills. The first water turbine dates back to the beginning 19th century, and at the end of this period the development of steam turbines commenced.

The progressive industrialization in the 18th and 19th centuries brought with it a growing demand for lighting. The development of light sources had a decisive impact on energy technology and economics. Oil and gas were used first as light sources and only later as sources of mechanical energy. The technology for high current and high electric transmission was also a result of spread of electric lighting. The milestones in the development were Hans Christian Oersted's (1820) discovery of electromagnetism, the discovery of electromagnetic induction by Michael Faraday (1831), and the demonstration of the dynamoelectric principle by Werner von Siemens (1866). Thomas Alva Edison introduced the first practical incandescent light bulb in 1879 and built the first public electric utility in the world in New York in 1882. This plant produced 500 kW at 100 V DC. Around 1900 Germany already had 94 power plants for illumination and electric motors in operation, with a total output of about 160 MW (1).

With the discovery of nuclear fission by Otto Hahn and Fritz Straßmann in December 1938, a completely new form of energy became available to mankind. In the fission of nuclei of heavy elements (thorium, uranium, plutonium, etc.) 1 kg of substance yields about 10^6 times as much energy as is liberated in the combustion of fossil fuel. In the atomic bomb, this liberation of energy is uncontrolled and takes place within a fraction of a millionth of a second, while the same amount of energy is controlled and liberated in an atomic reactor over a longer period of time. The first nuclear reactor (CP 1) was started up by Enrico Fermi on 2 December 1942 in Chicago.

Electricity can be obtained from all primary energy sources. This form of secondary energy is easily transported and handled, and is not liable to cause environmental damage. All electrical generating plants work on the same principle: heated steam drives a turbine which in turn drives the generator, which provides electrical power. In a conventional steam generating plant the required heat is produced by burning fossil fuels; in the nuclear power plant the heat comes from the energy liberated in the reactor core.

In a modern pressurized water reactor of the Biblis type (el. power ca. 1200 MW) about a ton per year of ^{235}U is converted into heat. In comparison, a coal-fired plant burns $2.5 \cdot 10^6$ tons of anthracite yearly, or 7000 tons per day, requiring a daily coal supply of 400 railroad cars.

2.2 *Physical and technical foundations*

2.21 Work, energy, power

The energy of a system is defined as its ability to perform work. A system can only acquire this ability to work, if work is done on it.

As an example, a constant force F operates on a body and moves it a distance l; the work done on the body is W:

$$W = F \cdot l \qquad (1)$$

(This holds if force and displacement have the same direction and the force is constant[1]). Hence a supply of work ability, that is to say energy, is stored in the body. If

---

[1] Work is defined as the scalar product of a force vector and a displacement vector. If the direction of the path and the amount and direction of force along it vary, then for individual path elements $d\vec{l}$ is : $dW = \vec{F}\,d\vec{l}$. The work expended over a finite path is the integral over all path elements : $W = \int \vec{F}\,d\vec{l}$. In a conservative force field or a conservative system, the work integral from location 1 to location 2 is independent of the path between the two.

one refers to the energy of a body – or more generally to the energy of a system – one means a certain amount of energy, or the energy content of the system. Energy is the ability to perform work. Thus energy and work are measured in the same unit, the Joule, where 1 J = 1 Ws.

The energy of a body in a certain state is equivalent to the amount of work required to place it into this state. Conversely, the body is able, by reversing the state, to transfer work to other bodies. Let us assume that an athlete lifts a 100 kg weight to a height of 2 meters. He has expended a certain amount of lifting energy.

If he lets the weight drop freely, it can now perform as much deformation work as the athlete previously expended in lifting it. This process demonstrates the conservation of energy. If a system is brought into a state of higher potential energy, work has to be added from the outside. The potential energy is converted to kinetic energy when the athlete lets go of the weight. Kinetic energy is converted to deformation work when the weight meets the ground.

Energy occurs in many forms. Some examples are: mechanical energy, heat energy, electrical energy, radiant energy, and nuclear binding energy. In other words, storing of work ability is not only possible in the form of mechanical energy (potential and kinetic energy), but also as heat energy (heat content of a body), in the form of distribution of electrical charges (electrical energy), in the form of a certain chemical state of a body (chemical energy), etc.

To denote that a certain amount of energy is given off or taken on by a system, one needs the concept of power. Consider two immersion heaters: one brings 1 liter of water to a boil in 5 minutes, the other needs 10 minutes. The amount of energy expended was the same in both units but the first was more powerful than the second. Performance, or power P is defined as the ratio of work W over time t:

$$P = \frac{W}{t} \tag{2}$$

provided work furnished in the same time interval is constant[1]. The unit of power is the Watt, W.

A person can exert the power of 1 kW for a short period of time. In order to do this, he must lift 102 kg to a height of 1 m within 1 second. Over a longer period, the human power output is more of the order of 100 W.

2.22 First and second laws of thermodynamics.

For the supply of available work, i.e. the energy of a body (system) – restricted at present to purely mechanical processes –, the total available energy, that is the sum

[1]) If the power furnished per time interval is not constant, the power is defined by the differential expression $P = dW/dt$. In the finite time t, the work furnished is $W = \int_0^t P dt$.

of potential and kinetic energy, in a closed system is constant. (Law of the conservation of energy).

If a closed mechanical system is to provide work without changing its energy stores, this can only be effected by adding energy from the outside. Otherwise the work furnished must result in a decrease of the energy of the system. (Law of the impossibility of a perpetuum mobile of the first kind: it is impossible to construct a machine which continously puts out work without adding energy from the outside).

Physical systems in which no other energy conversion occurs, other than changing potential into kinetic energy or vice versa, are called conservative systems; others in which conversion to other energy forms occur are called dissipative systems.

J. R. Mayer and J. P. Joule recognized that the law of the conservation of energy is not limited to mechanical processes but holds also for processes of converting mechanical energy into heat and vice versa. They recognized heat as an energy form and that it had to be included in the energy conservation law.

In the year 1842, J. P. Joule calculated the *mechanical heat equivalent* from the specific heat of gases, i.e. the quantitative relationship between the unit of calorie and the unit Joule. These two are one and the same thing, namely energy units. Joule arrived experimentally at $1\ J = 0.238\,845$ cal or $1\ kWh = 0.8598 \cdot 10^6$ cal. Mechanical work and heat are mutually interchangeable forms of energy.

If heat is included in the law of conservation of energy, a more comprehensive relationship is derived. The sum of the heat added to the system from the outside ΔQ and the work from the outside ΔW is equal to the increase in energy content:

$$\Delta U = \Delta Q + \Delta W \tag{3}$$

This is the first law of thermodynamics. (If the system is a homogeneous body, then $\Delta U = c_v m \Delta T$, where c_v is the specific heat capacity at constant volume, m the mass of the body and ΔT the increase in temperature).

Independently of J. R. Mayer, H. von Helmholtz formulated in 1846 the law of the conservation of energy, or in its most general form, the energy principle. Both recognized that in all fields of physics one can define units of the various types of energy, be it electrical, magnetic, chemical, or whatever.

The energy principle in its most general formulation can be stated as follows: In a closed system in which any processes whatever (mechanical, thermal, electrical, chemical) take place, the total energy remains unchanged. (A closed system is considered as one where energy is neither added nor removed). No energy form can be created out of nothing nor can any disappear into nothing. Energy can only be converted from one form to another. The total energy of a closed system never changes.

The energy principle can also be formulated as follows: A perpetuum mobile of the first kind cannot exist. This does not mean, literally, a contraption which remains continuously in motion without work added from the outside. Such a contraption would in fact be feasible if all friction were excluded, and would not be in contradiction with the energy principle. The motion of the planets around the sun is a very close approximation of such a system. However, it is not technically possible to build one on earth. A perpetuum mobile of the first kind is better described as an apparatus which furnishes work continuously without work added from the outside, and thus creates energy out of nothing. Such a contrivance is forbidden according to the energy principle.

The law of the conservation of energy cannot make any statement about the direction in which a natural process can occur. For example, it would be in agreement with the energy principle if a tile which previously had dropped from a roof were to jump back onto the roof by converting the heat content it had gained on impact into the kinetic energy required for the jump. Such a process has never been observed in nature. This means then, that not all processes do occur in nature which are possible on the basis of the law of the conservation of energy. There must be an additional reason for nature's preference of one process over the other. In another – more general – formulation: For a system which can exist in various states, there must be a reason why one state is the starting state only and the other only the final state, but not the reverse.

This problem, namely the direction in which a process proceeds from the starting state, is addressed by the second law of thermodynamics which, like the first law, is a formalized statement of experience. The second law also answers the question (which is essentially the same problem) why there are some states from which no process begins, i.e. equilibrium states.

A further example for a natural process which only proceeds in one direction is the following: A space filled with a gas is connected with an adjacent space which is a vacuum. Gas flows from the former space to the latter until the pressure is equalized. The gas never reverts to the one space and reestablishes vacuum in the other.

Likewise, if two unmixed gases are placed in a single vessel, they will interdiffuse, but never spontaneously separate.

A cube of sugar placed into a glass filled with water spontaneously dissolves – without any aid on our part – i.e. sugar molecules distribute themselves over the whole liquid volume. This process also operates only in one direction. It has never been observed that sugar molecules equipartitioned over a volume of water recombine spontaneously to form the sugar cube.

Such processes are termed irreversible processes. Irreversible processes cannot be reversed without making any lasting changes in the surroundings. All macroscopic processes in nature are irreversible processes. Or, strictly speaking, reversible processes do not occur in nature and in reality can only be approximated.

An irreversible process can of course be reversed at the expense of outside energy; yet this causes some irreversible change in the environment.

From the state parameters of a body, R. Clausius defined a new parameter, the entropy function S[1)], which is directly related to the one-sidedness of the processes occuring in nature.

An important special case is the one where the observed system is a closed one. For this case, the second law of thermodynamics can be formulated: The entropy of a closed system of bodies which mutually interact can only increase. In a closed system the entropy of the final state S (2) is always larger than the entropy of the starting state S (1):

$$S(2) > S(1) \tag{4}$$

Processes in a closed system are always accompanied by increases in entropy. This principle defines the direction of all processes. If the system is not closed, one can always make it a closed system by including all the bodies from its environment with which it is exchanging heat.

It can be shown that the existence of a periodically working machine which converts heat completely into work would violate the second law (see 2.23; law of the impossibility of a perpetuum mobile of the second kind). A perpetuum mobile of the second kind would thus be a machine which would satisfy the first but not the second law of thermodynamics.

2.23 Energy conversions

Any physical system in which processes consume and release energy is an energy exchanger. If the function of the system is to convert energy from one form to another, i.e. to deliver it in a certain form, it is called a machine. The degree of efficiency is the ratio of the energy furnished by the machine over the energy added in a different form to the machine. For example the degree of efficiency of an electrical generator η_G, is the ratio of the electrical energy W_e, supplied by the generator, over the rotational energy W_r, consumed by the generator:

$$\eta_G = \frac{W_e}{W_r} \tag{5}$$

The efficiency amounts to ca. 0.99.

The degree of efficiency of the total system, the generating plant η_K, is the ratio of the electrical energy W_e, supplied by the generator over the chemical energy W_c, of the consumed fuel:

$$\eta_K = \frac{W_e}{W_c} \tag{6}$$

---

[1] To each state of a system one can ascribe a function S, the entropy of the state. Its full differential dS in a reversible change is $dS = dQ/T$, where dQ is the amount of heat taken up at temperature T.

For fossil fuel-fired power generating plants this ratio is ca. 0.4.

Aside from the indicated energy forms which define the degree of efficiency, other energy forms may occur in the machine. For example, in a steam power plant, there is the chemical energy from the fuel, heat energy in the form of steam, rotational energy in the moving parts, and electrical energy. For all the steps in energy exchanges, partial degrees of efficiency are ascertained, and the total degree of efficiency of the power plant is the product of them.

Some further examples of degree of efficiency should be mentioned: For the Otto motor it is only 0.25. The efficiency of an incandescent light bulb is even smaller. It converts only 5% of electrical energy to visible light; a fluorescent bulb has an efficiency of ca. 20%. Degrees of efficiency can also be defined by performance data. If one divides numerator and denominator by the same interval of time, the degree of efficiency is not changed, but it is now expressed as the ratio of two performance parameters.

Historically, machines converting heat to mechanical energy have been termed heat engines. Up to the 18th century man had to rely on his own muscle power or on that of his domestic animals. James Watt built the first piston steam engine which converted heat to mechanical energy, and thus made available the energy contained in coal.

For the conversion of heat to mechanical energy, the only useful systems are those which cyclically repeat some sequence of processes, i.e. work periodically in a cyclic process.

Work can be converted completely into heat by friction, but heat cannot be completely converted into work (second law of thermodynamics). S. Carnot has shown that only a fraction of the available heat can be converted into work, even in an idealized cycle in which an ideal gas is supposed to be doing the work, and all heat losses by radiation, conduction or convection are neglected. The thermodynamic degree of efficiency of the so-called Carnot cycle is:

$$\eta = 1 \frac{T_2}{T_1} \qquad (7)$$

where T_1 and T_2, respectively are the temperatures at which heat is added and removed, respectively. (These temperatures must always be expressed in the Kelvin scale, which starts at absolute zero.)

Operating steam engines do not fulfil these conditions. For example, condensation occurs and the realizable degree of efficiency is thus much smaller than expected for the Carnot cycle. (A more correct comparative picture would be provided by the Clausius-Rankine process which will not be discussed here.) The fact remains that the higher T_1 of the steam, and the lower T_2 of the coolant water, the greater the efficiency of the process. Regarding T_2, it is not practical to go below the ambient temperature, and therefore one strives to raise efficiency by raising the steam temperature T_1. The upper limit on T_1 is set by the properties of the construction materials; it cannot be so high that loss of material strength or stability occurs.

If, for example, one chooses a starting temperature of the steam, $T_1 = 673$ K and assumes for $T_2 = 323$ K, one can calculate a degree of efficiency from equation (7) of $\eta \approx 52\%$. This means that even under ideal conditions the Carnot cycle engine can only convert heat to mechanical energy at 52%, and 48% is lost to the cooling water. This degree of efficiency is not reached under practical conditions.

We see that the heat machine cannot convert the energy of the fuel completely into a new energy form like mechanical energy. A certain portion of the mechanical energy is given off as heat energy and is termed conversion losses.

The degree of efficiency of a heat engine is then a measure of the usefulness of the energy conversion and is also a measure of the utility of the engine itself. It holds for operational conditions such as working at full load or partial load. The useful yield of a machine must be taken by observation over a longer period of time, with changing work load parameters. Here, also idling of the machine, down-time, influence of operator, and maintenance are taken into consideration.

The useful yield g, of a machine is the ratio of usable energy W_n to the total energy supplied W_z, or:

$$g = \frac{W_n}{W_z} \tag{8}$$

The difference between useful yield and degree of efficiency can be illustrated with the example of the diesel engine of a truck. Under normal conditions it has an efficiency of 40%. However, under practical conditions: slower traffic in towns, changing loads, different drivers etc., the useful yield is only ca. 25%. In contrast, the useful yield of a (stationary) oil heating plant is about 62%.

What we have discussed on hand of the steam engine example holds in principle for all energy conversion processes. Ultimately, every form of energy is converted to the heat of the environment, and every energy conversion goes in the direction of levelling all temperature differentials. For this reason we shall always need "high grade" sources of energy. This also explains why one refers to energy "generation" and energy "consumption" even though energy can never be generated or consumed, but only be converted or transformed.

Natural energy sources, such as coal, petroleum, or natural gas (fossil fuels), are called primary energy sources. Generally they cannot be used directly in heat engines but often need to be modified for use, e.g. by coke plants, or oil refineries. They then can be converted into secondary energy sources such as e.g. electrical energy, heating oil, gasoline, etc., which are much more easily transported and better suited for the end user. (Nobody in his right mind would use coal directly as a source of illumination, but rather "refined coal" in the form of electricity). Conversion always entails losses at any conversion step (compare Fig. 3-1).

The final energy form is the one furnished to the end-user. It embraces most secondary energy forms, but also such primary ones as natural gas or anthracite. Conversion losses also occur at the point of end use, in the form of heat, light, sound, etc. (compare Fig. 3-1).

2.3 The development of primary energy consumption in the world

2.31 Primary energy consumption and world population

Energy consumption in the world was very small over the span of many centuries. The reason was the low population density. About 6000 B.C. there were about $10 \cdot 10^6$ human beings on the globe, and at the time of the birth of Christ this number had grown to ca. $250 \cdot 10^6$. In the next 1650 years the world population doubled to a figure of ca. $500 \cdot 10^6$. In 1930 there were already $2 \cdot 10^9$ human beings, in 1970, $3.61 \cdot 10^9$, and in 1975, $3.97 \cdot 10^9$. The growth rate in 1650 was 0.3%, in contrast to 1950–1965 ca. 1.8%, and from 1965–1975 approximately 1.9%.

Assuming exponential growth, the relationship between doubling times t_d (in years) and growth rate p (in per cent per annum) is:

$$\begin{aligned} t_d\, p &= 100 \ln 2 \\ t_d\, p &\approx 70 \end{aligned} \tag{9}$$

A growth rate of $p = 1.9\%$ per year corresponds to a doubling time of $t_d = 37$ years.

Fig. 2-1 shows the growth of the world population. According to calculations made by the United Nations, the world population in the year 2000 will be $6.1 \cdot 10^9$ people (middle estimate). The highest and lowest estimates are $7.0 \cdot 10^9$ and $5.4 \cdot 10^9$, respectively. In the year 2075, this population avalanche is expected come to rest at $12.3 \cdot 10^9$ people (medium estimate); the highest and lowest estimates being $16 \cdot 10^9$ and $9.8 \cdot 10^9$ people, respectively.

The population growth curve may eventually prove to have a typical S – shape with an early slow growth rate increasing exponentially and leading to a saturation limit. The UN experts predict a turning point of the curve about 1980, when the growth rate will be at the maximum, then decrease again.

The proportion of people living in developing countries is still on the increase. In 1950, the number of the inhabitants of the developing countries was $1.6 \cdot 10^9$ people out of a total of $2.5 \cdot 10^9$ people or 65%. Around the year 2000, the figures are expected to be $5 \cdot 10^9$ people out of a total of $6.1 \cdot 10^9$ or 81%, and in 2075, the percentage of the population in developing countries should reach 84%.

In industrialized countries population growth has come to a halt or even reversed. Birth rates in the German Federal Republic in the past 10 years have markedly de-

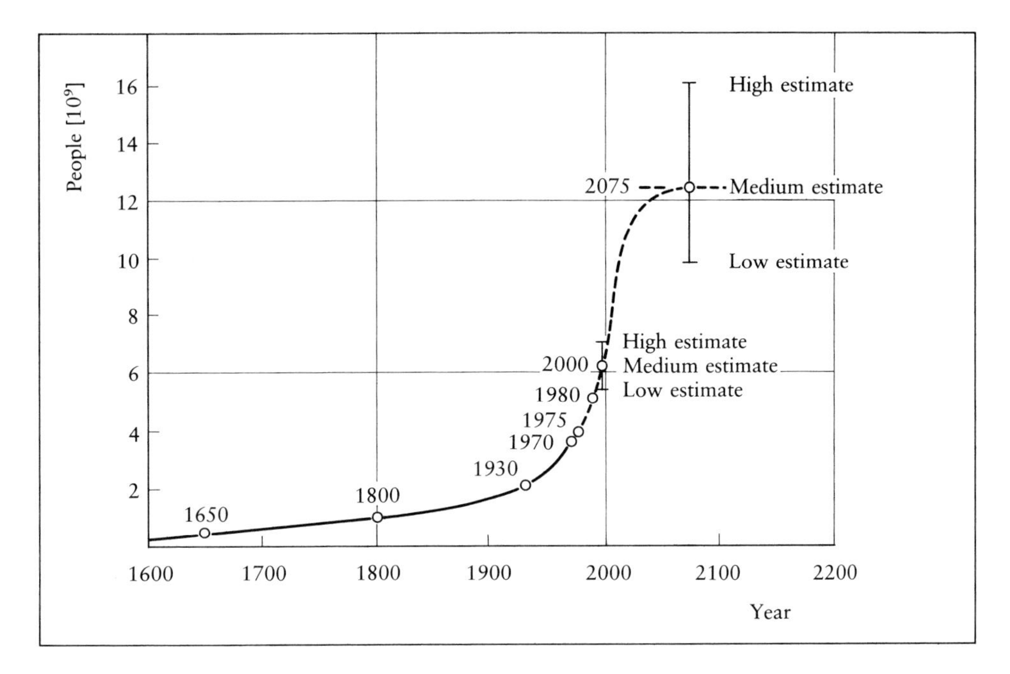

Fig. 2-1: Growth of the world population*
*The data are taken from various UN-publications

creased. Data from the German Statistical agency in Wiesbaden show 18.2 births per 1000 inhabitants in 1964, but only 10.2 per 1000 for the year 1973. Even in very populated countries like the People's Republic of China, the birth rate from 1970 to 1975 dropped from 1.85% to 1.18%. The reasons for this are manifold: different social conditions, birth control, welfare considerations and improved cultural opportunities.

During the world population conference of the UN in Bucharest in 1974, the USSR, China and nearly all Eastern bloc countries, as well as some not under the communistic umbrella, took the position that a population problem did not exist. The Chinese delegation argued persuasively, that the total population had been supplied adequately with food for some time, and that there were no food shortage problems. According to their statements, the population of the People's Republic of China had grown 60% since 1950, from $500 \cdot 10^6$ to $800 \cdot 10^6$ people, while in the same span of time grain production had more than doubled. The population had grown in this time span at 2% per year, while grain production had grown at a 4% annual rate.

Similar views have been developed by H. Kahn (5). Fig. 2-2 shows three different growth rates of the population A, B, and C, for the period from 1950 to 2010 under a) and for the period from 1776 to 2176 under b). The steep rise and the expected

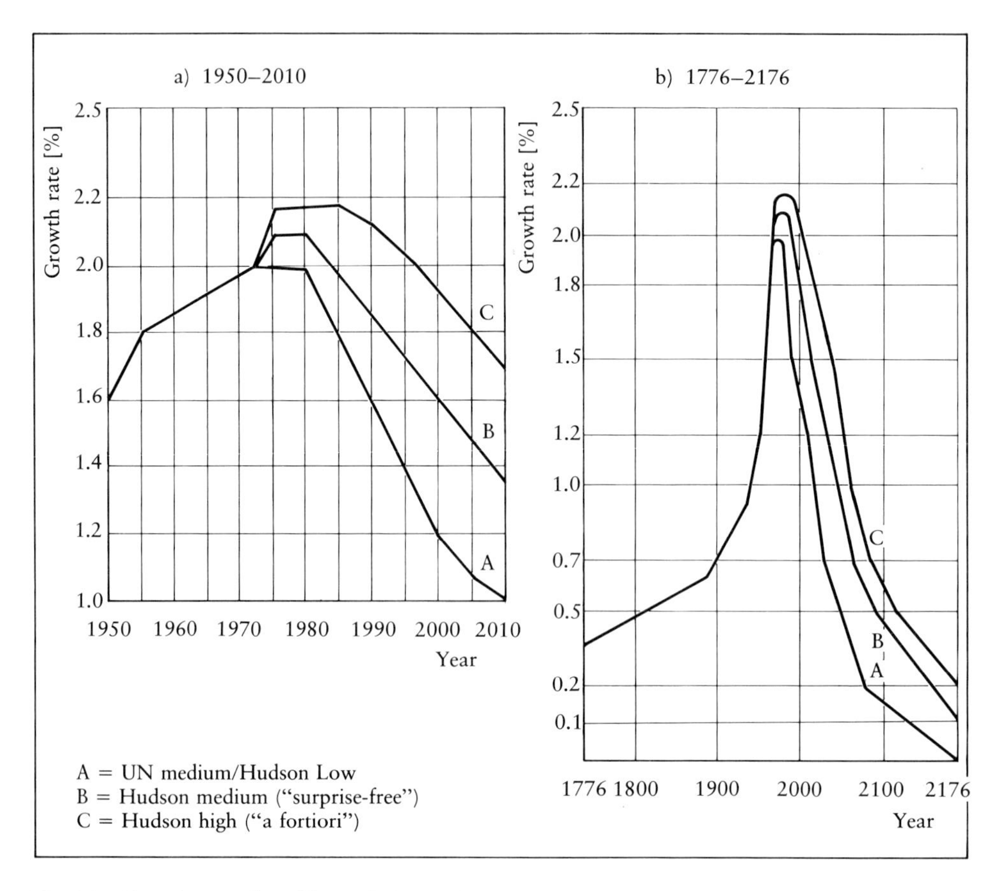

Fig. 2-2: Growth rate of world population

Source: H. Kahn et al., The next 200 years; A Scenario for America and the World, New York: William Morrow and Company 1976.

steep decrease is even more marked in the b) graph. "The slowest curve A is based on the United Nations Population Bureau's "medium" variant projection; the second curve B assumes higher rates and is used in many of our "surprise-free" projections[1], and the third curve C, with still higher rates is used in our "a fortiori" projections[2]. H. Kahn expects that the growth rate will slow up and that world population will reach a final figure of $15 \cdot 10^9$ people (compare Table 2-13).

---

[1] The "surprise-free" projection is one that assumes innovation and progress that would not be surprising in the light of past trends and current developments – that is, based on extrapolations of current or emerging tendencies and expectations.

[2] The "a fortiori" projection used here is one that is mostly based on current – or near current – technology and avoids the assumption of great future improvements, like those that have characterized past historical experience.

According to 1980 reports by the World Bank in Washington, D.C., the world population in the middle of 1978 was $4247 \cdot 10^6$ people, and the expected number in the year 2000 would be $6029 \cdot 10^6$ people. A stationary state population of $9771 \cdot 10^6$ people is predicted by the World Bank. The predicted development of the greatest population density countries in the world is shown in Table 2-1. Stationary state populations are expected for China in 2065, for India in 2150, for the USSR in 2095, and for USA in 2030.

Let us consider the world primary energy consumption from 1875 to 1975. During these 100 years the population increased from 1.2 to $3.97 \cdot 10^9$ people, i.e. it tripled. The world primary energy consumption during this time span grew from 250

Table 2-1: Primary energy consumption and gross national product per capita for the most populated countries in the world

Country[1]	Population mid – 1978 [10^6 people]	Projected population in 2000 [10^6 people]	Hypothetical population stationary state [10^6 people]	Primary energy cosumption per capita [kgce][2]	Gross national product per capita [US$]
China, P. R.	952	1251	1555	805	450
India	644	974	1645	176	180
USSR	261	310	360	5500	3700
USA	222	252	273	11374	9590
Indonesia	136	204	350	278	360
Brazil	120	201	345	794	1570
Japan	115	131	134	3825	7280
Bangladesch	85	143	314	43	90
Nigeria	81	153	425	106	560
Pakistan	77	139	332	172	230
Mexico	65	116	205	1384	1290
Germany, F. R.	61	61	61	6015	9580
Italy	57	61	63	3230	3850
United Kingdom	56	58	59	5212	5030
France	53	58	61	4368	8260
Vietnam	52	87	149	125	170
Phillipines	46	75	126	339	510
Thailand	45	68	103	327	490
Turkey	43	65	100	793	1200
Egypt	40	62	101	463	390
Total world	4247	6029	9771	2190	1300[3]

[1] Sequence of countries according to population data for mid-1978.
[2] kilograms of coal equivalent.
[3] This figure was for 1975.

Source: World Development Report 1980, World Bank, Washington D. C. 1980.

Table 2-2: Primary energy consumption in the world from 1950 to 1980 and the predicted primary energy demand for 1985 and for 2000 (in 10^9 tce[1])

Year	1950	1955	1960	1965	1970	1973	1975	1979	1980		1985		2000		2000	
from	(7)	(7)	(7)	(7)	(7)	(7)	(7)	(8)	(8)		(9)		(9)		(10)	
Energy Source										%		%		%		%
Petroleum	0.6	0.9	1.3	1.9	2.9	3.6	3.5	4.5	4.3	44	5.2	43	6.1	34	6.5	34
Natural gas	0.3	0.4	0.6	0.9	1.3	1.5	1.5	1.8	1.8	18	2.1	17	3.0	17	2.5	13
Coal	1.5	1.8	2.2	2.3	2.4	2.5	2.6	2.8	2.9	30	3.5	28	5.0	28	4.4	23
Hydropower and others[2]	0.1	0.1	0.1	0.1	0.1	0.1	0.2	0.6	0.6	6	0.7	6	1.4	8	2.7[2]	14
Nuclear energy	–	–	–	–	–	0.1	0.1	0.2	0.2	2	0.7	6	2.4	13	3.0	16
Total	2.5	3.2	4.2	5.2	6.7	7.8	7.9	9.9	9.8	100	12.2	100	17.9[3]	100	19.1	100

[1] tons of coal equivalent, 1 tce = $29.3 \cdot 10^9$ J.
[2] "other" sources, such as solar energy, wood, etc., are assumed as $1.6 \cdot 10^9$ tce for 2000.
[3] An estimate made by the Exxon Corporation, in 1981, predicted $15 \cdot 10^9$ tce for 2000; the percentage distribution of primary energy sources is similar.

Sources: United Nations Statistical Yearbook 1959–1977, New York 1960–1978 (7).
BP statistical review of the world oil industry 1980, London 1981 (8).
Energy Program of the Government of the Federal Republic of Germany, Ministry of Economics, Bonn, December 1977 (9).
World Energy Demand, Full Report to the Conservation Commission of the World Energy Conference, Guildford (UK), New York : IPC Science and Technology Press, 1978 (10).

to $7877 \cdot 10^6$ tce/a or increased by a factor of 32 (4). The world primary energy consumption is shown for some selected years in Table 2-2 (7, 8, 9, 10). The development is graphically shown in Fig. 2-3 and can be roughly divided into 3 phases: Up to about 1965: Energy consumption is primarily satisfied by coal. From 1965–1970: Petroleum becomes the primary energy source. From 1970: Beginning of nuclear energy.

Growth of world population and the intent on the part of governments to raise the living standard have brought about a much larger increase in energy consumption than population growth. According to UN statistics, world population grew between 1950 and 1975 from $2490 \cdot 10^6$ to $3920 \cdot 10^6$people, or ca. 66%. The primary energy consumption during the same time grew from $252 \cdot 10^6$ to $7877 \cdot 10^6$ tce/a or around 250%. This comes to a yearly growth rate of about 5% corresponding to a doubling time of about 14 years (compare Fig. 2-3).

It must be emphasized that in the year 1976, 57% of the world primary energy was consumed by only 18% of the world population: North America (6% of world population) consumed 32%, Western Europe (9% of world population) consumed 19%, and Japan (3% of world population), 6% of world primary energy. In contrast, 82% of the world population consumed 43% of world primary energy: Africa,

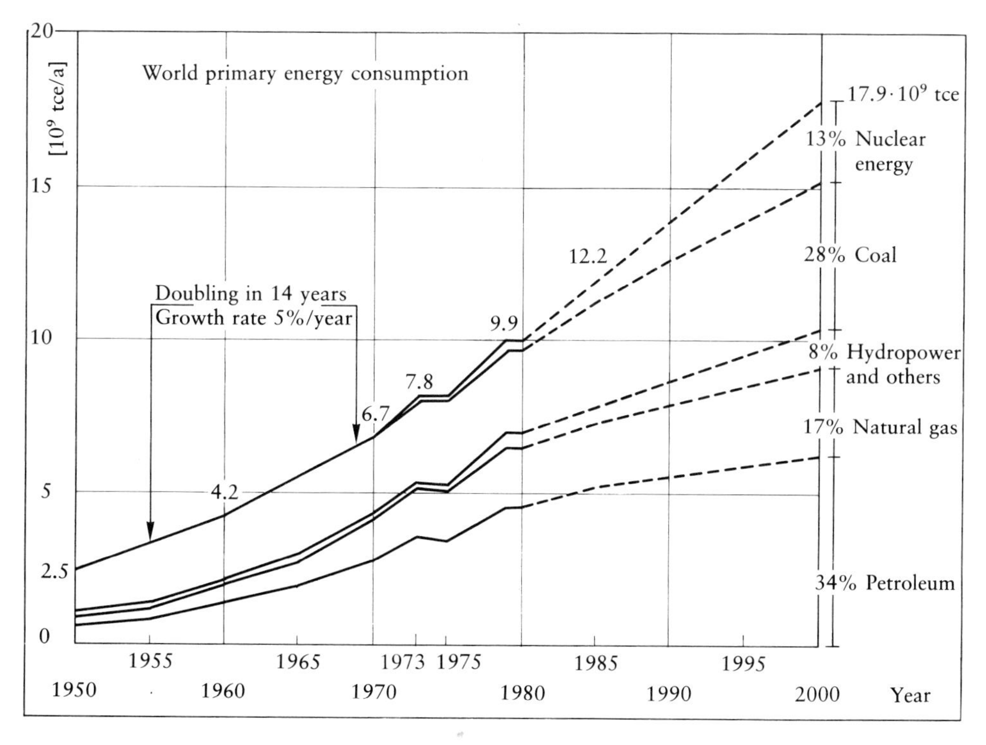

Fig. 2-3: World primary energy consumption according to energy sources*
*The data are taken from table 2-2 (7, 8, 9)

Asia, Australia, South America together have 52% of world population and used 13%; Eastern Europe, China, and USSR make up the remaining 30% and used 30% of world primary energy (10).

In highly industrialized countries, the growth rate of primary energy consumption has peaked already. For example, the US yearly growth rate of primary energy consumption between 1950 and 1974 was 3.6% per annum; for 1978 to 1985, the expected growth rate is less than 2% per annum. In the Federal Rep. of Germany, the annual growth rate of primary energy consumption between 1950 and 1974 was ca. 4.7%; for 1975 to 1985, and for 1985 to 1990, the expected growth rates are 3.6% and 2.1%, respectively (17).

The annual growth rate in most of the developing countries will be 5 to 6% for many years to come. However, this growth starts at a very low level, since the primary energy consumption per capita, in more than half of all countries on earth, is less than 500 kgce/a (see 2.32).

The world primary energy consumption, like population growth, will follow a sigmoid curve and will end in a saturation limit. As stated previously, about 81% of the world population will live in the developing countries about the year 2000, and this figure will increase in about 2075 to about 84%. There will be an enormous need for primary energy at this time, and the saturation limit of the S-curve will not be reached in the foreseeable future (see 2.333). The interdependence between food economics and energy economics will not be covered in this book (12-15). But it should be mentioned that the per capita level of energy consumption in the highly industrialized countries is not needed to alleviate hunger and poverty in the many developing countries.

2.32 Primary energy consumption and gross national product

If one considers the development of primary energy consumption of developed industrialized nations one observes a structural similarity with the world primary energy consumption (compare Fig. 2-3). Examples are furnished in Table 2-3, the primary energy consumption of the German Federal Republic (9, 16-19) and in Table 2-4, the primary energy consumption of the United States (8, 20, 21), for some selected years. Figures 2-4 and 2-5 show this development in graphic form. The predictions in all cases are based on the lower growth figures. The yearly growth rate increase for the Fed. Rep. of Germany amounted to ca. 4.7% from 1950 to 1974, and was thus quite close to the worldwide figure of 5%. For the USA the yearly growth rate for the time span of 1950 to 1974 was about 3% (21).

The per capita primary energy consumption has grown exponentially in several countries. In many countries it seems to be independent of the state of development or economic order. Examples are the USA, the USSR, the Federal Rep. of Germany, and India. Actual data for selected years are summarized in Tables 2-5 and 2-6; for regions and for countries.

Table 2-3: Primary energy consumption of the Federal Republic of Germany from 1950 to 1980 and predicted primary energy demand for 1985, 1990, and 2000 (in 10^6 tce[1])

Energy source	Year from	1950 (16)	1960 (16)	1970 (16)	1973 (16)	1975 (18)	1979 (17)	1980 (17)	%	1985 (18)	%	1990 (18)	%	2000 (18)	%	1990 (19)	%	2000 (19)	%
Petroleum		6.3	44.4	178.9	209.0	181.0	206.8	186.9	47	218	47	217	44	198	39	220	45	214	39
Natural gas		–	0.9	18.3	38.6	48.7	65.3	64.5	16	85	18	88	18	81	16	83	17	83	15
Anthracite		98.7	128.3	96.8	84.2	66.5	75.8	77.0	20	80	17	87	17	105	20	83	17	115–93	21–17
Lignite		20.7	29.2	30.6	33.1	34.4	38.1	38.7	10	37	8	36	7	36	7	40	8	33	6
Hydropower and others[2]		9.8	8.6	10.1	9.7	10.0	8.2	9.8	4	11	3	13	3	21[2]	4	15	3	17	3
Nuclear energy		–	–	2.1	4.0	7.1	13.9	14.1	3	33	7	52	11	72	14	49	10	88–110	16–20
Total		135.5	211.5	336.8	378.6	347.7	408.1	391.0	100	464	100	493	100	513[3]	100	490	100	550	100

[1] 1 tce = $29.3 \cdot 10^9$ J.
[2] "Other" sources, such as solar energy, wood etc., are assumed as 10^7 tce for 2000.
[3] An estimate made by the Shell Oil Company, in 1980, predicted $457 \cdot 10^6$ tce for 2000; the percentage distribution of primary energy sources is similar. In November 1981, the third revision of the German energy program predicted $460 \cdot 10^6$ tce to $497 \cdot 10^6$ tce for 1995.

Sources: M. Grathwohl, Zukunftsperspektiven der Energieversorgung, Naturwissenschaftliche Rundschau, Vol. 30,2 (1977) (16).
Energiebilanzen der Arbeitsgemeinschaft Energiebilanzen, Düsseldorf 1981 (17).
Shell Oil Company, Public Affairs and Information Department, Hamburg 1979 (18).
Exxon Corporation, Public Affairs and Information Department, Hamburg 1978 (19).

Table 2-4: Primary energy consumption in the United States (USA) from 1950 to 1980 and predicted primary energy demand for 1985, 1990, and 2000 (in 10^6 tce[1])

Energy source / Year from	1950 (20)	1960 (20)	1970 (20)	1973 (20)	1975 (20)	1979 (8)	1980 (8)	%	1985 (20)	%	1990 (20)	%	1985 (21)	%	2000 (21)	%
Petroleum	490	730	1070	1270	1180	1250	1140	43	1390	42	1580	42	1410	45	1160	33
Natural gas	220	460	800	810	750	719	708	27	710	22	700	18	620	20	360	11
Coal	470	370	460	480	460	570	590	22	840	26	1110	30	690	22	1160	33
Hydropower and others[2]	50	60	90	110	100	115	114	4	140	4	150	4	200[2]	6	360[2]	11
Nuclear energy	–	–	10	30	100	104	102	4	200	6	220	6	220	7	400	12
Total	1230	1620	2430	2700	2590	2758	2654	100	3280	100	3760	100	3140	100	3440[3]	100

[1] Values are rounded off.
[2] Contribution of "others" is estimated for 1985 at $100 \cdot 10^6$ tce and for 2000 ca. $230 \cdot 10^6$ tce. These sources include biomass and solar heat. However, contributions from wind energy, tidal energy, oceanic heat, solar electricity, and geothermal energy in the year 2000 are estimated at only $36 \cdot 10^6$ tce (21).
[3] An estimate made by the Exxon Corporation, in 1981, predicted $3200 \cdot 10^6$ tce for 2000; the percentage distribution of primary energy sources is similar.

Sources: BP statistical review of the world oil industry 1980, London 1981 (8).
Project Interdependence : US and World Energy Outlook through 1990. A Report printed by the Congressional Research Service, U. S. Government Printing Office, Washington, D. C., November 1977 (20).
E. T. Hayes, Energy Sources available to the United States, 1985–2000, Science, Vol. 203, 234 (1979) (21).

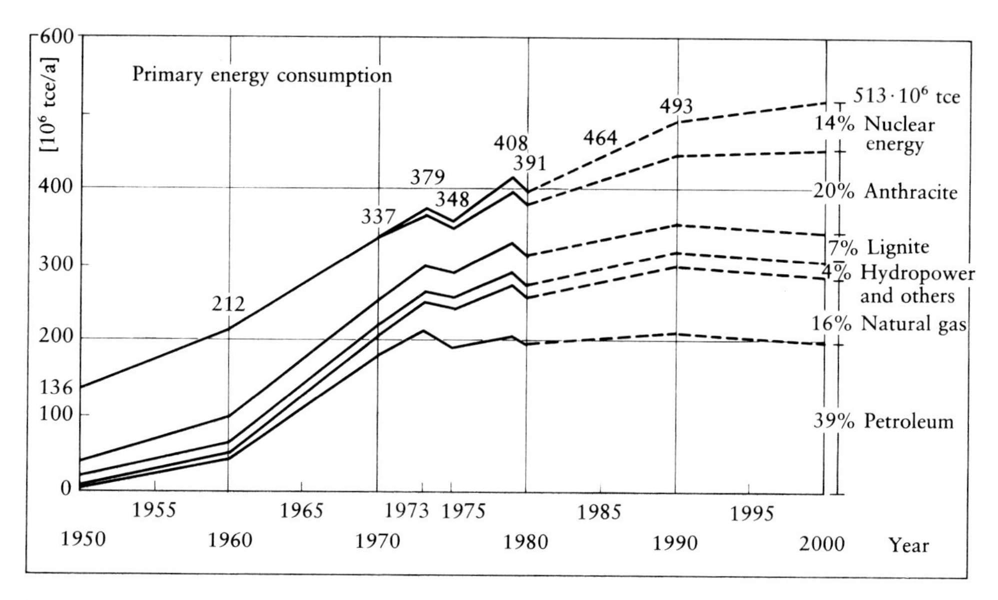

Fig. 2-4: Primary energy consumption in the Federal Republic of Germany*
*The data are taken from table 2-3

A relationship between primary energy consumption and gross national product exists and is shown for the United States for a span of over 40 years in Fig. 2-6 (22, 23). Even though this growth took place at a rather high level, the rate of increase in primary energy consumption was similar to the growth rate of the gross national product. The same holds for many other countries. Figure 2-7 shows this for the German Federal Republic (24).

In the time from 1960 to 1973, the Primary Energy Consumption (PEC) in the Federal Rep. of Germany increased ca. 0.98% per year, while the Gross National Product (GNP) increased by 1%. The energy-GNP elasticity coefficient k, defined as the percentage change in PEC divided by the percentage change in GNP hence was 0.98.

This connection is not based on any law of nature. The relationship between total economic growth and primary energy consumption is not a rigid one. It can be modified by technological progress, especially by energy conservation technologies and rational utilization (lesser conversion losses). In the industrialized nations, the elasticity coefficient has become smaller, owing to end-user resistance against the large price increases. This embraces industry, small users, households and transportation.

A similar relationship can apparently be drawn between the growth in electric power consumption W_S, and the growth in Gross National Product (GNP). This ratio k_S was 1.63 for the Federal Rep. of Germany between 1960 and 1973. In other words, electricity consumption grew at a faster rate than the gross national product, owing to greater affluence and increasing electrification of households. The pro-

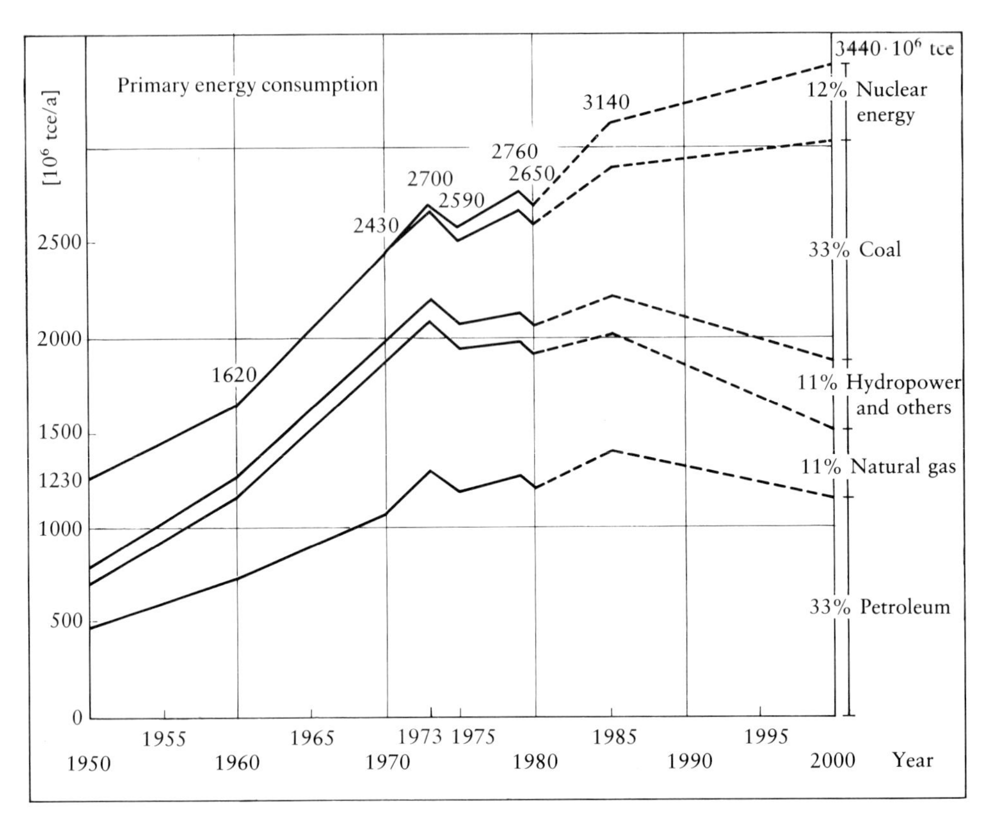

Fig. 2-5: Primary energy consumption in the United States*
*The data are taken from table 2-4

Table 2-5: Primary energy consumption per capita in some regions (in kgce)

Region / Year	1953	1963	1973	1976
Africa	186	256	353	397
America, North	7420	8403	11534	11395
America, Central	549	911	1259	1265
America, South	348	555	815	868
Asia, Middle East	311	439	977	1169
Asia, except Middle East	170	293	564	557
Europe, except Eastern Europe	2072	2904	4281	4289
Oceania	2358	3071	4524	4953
Centrally Planned Economies[1]	710	1329	1838	2030
World (average)	1109	1490	2041	2069

[1] Eastern Europe, China P. R., Mongolia, Democratic People's Republic of Korea, Vietnam, and USSR.

Source: United Nations Statistical Yearbook 1971, 1977 and 1978, New York 1972, 1978 and 1979.

Table 2-6: Primary energy consumption per capita in different countries (in kgce)

Country / Year	1955	1964	1973	1978
Canada	5279	7137	11237	9930
United States	7768	8772	11960	11374
Haiti	36	32	27	57
Mexico	642	1029	1355	1384
Venezuela	2123	3000	2818	2989
Argentina	982	1242	1908	1873
Brazil	289	364	566	794
France	2166	2933	4389	4368
Germany, F. R.	3257	4230	5792	6015
Italy	721	1659	2737	3230
Netherlands	2376	3278	6090	5327
Spain	606	996	1993	2405
Sweden	2744	4320	6110	5954
United Kingdom	4993	5079	5778	5212
Bangladesh	42	86	29	43
India	114	161	188	176
Indonesia	116	108	146	278
Iran	158	386	1086	1808
Israel	1124	1599	2868	2362
Japan	740	1660	3601	3825
Pakistan	42	86	149	172
Philippines	126	203	291	339
Saudi Arabia	222	320	1023	1306
Thailand	50	106	303	327
Turkey	224	333	625	793
Vietnam	37	63	274	125
Egypt	243	321	294	463
Nigeria	33	38	67	106
Rep. of South Africa	2387	2576	2815	2985*
Sudan	40	60	124	172
Germany, D. R.	3878	5569	6233	7121
Poland	2621	3513	4575	5596
USSR	2240	3430	4927	5500
China, P. R.	159	447	594	805
Australia	3545	4454	5956	6622

* This figure was for 1976.

Sources: United Nations, Statistical Yearbook 1959–1977, New York 1960–1978.
World Development Report 1980, World Bank, Washington D. C. 1980.

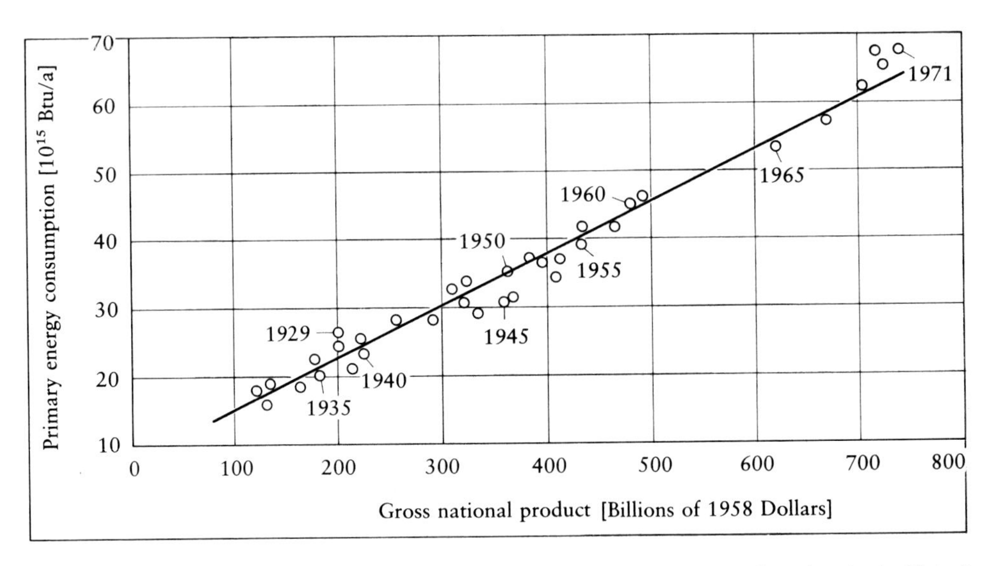

Fig. 2-6: Relationship between primary energy comsumption and gross national product in the United States

Source: H. Schneider, D. Schmitt, W. Pluge, Die Energie-Krise in den USA, Munich, Vienna: R. Oldenbourg 1974.

jected k and k_S values for the Federal Rep. of Germany for 1980–1985 are expected to be k = 0.7, and k_S = 1.55, and for 1985–1990 k = 0.6, and k_S = 1.43, respectively. Again a similar relationship between the growth of primary energy consumption and the growth of gross national product was observed for the OECD. Between 1960 and 1974 both parameters increased at the same rate resulting in a k = 0.99. The projection for 1974 to 1985 is k = 0.8 (see 2.333, (25) (26)).

Even though various countries vary greatly in their respective PEC per capita values, the relationship between PEC per capita and GNP per capita is a "linear" one (compare Fig. 2-8). (Some deviations may be due to climatic conditions). The PEC value for the USA is nearly twice that of the Federal Rep. of Germany. Obviously, there is room for economizing in energy consumption in the United States, as the quality of life between the two countries hardly differs by a factor of 2.

The differences between these nations are vast: Out of 187 nations in the UN in 1978, about 30 had a per capita PEC value of less than 100 kgce and a total of about 70 less than 500 kgce. About half of the nations have per capita consumptions of less than 500 kgce, which corresponds to 4% of the US per capita PEC. After the exorbitant price increase in raw materials, especially the primary energy source petroleum, it has become more meaningful to group the countries of the world into four groups: There are the industrialized nations with abundance in some raw materials (e.g. USA, Canada), then the raw-material-poor nations (e.g. Japan). The third

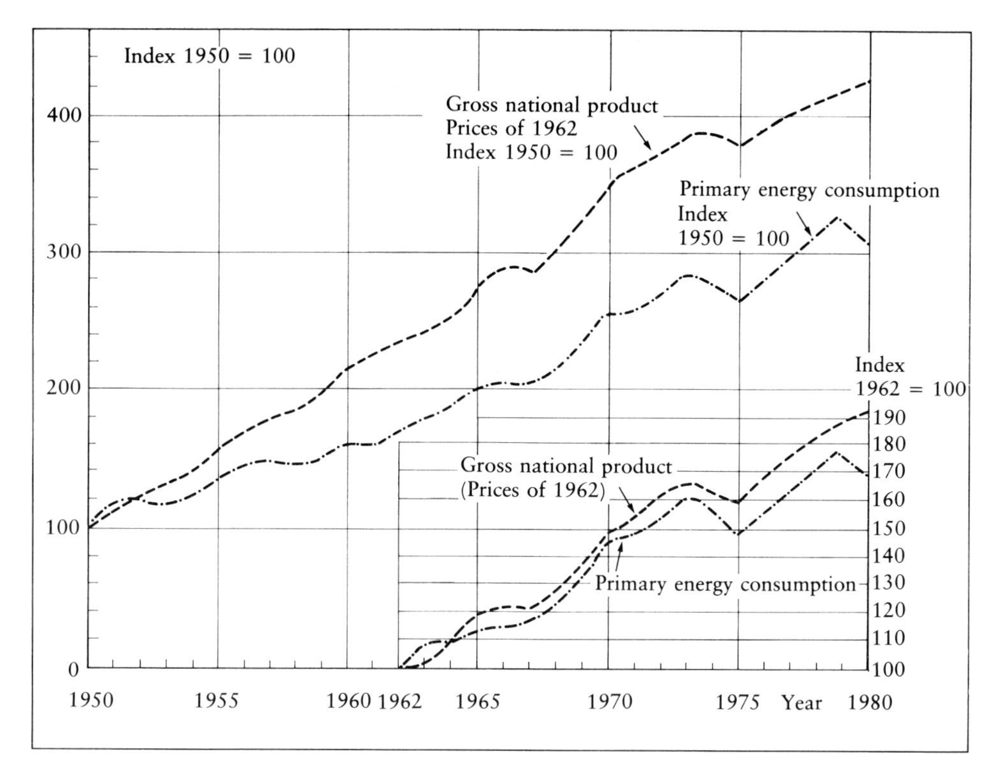

Fig. 2-7: Primary energy consumption and gross national product in the Fed. Rep. of Germany

Sources: E. Pestel et al., Das Deutschland-Modell, Stuttgart: Deutsche Verlags-Anstalt 1978.
The British Petroleum Company, BP, Hamburg 1981.

group consists of developing countries, rich in raw materials (e.g. OPEC nations, Mexico) (27-31). Nations of the fourth group have neither their own raw materials, nor enough money to buy them. These are the poorest countries in the world. They are called fourth world countries and contribute 3% to the world economics but 27% to world population. From a geographical view point, this "hunger belt" extends from the West coast of Africa, south of the Sahara desert eastwards and includes all of Southern Asia. Figure 2-9a depicts the poorest countries in the world (32, 33).

According to the World Bank report of 1980, these 39 countries, with a population of $1.2 \cdot 10^9$ people, in 1978 had a GNP per capita of up to \$ 360 (US-\$) corresponding to 4% of the North American GNP per capita. Within this group of states, there are wide differences. In 1978 in the population-dense Ethiopia, ($31 \cdot 10^6$ inhabitants) the GNP per capita was \$ 120. The figures are for Bangladesh ($85 \cdot 10^6$ people) \$ 90, for Burma ($32 \cdot 10^6$ people) \$ 150, for Zaire ($27 \cdot 10^6$ people) \$ 210, for India ($644 \cdot 10^6$ people) \$ 180, and for Pakistan ($77 \cdot 10^6$ people) \$ 230 (6).

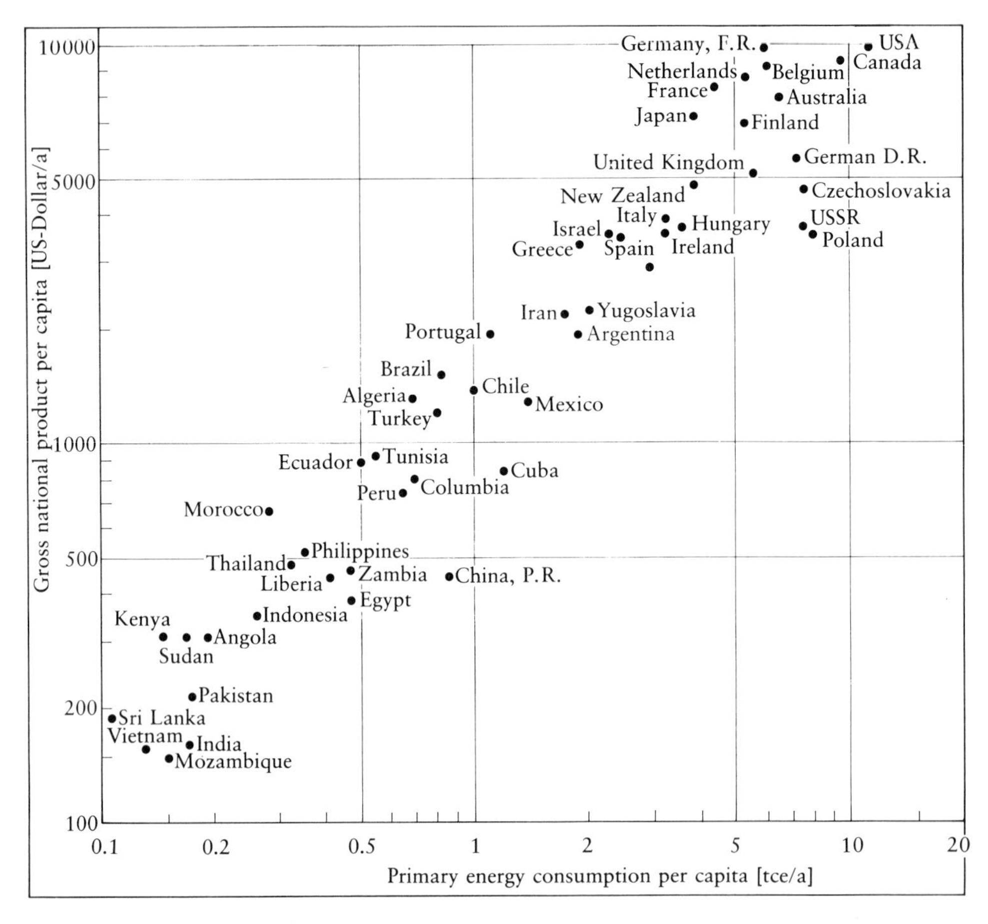

Fig. 2-8: Relationship between primary energy consumption per capita and gross national product per capita for different countries 1978*
*The data are taken from World Development Report 1980, World Bank, Washington D.C. 1980

Fig. 2-9a: The poorest countries in the world

Africa	11 Somalia	22 Mauretania	31 India
1 Mali	12 Uganda	23 Senegal	32 Sri Lanka
2 Guinea-Bissau	13 Rwanda	24 Togo	33 Bangladesh
3 Guinea	14 Burundi	25 Sudan	34 Burma
4 Sierra Leone	15 Zaire	26 Angola	35 Lao People's Dem. Rep.
5 Upper Volta	16 Kenya		36 Cambodia
6 Benin	17 Tanzania	*Asia*	37 Vietnam
7 Niger	18 Malawi	27 Afghanistan	38 Indonesia
8 Chad	19 Mozambique	28 Pakistan	
9 Central African Republic	20 Lesotho	29 Nepal	*America*
10 Ethiopia	21 Madagascar	30 Bhutan	39 Haiti

Source: World Development Report 1980, World Bank, Washington D.C. 1980.

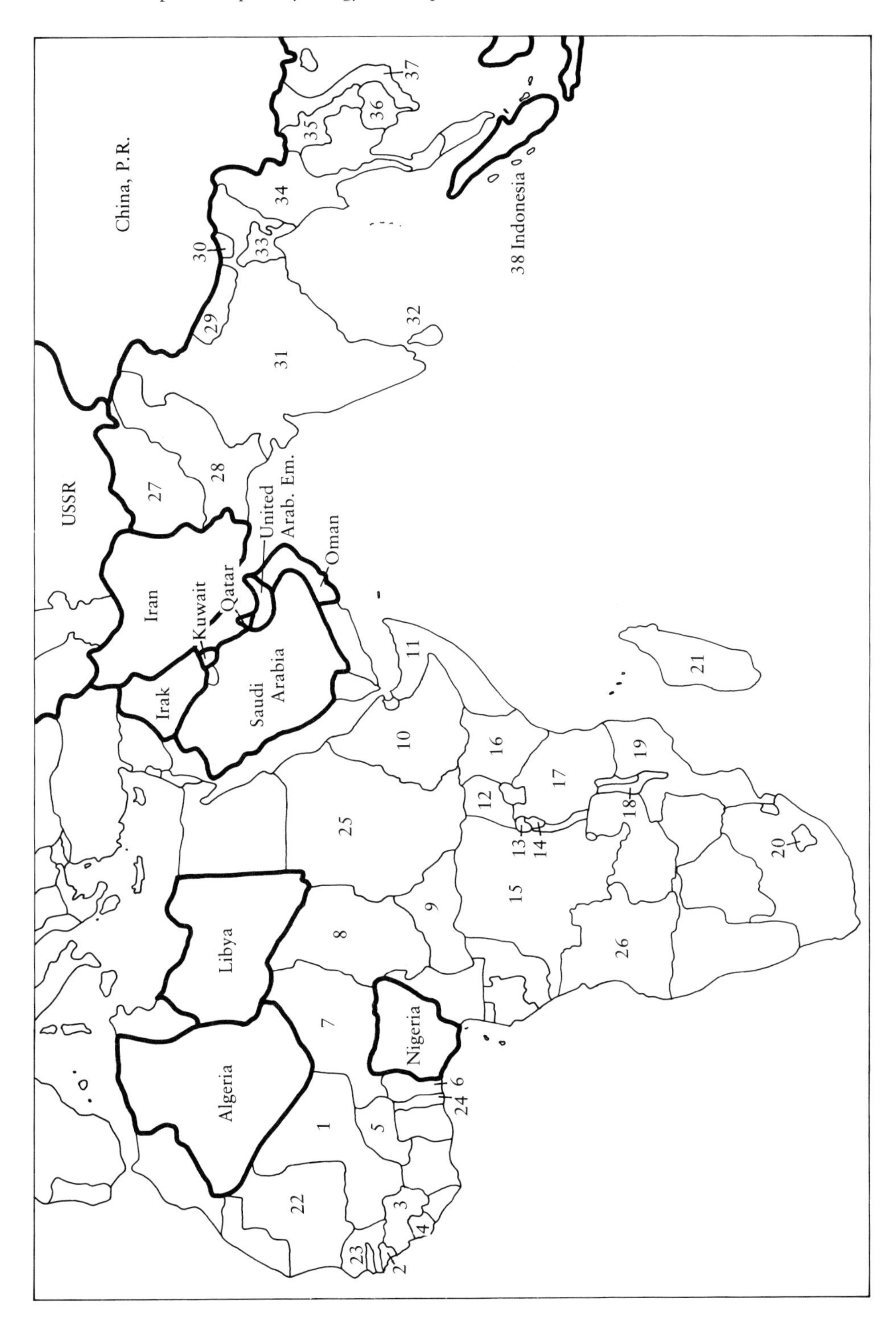

China, P.R.
USSR
Iran
Kuwait
Qatar
Irak
Saudi Arabia
United Arab. Em.
Oman
Libya
Algeria
Nigeria
38 Indonesia

One must stress the misery which is hidden in these figures: In the beginning of the '70s approximately $450 \cdot 10^6$ people (among them about half being children) suffered of malnutrition owing to lack of protein.

In contrast, the GNP per capita of all OECD countries in 1978 amounted to an average of about $ 6000, and in the OPEC countries about $ 5500. The differences within the OPEC countries are vast. At the top are the sparsely populated countries on the Persian gulf: Kuwait with $1.2 \cdot 10^6$ inhabitants has $ 14890 per capita, and Qatar ($0.2 \cdot 10^6$ people), $ 11400 (34–36). On the other hand, some oil producing countries like Indonesia ($136 \cdot 10^6$ people) have a low GNP per capita, with $ 360, as does Nigeria ($81 \cdot 10^6$ people) with $ 560 (6).

The North-South conflict is not a new problem (37, 38). However, political conditions have changed drastically. Owing to economical alliances of countries, and owing to improved communication the world has become "smaller". Countries which are economically weaker have succeeded, in the United Nations, in alerting the whole world regarding the vital problems that concern them. This overall process was favored by the growth in membership in the UN. Between 1959 and 1976 the membership of the UN grew from 83 to 145 nations. The number of Western industrialized countries and socialistic countries has not changed significantly since 1965, so that the status in September of 1976 indicates ca. 75% of UN member nations as developing countries (6).

These considerations show conclusively that most of nations in the world have a great need to catch up with the more affluent ones (39, 40). It would seem advisable to locate industrial production, and especially those industries which require large amounts of energy and raw materials, in countries rich in energy and raw material sources, and as much as possible, in fourth world countries. If one compares the geographical boundaries of the 15 largest petroleum producers in 1980 (Fig. 2-9b) with countries of the fourth world, one notes that most of the producers have common boundaries with one or more fourth world countries (41), although they are often boundaries between different social and economic systems as well.

With the exorbitant increase in petroleum prices, the developing countries in 1973 spent a considerable part of their export gains to cover their own energy needs by paying for crude oil and petroleum products. Petroleum technology, in contrast to the newer energy technology such as nuclear technology, is relatively simple and less capital-intensive. This will make it even more difficult for the developing countries to find substitutes for petroleum as energy sources (42). If industrialized nations refrain from developing nuclear energy, this will increase the demand for crude oil and intensify the worldwide competition for available petroleum. It certainly will affect many of the developing countries most severely (43).

The cost of crude oil will play an important role in the economy of many developing nations, particularly since their primary energy demand will increase with increased growth rates, in contrast to the industrialized nations. For this reason petroleum exploration in developing countries which do not as yet have a petroleum

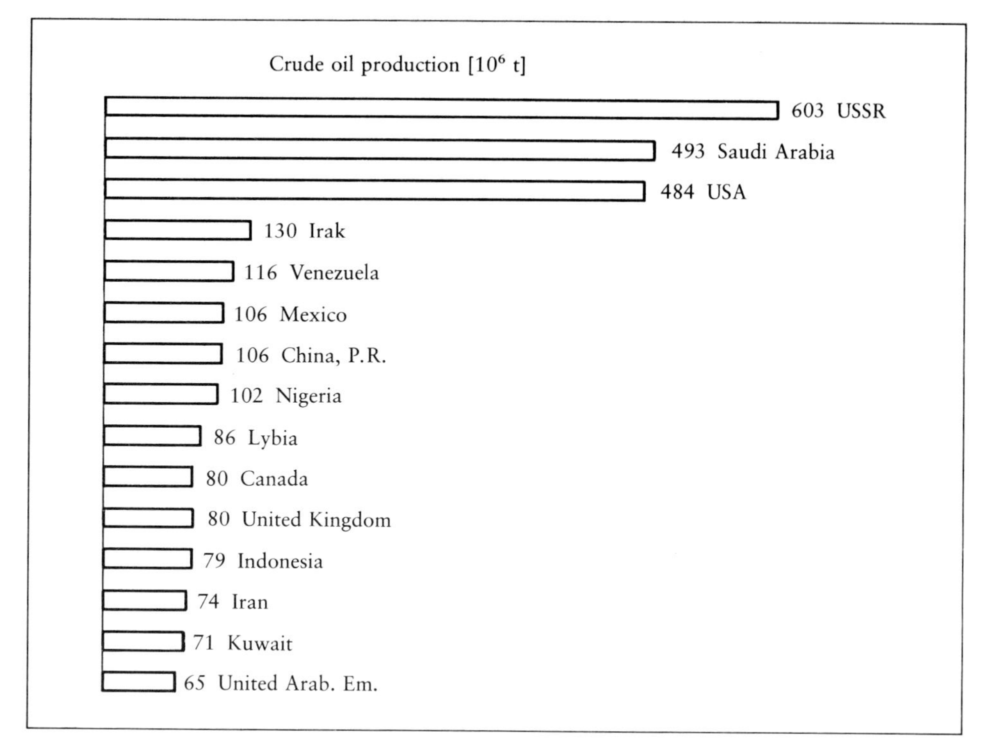

Fig. 2-9b: The largest crude oil producers in 1980*
*The data are taken from BP statistical review of the world oil industry 1980, London 1981

industry merits support. A study by the Exxon Corporation counts on oil finds or increased oil production in about 60 countries. Production costs are based on 1975 prices of US \$ 3.00 to \$ 6.00. This means a more economically attractive cost than the price of imported crude or offshore production of Alaskan or North Sea crudes (6, 44–48).

A considerable increase in petroleum production is expected in Mexico, Brazil, Egypt and India. Proved recoverable oil reserves in Mexico are $4058 \cdot 10^6$ tons (compare Table 3-14). Oil production rose from $47 \cdot 10^6$ t in the year 1976 to $106 \cdot 10^6$ t in 1980, and is expected to reach $115 \cdot 10^6$ t per year in 1982 (31).

Smaller increases in petroleum production are forecast for other countries, as e.g. Angola, Zaire, Chad, Pakistan, and Malaysia. There are assumed to be reserves in Bangladesh, Cambodia, and Sri Lanka. The World Bank predicts that petroleum production in the developing countries (excluding OPEC countries) will increase from the $3.7 \cdot 10^6$ barrels per day in 1976 to $8.3 \cdot 10^6$ barrels per day in 1985. Very optimistic predictions claim that some South American countries and some African nations, adjacent to the Middle East will become major oil producers at the end of this century (44).

The potential for natural gas production is supposed to be vast in the developing countries. Examples are again Mexico, Indonesia, and Malaysia. It will require much capital investment (6).

Other developing countries are striving to increase their coal production. This holds for India, Vietnam, Columbia, Mexico and Mozambique (6). Large increases in generating capacity for electricity are also expected in some countries. Argentina, Brazil, Pakistan, India and Turkey are planning major hydrothermal generating plants (6, 49). Electrical capacity will also be boosted by nuclear power plants (see 4.212).

Of further importance for the developing countries, particularly their rural areas, is the direct and indirect utilization of solar energy (e.g. use of bio-mass, bio-gas) (decentralized energy use)[1] (50–53).

2.33 Energy conservation potentials and predictions of future primary energy demand

2.331 Analysis of energy demand

Based on the connections between primary energy consumption, population, and gross national product, certain predictions regarding future primary energy demand can be made. In addition to these parameters several factors need to be taken into account to estimate future demands. Examples are the availability of primary energy sources, their cost, energy politics of individual states, or groups of states, attitudes of consumers, technological development, etc.

In order to ascertain the energy demand of a country, the energy consumption is broken down into categories such as consumption by industry, traffic, households, individual small consumers, etc. Assumptions about various growth rates have to be made as well as the fractional contributions of various energy sources. Fig. 2-10 shows the distribution of final energy consumption according to demand and end-users in the Federal Rep. of Germany for 1975. 75% of all energy demand there is required in the form of heat, of which 39% is for space heating purposes and 36% for furnishing warm water and process heat (35). Clearly the emphasis is on supplying heat while demand for light and power is placed in the background.

If one analyzes the distribution of final energy consumption within individual user sectors, it becomes evident that in the household and commercial sectors, the greater

[1] The annual average solar energy influx [W/m^2] is quite favorable in many fourth world countries (compare Figures 3-17a, b). Thus, these countries have a significant future energy source.

Fig. 2-10: Distribution of final energy consumption according to demand and end-users in the Fed. Rep. of Germany

Source: H. Schäfer, Energy Technology Research Institute, Munich 1980.

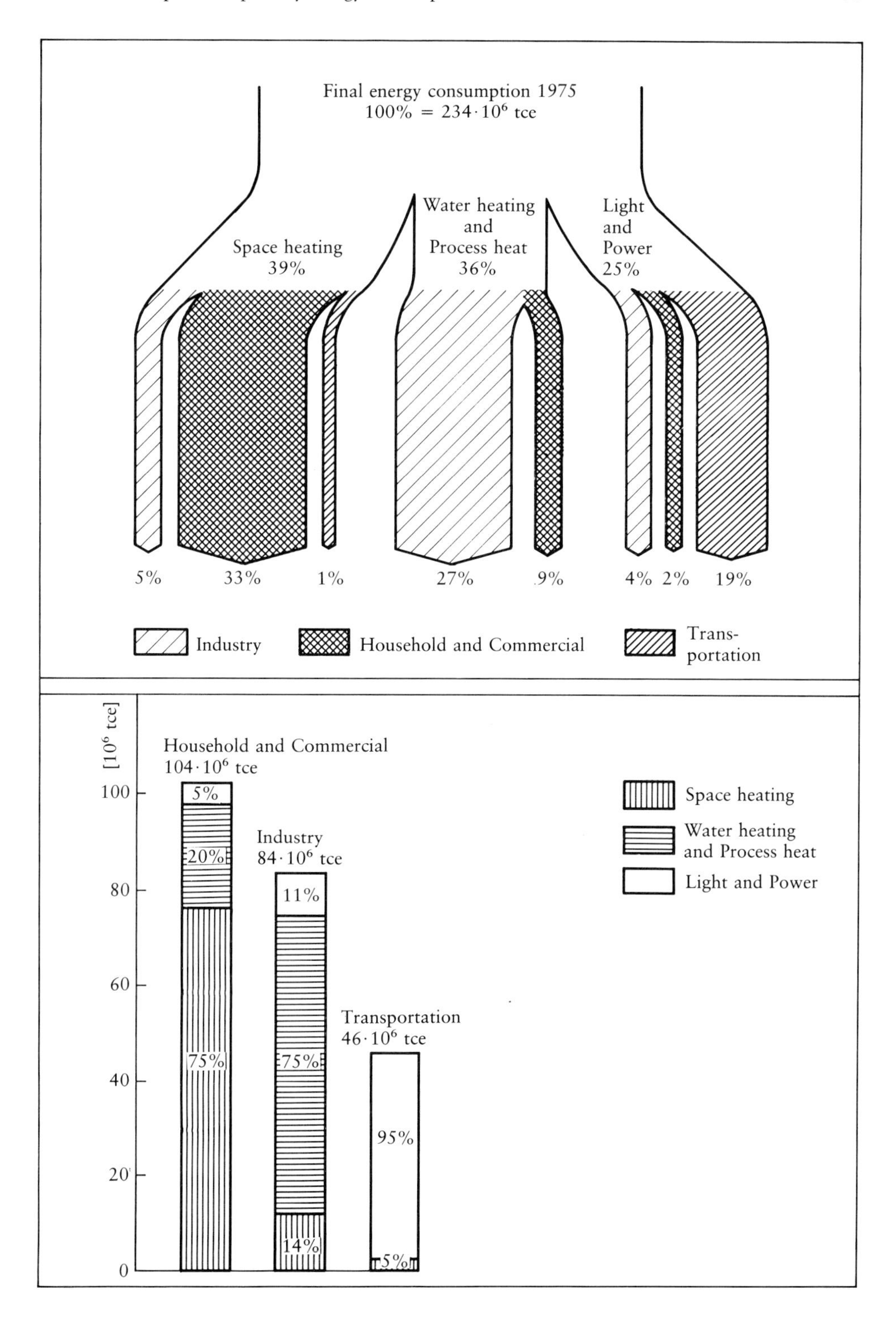

Final energy consumption 1975
100% = 234·10⁶ tce
Space heating
39%
Water heating
and
Process heat
36%
Light
and
Power
25%
5%
33%
1%
27%
9%
4%
2%
19%
Industry
Household and Commercial
Trans-
portation
[10⁶ tce]
Household and Commercial
104·10⁶ tce
Industry
84·10⁶ tce
Transportation
46·10⁶ tce
5%
20%
75%
11%
75%
14%
95%
5%
100
80
60
40
20
0
Space heating
Water heating
and Process heat
Light and Power

portion, 75%, is spent for space heating. In the industrial sector, the major portion, again 75%, is for heating water and for process heat, while in transportation 95% goes to light and power.

The heat consumed in industry is predominantly process heat, with a considerable portion as high temperature heat. Fig. 2-11 shows the heat demand (heat/degree, at temperature T) (54). The required heat demand is ascribed to the respective temperature level at which heat has to be furnished to the production process.

It is apparent that there are two areas, high temperature heat at about 1400°C, and low temperature heat at about 200°C. The industries typical for the low temperature area comprise textile, foodstuff, paper, and chemical industries, while the high temperature area is primarily the iron and steel industry.

Table 2-7a gives the final energy consumption for the Federal Rep. of Germany for 1950 to 1977 and the projected final energy demand for 1985, 1990 and 2000; and Table 2-7b, the distribution according to end-users for the various sectors (18, 56). Total consumption increased sharply in all sectors between 1950 and 1977, and is expected to increase further by 1985. In the household and commercial sectors, percentages of 35% in 1950 increase to 44% in 1977, while in the transportation area in the same time span the percentages grow from 17% to 21%. End energy use in toto, as well as in individual sectors, will tend to reach a saturation level, so that after the year 2000 only minor changes are to be expected.

The history of electricity demand shows a much more pronounced growth tendency. In 1977, gross electricity production in the German Federal Republic amounted to 335.5 TWh, and accounted for approximately 30% of all primary energy expen-

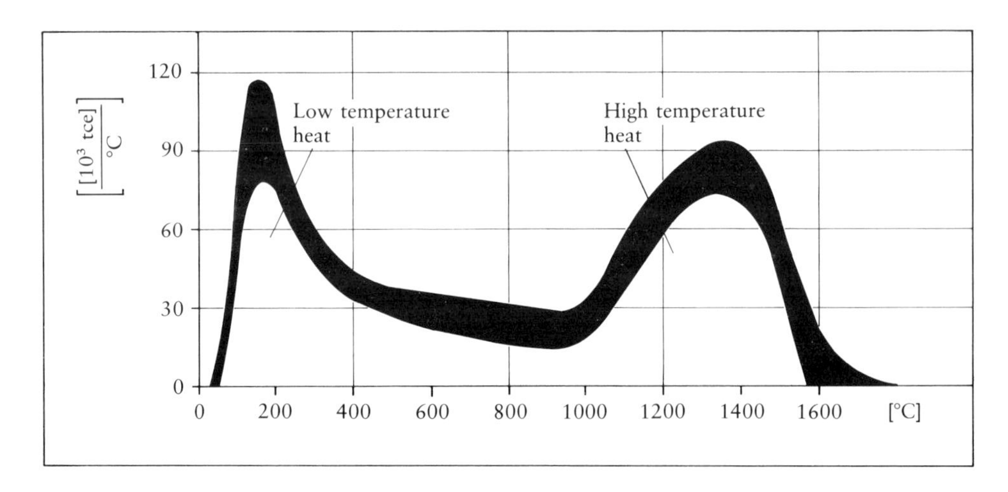

Fig. 2-11: Process heat demand of West German Industry

Source: Einsatzmöglichkeiten neuer Energiesysteme, Teil I, Federal Ministry of Research and Technology (Ed.), Bonn 1975.

Table 2-7a: Final energy consumption for the Federal Republic of Germany from 1950 to 1977 and predicted final energy demand for 1985, 1990 and 2000 (in tce*)

Year / Final energy consumption	1950	%	1960	%	1970	%	1977	%	1985	%	1990	%	2000	%
Industry	42	48	71	48	91	40	88	35	99	33	105	35	106	35
Household and Commercial	30	35	52	36	100	43	110	44	134	45	134	44	128	43
Transportation	15	17	23	16	39	17	51	21	64	22	65	21	65	22
Total	87	100	146	100	230	100	249	100	297	100	304	100	299	100
Non fuel uses	3		7		25		30		38		44		50	
Conversion losses	46		59		82		93		129		146		164	
Primary energy consumption	136		212		337		372		464		494		513	

* 1 tce = $29.3 \cdot 10^9$ J.

Sources: Energy balance-sheet of the Federal Republic of Germany, VWEW, Frankfurt 1979.
Shell Oil Company, Public Affairs and Information Department, Hamburg 1979.

Table 2-7b: Distribution of final energy consumption according to end-users in the Federal Republic of Germany

End-user	Source	1977 [10^6 tce]*	1977 %	2000 [10^6 tce]*	2000 %
Industry	Petroleum	29	33	20	19
	Natural gas	19	22	26	25
	Coal	22	25	20	19
	Electricity	18	20	29	27
	Renewable Sources	–	–	1	1
	Others	–	–	10	9
	Final energy consumption	88	100	106	100
Household and Commercial	Petroleum	67	61	45	35
	Natural gas	15	14	32	25
	Coal	11	10	4	3
	Electricity	17	15	38	30
	Renewable Sources	–	–	9	7
	Final energy consumption	110	100	128	100
Transportation	Gasoline	32	63	31	48
	Diesel fuel	14	28	25	38
	Others, Coal	4	7	8	12
	Electricity	1	2	1	2
	Final energy consumption	51	100	65	100

* 1 tce = $29.3 \cdot 10^9$ J

Source: Shell Oil Company, Public Affairs and Information Department, Hamburg 1979.

diture. The contributions of primary energy sources are 57% coal (29% lignite, 28% anthracite), 15% natural gas, 11% nuclear energy, 9% hydrothermal and other sources, 8% fuel oil. The electricity production predicted for the year 2000 is 648 TWh, corresponding to a yearly growth rate of 2.8%. The primary sources would be 47% coal (30% anthracite, 17% lignite), 36% nuclear energy, 8% natural gas, 6% hydrothermal and others, and 3% petroleum (18) (see also Table 4-4).

In analyzing the structure of end consumption for the individual sectors (compare Table 2-7b), one must take into account that the position of petroleum, which presently dominates at 33%, will probably decrease by the year 2000 to a share of 19%, being replaced in part by natural gas and electricity. At this future time natural gas will have a share of 25% and electricity of 27%. In the household and commercial sectors, the predominant part goes to heating and low temperature hot water production, where petroleum products can be replaced more easily than in the transport sector or the chemical industry, where they serve as raw materials. The petroleum-derived share will probably drop from 61% to 35%. Coal consumption likewise will shrink with marked concommitant growth in natural gas and electricity. The household sector will be the largest user of electricity, with 30% in the year 2000. Regenerative sources will provide 7% of this demand in 2000. A steady increase is predicted in the transportation sector up to 1985. The trend towards more energy-effective vehicles will persist and the number of diesel-powered cars will grow, but petroleum products will maintain their dominant position in this sector.

This structure of end energy demand is typical for all highly industrialized nations with a comparable climate. In other words, the statements hold for all other industrialized countries of Western Europe and – with only minor modification – also for the United States and Japan (59–63).

Table 2-8a shows the final energy consumption in the United States for 1950 to 1975 and the predicted final energy demand for 1990. Table 2-8b gives the distribution of final energy consumption according to end users in the United States for the individual sectors (20). The total consumption as well as consumption in individual sectors grew considerably from 1950 to 1975, and is expected to continue to increase up to 1990 (compare Fig. 2-5). The percentage contribution of the household and commercial sectors increased from 28% in 1950 to 32% in 1975; in the trans-

Table 2-8a: Final energy consumption for the United States from 1950 to 1975 and predicted final energy demand for 1990 (in 10^6 tce)[1)]

Year Final energy consumption	1950	%	1960	%	1970	%	1975	%	1990	%
Industry	430	42	520	39	710	37	690	35	1040	42
Household and Commercial	290	28	420	31	620	32	630	32	650	26
Transportation	310	30	390	30	590	31	670	33	800	32
Total	1030	100	1330	100	1920	100	1990	100	2490	100
Non fuel uses	40		60		110		110		230	
Conversion losses	160		230		400		490		1040	
Primary energy consumption	1230		1620		2430		2590		3760[2)]	

[1)] Values are rounded off; 1 tce = $29.3 \cdot 10^9$ J.
[2)] Percentage distribution of final energy is probably similar, based on a predicted lower growth rate (compare Table 2-4).

Source: Project Interdependence: U. S. and World Energy Outlook throught 1990, A Report printed by the Congressional Research Service, U. S. Government Printing Office, Washington D. C., November 1977 (20).

portation sector it grew from 30% in 1950 to 33% in 1975. The percentage share of industry decreased from 42% to 35% in the same time span. All three sectors shared end energy consumption roughly in thirds in 1975 – disregarding minor differences (64). If one compares the end energy consumption among like sectors for the United States and the Federal Rep. of Germany, the higher figures in the U.S. transportation consumption are astonishing (65). The household and commercial consumption is expected to reach saturation levels, but there are large increases in the conversion area – consumption of electricity has grown at a much faster rate than overall energy consumption – and in non-energy consumption. The trend in both the latter areas is going to continue for some time to come (62).

Table 2-8 b: Distribution of final energy consumption according to end-users in the United States

End-User	Source	1972 from (62) $[10^6$ tce$]$[1]	%	1990 from (20) $[10^6$ tce$]$	%
Industry	Petroleum	120	15	140	14
	Natural gas	420	50	460	44
	Coal	160	20	260	25
	Electricity	120	15	180	17
	Final energy consumption	820	100	1040	100
Household and Commercial	Petroleum	200	33	230	35
	Natural gas	270	45	170	27
	Coal	10	2	–	–
	Electricity	120	20	250	38
	Final energy[2] consumption	600	100	650	100
Transportation	Petroleum	620	100	800	100
	Natural gas	–	–	–	–
	Coal	–	–	–	–
	Electricity	–	–	–	–
	Final energy consumption	620	100	800	100

[1] 1 tce = $29.3 \cdot 10^9$ J

[2] According to recent investigations, solar energy may provide up to $70 \cdot 10^6$ tce for this end use, in the year 2000, via solar heating (21).

Sources: Energy R & D, OECD, Paris 1975 (62).
Project Interdependence: U. S. and World Energy Outlook through 1990, A Report printed by the Congressional Research Service, U. S. Government Printing Office, Washington D. C., November 1977 (20).

Gross production of electricity in the United States was 2251.2 TWh in 1976. 77% came from fossil fuels (coal, gas, oil), 14% came from water and other sources, and 9% from nuclear energy. For the year 1990, an electricity generation of 3800 TWh is anticipated, 58% from coal 15% from nuclear energy, 13% from petroleum, 10% from water and other sources, and 4% from natural gas (20).

The predicted pattern of primary energy sources for the various user sectors follows Table 2-8b: In industry natural gas will retain its dominant position in 1990 with a 44% share. The share of coal will probably increase to 25%. The use of electricity in the household and commercial sectors will continue to grow. Increased use of electrical household appliances in 1950 to 1975 brought about a six-fold increase in electricity consumption. This sector is expected to become the largest user by 1990, accounting for about 38%.

If these predictions come true, the household and commercial sectors of the industrialized nations will be the largest consumers of electricity. In the industrial sector, natural gas will have a larger share than petroleum, and conversely, the share of petroleum will be larger in households and commercial uses. Coal will play practically no role there; its use will be exclusively in industry and electricity generation. Around the turn of the century electricity will be a predominant source with strong growth also in nuclear generating plants. In transportation, petroleum derived products will remain the preponderant source of energy even around the year 2000.

2.332 Energy conservation possibilities

Since 1973, experience has clearly shown that rising petroleum prices and energy consciousness in the major consumer countries leads to many economic and sociological changes. The predicted development of primary energy demand was revised downward for individual countries, and consequently for the whole world. For example, for the Federal Rep. of Germany, in 1973 – shortly before the oil crisis – the primary energy demand in 1985 was estimated at $610 \cdot 10^6$ tce. Owing to a changed energy-political situation, one year later – in the first formulation of an energy program – the 1985 estimate was reduced to $555 \cdot 10^6$ tce. The government of the German Federal Republic published in March 1977 a new version of the energy program, called "Grundlinien und Eckwerte", where the primary energy demand was scaled down another 10% to $496 \cdot 10^6$ tce (11). A second revision, published in December 1977, predicted a 1985 figure of $483 \cdot 10^6$ tce (9). Finally, an estimate made by the Shell Oil Company, in 1979, predicted $464 \cdot 10^6$ tce for 1985 (18).

In long range forecasts, e.g. for the year 2000, the deviations are even larger. Prior to the oil crisis in 1973, the German Federal Republic figure for 2000 was $888 \cdot 10^6$ tce. The December, 1977 second revision of the German energy program reduced it by a third to $600 \cdot 10^6$ tce (9). An estimate made by the Shell Oil Company, in 1979, predicted $513 \cdot 10^6$ tce (18), and in 1980, the Shell Oil Company predicted $457 \cdot 10^6$ tce.

Similar predictions for future primary energy demands are available for the United States (66–72). In a report entitled "U.S. Energy through the Year 2000", first published in 1972, the predicted U.S. primary energy consumption in 2000 was estimated at $4200 \cdot 10^6$ tce ($116.1 \cdot 10^{15}$ Btu) (66). In January 1974, shortly after the 1973 oil crisis, president Nixon proclaimed the "Project Independence", aimed at making the United Sates energy self-sufficient. The 1985 figure was estimated at 3700 to $3900 \cdot 10^6$ tce (102.9 to $209.1 \cdot 10^{15}$ Btu) (67). The second revision of this report: "U.S. Energy Through the Year 2000" published in 1975 predicted a 1985 demand of $3750 \cdot 10^6$ tce ($103.5 \cdot 10^{15}$ Btu) (66). The report "Project Interdependence: U.S. and World Energy Outlook through 1990" of November 1977 expects a 1985 primary energy demand of $3280 \cdot 10^6$ tce ($91.2 \cdot 10^{15}$ Btu) (20). E. T. Hayes, in 1979 used a value of $3140 \cdot 10^6$ tce ($86.7 \cdot 10^{15}$ Btu) (21). In other words, the primary energy demand figures were revised downward from 1972 to 1979 by $1060 \cdot 10^6$ tce ($29.4 \cdot 10^{15}$ Btu), a reduction of ca. 25%.

In the long range prediction, e.g. for the year 2000, deviations – just like those for other countries – are large. The report: "U.S. Energy Through the Year 2000", in the first edition in 1972, predicted for the U.S. in 2000 $6900 \cdot 10^6$ tce ($191.9 \cdot 10^{15}$ Btu) (9), and in its second edition of 1975 $5900 \cdot 10^6$ tce ($163.4 \cdot 10^{15}$ Btu) (21). E. T. Hayes, in 1979, estimates $3440 \cdot 10^6$ tce ($95.0 \cdot 10^{15}$ Btu) (21). An estimate made by the Exxon Corporation, in 1981, predicted $3200 \cdot 10^6$ tce ($89 \cdot 10^{15}$ Btu). It is evident that predictions for the year 2000 during the years 1972 to 1979 were revised downward by $3500 \cdot 10^6$ tce ($96.9 \cdot 10^{15}$ Btu); the predicted demand for the year 2000 estimated in 1979 is 1/2 of the 1972 value.

The examples given for the United States and for the Federal Rep. of Germany hold for other industrialized nations and groups of nations such as the European Community (EC), and the OECD countries (73–77).

The same downward revision is seen in estimates of world demand (78). Prior to the oil crisis in 1973, the world primary energy demand for the year 2000 was estimated at $30 \cdot 10^9$ tce ($829 \cdot 10^{15}$ Btu) (62). In 1976, H. Kahn predicted $22 \cdot 10^9$ tce ($608 \cdot 10^{15}$ Btu) (5).

The Conservation Commission of the World Energy Conference, in 1977 estimated the world primary energy demand for 2000 at $19.1 \cdot 10^9$ tce ($528 \cdot 10^{15}$ Btu) (26). Finally, also in 1977, the Federal Republic of Germany proposed a year 2000 figure of only $17.9 \cdot 10^9$ tce ($494 \cdot 10^{15}$ Btu), and an estimate made by the Exxon Corporation, in 1981, predicted $15 \cdot 10^9$ tce ($417 \cdot 10^{15}$ Btu) (compare Table 2-2) (9). The prime reason for the repeated downward revisions of the energy demand estimates is the fact that potential savings had been largely underestimated (79–83).

There are a number of reasons to conserve energy, or to utilize energy more efficiently, e.g. by improving the efficiency of conversion processes. Some of the advantages are lesser dependence of a country on imported primary energy, conservation of non-renewable sources, avoidance of ecological damage or crises, and extension of existing energy reserves, thus also extending the time available for the development of new energy technologies.

Decreasing the dependence of a country on primary energy imports, especially petroleum, increases the assurance of available supplies. Furthermore, a saving strategy on the part of the principal user countries can ease the world petroleum market tension and afford some leeway for potential price hikes, thus keeping the oil producing countries within limits.

Preservation of non-renewable energy sources – this holds particularly for petroleum – is based not only on the necessity that they not be consumed as energy sources, but on the fact that they constitute a raw material source which is difficult to replace. It is also a moral question whether it is justifiable to squander, in a relatively short time span, raw materials which took millions of years to form.

With the present world-wide growth rates, the growth in the use of fossil fuels creates a global ecological liability: an excessive CO_2 content in the upper atmosphere will influence the climate (see 5.3). The reduction of the energy demand by conservation energy brings about a slowing of the growth rate, i.e. extends the availability of the primary energy resources. Owing to the uneven geographical distribution, this in turn will tend to lessen international tensions. Long range solutions with long range advantages will be preferable to short range solutions with only short range advantages.

The primary energy consumption per capita in the United States is almost twice that of the Germany Federal Republic. As both countries enjoy a comparable quality of life, it is clear that important energy consumers can use energy more efficiently (84–86). The comparison also indicates that less primary energy consumption does not necessarily mean giving up of goods and services or a decrease in standard of living (87–89).

Even though in German Federal Republic – in contrast to other highly industrialized countries – a relatively high gross national product per capita was realized in 1978 with only a 33% exprense of primary energy, ca. 67% of this primary energy was lost in conversion, in use and transportation (56,80). This may be related to the fact that the type of primary energy used in the individual sectors has not been chosen on the basis of the form in which it is consumed. The energy supply systems in many countries do not take advantage of the fact that the major portion of the end use is low-grade heat. Of course, some of the losses are unavoidable and are dictated by the laws of nature (see 2.2), but others can be avoided or reduced by better technology.

Detailed investigations show that there is ample room for energy saving in the German Federal Republic. The relative final energy saving potential for the industrial sector in 1985 – the ratio of the (hopefully) conserved final energy over the predicted final energy consumption without any saving – is estimated at ca. 11%. A corresponding final energy conservation potential in the transportation sector is said to be ca. 10% and in the household and commercial sector ca. 15% (89).

Furthermore, energy can be conserved without any sacrifice in comfort. The same overall effect can be achieved with less energy if the proper measures are taken (91–95). The same room temperature can be attained with more added heat and less insulation, or with less heat input and better insulation. In other words, comfort in heated rooms can be achieved by a combination of heat and insulation; i.e. more efficient energy utilization. In this way one can consider energy conservation as a new "energy source" (89).

All energy conservation methods are characteristically capital intensive (96, 97). One can reason, however, that substitution of saving measures (capital) for energy expenditures, i.e. achieving a particular end with a lower energy consumption does not necessarily reduce the gross national product (89).

Some characteristics of specific energy conservation possibilities in the most important use sectors (industry, household and commercial, transportation) in the OECD countries will be discussed. According to a Shell Oil Company study of the year 1979, the primary energy consumption per unit of gross OECD product decreased by about 7% between 1973 and 1978. The following factors were responsible: The proportion of low energy intensive production of consumer goods increased, and new producers, especially in heavy industry were more energy efficient (87). The industrial energy demand varies widely from sector to sector. Heavy industry (iron and steel, nonferrous metals, glass, minerals, cement), and the chemical industry are particularly energy intensive (87). This type of industry accounts only for 20% of industrial production in the OECD countries, yet consumes more than half of all industrial energy. For this reason, rational utilization of energy in these areas is of prime importance. In heavy industry, as a rule, the plants are large and capital-intensive with a long useful life (low replacement rate). For this reason, the short and medium range possibilities for rational use of energy will largely consist of introducing energy saving components into existing plants – combined with improved process controls – and the use of different fuels and recovery of waste heat. Savings can often be effected in steam generation (e.g. by better measurement and regulation devices) and in space heating (e.g. better utilization of waste heat).

Frequently, integration of energy demand over several segments of an industrial complex permits a more rational energy utilization. For example, waste heat from a high temperature process can be used, via heat exchangers, as a heat supply for a process which operates at a low temperature (compare Fig. 2-11). Wherever applicable, the principle of heat/power coupling, or heat recovery should be employed (see 4.62). Arbitrary barriers which obstruct the exchange of (electrical) energy between individual companies, or with a public utility grid should be abolished.

The major portion of energy consumed in the household/commercial sector is spent for space heating (compare Fig. 2-10). Residential buildings have a relatively long useful life – up to 80 years. Hence more rational use of energy in existing structures (energy saving) is important, particularly since replacement by erecting new buildings would require considerable time. Even if the energy consumption of existing units could be cut by 20% by 1985, and energy saving investments and building codes for buildings erected after 1980 saved 50%, only half of the realizable energy saving potential in the building area would be achieved by the year 2000 (87).

Studies have shown that people all over the world prefer to live and work in a narrow temperature confine, namely between 22°C and 24°C. However, this temperature range can be affected by clothing, by ventilation and air humidity. This preferred range could be lowered substantially if people were willing to wear warmer clothing. In Western Europe, each additional degree of interior temperature for indoor heating requires a 5 to 10% energy increment. Hence, considerable energy can be saved by lower thermostats settings in homes and offices.

Apartment buildings can be divided into those supplied by a central heating plant and those heated by individual units. Capital investment savings and greater fuel flexibility argue for centralized heating plants. However, heat provided by the central unit should be metered out individually. Only this way can the renter monitor his own consumption and be motivated to save. Concerning the heating plant of an apartment house, there is a dichotomy of interest on the part of the owner or landlord and the tenant. The installation of a fuel-saving (or different fuel-fired) heating plant demands capital outlay from the owner, but benefits the tenant in lower heating costs. There are signs that these inherent contradictions have been recognized and that appropriate legislative steps are being taken.

The transportation sector is the largest user of petroleum products. In Japan, ca. 20% of petroleum imports go to the transportation area, in Western Europe, the average is 27% (in the Federal Rep. of Germany it accounts for 30%). In the United States, which has the largest traffic in individual cars, this figure is 50%. (In 1978, the U.S. per capita consumption of gasoline was 1281 liters, or about 4 times that of West German consumption of 289 liters). Land, sea and air traffic are almost 100% dependent on petroleum. In the area of freight traffic, energy accounts for 30% of all cost – which has been recognized as a major cost factor for some time and hence was a target for saving measures. The various modes of transportation (shipping, truck traffic, rail, air) already have specialized in the areas of their competitive advantages. Thus substituting one mode of freight movement for another has its limitations (87).

Efficiency of land passenger traffic is dependent on the load factor. Railroads and buses are more efficient at full capacity operations. The load factor in turn is a function of the public transportation grid. A dense public transportation grid will necessarily be operated at lesser loads, and a less dense grid most likely will operate at an increased degree of load, or greater energy utilization. However, the inconvenience of a sparse public grid will act as a deterrent to riders who will tend to make less use of public transportation (87).

Land passenger traffic will improve in the next few years with more efficient vehicles, especially fuel-saving engines leading to considerable fuel conservation. (Fuel savings up to 35% are possible for the same distance or amount of travel).

The Conservation Commission of the World Energy Conference has studied the world wide energy saving potential up to the year 2020. The growth of the gross world product (GWP) is assumed at 4.6% between 1975 and 2000 and then should decrease to a 4.1% annual rate between 2000 and 2020; this would correspond to an averaged growth of the GWP over the total period of 4.37% per year. The studies are based on the present state of technology (e.g. present specific energy consumption for industrial production, or energy spent per product unit), and the expected technological development (e.g. expected development of specific energy consumption for industrial production). As a frame of reference one uses knowledge as to how much energy has already been saved over the originally forecast energy demand (26). The world primary energy demand without any saving, i.e. constant elasticity of demand ratio or reference case (ED-ratio = the percentage change in energy demand divided by the percentage change in GWP; 1975 value = 1.00), is estimated at $63.1 \cdot 10^9$ tce for 2020. With the anticipated energy saving it is predicted at $34.1 \cdot 10^9$ tce (This is based on an ED-ratio of 0.54). In other words, the energy conservation potential for the year 2020 is estimated at a maximum of 46%. The energy conservation potentials in the various sectors are estimated as follows: Industry 41%, household and commercial 76%, transporation 40%, energy sector 33% (26).

Corresponding energy conservation potentials are 9% for 1985 and 33% for 2000 (26). Similar investigations of future energy conservation potentials from other sources arrive at similar conclusions (87, 95).

2.333 Probable development of the world primary energy demand

It is a very difficult task to predict the probable development of the primary energy demand of individual countries or groups of states, owing to many interlocking factors.

The price increases and trade-curtailing practices of the OPEC countries after October 17, 1973, caused a world-wide reduction in oil consumption in the years 1974, 1975, and 1980, 1981 (98). Furthermore – as already discussed – based on the tremendous price increases and the ensuing changes in user attitudes, the primary energy demand estimates, predicted for individual countries, were revised downwardly several times (see 2.332).

Fundamental aspects of the energy programs of individual countries have been repeatedly revised, which shows how difficult it is for governments to develop appropriate energy concepts for their respective countries. This is exemplified by comparing the U.S. energy program "Project Independence", proclaimed by president R. Nixon on 23 January 1974 and aimed at making the United States energy self-sufficient (99), with the message of president J. Carter of April 19, 1977, where the concept was an admonition to the American population to conserve energy (100–103).

Similar comments on the development of energy concepts can be made for the Federal Republic of Germany and for other industrialized states (9, 11, 104). Not only were quantitative estimates revised, it also became apparent that hopes to replace the petroleum component in the energy demands of the major consumers by other sources (e.g. coal, nuclear energy, solar energy) at least in part, could not be realized to the degree originally hoped for (see 3.313) (105, 106).

The possibilities for a rapid conversion to nuclear energy in many developed industrialized countries have been much overestimated in the past years. Resistance on the part of the population against certain developments such as breeder reactors is very strong in some countries (107–110). Discussions in the United States, in Austria, in Sweden, in Switzerland, and in the German Federal Republic are worth consideration (111–115). On the other hand, there seems to be no great resistance on the part of the French population towards an expanded nuclear energy development.

The United States Atomic Energy Commission, in 1973, predicted a nuclear energy derived electrical output of 240 GW for 1985; in the years 1977 and 1978, these figures were revised to 163 GW and 110 GW, respectively. Actually installed nuclear capacity in the second quarter of 1978 was 48 GW. Present estimates of 1985 installed nuclear capacity are about 100 GW. This corresponds to a primary energy share of $220 \cdot 10^6$ tce ($6 \cdot 10^{15}$ Btu) or 7% (21). Deviations in long range predictions are even larger. Recent estimates for nuclear capacity in the year 2000 are only 185 GW, corresponding to a share in primary energy of $400 \cdot 10^6$ tce ($11 \cdot 10^{15}$ Btu) or 12% (21).

In the year 1973, predictions for the German Federal Republic estimated a gross nuclear electrical output of ca. 55 GW annual capacity for 1985. Downward revisions in March 1977 and in December 1977 arrived at figures of 30 GW and 24 GW, respectively. The last figure is equivalent to a primary energy share of $50 \cdot 10^6$ tce or 10% for 1985 (9, 11). Again, discrepancies in the predictions for the year 2000 are even larger. In 1973, the nuclear derived capacity estimate for 2000 was 170 GW, but in December 1977, the government of the German Federal Republic estimated 75 GW (9). The 1979 estimate of the Shell Oil Company for 2000 was ca. 35 GW, corresponding to a primary energy share of $72 \cdot 10^6$ tce or 14% (87).

The OECD Nuclear Energy Agency (NEA), in 1975, predicted a total world nuclear generated electricity of 2000 GW for the year 2000. This value was scaled down to 1000 GW in December 1977.

France will probably have a nuclear capacity of 30 GW in 1985. Their large share of petroleum-derived electricity production, ca. 51% is expected to be reduced drastically. (In the United Kingdom, the petroleum share of electrical power production in 1978 was ca. 25%, in the German Federal Republic, ca. 8%).

The predicted primary energy consumption for the German Federal Republic up to 2000 and its structural development is shown in Fig. 2-4 and Table 2-3, respectively; the comparable data for the United States are shown in Fig. 2-5 and Table 2-4; and world wide predictions are summarized in Fig. 2-3 and Table 2-2.

Even fairly recent predictions contain large discrepancies, as already discussed in section 2.332. (Later predictions as a rule give lower values).

There is a structural similarity in the probable developments of primary energy demand in different highly industrialized nations, and fact, in the world as a whole. All the predictions for the year 2000 have in common that fossil fuel sources – petroleum, natural gas, and coal remain the basic energy sources. Petroleum will maintain its predominant position up to 2000. Coal consumption will increase considerably and nuclear energy will show the strongest growth.

Experience has shown that accurate prediction of future primary energy demand is almost impossible even for a single country and only for one or two decades. This explains the difficulty of predicting world wide energy demand data up to the turn of the century. There are many interrelated factors which are different from country to country and from region to region.

Forward planning in the energy field requires tremendous lead times. To build a nuclear power plant will take ca. 10 years from the time the decision to do so is reached until the start of operation. If one assumes a useful life of 20 years, the forward planning is of the order of 20 years. The time horizon moves further into the future if a new, untried technology is involved. According to a study made by the International Institute for Applied Systems Analysis (IIASA) in Laxenburg/Vienna, the time required for gain (or loss) of a 50% market share when one energy source is succeeded by another (series: wood – coal – oil – gas – nuclear) is in excess of 50 years (117). These facts re-emphasize that in order to guarantee long term energy supplies for mankind, accurate predictions of world wide primary energy demand and supply data are urgently needed. For these reasons, predictions of the Conservation Commission of the World Energy Conference about the probable development of the world primary energy supply are here discussed. By co-operation of 76[1)] nations from all the four corners of the globe, certain nationally and certain regionally conditioned influences are eliminated.

[1)] The National Committees of the World Energy Conference are:

Algeria
Arab Republic of Egypt
Argentina
Australia
Austria
Bangladesh
Belgium
Brazil
Bulgaria
Canada
Chile
Colombia
Costa Rica
Cuba
Czechoslovakia
Denmark
Ecuador
Ethiopia
Finland
France
Germany (Democratic Republic of)
Germany (Federal Republic of)
Ghana
Great Britain
Greece
Hungary
Iceland
India
Indonesia
Iran
Ireland
Israel
Italy
Ivory Coast
Japan
Jordan
Korea (Republic of)
Liberia
Luxembourg
Malaysia
Mexico
Morocco
Nepal
Netherlands
New Zealand

Table 2-9: Probable development of world primary energy demand according to regions

Year	OECD [10^9 tce]*	%	Centrally planned economies [10^9 tce]	%	Developing nations [10^9 tce]	%	World [10^9 tce]	%
1980	6.1	57	2.9	28	1.6	15	10.6	100
1990	7.2	51	4.1	28	2.9	21	14.2	100
2000	8.2	43	5.7	30	5.2	27	19.1	100
2010	8.9	35	7.9	31	8.6	34	25.4	100
2020	9.5	27	11.1	33	13.5	40	34.1	100

* 1 tce = $29.3 \cdot 10^9$ J.

Source: World Energy: looking ahead to 2020. Report by the Conservation Commission of the World Energy Conference, Guildford (UK) and New York: IPC Science and Technology Press 1978.

The countries of the world are divided into three major groups: The OECD countries (North America, Western Europe, Japan, Australia and New Zealand), the Centrally planned economies (USSR, Eastern Europe, P.R. of China, the communistic countries of Asia), and the Developing nations (OPEC, Latin America, Middle East, Africa, South and East Asia).

Table 2-9 summarizes the probable world primary energy demand according to regions up to the year 2020 and Fig. 2-12 graphically depicts this development. The primary energy demand is expected to increase from $9.8 \cdot 10^9$ tce in 1980 to $34.1 \cdot 10^9$ tce in 2020 (26). The share of the OECD will decrease to 27% by 2020 while the share of the developing nations will increase to 40%. The $13.5 \cdot 10^9$ tce demand predicted for the developing nations in 2020 would give them an average per capita energy consumption equal to the present world average (compare Table 2–5).

Estimates of the future world primary energy demand depend critically the estimate of the future development of the energy-GNP elasticity coefficient k (see 2.32). This quantity k is at present about 0.8 in the OECD countries and 1.3 in developing

Nigeria
Norway
Paraguay
Pakistan
Peru
Philippines
Poland
Portugal
Romania
Senegal (Republic of)
Sierra Leone
South Africa (Republic of)
Spain
Sri Lanka
Sudan (Democratie Republic of the)
Sweden
Switzerland
The Territories of Taiwan, Kinmen, Matsu, and Penghu of the Republic of China
Tanzania
Thailand
Trinidad and Tobago
Tunisia
Turkey
Uganda
Union of Soviet Socialist Republics
United States of America
Uruguay
Venezuela
Vietnam
Yugoslavia
Zambia

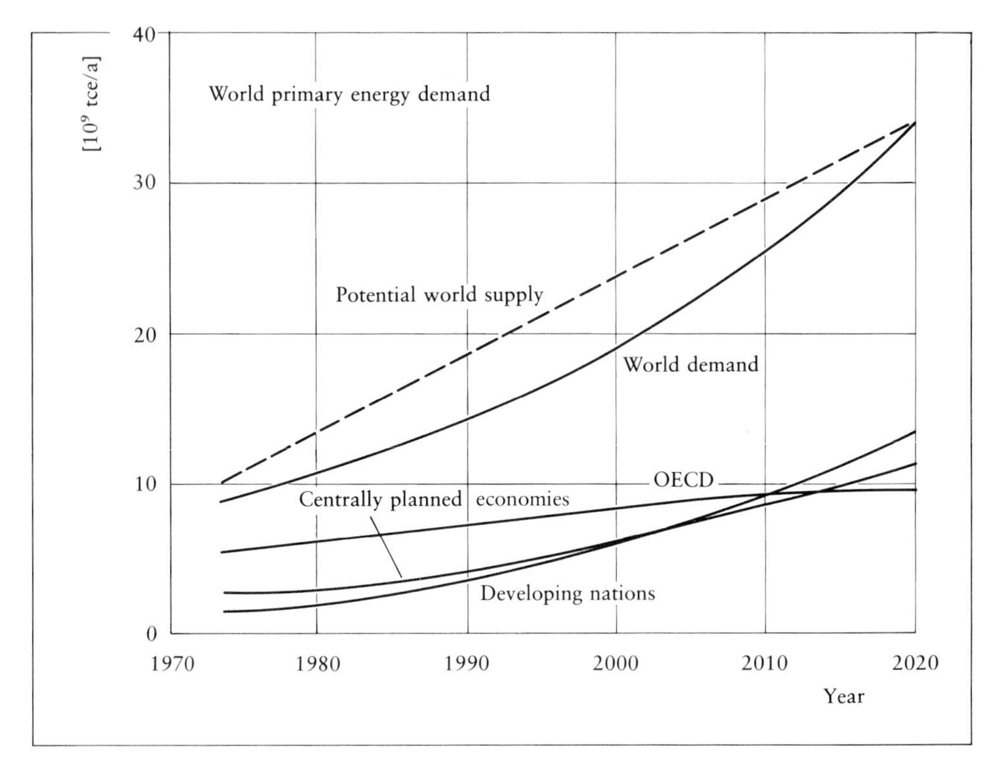

Fig. 2-12: Projected development of world primary energy demand

Source: World Energy: looking ahead to 2020. Report by the Conservation Commission of the World Energy Conference, Guildford (UK) and New York: IPC Science and Technology Press 1978.

countries. With increasing industrialization, k should decrease. In other words, structural changes and adaptation in economy, saturation phenomena and more rational energy consumption will tend to decrease energy expense per unit of GWP. The Conservation Commission expects that the energy-GNP elasticity coefficient in the OECD countries will decrease from its present k = 0.8 to k = 0.4 in 2020; and that the present k = 1.3 value of developing countries will drop gradually to k = 0.9 by 2020 (26).

The Conservation Commission has also taken into account the influence of future energy costs on consumption by means of the price elasticity coefficient. This price-elasticity coefficient is defined as the ratio of the percentage change in energy consumption to the percentage change in energy price. For example, for the OECD countries a price-elasticity coefficient of 0.4, and for the developing countries of 0.3 is assumed. This means that a 10% rise in energy costs in the OECD countries leads to a reduction of energy consumption of 4%, in the developing countries of 3%.

The question remains whether and how the future world primary energy demand can be satisfied. Based on the extensive investigations of the Conservation Commission, the world primary energy demand was met, up to the year 2020, as is shown in Tables 2-10 to 2-12. Tables 2-10 and 2-11 for the existing energy potentials for petroleum and natural gas, indicate the distributions over the various regions (see 3.32, 3.33). Possible further extension of fossil fuel sources, while taking into account the data in Tables 2-10 to 2-12, is shown graphically in Fig. 2-13. Crude oil production will reach its peak between 1985 and 1995, then decline. Natural gas production will peak much later, about the turn of the century, and remain at a relatively high level for some time. The future petroleum share of the OECD countries will range

Table 2-10: Potential energy from oil resources

Year	OECD [10^9 tce]*	%	Centrally planned economies [10^9 tce]	%	Developing nations [10^9 tce]	%	World [10^9 tce]	%
1972	0.8	21	0.9	24	2.1	55	3.8	100
1980	1.0	15	1.5	23	4.1	62	6.6	100
1990	1.2	16	1.6	21	4.7	63	7.5	100
2000	1.0	15	1.8	27	3.8	58	6.6	100
2010	0.8	16	1.3	26	2.9	58	5.0	100
2020	0.7	19	0.8	22	2.1	59	3.6	100

* 1 tce = $29.3 \cdot 10^9$ J.

Source: World Energy: looking ahead to 2020. Report by the Conservation Commission of the World Energy Conference, Guildford (UK) and New York: IPC Science and Technology Press 1978.

Table 2-11: Potential energy from natural gas resources

Year	OECD [10^9 tce]*	%	Centrally planned economies [10^9 tce]	%	Developing nations [10^9 tce]	%	World [10^9 tce]	%
1972	1.0	63	0.4	25	0.2	12	1.6	100
1980	1.0	50	0.6	30	0.4	20	2.0	100
1990	1.1	35	1.2	39	0.8	26	3.1	100
2000	1.3	27	2.0	41	1.6	32	4.9	100
2010	1.0	21	1.7	35	2.1	44	4.8	100
2020	0.6	14	1.2	28	2.5	58	4.3	100

* 1 tce = $29.3 \cdot 10^9$ J.

Source: World Energy: looking ahead to 2020. Report by the Conservation Commission of the World Energy Conference, Guildford (UK) and New York: IPC Science and Technology Press 1978.

Table 2-12: Potential world primary energy production (in 10^9 tce)*

Resource / Year	1972	1985	2000	2020
Petroleum	3.8	7.4	6.6	3.6
Natural gas	1.6	2.6	4.9	4.3
Coal	2.3	3.9	5.8	8.7
Hydropower	0.5	0.8	1.2	1.9
Nuclear Energy	0.1	0.8	3.0	10.7
Unconventional oil and gas	–	–	0.1	1.4
Renewable (solar, geothermal, biomass etc.)	0.9	1.1	1.9	3.4
Total	9.2	16.6	23.6	34.1

* 1 tce = $29.3 \cdot 10^9$ J.

Source: World Energy: looking ahead to 2020. Report by the Conservation Commission of the World Energy Conference, Guildford (UK) and New York: IPC Science and Technology Press 1978.

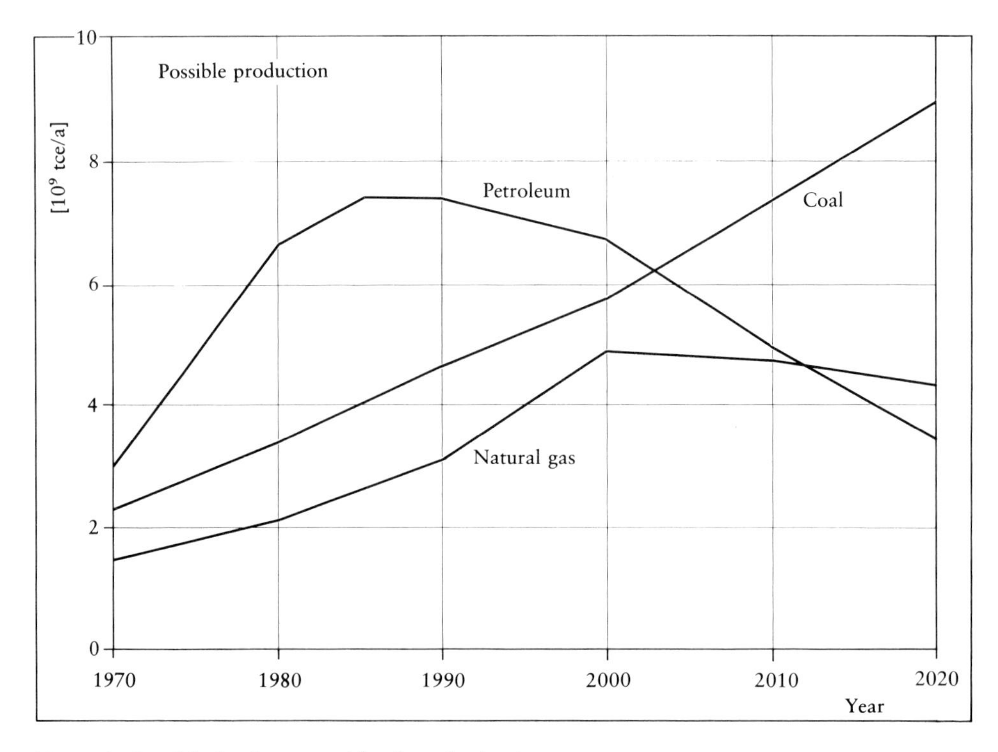

Fig. 2-13: Possible development of fossil production*
*The data are taken from tables 2-10 to 2-12

between 15% and 19% in the envisioned time span; that of the developing countries between 59% and 63%. Regarding the distribution of natural gas, the OECD share will decrease from the present ca. 50% to 14% in 2020, with a converse gain in the developing countries from 20% to 58%.

It should be stressed here again that political developments may cause crude oil starvation in some important industrialized nations, long before the oil wells of the owner countries run dry.

Fig. 2-13 shows that coal production in the contemplated time span – and further (see 3.31) – has a pronounced growth tendency. The proved coal reserves are adequate to allow this growth provided that development of production technology continues vigorously. It is significant that a large portion of this primary energy source is located in the OECD nations (see 3.31). The increase in West European production between 1975 and 2020 will be relatively small; from $300 \cdot 10^6$ tce to $400 \cdot 10^6$ tce. The greatest contributions for this time span are anticipated for the United States, $581 \cdot 10^6$ tce going to $2400 \cdot 10^6$ tce; for the USSR, $614 \cdot 10^6$ tce going to $1800 \cdot 10^6$ tce; and for China, $349 \cdot 10^6$ tce going to $1800 \cdot 10^6$ tce (see 3.312) (26).

As the coal reserves are located in these important industrialized nations, it is anticipated that they will be used to cover their own needs and the world coal trade, for the immediate future, will remain quite small in comparison with crude oil trade.

The world coal trade volume presently amounts to $240 \cdot 10^6$ tce, of which, however, about a third is a kind of domestic trade within the economic blocs EC, COMECON, and USA/Canada (see 3.312). World trade is expected to grow by 2020 to $800 \cdot 10^6$ tce, about 30% of the present day crude oil trade.

The share of the fossil fuels petroleum, natural gas and coal in the present day primary energy consumption of the world is about 90%. Given the probable developments in petroleum and natural gas, mankind will be forced to substitute other energy sources for these hydrocarbons. The Conservation Commission of the World Energy Conference came to the conclusion that the potential world primary energy production, shown in Table 2-12 will cover the probable world primary energy demand for the time being, up to 2020 (compare Table 2-9 and Fig. 2-12).

After the year 2020, the fossil energy sources (petroleum, natural gas, coal) can only provide $16.7 \cdot 10^9$ tce/year or about half of the primary energy demand of $34.1 \cdot 10^9$ tce/year. The other half would have to be covered by hydropower, nuclear energy, solar energy, biomass etc.

Hydropower is expected to grow about 4-fold in the non-industrialized nations to ca. $1.9 \cdot 10^9$ tce/year. The major portion, based on this study, will have to come from nuclear energy with $10.7 \cdot 10^9$ tce/year, which would make it the main source. Regenerative sources are expected to account for $3.4 \cdot 10^9$ tce/year. Finally some presently unavailable oil and gas reserves, including future deep ocean wells, future strikes in the arctic, oil shale and oil sands, will have to provide the balance up to $1.4 \cdot 10^9$ tce/year.

It was shown in 2.332, that realizable energy conservation potentials have been repeatedly underestimated. One reason for the discrepancies may be that the efforts to develop new energy sources particularly for the main consumer countries were enormous, while the efforts to use energy more rationally and practice conservation began only in the mid 1970's. Developments in the past years also show decisively that the rate of nuclear energy development has been repeatedly exaggerated. However, even enough nuclear energy may continue to grow at the highest rate, the fraction of the primary energy demand met by it will only grow moderately.

With further increases in energy costs, the trend for more rational energy use will persist – particularly in the major consumer countries. A slower increase in demand in many countries is presently expected as a logical result.

It is generally agreed that the more uncommon energy sources: tidal, geothermal, and some forms of solar energy, such as wind, will acquire at most local, but certainly not global importance.

There are some signs, however, that opinions are changing with respect to solar energy utilization, particularly for providing low temperature heat. Many countries e.g. the United States, Japan, and the European Community (EC) countries now believe that solar energy may supply a significant amount of low temperature heat sooner than hitherto expected (see 2.331). The Energy Research and Development Administration (ERDA) estimates a primary energy share for solar energy in the USA in the year 2000 of 6%, increasing to 25% by 2020 (118, 119). Recent projections are even higher (120–122). For the EC bloc, 5–10% of primary energy is expected to come from the sun in the year 2000 (123). Japan has a government sponsored program entitled "Sunshine Project" (124, 125).

Contributions to the primary energy demand in the future may come from presently unavailable oil and gas finds (deep ocean wells, and wells in arctic regions) as well as from oil shale and oil sands, sources referred to as unconventional oil and gas, which have been underestimated so far (126–129). The reason for this development is that since 1950, the trend in development of secondary sources in the industrialized countries is to provide easily transportable gas, liquid, or electrical energy. This trend is likely to persist.

As a result Europe, North America and other parts of the world have widely branched pipeline nets, which are supplemented by low pressure grids for individual consumers in many countries. (In the German Federal Republic, the gas pipeline net extends ca. 114000 km; in the USA, natural gas pipelines extend over 640000 km). The petroleum industry has its own pipeline nets, tanker and truck fleets, built at great capital investment cost. (The North American pipelines for transporting crude oil are about 370000 km long). In other words, the present investment in transportation grids gives liquid and gaseous energy forms great advantages over solid sources (see 3.313, 3.323, 3.333). It is very likely that as petroleum and natural gas become less available, they will probably be replaced by synthetic liquid and gaseous fuel from coal (see 4.64) (130). Natural wells will be replaced by "synfuel" wells.

Owing to the many factors in estimating future demand for primary energy sources, it is clear that predictions beyond the year 2020 are extremely tenuous and difficult. W. Häfele and W. Sassin extrapolate to the year 2030 with roughly $38 \cdot 10^9$ tce (131, 132). H. Kahn predicts a primary energy consumption in the year 2176 of $3.6 \cdot 10^3$q (1q $= 3.62 \cdot 10^7$ tce $= 10^{15}$ Btu $\approx 10^{18}$ J) (compare Table 2-13). This presupposes a new "energy awareness" so that conversion and utilization of energy become more and more effective. Energy demand will not grow as fast as the GWP. In other words, the primary energy demand in the next 200 years will grow 15-fold, while the GWP will grow 60-fold.

To understand the vastness of the figures in connection with future primary energy demand, a simple model is developed. In 1978 the per capita primary energy consumption in Japan amounted to ca. 3.8 tce. Other nations on the globe go through similar economic developments and one can project a (lower limit) total world population of ca. $10 \cdot 10^9$ people. This equates to a world primary energy consumption of ca. $38 \cdot 10^9$ tce (= 1020 q/year = $1.02 \cdot 10^{18}$ Btu/year). One can argue that all these are hypothetical considerations, as the primary energy consumption of highly developed industrialized countries must still increase, even though at a lesser rate. The tenuous nature of the predictions is related to the fact that by the middle of the coming century, over 80% of the world population will live in the developing countries, and that a large amount of their energy demand will not have to be spent for space heating as e.g. is now the case for Western Europe. A comparable living standard in these regions will be realized with a much lower primary energy consumption per capita.

Table 2-13: Estimates of world primary energy consumption

Year	Population [10^9 people]	GWP[1)]/ capita [1975 US-Dollar]	Annual consumption [10^3q][2)]	Cumulative consumption from 1975 [10^3q]
1975	4.0	1300	0.25	–
1985	5.0	1700	0.35	3
2000	6.6	2600	0.60	10
2025	9.3	5600	1.20	30
2076	14.6	10400	2.40	115
2126	15.0	15200	3.20	240
2176	15.0	20000	3.60	400

[1)] GWP: Gross World Product.
[2)] 1q $= 3.62 \cdot 10^7$ tce $\approx 10^{18}$ J.

Source: H. Kahn et al., The Next 200 Years, A Scenario for America and the World, New York: William Morrow and Company, Inc. 1976.

2.4 Aspects of energy economics

2.41 Development of energy economics

The economic development in the past century was made possible by the invention of engines which were independent of time or location, and by the availability of large amounts of cheap energy. Large industries frequently located near the geographical site of the available primary energy.

In the second half of the 19th century and in the beginning of the 20th, the basis for the industrialization of the highly developed nations was the technology connected with anthracite. As a reducing agent for iron ore in high temperature furnaces, it created an iron producing industry in Western Europe, Japan, the United States, and the USSR. Anthracites plays a decisive role in the industrialization process of China, India, Poland and Czechoslovakia. Heavy industry is often found near the location of large anthracite mines. Examples are: the Ruhr and Saar basins of the German Federal Republic, Yorkshire, Nottinghamshire and Durham of Great Britain, Alsace-Lorraine of France, Beuthen and Cattowice of Poland, the Donets basin, Kusnezk, and Karaganda of the USSR, Pennsylvania, Virginia, and Kentucky in the United States, Chikugo, Ishikari in Japan, and Bihar and Bengal in India (133).

As industrial nuclei develop, their energy appetites grow. New finds, adjacent to the older locations are made, as older mines become exhausted, or non-economical to operate, or no longer are able to satisfy the demands of heavy industry. As production has to move from the industrial sites, it frequently entails relocation of energy-intensive industrial branches. They, in turn, attract complementary industries, and new industrial sites, removed from the original industrial center, come into their own.

The above described process can be traced in many instances. For example, the industrialization of the Moscow region (nucleus) began with the operation of the local mines. The first satellite industry was the Donets basin which grew into a major nucleus of its own. The far satellite, the Urals where metallurgical industry based on charcoal had developed prior to the revolution, in turn found its complementary region in the 200 km distant Kutznek basin. This was followed by utilization of Karaganda coal, of the Fergana deposits, etc. (134). Similar steps can be outlined for the USSR petroleum sources. With geographical constraints, the same process can occur in smaller countries in a much smaller space[1].

The fossil fuels, coal, petroleum, and natural gas, were used as raw materials in ever increasing amounts. In the early stages of industrialization, the chemical industry had its beginnings near anthracite and lignite deposits. Then petroleum and its

[1] In the Federal Republic of Germany (Lower Saxony), in Eastern France, in Corby in Great Britain, and in Sweden, ore deposits became the basis of smelters, which owed their successful existence to the smelting of low cost regional raw materials.

products became raw materials for entirely new industries. The invention of many new chemical products led to replacement of one raw material by another, for example, metal ores (metals) by petroleum (plastics). This development will persist for some time to come, and for reasons not discussed here, it will be much more difficult for mankind to replace petroleum as a raw material than as an energy source.

The availability of low cost energy was also often responsible for bringing other industries into being. An example is the electrical furnace production of aluminum metal, as well as other metals by the energy-intensive electrometallurgical industry, which is often located near hydroelectric power sources.

Changes in energy technology and in supply and demand of available energy led to technological developments which were of much greater importance than geographical proximity. The more important factors were transportation, climatic, demographic and ecological circumstances.

For example, an important steel industry came into being in Japan, even though this industry is very energy-intensive and Japan is forced to import most of the iron ore and coal for coking from overseas. This was the reason for locating the Japanese steel industry on the coast line, reducing transportation costs of import and export.

The problem of furnishing energy over long distances was solved by electrical power grids of national and international scope. In the same manner, international pipelines for crude oil and natural gas were constructed and supertankers for crude oil and liquified natural gas were built. This meant that the previously existing industrial areas could develop much father than the local sources would have allowed. (The economical transportation of oil and gas in pipelines is largely responsible for the displacement of coal by petroleum and natural gas as energy sources in many areas (see 3.232 and 3.333)).

The facile transport of crude oil also led to the fact that in many countries, refineries were no longer built adjacent to the oil wells (raw material orientation), but near the user centers (use orientation). This tendency was strengthened by the rapid growth of consumption of petroleum products and heating oil, and could be observed in countries as different as the Federal Republic of Germany and the USSR. Owing to the smaller distances and the geographical size of the German Federal Republic vis-à-vis the USSR, conditions there were most favorable. For example, no petroleum refining was done south of the Main river prior to 1962. In 1967, however, 44% of the West German refining capacity was located south of the Main. Thus, no end-user in the Federal Republic of Germany is further than 120 km from the nearest refinery.

Due to differences in geological conditions, the production costs for coal as well as for the hydrocarbon fuels petroleum and natural gas can vary enormously from site to site. For example, there are vast coal reserves in the American Middle West which are located close to the earth's surface and where coal can be obtained with little effort (strip mining). In Western Europe, on the other hand, the expense of mining anthracite is quite high. Importing energy material is often cheaper than the opening of new deposits or sinking very deep mine shafts. For example, in the Federal Republic of Germany, in spite of temporary excess production, and capacity reductions – in the years 1957 to 1974, a total of 105 coal mines were abandoned – anthracite was imported from the United States, which was particularly economical for coastal regions. In 1976, the German Federal Republic imported 13% of its anthracite requirement. The situation for petroleum is similar. The cost for producing oil from the North Sea fields is many times larger than the cost of Near East imports.

As the industrial nations, prior to the oil crisis in the Fall of 1973, could import cheap crudes in large amounts, those of their own deposits which were uneconomical for geological reasons were largely disregarded. The oil price politics of the OPEC countries have changed all this. To overcome adverse geological conditions, new technologies (off-shore drilling) and new mining operations (secondary, tertiary production) had to be developed. The temporary deprivation politics of the OPEC countries during the height of the oil crisis contributed to and speeded up the development of new technologies in the industrialized nations (see 3.323).

A number of nations own vast primary energy resources in unfavorably located regions – on the Northern part of the globe. Their exploration is planned or they are being actively developed. The nations are the United States, with large petroleum deposits in Alaska, Canada, the USSR (136), and the Skandinavian countries. Norway has begun the exploration of its continental shelf in the Barents sea. The Northern parts of Alaska, Canada, and the USSR are sparsely populated or, in some areas not populated at all (demographic factors). As regards Northern Canada and Alaska, one tries to keep the human work element at a minimum and to automatize production. This confines complementary industry to the bare necessicities. Materials needed in production, i.e. machines and pre-fab housing for workers are flown in, and repair shops operate by exchanging machine parts. In contrast, in the USSR, complementary industries such as home building are started in the northern exploration regions.

The excesses of regionally concentrated industrial areas led to an ever-increasing global awareness of ecological factors. The energy industry is particularly affected. Examples are power plants, be they convertional fossil fueled, or nuclear fueled (137). Ecological considerations are important not only in the most diverse refining processes, but also in the production and transportation of primary energy materials, particularly crude oil. Off-shore technology has particularly deleterious effects, which are caused by spills during exploratory drilling, loading of tankers, and transportation through underwater pipelines.

The factors relevant to the development of energy economies are not independent of each other. Many areas with large deposits of primary energy materials have inadequate traffic or transportation systems. They may have inclement climate, or an uneven demographic distribution. Examples are the Asian portion of the USSR and the upper North American continent. In some instances, however, technological developments have made the working of primary energy deposits possible, even under adverse conditions.

Other non-economical factors may affect energy economics. Owing to the fundamental importance of insuring the energy supply for the national economy, national energy programs and laws have always favored domestic energy sources. In the Federal Republic of Germany we can cite as examples the law for furthering the use of anthracite (1. Verstromungsgesetz) of August 1965; the law to assure the use of anthracite in electric power generation (2. Verstromungsgesetz) of September 1966; the law for further assurance of anthracite use in the European Community nations for electric power generation (3. Verstromungsgesetz) of 1974; the law to assure providing of energy in cases of threatened or actual interruption in the importation of crude oil or natural gas (Energiesicherungsgesetz) of November 1973. Similar measures were taken in the United States in order to assure this country's continued energy supplies (20, 99–101, 114, 130). Strategic factors for energy political decisions may also play a role, as shown in the example of the USSR.

Since energy problems can only be solved by international cooperation, several international (intergovernmental) organizations are involved in the problems of providing energy. In the Western industrial nations, for example, the European Community (EC), the International Energy Agency (IEA), and the Nuclear Energy Agency (NEA) of the Organization for Economic Cooperation and Development (OECD); within the Centrally planned economies, for example the Council for Mutual Economic Assistance (COMECON), are concerned with energy provision. Furthermore, there are international (intergovernmental) organizations in which all nations independent of their political or economic affiliation, cooperate in solving the energy supply problems. Examples for these are the International Atomic Energy Agency (IAEA), an organization of the United Nations with headquarters in Vienna, and the International Nuclear Fuel Cycle Evaluation (INFCE), an expert comission created by the governments of important nations to evaluate established and alternative fuel cycles (see 5.83).

Factors which are important in energy economics are fundamentally independent of the respective economical system, although the degree of priority accorded them differs from country to country. Environmental burdens have become acute both in highly developed free market economies and in centrally planned economies. In a free market economy, industry is profit-oriented, and thus strives to keep production costs as low as possible. For a long time, the costs ascribable to ecological damage were grossly disregarded and not considered in the general pricing structure. Even in the planned economies, where the profit motive is not predominant, industry still has to minimize costs. Under the pressure to achieve ambitious production goals, ecological impacts were mostly overlooked and disregarded. They were considered as restricting production and standing in the way of industrial growth. Admittedly, the ecological burdens are much less pronounced in the East than in the Western industrialized nations. But the reason for this is the considerably lower living standard.

In the densely populated countries of Western Europe, with a high energy consumption per capita and a high degree of motorization, air and water suffer a much greater ecological burden than in the thinly populated states in the East, where the energy consumption per capita is less, and the automobile is not an indispensible part of daily life.

Many ecological problems in the industrialized nations have been recognized, albeit rather late. Industry can only create an ecological burden as long as the government tolerates it. If the government prevails in laws, regulations or sanctions, industry must bear the cost of the ecological burden and calculate it into the price structure (138–140). This principle of forcing the responsible party to bear the cost is the goal of many industrialized nations of the free world. Beyond that, there is clearly a need to develop technologies compatible with the ecology of the surroundings (142–146).

Many problems of the environment, especially in Western Europe, have an international character. Wind, oceans and rivers do not recognize national boundaries. International cooperation is indispensable in order to solve these problems. This fact has been recognized by NATO. This organization commands the means and the organizational structure, and as an international body, in which most large industrial nations are members, possesses the most favorable qualifications for the role of environmental coordinator. In signing the North Atlantic Treaty Organization on April 4, 1949, 12 European and North American nations – today there are 16 member nations – obligated themselves to consider an act of aggression against one nation as aggression against all. Aside from regulations dealing with the threat of war and with the solving of international problems, article 2 of the pact covers a peace program. It is stated there that NATO has to be more than a military alliance. Twenty years after the original signing of the pact, NATO decided on November 6th 1969 on the formation of a Committee on the Challenges of Modern Society (CCMS).

The task of this NATO committee is based on the following fundamental theses:

1. In each concrete objective, one country directs and is responsible for each project undertaken by the environment committee.
2. The committee is oriented toward concrete measures and not research. Available data are collected. The member nations are apprised of the facts and are given recommendations which are then supposed to lead to concrete action.
3. The results of the CCMS committee are available to everyone; the documents are not classified. For example, the United States invited the USSR and other nations to participate in some of the sessions.

The increasing pollution of the oceans, particularly by oil spills, has reached dangerous proportions. It is estimated that an amount of the order of magnitude of 10^6 t of oil, stemming from sewers, underwater wells, and leaks (and backwashing) of tankers is spilled into the oceans yearly. When the oil tanker "Terrey Canyon" foundered on the 18th of March 1967 on the southwest coast of Great Britain, an estimated 110000 tons of crude oil flowed into the ocean. Other mishaps, resulting in oil spills of the same magnitude, happened in May 1976 off the northern coast of Spain (tanker "Urquiola"), in December 1976 near Philadelphia (tanker "Olympic Games"), and in March 1978 near Brest (tanker "Amoco Cadiz").

Owing to the great discrepancies between production and consumption of petroleum, the world trade routes are of utmost importance to the free world, which is the major consumer. Roughly every other barrel of crude oil is exported by the producing country and transported to a consumer center (compare Fig. 3-7).

The off-shore catastrophies of the Ekofisk oil field in April of 1977 where hundreds of thousands of gallons of crude oil seeped into the North Sea, and the accident at a well on the South Mexican coast (near Ciudad del Carmen) in June 1979, where ca. 10^6 liter of crude oil flowed into the Mexican gulf, have demonstrated the ecological risks of utilizing off-shore reserves. The world-wide share of the offshore reserves, presently 17% is expected to grow to 50% by the year 2000. This danger potential must be given more attention in the future, particularly as the North Sea is estimated as the second largest off-shore find of the world, ranking only after the Persian gulf (148).

If one considers that ca. 80% of all the crude oil transported in the world is shipped in tankers belonging to the NATO countries and that the producers of North Sea crude are also NATO member countries, one realizes that cooperation of these countries to assure safety measures is a gigantic step on the road to the solution of global ecological problems.

2.42 Future investment needs in energy economics

2.421 Investment needs in the petroleum industry

The quadrupling of crude oil prices in ca. 2 years after the oil crisis of October 1973 and the oil deprivation politics of the OPEC countries have led the industrial countries which are dependent on imported oil to assign the assurance of the energy supply a high priority. All the subsequently developed energy concepts had the following goals: To use energy more efficiently and to conserve it, to carry out research and development of alternate energy sources, to strive for geographical diversification regarding energy suppliers, and to open up new fossil energy reserves, particularly petroleum, in regions which are considered politically safe.

From the prediction of future energy demand development, some of the investment needs to provide new energy supplies can be estimated. It must be borne in mind that the greater difficulties of opening up new deposits and the concomitant expenses of production processes will result in enormous cost increases. For example, new technologies for obtaining new energy furnished from conventional raw materials must be developed, such as obtaining petroleum from deep wells or the arctic regions.

Owing to the limited oil reserves, large scale technical production of synfuels must also be achieved, for example the production of liquid and gaseous fuels from coal as raw material. Present estimates indicate that synfuels will satisfy only about 1% of the world primary energy demand by 1990.

As has been discussed before, all these predictions are based on the assumption that petroleum will remain the main energy source up to the turn of the century. Petroleum production will require more and more cost-intensive processes (see 3.323). According to a publication of the Chase Manhattan Bank, which traditionally deals with energy problems, the capital investment of the world-wide petroleum industry, in 1972, amounted to $ $25 \cdot 10^9$. To this one must add $ $1.5 \cdot 10^9$ for exploration costs, bringing the total investment to $ $26.5 \cdot 10^9$ (In 1960, this figure was $ $1.6 \cdot 10^9$). In 1975 the total investment had climbed to $ $52 \cdot 10^9$. The bank estimates that the total capital outlay between 1970 and 1985 will amount to a total of $ $810 \cdot 10^9$. (This is 4 times the amount which the petroleum industry spent between 1955 and 1970). To the capital needs and the exploration costs, one must add debt service and interest on capital investment. The predicted total investment of the world petroleum industry from 1970 to 1985 will thus come to $ $1350 \cdot 10^9$.

The various stages from the crude oil well to the consumer require capital investments of vastly different magnitudes. In 1972, the sum for exploration and production was roughly $ $11.6 \cdot 10^9$. For the period from 1970 to 1985 it is estimated to be $ $450 \cdot 10^9$. (For the 15 year period from 1955 to 1970 the comparable figure was $ $104 \cdot 10^9$) (155). The increases reflect the fact that exploration and production will have to move less accessible and less favorable regions.

Investment in the transportation sector in 1972 amounted to $ $5 \cdot 10^9$, of which $ $3.8 \cdot 10^9$ was for tankers and $ $1.2 \cdot 10^9$ for pipelines. In the future, pipeline investment costs will rise much more steeply, as they will have to be built in inaccessible and climatically inclement regions. For example, offshore pipeline costs 2 to $3 \cdot 10^6$ $/km, while comparable land piplines cost 0.4 to $0.8 \cdot 10^6$ $/km. The Alaska pipeline, completed in 1977, cost $5.5 \cdot 10^6$ $/km or, at a length of 1276 km, $ $7 \cdot 10^9$.

Investment in the refinery sector in 1972 was ca. $ $6.3 \cdot 10^9$. (In 1960: $ $1.6 \cdot 10^9$). The reason for this 4-fold increase was caused by the enlargement of refinery capacity in the Western world which grew from $1.1 \cdot 10^9$ t to $2.5 \cdot 10^9$ t. Also one must consider the expenses for protecting the environment, which are considerable, especially in the industrial countries, and contributed much to the overall costs. In the construction of a modern refinery, the protection costs amount to ca. 20%. This percentage is exceeded only in the construction of new power plants.

In the marketing sector, the final phase from oil well to consumer, investment is relatively small. Total investment in 1972 was about $ $2.8 \cdot 10^9$.

If one considers the regional distribution of the petroleum industry in the Western world, the United States rank first before Western Europe. This is understandable, since the United States are not only the starting point for the world petroleum industry, but also by far the largest consumer country; the USA is still an important producer (compare Fig. 2-9b). According to a study by the Irving Trust Company, New York, 85% of the investment volume (except the electric power industry) in the USA will be for petroleum and natural gas (including construction of tankers and pipelines) up to the year 1985.

The world petroleum industry investment in Western Europe has been concentrated on the refinery operation. The off-shore finds of oil and gas brought about a shifting in favor of the exploration and production sector, particularly as the North Sea fields, compared with other off-shore locations like those in the Middle East, are unfavorably located reserves (see 3.323). The approximate investment in the North Sea finds up to 1975 amounts to ca. $\$ 1.5 \cdot 10^9$, and towards the end of the 1980s, $\$ 40 \cdot 10^9$ (in prices of 1975) will be required in order to reach the production goal of 150 to $200 \cdot 10^9$ t/per year (156).

2.422 Investment needs in the entire energy industry

The highly developed industrial nations still face the necessity of making even greater efforts to find alternatives for petroleum as an energy source. According to all projections, only coal and energy from nuclear fission are available as substitute energy forms during the transition from the petroleum phase to the fusion and/or solar energy phase, to guarantee a sufficient supply to meet future energy demand. A part of this primary challenge will be to develop a secondary energy source which is environmentally sound, and to construct a supply system for it. The investments required in the individual energy sectors, and consequently the total anticipated investment, can be estimated on the basis of the predicted demand structure. One must keep in mind, however, that following the exorbitant rise in oil prices, the prices for the other primary energy materials, such as natural gas, coal, and uranium ore have also gone up.

According to data published by Citibank, New York, world wide energy investment (without the Eastern bloc) in 1970 amounted to $\$ 74 \cdot 10^9$. Of this, 56% went towards electricity generation, 28% to the petroleum industry, 13% to providing gas, and 3% to others, which includes nuclear energy. The figure for 1980 is $\$ 162 \cdot 10^9$, corresponding to a yearly growth rate of 8.2%. If one assumes the same continued growth rate for the free world, the period from 1970 to 1985 will require an investment of ca. $\$ 2000 \cdot 10^9$, half of which will be needed in the 1980's (155).

Estimates of the Chase Manhattan Bank are somewhat higher. According to this bank, the energy consumption of the Western world between 1970 and 1985 will require ca. $\$ 2700 \cdot 10^9$. (This is six times the GNP of the Federal Republic of Germany in 1975).

W. Häfele and W. Sassin assume a world-wide primary energy demand for the year 2030 of $38 \cdot 10^9$ tce, to satisfy which an investment of ca. $\$ 40 \cdot 10^{12}$ (in 1975 prices) will be needed (131, 132).

If one investigates the regional distribution of the total anticipated investment sum, the United States is in the lead position. Among the most important reasons for this is that the petroleum share of primary energy consumption for the USA in 1980 was 43%, of which about 40% came from imported crudes. This petroleum dependence is a mortal danger for the most powerful nation on earth. The second greatest power, the USSR, and the third potential great power, the Peoples' Republic of China, are independent of the world primary energy market. According to C. F. von Weizsäcker, these three nations are in fact world powers owing to their relative energy independence (157).

Even though the United Sates is economically the most powerful nation achieving complete independence from the petroleum phase cannot be realized, as long as petroleum is the major prime energy source (158). For this reason, the U.S. energy program "Project Independence", which was promulgated by president R. Nixon on 23. January 1974, and which was aimed at making the USA energy-independent, was doomed to fail from the beginning (99). Furthermore, investment requirements for this plan up to 1990 would have come to \$ $500 \cdot 10^9$, according to estimates of the National Petroleum Council. If one recalls that the Manhattan Project[1)] cost \$ $2.5 \cdot 10^9$ and that the realization of Project Apollo required ca. \$ $25 \cdot 10^9$ the vast differences in the financial orders of magnitude become quite apparent. In contrast to the foundered "Project Independence", president J. Carter presented to the American people a concept, on the 19th of April 1977, which centered around the saving and conservation of energy. According to this, the present growth rate of energy consumption of 3–4% per year should be reduced to less than 2% by 1985. Domestic coal production will be stepped up drastically, and research and development to utilize solar energy is to be scaled up (100–103).

The discrepancy between production and consumption of primary energy material is most pronounced in Western Europe. In the year 1980 Western Europe had to import ca. 90% of its petroleum. This dependence on the world market of primary energy sources will remain for some time to come, based on the present scientific and technical state of the art. It obviously applies to petroleum, in the major part also to nuclear energy based on fission, and owing to high production costs, in a certain measure also to coal. If nuclear energy based on fusion can be realized in the near future, its contribution to the energy supply must not be expected before 40 years at the earliest. This energy source could make Western Europe independent of the world energy market (see 3.36). The basic technology for solar energy utilization, in contrast to fusion technology, is known. Use of solar energy in these latitudes may ease the dependence from energy imports, particularly in the generation of low temperature heat. Natural conditions for large scale solar energy use (electric power, production of hydrogen) are more favorable for North Africa than for European latitudes (see 3.37) and Western Europe may have to depend on energy imports from

[1)] Code name for the first US atomic bomb project

North African countries. West European nations should take account of these potentialities when arranging to safeguard future energy supplies in this area.

Investments of the EC countries in the energy sector between 1975 and 1985 will come to 2 to 2.5% of the GNP of the EC bloc, as contrasted with 1.5% for 1965 to 1970. Total investment in the energy sector for 1975 to 1985 will amount to \$ $300 \cdot 10^9$ (in 1973 prices). Of this, ca. \$ $150 \cdot 10^9$ is allocated to electricity generation, inclusive of nuclear power plants, \$ $110 \cdot 10^9$ to petroleum and natural gas industries, and roughly \$ $6 \cdot 10^9$ to coal (155, 159).

It has already been pointed out that measures for energy conservation are connected with a substantial capital outlay. Nevertheless, there are many cogent reasons to save energy or to employ energy as efficiently as possible (see 2.332). Conversion processes in the EC with the use of energy saving technologies will result in an estimated investment of ca. \$ $500 \cdot 10^9$ by 1990 (160).

The Federal Republic of Germany counts on an investment need for the span from 1975 to 1985 of ca. DM $250 \cdot 10^9$ for the total energy sector. Of this ca. 30–40% will go to the production of electricity (new power plants, new distribution networks, fuel costs), 20–30% to the petroleum industry (new refineries, new storage facilities, pipelines, distribution, new tankers, production costs), 15% to natural gas, and 10% to the anthracite – lignite sector.

3. The World's Energy Potential

3.1 The basic types of available energy

There are basically two types of energy source available to meet the energy demands of humanity: those which are found in the earth's crust (fossil and nuclear fuels) and are not renewable, and those which are continuously available (solar, tidal and geothermal energy), and are thus classified as renewable sources (compare Fig. 3-1). Over the time span covered by human civilization, these sources can be considered constant (see 3.37, 3.38 and 3.39).

Fossil fuels include peat, brown (lignite) coal, anthracite coal, petroleum, natural gas, oil shales and oil sands. The fissionable nuclear fuels are uranium and thorium, and the fuels for fusion are deuterium and lithium.

Solar energy is available in many forms; even fossil fuels are none other than solar energy which has been stored for millions of years. (It is nonetheless conventional to regard these nonrenewable fuels separately from the other forms of solar energy). Radiant energy from the sun warms the earth and provides energy for the photosynthesis of biomass and free oxygen, and thus makes life possible on earth. Without solar energy, there would be no water power, wave energy, ocean heat or currents, or wind energy. Sunlight can be converted to other forms, for example by solar thermal, photoelectric or photochemical conversion. Ignoring, for the moment, technological and ecological problems, we can see that the long-term options for energy supply derive from the division of energy sources into renewable and non-renewable types. Solar and geothermal energy, as continuous flows of energy, are two options for an unlimited supply of energy. (The flow of energy in the tides is small compared to the previous two; compare Table 3-3 and Fig. 3-11). In the sun, 10^{38} helium nuclei are formed from protons each second. However, since the sun contains $2 \cdot 10^{33}$g of protons, it can maintain the radiation production on which we depend for life for another 10^{11} years, in spite of the consumption of $7 \cdot 10^{14}$ g of protons per second.

Fig. 3-1: Primary energy and its conversion to utilized energy

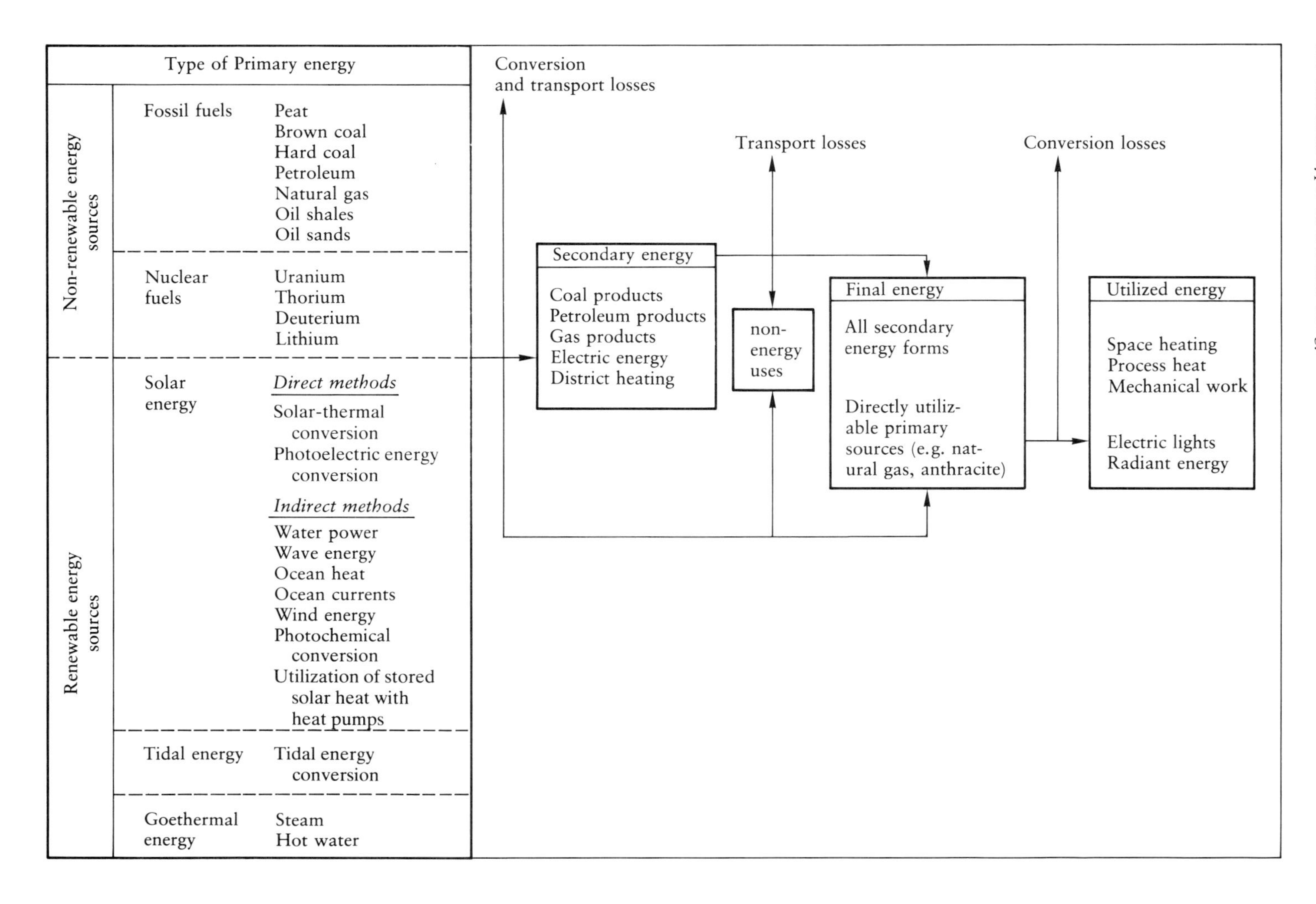
Type of Primary energy
Non-renewable energy sources
Fossil fuels
Peat
Brown coal
Hard coal
Petroleum
Natural gas
Oil shales
Oil sands
Nuclear fuels
Uranium
Thorium
Deuterium
Lithium
Renewable energy sources
Solar energy
Direct methods
Solar-thermal conversion
Photoelectric energy conversion
Indirect methods
Water power
Wave energy
Ocean heat
Ocean currents
Wind energy
Photochemical conversion
Utilization of stored solar heat with heat pumps
Tidal energy
Tidal energy conversion
Goethermal energy
Steam
Hot water
Conversion and transport losses
Secondary energy
Coal products
Petroleum products
Gas products
Electric energy
District heating
Transport losses
non-energy uses
Final energy
All secondary energy forms
Directly utilizable primary sources (e.g. natural gas, anthracite)
Conversion losses
Utilized energy
Space heating
Process heat
Mechanical work
Electric lights
Radiant energy

If controlled nuclear fusion can be achieved, the world's lithium an deuterium reserves are large enough to provide humanity with another practically unlimited energy source (compare Table 3-2b; see 4.22).

3.2 World reserves and lifetimes of primary energy carriers

If one compares the predicted development of the world demand for primary energy (see 2.333) with the reserves of primary energy sources, one can estimate the lifetime of various primary energy sources – that is, one can predict when they will become scarce. The exact results of such estimates are not so important as the fact that within the foreseeable future, the presently used primary energy resources will not suffice to cover a growing portion of the world's energy demand (2).

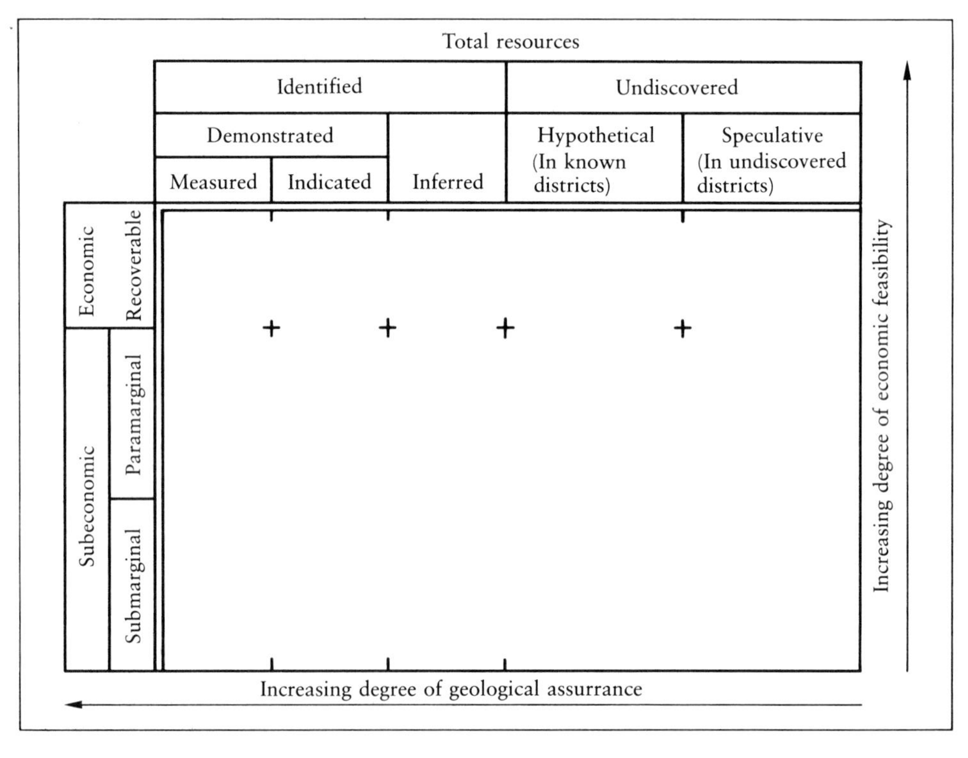

Fig. 3-2: Joint US-Geological Survey-Bureau of Mines classification scheme for mineral resources and reserves 1975

Source: L. Bauer, G. B. Fettweis, W. Fiala, Classification Schemes and their Importance for the Assessment of Energy Supplies, 10th World Energy Conference, Istanbul 1977.

With a few exceptions, the published estimates of the primary energy reserves/resources differ only slightly from each other. This is due in part to the fact that progress has been made in the last few years in the development of internationally uniform principles for the classification of mineral raw materials (3–7).

Fig. 3-2 shows the Joint US Geological Survey-Bureau of Mines classification scheme for mineral resources and reserves (6). The scheme, called a McKelvey Diagram, is based on a matrix in which the *degree of geologic assurance* increases from right to left, from *speculative in undiscovered districts* to *measured*. The *degree of economic feasibility* increases from bottom to top (8).

These guidelines distinguish between *resources* and *reserves*. According to the World Energy Conference: Survey of Energy Resources 1974, the definitions are: "In the broadest sense resources of nonrenewable raw materials are the total quantities available in the earth that may be successfully exploited and used by man within the foreseeable future. Reserves, however, are the corresponding fraction of resources that have been carefully measured and assessed as being exploitable in a particular nation or region under present local economic conditions using existing available technology; recoverable reserves are that fraction of reserves-in-place that can be recovered under the above economic and technical limits."[1)]

[1)] *Geological resources:* A concentration of naturally occuring solid, liquid, or gaseous materials in or on the earth's crust in such form, that economic extraction of a commodity is currently or potentially feasible.

Identified resources (measured, indicated, inferred): Specified resources whose location, quality, and quantity are known from geologic evidence supported by engineering measurements with respect to the demonstrated category.

Undiscovered resources (hypothetical, speculative): Unspecified resources surmised to exist on the basis of broad geologic knowledge and theory.

Reserve: That portion of the identified resource that can be economically and legally extracted at the time of determination – also referred to as technically and economically recoverable reserve. The reserve is derived by recoverability calculations from that component of the identified resource designated as the reserve base.

Identified-Subeconomic resources: Resources that are not Reserves, but may become so as a result of changes in economic and legal conditions.

Paramarginal: The portion of Subeconomic Resources that (a) borders on being economically producible or (b) is not commercially available solely because of legal or political circumstances.

Submarginal: The portion of Subeconomic Resources which would require a substantially higher price (more than 1.5 times the price at the time of determination) or a major cost reducing advance in technology.

Hypothetical resources: Undiscovered resources that may reasonably be expected to exist in a district under known geologic conditions. Exploration that confirms their existence and reveals quantity and quality will permit their reclassification as a Reserve or Identified-Subeconomic resource.

Speculative resources: Undiscovered resources that may occur either in known types of deposits in a favorable geologic setting where no discoveries have been made, or in as yet unknown types of deposits that remain to be recognized. Exploration that confirms their existence and reveals quantity and quality will permit their reclassification as Reserves or Identified-Subeconomic resources.

Measured: Identified resource components for which estimates of the quality and quantity have been computed, within a margin of error of less than 20 percent, from sample analyses and measurements from closely spaced and geologically well-known sample sites.

Table 3-1 shows the world reserves of fossil primary energy sources (2, 10–12). It is clear that the coal reserves constitute the majority of the fossil energy sources in each category. The fraction of hydrocarbons is considerably lower. In view of the enormous coal reserves, a "coal renaissance" seems very probable, especially in connection with the development of the High Temperature Gas Reactor (HTGR) (13, 14).

The Federal Institute for Geosciences and Natural Resources, Hannover, Federal Republic of Germany, estimates the *proved recoverable reserves* of solid fossil fuels to be $548 \cdot 10^9$ tce and the *geological resources* to be $9890 \cdot 10^9$ tce (2). The World Energy Conference (WEC) 1980 estimates the *proved recoverable reserves* of solid fossil fuels to be $693 \cdot 10^9$ tce and the *geological resources* to be $11\,184 \cdot 10^9$ tce. The estimates of the coal reserves tend to rise, because in many regions of the world exploration for coal is only beginning, and it will become technically and economically feasible to extract an increasing fraction of the coal (10, 15).

The current *proved recoverable reserves* of crude oil[1)] (conventional oil) are estimated to be $128 \cdot 10^9$ tce ($\triangleq 89 \cdot 10^9$ t oil) by the World Energy Conference 1980. The *geological resources* are estimated to be $890 \cdot 10^9$ tce ($\triangleq 619 \cdot 10^9$ t oil) (oil in place) (2). L. Auldrige, in the Oil and Gas Journal (Worldwide Issue), estimated the technically and economically recoverable reserves (estimated proved reserves) (as of 1/1/1978) as $126 \cdot 10^9$ tce ($\triangleq 87.3 \cdot 10^9$ t oil) (11).

The total oil content of all the oil shale reserves in the world, including those with very low oil contents, has been estimated by the World Energy Conference (WEC) (1974) as about $75\,000 \cdot 10^9$ tce, and by the Colorado School of Mines Research Institute (1975) as $450\,000 \cdot 10^9$ tce. Although the reserves which are at present *proved recoverable reserves* amount to only $66 \cdot 10^9$ tce, these reserves represent an enormous potential source of energy for humanity.

The *proved recoverable reserves* of natural gas (conventional gas) have been estimated by the World Energy Conference (WEC) 1980 to be $98 \cdot 10^9$ tce ($\triangleq 74 \cdot 10^{12}$ m^3), and the *geological resources* to be $353 \cdot 10^9$ tce ($\triangleq 266 \cdot 10^{12}$ m^3) (2). In the Oil and Gas Journal (Worldwide Issue), L. Auldridge estimated the technically and economically recoverable reserves (estimated proved reserves) (as of 1/1/1978) as $94 \cdot 10^9$ tce ($\triangleq 71 \cdot 10^{12}$ m^3) (11).

Improved techniques for recovering oil could considerably increase the lifetimes of known reserves. The recoverable fraction of a deposit, i.e. the fraction which can be

---

Indicated: Identified resource components for which estimates of the quality and quantity have been computed partly from sample analyses and measurements and partly from reasonable geologic projections.

Demonstrated: A collective term for the sum of measured and indicated resources.

Inferred: Identified resource components in unexplored extensions of Demonstrated resources for which estimates of the quality and size are based on geologic evidence and projection.

[1)] These comprise the *demonstrated reserves (measured and indicated reserves)* (compare Fig. 3–2) of the US Geological Survey (USGS) or the *proven recoverable reserves* in international usage (compare Fig. 3–2, 3–9).

extracted, is reported by the oil companies to be relatively low. If enhanced oil recovery (EOR) processes are used, however, the recoverable fraction, and thus the recoverable reserves can be increased, in some cases considerably. At present the aver-

Table 3-1: World supplies of fossil fuels[1)]

Energy source	proved recoverable reserves		additional resources		geological resource	
	[10^9 tce]	%	[10^9 tce]	%	[10^9 tce]	%
Hard coal[2)]	488	46.8	6161	55.0	6936	51.2
Brown coal[3)]	200	19.2	3840	34.2	4126	30.4
Peat	5	0.4	101	1.0	122	0.9
Solid fossil fuels	693	66.4	10102	90.2	11184	82.5
Crude oil	128	12.3	305	2.7	890[4)]	6.6
Natural gas	98	9.4	255[5)]	2.3	353	2.6
Oil sands	58	5.6	110	1.0	487	3.6
Oil shales[6)]	66	6.3	422	3.8	642	4.7
Oil + Gas	350	33.6	1092	9.8	2372	17.5
Fossil fuels (total)	1043	100.0	11194	100.0	13556	100.0

1) The method of resource classification practised in the Survey of Energy Resources 1980 of the World Energy, Conference 1980 has become virtually traditional. It lists resources into the two categories of *known reserves* and *additional resources*, where in the case of known reserves it is differentiated between reserves in place and the amount that could be recovered. Based on the definitions used for hydrocarbons, the following criteria were chosen to prepare the compilation submitted here: *Proved reserves* (in place/in situ) represent the fraction of total resources that has not only been carefully measured but has also been assessed as being exploitable in a particular nation or region under present and expected local economic conditions (or at specified costs) with existing available technology. *Proved recoverable reserves* (measured and indicated) are the fraction of proved reserves in place that can be recovered (extracted from the earth in raw form) under the above economic and technological limits (compare Fig. 3-2; Fig. 3-9).
Additional resources (in place/in situ) embrace all resources, in addition to proved reserves, that are of at least foreseeable economic interest. The estimates provided for additional resources reflect, if not certainty about the existence of the entire quantities reported, at least a reasonable level of confidence (see Fig. 3-9). Resources whose existence is entirely speculative are not included. This system is not quite as detailed as the classification systems used by some countries, but it has the advantage of being internationally compatible.
Geological resources see footnote 1, P. 67.
2) Bituminous coal and anthracite.
3) Subbituminous coal and lignite.
4) Oil in place.
5) One of the advantages of natural gas production is that it is usually possible to extract 75% of the gas in place without additional measures, and that some of the remaining 25% can be obtained by using additional measures.
6) Oil shales with > 40 liters shale oil/t rock, oil content.

Source: World Energy Conference 1980: Survey of Energy Resources 1980, Munich, September 1980.

age recoverable fraction, worldwide, is about 32%; an increase of 1% would release on the order of a year's supply of petroleum at the current worldwide rate of consumption.

Table 3-2a: World reserves of nuclear fuels (nuclear fission; uranium and thorium)

Fuel / Reserves	Uranium t	Uranium [10^9 tce]	Thorium t	Thorium [10^9 tce]	Uranium and Thorium [10^9 tce]
Economically recoverable[1] (from (12))	$7073 \cdot 10^3$	313	$3974 \cdot 10^3$	176	489
Economically recoverable plus possible and speculative potential [2] (from (12))	$13373 \cdot 10^3$	592	–	–	–
Economically recoverable plus possible and speculative potential plus low-grade ores[3] (from (2, 12))	$52 \cdot 10^6$	2300	$8800 \cdot 10^3$	389	2689
Existent[4]					
– Sea water (from (2, 12))	$4000 \cdot 10^6$	$177 \cdot 10^3$	–	–	–
– Potentially minable areas to 3000 m depth (from (2, 12))	$3298 \cdot 10^9$	$146 \cdot 10^6$	$1976 \cdot 10^9$	$87.3 \cdot 10^6$	$233.3 \cdot 10^6$

[1] The Nuclear Energy Agency (NEA) has developed a special system for the classification of uranium and thorium reserves. It is based on the estimated cost of mining and refining the reserves, given present and foreseeable technologies. Uranium *(reasonable assured resources + estimated additional resources)* in the cost category up to 130 $/kg U; Thorium in the cost category up to 75 $/kg Th (compare 3.351).
All tce estimates are based on the degree of utilization achieved in thermal reactors. Complete utilization can only be achieved in breeder reactors (the theoretical energy content of 1 t uranium or thorium is about $2.95 \cdot 10^6$ tce); thermal reactors utilize only about 1.5% of the theoretically possible energy.

[2] Uranium as under footnote 1 plus possible and speculative potential in the cost category up to 130 $/kg U.

[3] Uranium as under footnote 2 plus the low-grade ores (50 to 150 g Uranium/t), cost category > 130 $/kg U; thorium without cost estimates.

[4] Present in sea water and minable areas down to 3000 m depth.

Sources: World Energy Conference 1980: Survey of Energy Resources 1980, Munich, September 1980 (12).
Federal Institute for Geosciences and Natural Resources, Hannover, Federal Republic of Germany, 1976 (2).

The earth also has a large energy potential in the form of nuclear fuels. Table 3-2 a shows the world reserves of nuclear fuels for fission (uranium and thorium) (2, 16) and Table 3-2 b, the world reserves of fuel for fusion (lithium and deuterium)

Table 3-2 b: World reserves of nuclear fuels (nuclear fusion; lithium[1] and deuterium)

Lithium in the landmass (total resources)

		Identified			Undiscovered	
		Demonstrated			Hypothetical (In known districts)	Speculative (In undiscovered districts)
		Measured	Indicated	Inferred		
Economic	Recoverable	$1.4 \cdot 10^6$ t $35 \cdot 10^3$ q to $123 \cdot 10^3$ q [2]				
Subeconomic	Paramarginal		$5.2 \cdot 10^6$ t $(128 \text{ to } 455) \cdot 10^3$ q		$4.3 \cdot 10^6$ t $(104 \text{ to } 370) \cdot 10^3$ q	
	Submarginal				total resources: $1.2 \cdot 10^8$ t $(0.29 \text{ to } 1.0) \cdot 10^7$ q	

Increasing degree of economic feasibility (↑)

Increasing degree of geologic assurance (←)

Amount of lithium in sea water: $1.1 \cdot 10^{11}$ t $\triangleq$ $(2.7 \text{ to } 9.5) \cdot 10^9$ q
Amount of deuterium in sea water: $\approx 4.6 \cdot 10^{13}$ t[3]

1) In all probability, it will be easier to utilize the D-T reaction than the D-D reactions. Therefore lithium as well as deuterium is required as a fuel for the D-T reactor. Since more lithium than deuterium is needed, and the available amounts of lithium are smaller than those of deuterium, lithium is the determining factor in the amount of energy which can be released.

2) The energy equivalents depend on the type of reactor. The lower value is for a reactor using solid lithium compounds as breeder materials, while the higher value is for a reactor using liquid lithium as breeder material and coolant (17). (1 q $\triangleq$ $3.62 \cdot 10^7$ tce).

3) This amount of deuterium would give an energy equivalent of $15 \cdot 10^{12}$ q in a D-D reactor (18).

Source: Max-Planck-Institut für Plasmaphysik, Garching/Munich ASA-ZE/09/78 (1978).

(McKelvey Diagram) (17–19). Nuclear fuels are already being used for nuclear fission in large-scale and commercial power plants. Fusion reactors, however, have not yet been developed.

In addition to these non-renewable energy sources, there are the renewable sources: solar, geothermal and tidal energy. Table 3-3 shows the potential of renewable sources (2, 15, 20). The utilization of these energy resources, for which the basic technology already exists, in contrast to nuclear fusion, depends on a number of factors, such as technological developments, price trends, availability of conventional sources of energy, and local conditions.

Ignoring for the moment such factors as the different demand structures for the various fuels, ecological problems, availability, etc., one can calculate from the predicted world demand for primary energy that the reserves of fossil fuel technically and economically recoverable by the presently available technology would supply the total world energy for the next 50 years[1)] (semidynamic lifetime, i.e. increasing consumption as in Fig. 2-12 without increase in the reserves.) The technically recoverable reserves (reserve base), however, would suffice to meet a growing demand for about one hundred years.

Realistically, however, the established world-wide consumption patterns for fossil fuels must be taken into account, along with the foreseeable future developments. At present, petroleum and natural gas supply about two thirds of the energy consumed, and the prediction is that the annual consumption of petroleum or natural gas will increase by the end of this decade. It must also be assumed that an increasing fraction of the technically recoverable reserves (reserve base) of petroleum and natural gas will be economically recoverable (10, 15). Based on the technically recoverable reserves of petroleum and natural gas, the following production would be possible: for oil, a 2% annual increase in production until the year 2000, thereafter constant production for 35 years; for natural gas, an annual increase in production until the year 2000, and thereafter 55 years of constant production (2). (Larger annual rates of production, given constant reserves, mean shorter lifetimes.)

Some other studies have come to still more optimistic results (see 2.333). Fig. 3-3 shows a possible model of the development of fossil fuel production and the corresponding lifetimes by M. Grenon, International Institute for Applied Systems Analysis, Laxenburg/Vienna (10). The model assumes a world consumption of primary energy of $38 \cdot 10^9$ tce in the year 2030. The recoverable fossil fuels on which the model is based are coal (technically and economically recoverable reserves), $640 \cdot 10^9$ tce; oil (conventional oil, ultimately recoverable resources), $450 \cdot 10^9$ tce (about $300 \cdot 10^9$ t oil, recovery costs up to about $ 20/bbl in 1976 dollars); gas (con-

[1)] For practical purposes it is useful to give the lifetime (depletion time) T of a raw material: $T = 1/P \ln (PX/X_0 + 1)$. X is the total reserve of the raw material, X_0 is the amount consumed in the starting year, and P is the growth rate of the consumption (in percent per year). It can be shown that the lifetime of a raw material is increased by rising prices, partly because less of it is consumed and partly because an increasing fraction of the reserve can be economically recovered.

Table 3-3: Potential for energy from regenerative sources

Source of energy: Source	Solar energy[1] Amount from (12, 15)	Source	Geothermal energy[2] Amount from (2, 12, 20)	Source	Tidal energy[3] Amount from (12, 15, 20)
Energy falling on the entire earth	$192 \cdot 10^{12}$ tce/y[4]	Total heat content	$\approx 4 \cdot 10^{20}$ tce[5]	Total power potential of the tides	$3.2 \cdot 10^{9}$ tce/y
Energy from utilization of solar radiation on 0.2% of the land surface at 20% efficiency	$15 \cdot 10^{9}$ tce/y	Heat content below the continents (about $150 \cdot 10^{6}$ km[2])		Technically utilizable tidal power potential	$60 \cdot 10^{6}$ tce/y
		– to 5000 m depth	$\approx 8 \cdot 10^{14}$ tce		
		– to 10000 m depth	$\approx 90 \cdot 10^{14}$ tce		
Technically usable hydroelectric potential	$2200 \cdot 10^{3}$ MW	Contribution of geothermal energy of the electric capacity of the world		Contribution of tidal energy to the electric capacity of the world	
– in 1978 (installed)	$372 \cdot 10^{3}$ MW	– in 1978 (installed)	1325 MW	– in 1978 (installed)	240 MW
– in 2000 (estimated)	$840 \cdot 10^{3}$ MW	– in 2000 (estimated)	$100 \cdot 10^{3}$ MW	– in 2000 (estimated)	$60 \cdot 10^{3}$ MW to $100 \cdot 10^{3}$ MW

[1] see 3.371 [2] see 3.39 [3] see 3.38 [4] 1 tce/year $\triangleq$ 0.928 kW [5] 1 tce $\triangleq$ $8.140 \cdot 10^{3}$ kWh

Sources: World Energy Conference 1980: Survey of Energy Resources 1980, Munich, September 1980 (12).
World Energy: looking ahead to 2020. Report by the Conservation Commission of the World Energy Conference, Guildford (UK) and New York: IPC Science and Technology Press 1978 (15).
Federal Institute for Geosciences and Natural Resources, Hannover, Federal Republic of Germany, 1976 (2).
Essam El-Hinnawi, Energy, Environment and Development, 10th World Energy Conference, Istanbul, September 1977 (20).

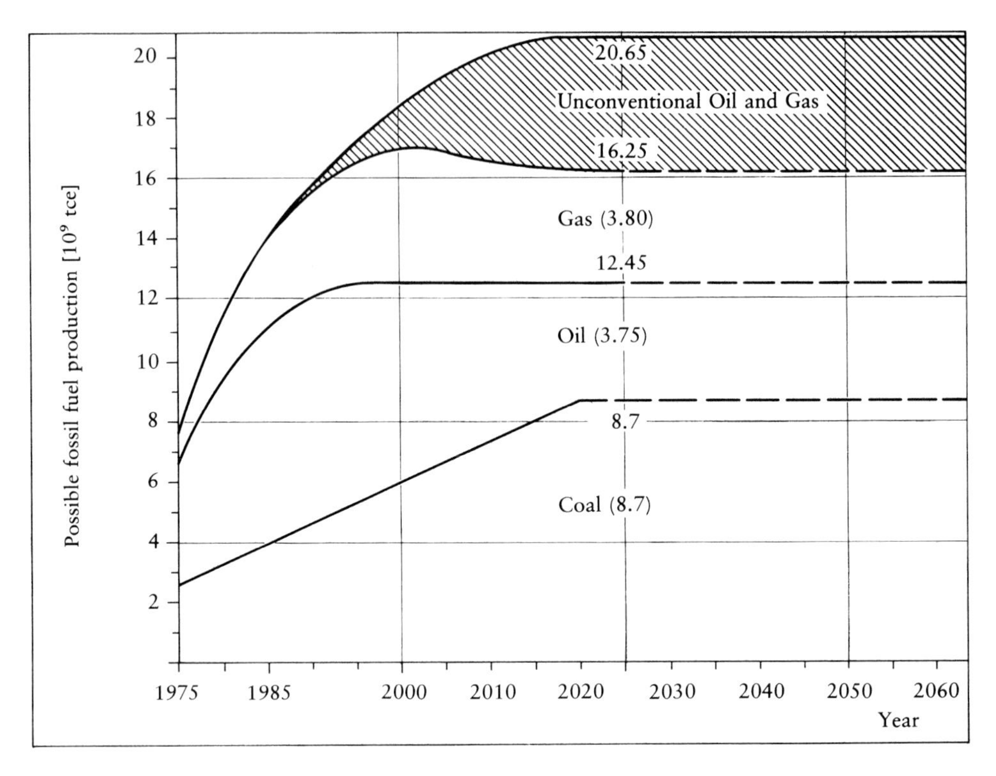

Fig. 3-3: Total fossil fuels: Possible production and lifetimes

Source: M. Grenon, On Fossil Fuel Reserves and Resources, International Institute for Applied Systems Analysis, Laxenburg/Vienna, RM-78-35, June 1978.

ventional gas, ultimately recoverable resources), $333 \cdot 10^9$ tce (about $250 \cdot 10^{12}$ m^3, recovery costs up to about \$ 20/bbl oil equivalent at 1976 prices). Another 200 to $300 \cdot 10^9$ t oil from nonconventional sources could be utilized, starting about 1990. Nonconventional sources of oil generally include enhanced oil recovery (EOR) from old fields, deep offshore and polar deposits, oil shales, oil sands and heavy-oil deposits, and synthetics from coal or biomass. The recovery of these nonconventional oils should cost \$ 20 to \$ 25/bbl (1976 dollars). In addition, it is hoped that gas can be recovered from nonconventional sources such as coal beds, shales, tight formations, geopressured gas, coal and biomass conversion (15).

As shown in Fig. 3-3, coal production is expected to rise from $2.6 \cdot 10^9$ tce in 1975 to $8.7 \cdot 10^9$ tce in 2020. Given coal reserves of $640 \cdot 10^9$ tce, this production could be maintained until about 2065. It is expected, however, that an increasing fraction of the technically recoverable reserves (reserve base) can be economically recovered, i.e. that one can assume a technically and economically recoverable reserve of about $1200 \cdot 10^9$ tce. This would mean that the production of $8.7 \cdot 10^9$ tce per year could be maintained until 2130. The production of petroleum will probably reach its peak about 1990 with an annual production of somewhat more than $7 \cdot 10^9$ tce (about $5 \cdot 10^9$ t oil). Thereafter it will drop back to about $3.75 \cdot 10^9$ tce (about $2.5 \cdot 10^9$ t oil) in 2020. This level of production could be maintained until about 2065. The production of natural gas will probably reach its peak of about $5 \cdot 10^9$ tce (about $3.8 \cdot 10^{12}$ m^3) in the year 2000, and decline thereafter until it reaches about $3.8 \cdot 10^9$ tce (about $2.8 \cdot 10^{12}$ m^3) in 2020 (21). This level of production could be maintained until 2065. The amount of nonconventional oil and gas could, beginning about 1990, rise to about $4.75 \cdot 10^9$ tce in the year 2020 (10).

This means that the fossil fuels could, without nonconventional oil and gas, meet an increasing demand up to about $17 \cdot 10^9$ tce; thereafter, an amount of somewhat more than $16 \cdot 10^9$ tce, could be produced for almost half a century (compare Fig. 3-3). (Coal, of course, will be available for much longer, even at a considerably higher rate of production, such as $12 \cdot 10^9$ tce annually.) These predictions agree quite well with the results of the Conservation Commission of the World Energy Conference (see 2.333). It is probable, however, that the costs of producing petroleum from the known reserves will rise so severely in the coming century that further production will be less and less economically feasible. It is in fact very likely that the costs of recovering the remaining known reserves of petroleum will become prohibitively high. This will also occur with natural gas, beginning about two decades later. Furthermore, it must be remembered that the discussion to this point has ignored important aspects of the problem, such as the geographical distribution of the reserves and political factors. As the past has shown, oil may well become an increasingly important political instrument, i.e. the important oil-producing countries may limit production before the oil wells are exhausted (22–25).

According to a study of the International Institute for Applied Systems Analysis (IIASA) in Laxenburg (Vienna), it has taken each of the primary sources of energy (wood, coal, oil, gas, nuclear fission) more than 50 years to gain (or lose) a 50% share of the market (26). In view of the lifetime of the oil reserves, this can only mean that the time remaining to develop alternative sources of energy is so short that humanity must make every possible effort to master this task of the century. If one remembers that more than 40% the primary energy consumption of the world (not including the Eastern block), will, in all probability, be supplied by petroleum well into the 1990's, one must realize that competition for the world supply of oil will become ever sharper (27).

The present situation in the energy sector differs, however, from the earlier transitions in one crucial respect, namely that in earlier transitions from one major fuel to the next, the new fuel was abundant and cheaply obtained, and had definite advantages compared to its predecessor. The result was that there were few problems in introducing it. This was especially true in the transition from coal to petroleum. In contrast to this, it seems unlikely that another period of cheap and abundant energy, such as the petroleum era of the 60's, will occur. Furthermore, in all probability, the future will not see one single major fuel. Instead, there will be several sources of energy, whose type and relative importance will vary in different parts of the world.

Given the magnitude of the world's energy consumption, the use of the various energy sources will grow to such dimensions that some effects which could be ignored when smaller amounts were consumed (e.g. the CO_2 problem), will become essential criteria in the choice of fuels to obtain an optimal combination. It should be remembered that any fuel which is used in sufficiently large amounts will have some effects on the environment (see chapter 5.).

In the choice of resources for the long-term provision of energy, the main problem with the options discussed above is not the limits on the available fuels, but the specific effects of each fuel on the environment. In addition, the second law of thermodynamics tells us that the heat produced by every energy-releasing process is the same. A primary energy consumption of $2.40 \cdot 10^3$ q/a which is about 10 times the present level, corresponds to an artificial additional release of energy equivalent to about 0.06% of the mean solar energy (232 W/m^2) falling on the entire surface of the earth, and to 0.2% of the solar energy falling on land. (In heavily industrialized areas, like the Ruhr Valley or New York City, the energy released artificially is equivalent to about 10% of the solar supply.) People are justifiably concerned that if the artificial release of energy approaches a small percentage of the natural energy supplied by the sun, it could produce climatic changes (see chapter 5.).

3.3 *Primary energy carriers*

3.31 Coal

3.311 Geographical distribution of the coal reserves

An adequate supply of primary energy resources for countries and regions depends not only on whether there are sufficient reserves on the earth, but on their geographical distribution. This may be the deciding factor in the choice of a primary energy source or combination of fuels, especially since the negative experiences of many countries during the oil crisis has led them to give the security of their energy supplies a high priority.

It has already been mentioned that the coal reserves make up the majority of the fossil fuel reserves, and that the reserve situation for the western industrial countries (OECD) is much more favorable for coal than for other fuels. Table 3-4 shows the total resources of solid fuels; Table 3-5a shows the *proved recoverable reserves* of coal and peat according to continents and economic-political groups and Table 3-5b shows the *additional resources* of coal and peat (12). The countries with the most solid fossil fuels are listed in Table 3-6. According to present estimates,

Table 3-4: Total resources of solid fuels

Classification[2]	COAL[1]						PEAT		Total	
	Bituminous coal and Anthracite		Subbituminous coal		Lignite					
	[10^9 tce]	%	[10^9 tce]	%	[10^9 tce]	%	[10^9 tce]	%	[10^9 tce]	%
1. Proved recoverable reserves	487.7	70.4	111.6	16.1	88.1	12.7	5.8	0.8	693.2	100.0
2. Proved reserves (in situ)	774.6	71.6	172.8	16.0	113.2	10.5	20.7	1.9	1081.3	100.0
3. Additional resources (in situ)	6161.4	61.0	2991.4	29.6	848.3	8.4	101.3	1.0	10102.4	100.0
Total of 2 + 3	6936.0	62.0	3164.2	28.3	961.5	8.6	122.0	1.1	11183.7	100.0

[1] Conversion factors for coal equivalents (12):
Bituminous coal/Anthracite = 1
Subbituminous coal = 0.78
Lignite = 0.3 (Turkey, German Dem. Rep., Germany FR, Poland, Australia)
= 0.33 (all African countries, all Asiatic countries except Turkey, USSR, France, Greece, Hungary, Italy, Romania, Spain)
= 0.5 (Canada, USA, Albania, Austria, Bulgaria, Portugal, New Zealand, Yugoslavia)
= 0.6 (Czechoslowakia)
Peat = 0.43 (all countries except Germany FR, (= 0.49); Ireland, Rwanda/Burundi (= 0.50); USSR (= 0.33))

[2] See Table 3-1

Source: World Energy Conference 1980: Survey of Energy Resources 1980, Munich, September 1980.

Table 3-5a: Proved recoverable reserves of coal and peat according to continents and economic-politcal groups

Continent or economic-political groups	proved recoverable reserves									
	Bituminous coal and Anthracite		Subbituminous coal		Lignite		Peat		Total	
	[10^9 tce]	%	[10^9 tce]	%	[10^9 tce]	%	[10^9 tce]	%	[10^9 tce]	%
Africa	32.5	6.7	0.1	0.1	0.0	-.-	-.-	-.-	32.6	4.7
America	111.4	22.8	75.3	67.4	13.2	15.0	0.3	5.2	200.2	28.9
Asia	113.9	23.4	0.8	0.7	1.4	1.6	-.-	-.-	116.1	16.7
USSR	104.0	21.3	32.8	29.4	28.7	32.6	3.6	62.0	169.1	24.4
Europe	100.5	20.6	1.3	1.2	35.1	39.8	1.9	32.8	138.8	20.0
Oceania/Australia	25.4	5.2	1.3	1.2	9.7	11.0	-.-	-.-	36.4	5.3
Total	487.7	100.0	111.6	100.0	88.1	100.0	5.8	100.0	693.2	100.0
EC	70.0	14.4	0.0	-.-	10.6	12.0	0.5	8.6	81.1	11.7
OECD	205.9	42.2	74.6	66.8	34.9	39.6	2.1	36.2	317.5	45.8
COMECON	134.2	27.5	32.7	29.3	45.1	51.1	3.6	62.1	215.6	31.1
Developing countries	22.5	4.6	2.8	2.5	1.3	1.5	0.0	-.-	26.6	3.8
OPEC	0.4	0.1	0.2	0.2	0.2	0.2	-.-	-.-	0.8	0.1

Source: World Energy Conference 1980: Survey of Energy Resources 1980, Munich, September 1980.

Table 3-5b: Additional resources of coal and peat according to continents and economic-political groups

Continent or economic-political groups	additional resources (in situ)									
	Bituminous coal and Anthracite		Subbituminous coal		Lignite		Peat		Total	
	[10^9 tce]	%	[10^9 tce]	%	[10^9 tce]	%	[10^9 tce]	%	[10^9 tce]	%
Africa	144.4	2.3	0.8	0.0	0.0	-.-	1.3	1.3	146.5	1.5
America	1181.2	19.2	1334.1	44.5	406.8	48.0	47.9	47.3	2970.0	29.4
Asia	1423.2	23.1	2.2	0.1	19.5	2.3	9.2	9.0	1454.1	14.4
USSR	2480.0	40.2	1570.9	52.5	381.5	45.0	37.4	37.0	4469.8	44.2
Europe	429.5	7.0	1.1	0.0	12.1	1.4	5.5	5.4	448.2	4.4
Oceania/Australia	503.1	8.2	82.3	2.8	28.4	3.3	0.0	-.-	613.8	6.1
Total	6161.4	100.0	2991.4	100.0	848.3	100.0	101.3	100.0	10102.4	100.0
EC	335.4	5.4	0.0	-.-	0.0	-.-	2.5	2.5	337.9	3.3
OECD	2007.3	32.6	1399.6	46.8	432.5	51.0	51.4	50.8	3890.8	38.5
COMECON	2572.3	41.7	1570.9	52.5	391.4	46.1	39.5	39.0	4574.1	45.3
Developing countries	214.4	3.5	19.0	0.6	9.3	1.1	9.9	9.8	252.6	2.5
OPEC	4.7	0.1	4.6	0.2	5.9	0.7	8.3	8.2	23.5	0.2

Source: World Energy Conference 1980: Survey of Energy Resources 1980, Munich, September 1980.

Table 3-6: The countries with the most solid fossil fuels (in 10^6 tce)

Country[1]	Bituminous coal and Anthracite		Subbituminous coal and Lignite		Peat	
	proved recoverable reserves	additional resources (in situ)	proved recoverable reserves	additional resources (in situ)	proved recoverable reserves	additional resources (in situ)
1. USA	107183	1072000	83707	1447200		9890
2. USSR	104000	2480000	61470	1952400	3594	37445
3. China, P. R.	99000	1326000		13365		473
4. Great Britain	45000	145000	3600	7200		1359
5. Poland	27000	84000				
6. Australia	25400	503000	10902	108600		
7. Rep. of South Africa	25290	33762				
8. Germany, F. R.	23991	186300	10545		441	
9. India	12610	91139	524	93		
10. Botswana	3500	100000				
11. Czechoslovakia	2700	5500	1716	972		30
12. Swaziland	1820	3000				
13. Canada	1607	93413	2761	272802	219	37710
14. Mexico	1200	1300	300	390		
15. Japan	1050		6			
16. Colombia	1010	7200	19	526		
17. Zimbabwe-Rhodesia	734	5820				
18. Zaire	600					
19. France	550	200	24	27		132
20. Belgium	440	2617				11
Spain	398	2375	238	884		
Korea, D. P. R.	300	2700	234	1716		
Yugoslavia	70	22	8670	1965		15
German D. R.			7500			460

[1] The countries are listed in order of the size of their proved recoverable reserves (Bituminous coal and Anthracite).

Source: World Energy Conference 1980: Survey of Energy Resources 1980, Munich, September 1980.

the geological resources of coal amount to about $11\,000 \cdot 10^9$ tce, of which only $3060 \cdot 10^9$ tce have been identified (28–33).

In comparison to other primary fuels, the estimates of the coal reserves have not been significantly increased in the past. They were already estimated to be about $6400 \cdot 10^9$ tce in 1913 and $7000 \cdot 10^9$ tce in 1936 (3). The World Energy Conference (WEC) in 1980 estimated the *proved recoverable reserves* as $693 \cdot 10^9$ tce and the geological resources as $11\,184 \cdot 10^9$ tce (15). The increase is essentially due to the inclusion of reserves at greater depths or in narrower seams. The vast majority of the identified coal reserves lie in North America, the East block and Western Europe. Except for Indonesia and Venezuela, the OPEC have practically no coal, and the same applies to the developing countries with the exception of India, Columbia, Mexico, Brazil, the Republic of Botswana and Swaziland.

The coal situation is favorable for the countries of the European Community (EC), especially for the United Kingdom and the Federal Republic of Germany (see Tables 3-5a, 3-5b).

3.312 Centers of coal production and consumption

In 1979, the world production of hard coal was $2792 \cdot 10^6$ t, and of soft coal, $956 \cdot 10^6$ t. The largest producers are given in Tables 3-7a and 3-7b (36). About two-thirds of the total was mined by the three largest producers (USA, USSR and the Peoples Republic of China), and about 90% by the ten largest producers.

The figures for soft coal are similar, i.e. the three largest producers mined about two-thirds of the total. It is to be expected that the soft coal production of North America, which at $42 \cdot 10^6$ t has in the past been of secondary importance, will rise considerably in the future. This will be possible, given the reserves (compare Table 3-6).

The China P.R. leads the world in the production of anthracite coal, followed by the United States, the USSR, and Poland.

The total world production of anthracite coal had already reached $1997 \cdot 10^6$ t in 1964, and although the total world consumption of primary energy increased between 1964 and 1980 by about 60%, the world coal production rose only about 20% in this time. The EC countries produced about 9% of the world's anthracite coal in 1979; the OECD countries about 33%, and the entire western world, about 45%.

The future of coal production depends on a number of factors which are still uncertain. A few of the important ones are the economic growth to be expected, matters of national security in the procurement of energy, the possible development of other, cheaper and more abundant sources of energy with a more certain future, the possible perfection of coal technologies which are less detrimental to the environment and more convenient for the consumer, and the costs of mining. The USA and

Table 3-7a: The most important producers of anthracite coal (production in 10^6 t)

Country[1] Year	1964	1970	1973	1977	1979
1. China, P. R.	290	360	410	550	663
2. USA	455	550	530	606	655
3. USSR	409	474	512	556	554
4. Poland	117	140	157	186	200
5. United Kingdom	197	145	130	121	121
6. India	62	74	77	100	105
7. Rep. of South Africa	45	55	62	86	96
8. Germany, F. R.	142	111	104	91	93
9. Australia	28	50	62	77	84
10. Korea (North + South)	19	30	38	63	63
11. Czechoslovakia	28	29	28	28	29
12. Canada	8	12	17	23	28
13. France	53	37	26	21	19
14. Japan	51	40	22	18	18
15. Spain	12	11	10	12	11
EC	426	310	270	240	239
Western Europe	439	321	281	252	250
North America	463	562	547	625	683
World	1997	2176	2238	2538	2792

[1] These were the 15 largest producers in 1979.

Source: Statistik der Kohlenwirtschaft, Essen und Köln, September 1980.

the USSR plan considerable increases in coal production by the year 2000 (compare Table 3–10b).

Summary: The coal reserves are large enough to permit a considerable increase in production. The expected lifetimes are listed in Table 3-8 (static lifetime for a few selected countries or groups of countries).

The differences in production costs in different countries can be expected to play an important role in the future importance of coal as a primary energy source in those countries (compare Table 3-9). In North America, South Africa, India and Australia, production costs for bituminous coal and anthracite were nearly always under \$ 30/t. Solely in the USA and Canada are there considerable quanties of recoverable reserves with production costs lying above this, between \$ 30/t and \$ 60/t. In contrast, in Europe the production costs are always more than \$ 30/t, and in fact the costs are mostly more than \$ 60/t. The costs in Japan and South Korea are similarly high. However one can still find minimal production costs in Zimbabwe-Rhodesia, Swaziland, Botswana, and in some regions of Columbia.

Table 3-7b: The most important producers of brown coal (production in 10^6 t)

Country[1] Year	1964	1970	1973	1977	1979
1. German Dem. Rep.	257	261	246	254	255
2. USSR	145	145	153	160	165
3. Germany, F. R.	111	108	119	123	131
4. Czechoslovakia	76	82	82	93	96
5. Yugoslavia	28	28	32	39	42
6. Poland	20	33	39	38	38
7. USA	3	5	13	26	37
8. Australia	19	24	25	30	32
9. Bulgaria	24	29	26	25	28
10. Greece	4	8	13	23	23
EC	117	112	124	128	135
Western Europe	129	126	144	161	197
North America	4	9	17	32	42
World	750	792	826	902	956

[1] These were the 10 largest producers in 1979.

Source: Statistik der Kohlenwirtschaft, Essen und Köln, September 1980.

Table 3-8: Lifetime of the Bituminous coal/Anthracite and of the Subbituminous coal/Lignite (proved recoverable reserves)[1]

Country	Bituminous coal/Anthracite lifetime (static)[2] [years]	Subbituminous coal/Lignite lifetime (static) [years]
USA	165	2260
USSR	190	370
China, P. R.	150	–
United Kingdom	370	–
Poland	135	–
Australia	300	340
Rep. of South Africa	265	–
Germany, F. R.	260	80
India	120	–
EC	290	75
OECD	220	460
World	190	210

[1] Based on the production and reserves in 1979 (compare Tables 3-7 a, b; 3-6 and Table 3-5 a).
[2] Static lifetime, that is, assuming constant consumption and no growth of the reserves.

Source: Author's calculations.

Table 3-9: Recoverable reserves (in 10^6 t) according to production costs (except for countries with centrally planned economies)

	Country	< 15	15–30	30–60	> 60
		US $ / t			
		BITUMINOUS COAL / ANTHRACITE			
Africa	Rep. of South Africa	22 090.0	3 200.0		
	Zambia		11.7	3.6	
America	Canada			1 607.0	
	USA	15 627.0	57 614.0	33 942.0	
	Brazil	123.0	66.0		
	Mexico		1 200.0		
	Venezuela		134.0		
Asia	India		12 610.0		
	Indonesia			10.9	
Europe	Belgium	31.0	62.0	62.0	174.0
	France				550.0
	Federal Republic of Germany				23 991.0
	Great Britain				45 000.0
	Netherlands				130.0
	Norway				18.0
	Spain			247.0	151.0
Oceania/Austr.	Australia		25 400.0		
		SUBBITUMINOUS COAL			
America	Canada	2 182.0			
	USA	91 676.0			
	Argentina			100.0	
	Brazil	600.0	324.0		
	Mexico		384.0		
	Venezuela		2.5	3.8	
Asia	Indonesia		108.4		
	Taiwan			140.0	
Europe	France				10.0
	Spain		123.0		
Oceania/Austr.	Australia		1 500.0		
		LIGNITE			
America	Canada	2 117.0			
	USA	24 400.0			
Asia	Indonesia		420.0		
	Thailand	103.0			

Table 3-9 (continued): Recoverable reserves (in 10^6 t) according to production costs (except for countries with centrally planned economies)

	Country	< 15	15–30	30–60	> 60
		US $ / t			
		LIGNITE			
Europe	Federal Republic of Germany	10000.0			
	Italy	6.0	15.0		
	Spain	430.0			
Oceania/Austr.	Australia	32440.0			
		PEAT			
Europe	Finland	2340.0			
	Federal Republic of Germany	900.0	12.0		
	Ireland	99.0		77.0	

Source: World Energy Conference 1980: Survey of Energy Resources 1980, Munich, September 1980.

Mining of subbituminous coal nearly always costs less than $ 30/t, and in Canada and the USA it is cheaper than $ 15/t. In the most important countries, the mining costs for soft brown coal and lignite, both of which are mined more or less exclusively open cast, do not exceed $ 15/t, and in fact in some situations the costs can be considerably less than this. However, in the Federal Republic of Germany only 10 Gt of soft brown coal can be mined for this price. (Peat has a very low caloric value, and can only be recovered economically by open cast mining). The high mining costs in the EC countries were one of the causes for the decline in production of bituminous coal (compare Table 3-7 a) and a reason for the substitution of cheaper imported coal.

Unlike brown coal, hard coal (anthracite) is an internationally traded commodity. Because the centers of production and consumption are fairly evenly distributed, with a few exceptions like Japan, France and Italy, the international trade in anthracite is far less significant than in petroleum and natural gas. The deficit in Asian production is covered primarily by coal from the USA, Australia and Canada, while that of Western Europe is supplied from the USA and Poland. The total international volume of trade in 1978 was about $230 \cdot 10^6$ t. The majority of this was traded within the EC, and COMECON, or between the USA and Canada. In contrast, every second ton of crude petroleum was shipped by sea from the producing to the consuming country.

Table 3-10 a shows the countries exporting and importing anthracite and anthracite coke in 1978 (36). The exports from the United States went primarily to Cana-

Table 3-10a: Exporters and importers of anthracite coal and coke in 1978 (in 1000 t)

To: \ From:	Germany, F. R.	France	Italy	Netherlands	Belgium	Great Britain	Ireland	Denmark	EC	Czechoslovakia	German, D. R.	Greece
Germany, F. R.	x	391	–	418	143	772	–	9	1733	227	–	–
France	8542	x	27	145	200	892	–	–	9806	–	–	–
Italy	2564	116	x	–	–	38	–	–	2718	–	–	–
Netherlands	1933	25	–	x	19	290	–	–	2267	154	–	–
Belgium	4396	214	–	273	x	203	–	–	5086	–	–	–
Luxembourg	2624	36	–	40	154	12	–	–	2866	–	–	–
Great Britain	258	2	–	10	20	x	53	–	343	–	–	–
Ireland	7	–	–	–	1	227	x	–	235	–	–	–
Denmark	997	84	–	–	–	154	–	x	1235	–	–	–
EC	21321	868	27	886	537	2588	53	9	26289	381	–	–
Bulgaria	–	–	–	–	–	–	–	–	–	–	–	–
Czechoslovakia	–	–	–	–	–	–	–	–	–	x	–	–
German Dem. Rep.	354	–	–	–	–	–	–	–	354	854	x	–
Finland	239	–	–	–	–	31	–	–	270	–	–	–
Yugoslavia	203	–	16	–	–	12	–	–	231	626	–	–
Norway	104	96	–	3	11	321	–	31	566	–	–	–
Austria	623	8	31	–	–	–	–	–	662	1175	–	–
Poland	–	–	–	–	–	–	–	–	–	–	312	–
Rumania	418	–	126	–	–	–	–	–	544	1140	–	–
Sweden	551	25	–	–	4	187	–	34	801	133	–	–
Spain	467	37	79	23	21	13	–	–	640	–	–	–
Hungary	67	–	–	–	–	–	–	–	67	940	–	–
Rest of Europe	222	74	38	39	9	61	–	–	443	31	–	–
Europe (excl. USSR)	24569	1108	317	951	582	3213	53	74	30867	5280	312	–
USSR	–	–	–	–	–	–	–	–	–	–	–	–
Japan	331	–	–	–	–	–	–	–	331	–	–	–
Rep. of South Africa	–	–	–	–	–	–	–	–	–	–	–	–
Canada	595	–	–	–	–		–	–	595	–	–	–
USA	4299	–	250	245	–	277	–	–	5071	–	–	–
Australia + New Zealand	–	–	–	–	–	–	–	–	–	–	–	–
Other	723	416	245	–	22	29	5	8	1448	1066	–	–
Total	30517	1524	812	1196	604	3519	58	82	38312	6346	312	–

Source: Statistik der Kohlenwirtschaft, Essen und Köln, September 1980.

Norway	Austria	Poland	Sweden	Spain	Hungary	Rest of Europe	Europe (excl. USSR)	USSR	Japan	Rep. of South Africa	Canada	USA	Australia + New Zealand	Other	Total
95	1	2069	30	–	–	–	4155	120	–	1112	438	1374	763	175	8137
–	–	4752	–	–	–	–	14558	853	–	6834	–	1499	1771	168	25683
–	1	3246	–	–	–	13	5978	1073	–	961	164	2649	1320	43	12188
20	–	640	4	–	–	–	3085	57	–	309	59	679	1490	106	5785
–	–	497	–	15	–	–	5598	279	–	621	174	833	210	83	7798
–	–	–	–	–	–	–	2866	50	–	124	–	1	–	–	3041
–	–	416	–	–	–	–	759	106	–	26	–	422	1025	33	2371
–	–	600	–	–	–	–	835	–	–	–	–	–	–	17	852
–	–	3078	9	–	–	–	4322	528	–	869	309	2	177	50	6257
115	2	15298	43	15	–	13	42156	3066	–	10856	1144	7459	6756	675	72112
–	–	–	–	–	–	18	18	6586	–	–	–	–	–	–	6604
–	–	2412	–	–	–	–	2412	3226	–	–	–	–	–	–	5638
–	–	1915	–	–	–	–	3123	5503	–	–	–	–	–	648	9274
–	–	4000	74	–	–	–	4344	1654	–	–	–	–	–	–	5998
–	–	156	–	–	5	–	1018	1980	–	–	–	177	–	3	3178
x	–	153	–	–	–	–	719	–	–	–	–	108	–	49	876
–	x	987	–	–	–	–	2824	727	–	–	–	–	–	5	3556
–	–	x	–	–	–	–	312	762	–	–	–	–	–	–	1074
–	79	144	–	–	–	104	2011	1446	–	–	226	1074	–	373	5130
–	–	246	x	–	–	–	1180	470	–	–	154	263	–	26	2093
–	–	1439	–	x	–	–	2079	196	–	114	68	805	454	25	3741
–	–	826	–	–	x	–	1833	1180	–	–	–	–	–	12	3025
–	–	280	–	–	–	x	754	12	–	23	–	262	207	44	1302
115	81	27856	117	15	5	135	64783	26808	–	10993	1592	10148	7417	1860	123601
–	–	10022	–	–	–	–	10022	x	–	–	–	–	–	–	10022
–	–	–	–	–	–	–	331	2480	x	2520	10960	8870	25180	1836	52177
–	–	–	–	–	–	–	–	–	–	x	–	–	–	–	–
–	–	226	–	–	–	–	821	–	–	–	x	12997	–	–	13818
104	–	599	–	–	–	–	5774	–	–	996	142	x	933	1031	8876
–	–	–	–	–	–	–	–	–	–	–	–	–	x	–	–
–	–	4209	32	3	32	59	6849	1839	–	880	1072	5414	5150	x	21204
219	81	42912	149	18	37	194	88580	31127	–	15389	13766	37429	36680	4727	229698

da, Japan and the European Community (EC). The second most important exporter was Poland, which supplied the USSR, France, Italy, Denmark and Finland. The Federal Republic of Germany exported mainly to France, Belgium, Luxembourg and Italy. The USSR exported primarily to the Eastern block countries (Bulgaria, German Democratic Republic, Czechoslovakia and Romania) and to Japan.

The largest importer was Japan, whose most important suppliers were Australia and New Zealand, the USA, Canada and the USSR. France imported mainly from the Federal Republic of Germany, Poland and the USA. Canada imported almost entirely from the USA. (Although Canada has its own large reserves, the production costs in the USA are much lower).

It was mentioned above that the production of coal is increasing especially in the USA, the USSR, China, and to a lesser extent in a few Western European countries. However, since the coal reserves, in contrast to the petroleum, lie primarily in the industrial countries which use it to cover their own needs, there will probably be only a slight increase in the international coal trade in the future. Table 3-10b shows the possible development of coal production in the main consumer countries until 2020, along with the projected fraction to be exported (15). This volume of world trade of about $800 \cdot 10^6$ tce would, however, be only 30% of the volume of petroleum traded at that time. (At present, the world trade in coal is only 5% of the volume of petroleum traded.) The realization of this production will only be possible, however, if

Table 3-10b: Possible development of coal production and exports in the largest producers (in 10^6 tce)

Year	1985		2000		2020	
Country[1]	Production	Export	Production	Export	Production	Export
USSR	851	37	1100	50	1800	60
USA	842	68	1340	90	2400	145
China, P.R.	725	7	1200	30	1800	50
Poland	258	45	300	50	320	50
Australia	150	60	300	180	400	240
United Kingdom	137	10	173	10	200	10
India	135	7	235	13	500	32
Germany, F.R.	129	25	145	30	155	30
Rep. of South Africa	119	23	233	55	300	60
Canada	35	15	115	40	200	65
Japan	20	–	20	–	20	–
Other countries	483	6	619	34	751	46
Total	3884	303	5780	582	8846	788

[1] The countries are listed in order of possible production in 1985.

Source: World Energy: looking ahead to 2020. Report by the Conservation Commission of the World Energy Conference, Guildford (UK) and New York: IPC Science and Technology Press 1978.

the countries involved make an enormous effort. For example, the expansion of production in the USA to this level would require investments on the order of $\$ 100 \cdot 10^9$ (1977 dollars) by the year 2000 (38).

For the sake of completeness, it should be mentioned that wood, and especially peat, are of essentially no importance as fuels. Although wood was practically the most important fuel in the Middle Ages, its contribution to the total energy production in Western Europe is now less than 1%. Today wood is an important fuel in developing countries. In 1972, the firewood consumption of the world, including charcoal, was $370 \cdot 10^6$ tce, and was concentrated in developing nations, such as Brazil.

3.313 Special aspects of coal technology

Because the reserves of coal are large, the coal economy faces not only the task of transforming coal into secondary fuels which are more convenient for the consumer and less destructive to the environment, but also the necessity of remaining economically competitive through improvements in the individual phases of coal technology, from locating deposits to final consumption. As a solid, coal is both technologically and physical-chemically at a disadvantage with respect to liquid and gaseous fuels. This is the reason for the preferential utilization of the hydrocarbons.

The discovery of coal deposits is easier than locating petroleum and natural gas, and for this reason the coal reserves are in general better known. Three important aspects in the evaluation of a deposit are the quality and quantity of coal in relation to the purpose for which it is to be used, the physical factors which influence mining the deposit and thus the cost, and the geographic location of the deposit (37). The primary requirement is that the deposit contain a sufficient amount of the desired quality; this depends on the intended use of the coal. For example, the heat content and the amount of gaseous components are deciding criteria for coal to be used in electrical generators. The important physical factors include the depth and thickness of the seam and the amount of groundwater. The greater the depth, the more expensive it is to open the mine and to remove the coal. In addition, the temperature is higher at greater depths, which increases the amount of ventilation needed. (In Western Europe, the temperature rises about 3K for each 100 m depth). If there is a large amount of ground water, both cutting and removing the coal from the mine are considerably more difficult than in a dry mine. The importance of the geographical location of a primary fuel has already been discussed under 2. 4.

The expense of opening either a new coal mine or a new oil field can vary greatly. The coal fields in the American Midwest, for example, were opened relatively cheaply, compared to those in Europe. (Analogously, the opening of the petroleum fields in the Near East was very inexpensive, compared with the North Sea field.) A minable seam can be extracted either by strip or by underground mining, depending on how it lies. In strip mining, the seam is exposed to the air by removing the rock layers above it. In the USA, the USSR and South Africa, anthracite is obtained from strip mines, while in the Federal Republic of Germany, it is lignite that is so obtained. Underground mines are reached by shafts sunk in the ground; this is the type of mine used in Western Europe to reach anthracite coal (3).

The subsequent phases of coal technology are separation of the coal from the surrounding rock, removal from the mine and processing (e.g. according to size of lump), to make a salable or technologically usable product from the raw mined coal. It is in these phases that the differences between coal and oil (or gas) technology are greatest, due to the difference in physical state. Petroleum and natural gas can be pumped, so that they can be mined with very little effort (primary recovery). The advantage is only lessened when it is necessary to go to secondary or tertiary recovery methods. The technology for recovery of oil from oil shales or sands is similar to that for coal, i.e. a solid material must be mechanically loosened and transported. In comparison to oil recovered from such sources, coal is at less of a disadvantage.

The possibility of stockpiling and the transport costs are other important criteria in the evaluation of a primary source of energy. Coal has the advantage with respect to storage, because all that is required is a sufficiently large area for a coal dump. For oil or gas, by contrast, one must have containers or underground caverns whose preparation requires considerable effort. The fact that coal is a solid makes its transport more expensive than that of petroleum or gas, which can be cheaply moved in pipelines or tankers (natural gas in liquified form). An increased use of coal will require, even in the producing countries, that it be more cheaply transported than at present. One promising method is the hydraulic solid transport (a slurry of large particles in water) in pipelines, which has certain advantages over conventional means of transport like ships, railroads or trucks. It is less detrimental to the environment, independent of weather and possible even in difficult terrain. In the USSR, for example, construction of a 250-km pipeline to bring coal to a thermal power plant is supposed to be starting (40).

Compared to the world-wide network of pipelines for oil and gas transport, however, hydraulic solid transport is still in the research or pilot stage. In North America alone, the pipeline network for petroleum transport is about 370000 km long, and the natural gas system includes about 640000 km of pipes.

In the USSR, there were already about 72000 km of gas and about 50000 km of oil pipelines in 1975. The longest crude oil pipeline in the world begins in the Volga-Urals area at Knibyschev and goes west to Mozyr. This 1350 km trunk branches at Mozyr into two lines, one leading to Czechoslovakia and Hungary, the other to Poland and the German Dem. Rep. The total length of the pipeline is 5300 km. The Transsiberian Pipeline, which begins in Tujmasy, also in the Volga-Urals area, and goes to Irkutsk via Omsk, is 3800 km long. The construction of another pipeline is planned.

3.32 Petroleum

3.321 Geographical distribution of the petroleum reserves

Petroleum is at present the main fuel used in the entire world, but it is also the energy source which will first be exhausted. This situation is exacerated by the very uneven distribution of the petroleum reserves, especially those which are at present technically and economically recoverable. For the OECD countries, the reserve situation with respect to oil is especially unfavorable compared to coal, natural gas and uranium. Furthermore, in spite of its greatest efforts, humanity seems to have no genuine alternative to petroleum in the short and medium term.

Several attempts have been made to estimate the total available and recoverable petroleum reserves of the earth (41–48). Table 3-11 and Fig. 3-4 show the estimates of ultimate recoverable resources of crude oil (10, 41). The older estimates have proven to be far too low. In 1920, the recoverable world reserves were estimated to be $5.9 \cdot 10^9$ t petroleum. Since then, however, about $53 \cdot 10^9$ t petroleum have been produced, and the not yet recovered *proved recoverable reserves* are now estimated (with slight variations) at $100 \cdot 10^9$ t.

In recent years the majority of estimates made for the total amount of recoverable oil (cumulative production, proved recoverable reserves and additional recoverable resources) lie around $240 \cdot 10^9$ t to $360 \cdot 10^9$ t excluding Natural Gas Liquids (NGL). It is now estimated that the total quantity of recoverable oil was $354 \cdot 10^9$ t, and $53 \cdot 10^9$ t (15%) have already been recovered; on 1. 1. 1979 the proved recoverable reserves were approx. $89 \cdot 10^9$ t (25%) and the additional recoverable resources were around $212 \cdot 10^9$ t (60%) (12).

The *proved recoverable reserves* of petroleum (conventional oil) are given as $98 \cdot 10^9$ t ($\triangleq 141 \cdot 10^9$ tce) by the Federal Institute for Geosciences and Natural Resources, Hannover. It is assumed here that an average of 32% of the oil can be recovered. L. Auldridge, writing in the Oil and Gas Journal (Worldwide Issue), estimates this value as $87.3 \cdot 10^9$ t ($\triangleq 126 \cdot 10^9$ tce) as of 1/1/1978 (11).

The "oil in place" is estimated by the Federal Institute for Geosciences and Natural Resources, Hannover, to be $725 \cdot 10^9$ t oil (2). The majority of the estimates are comparable, but they range up to $1000 \cdot 10^9$ t "oil in place" (45, 49).

Although it will take a few years to evaluate these estimates in detail, it can be assumed that they are very close to the right order of magnitude. One disadvantage, however, is that the methods by which the data were assembled are often not given in detail, and in addition, the data are not given for individual countries.

It should be mentioned, however, that the exploration for petroleum is only beginning in many regions of the earth. A study by the Exxon Corporation indicates that there are about 60 countries in which oil production can be expected to be begun or increased (50). These include parts of Latin America (e.g. Mexico and Brazil), Africa (e.g. Egypt, Angola, Zaire, Chad) and Asia (e.g. Pakistan, India, Bangladesh,

Table 3-11: Estimates of the total recoverable oil (excluding oil from oil shales and bituminous sands)

	Year	Source	[10^9 t]
PWS	1942	Pratt, Weeks, Stabinger	82
D	1946	Duce	55
P	1946	Pogue	76
We	1948	Weeks	183
Le	1949	Levorsen	205
We	1949	Weeks	138
McN	1953	Mac Naughton	136
Hu	1956	Hubbert	171
We	1958	Weeks	205
We	1959	Weeks	273
He	1965	Hendricks (USGS)	338
Ry	1967	Ryman (Esso)	285
Sh	1968	Shell	246
We	1968	Weeks	300
Hu	1969	Hubbert	184–286 (235)
Mo	1970	Moody (Mobil)	246
Wa	1971	Warman (BP)	164–273 (218)
We	1971	Weeks	312
US	1971	US National Petroleum Council	364
Li	1972	Linden	402
We	1972	Weeks	498
Mo	1972	Moody, Emerick (Mobil)	246–259 (252.5)
WEC	1973	WEC (USGS)	184–1840
Ad	1975	Adams and Kirby (BP)	273
Mo	1975	Moody (Mobil)	277
BGR	1975	BGR	336
Kle	1976	Klemme (Weeks)	259
WEC	1978*	WEC	127–950
Ha	1979	Halbouty (Moody)	304
Mey	1979	Meyerhoff	300
Roo	1979	Roorda	330
WEC	1980	WEC	354

* WEC 1978: The mean statistic used by the WEC from this range was $350 \cdot 10^9$ t which included
$260 \cdot 10^9$ t conventional
$40 \cdot 10^9$ t non-conventional and
$50 \cdot 10^9$ t already produced

$350 \cdot 10^9$ t

Source: World Energy Conference 1980: Survey of Energy Resources 1980, Munich, September 1980.

Cambodia, Sri Lanka, and China P.R.) (51–60). And in the countries with large petroleum reserves, including Saudi Arabia, the USA and the USSR, the exploration for oil continues (61–67).

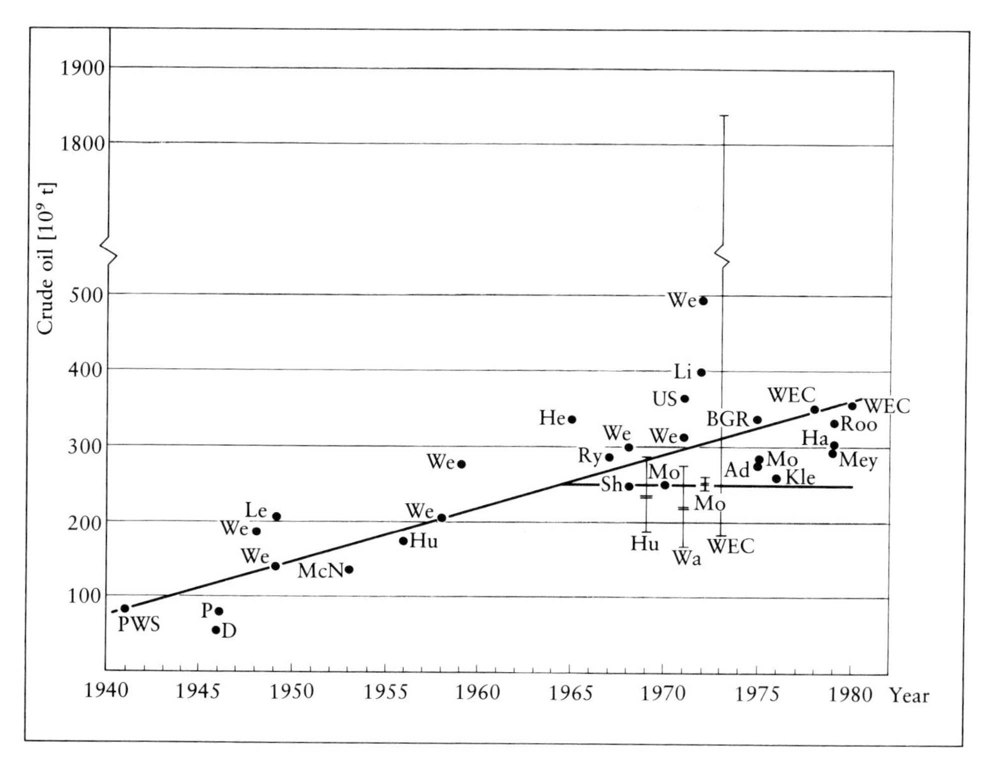

Fig. 3-4: Development of "ultimate recovery" estimates for crude oil (excluding oil from oil shales and bituminous sands)

Source: World Energy Conference 1980: Survey of Energy Resources 1980, Munich, September 1980.

Table 3-12a shows the cumulative production, proved recoverable reserves, estimated additional resources and ultimated recovery for crude oil according to region. Even ahead of the Middle East, North America is the region with the greatest cumulative production up till now. However the Middle East has 57% of the proved recoverable reserves and is therefore by a wide margin the region with the greatest reserve margin. The majority of the additional recoverable resources are expected in the USSR, P.R. of China and Eastern Europe with jointly 30%, and the Middle East with 24% of the world total (12). (Estimates of the additional recoverable resources are based on the assumption that the recovery factor of about 0.3 will increase to 0.4 in the future (12)). Table 3-12b shows the reserves and resources of natural gas liquids (NGL) and crude oil (12). (Gas condensates are mixtures of hydrocarbons that are liquid under normal conditions but gaseous under the conditions found in the deposit. They arise during production of natural gas or associated gas as a result of expansion and cooling which leads to condensations or absorption).

Table 3-12 a: The cumulative production, reserves, resources, and ultimate recovery for oil according to region

Region	cumulative production up to 1. 1. 1979		proved recoverable reserves[1] on 1. 1. 1979		estimated additional recoverable resources[2]		ultimate recovery	
	[10^6 t][3]	%	[10^6 t]	%	[10^6 t]	%	[10^6 t]	%
Africa	3750	7	8040	9	34000	16	45790	13
North America	17520	33	4480	5	24000	11	46000	13
Latin America	7040	14	7770	9	12000	6	26810	8
Far East/Pacific	1720	3	2390	3	12000	6	16110	4
Middle East	14680	28	51050	57	52000	24	117730	33
Western Europe	560	1	2710	3	10000	5	13270	4
USSR, China P. R., Eastern Europe	7530	14	12700	14	64000	30	84230	24
Antarctic	–	–	–	–	4000	2	4000	1
Total	52800	100	89140	100	212000	100	353940	100

[1] *Proved reserves* represent the fraction of total resources that has not only been carefully measured but had also been assessed as being exploitable in a particular nation or region under present and expected local economic conditions (or at specified costs) with existing available technology.
Proved recoverable reserves are the fraction of proved reserves in-place that can be recovered (extracted from the earth in raw form) under the above economic and technological limits.

[2] *Estimated additional recoverable resources* embrace all resources, in addition to proved reserves that are of at least foreseeable economic interest. The estimates provided for additional resources reflect, if not certainty about the existence of the entire quantities reported, at least a reasonable level of confidence. Resources whose existence is entirely speculative are not included.
This system is not quite as detailed as the classification systems used by some countries, but it has the advantage of being internationally compatible.

[3] 1 t oil ≙ $42.2 \cdot 10^9$ J.

Source: World Energy Conference 1980: Survery of Energy Resources 1980, Munich, September 1980.

Table 3-12b: The reserves and resources of natural gas liquids (NGL) and crude oil

Region	proved recoverable reserves		estimated additional recoverable resources		estimated additional resources crude oil in place
	natural gas liquids	crude oil	natural gas liquids	crude oil	
	[10^6 t]	[10^6 t]	[10^6 t]	[10^6 t]	[10^6 t]
Africa	494	8040	450	34000	85000
North America	510	4480	1100	24000	60000
Latin America	382	7770	250	12000	30000
Far East/Pacific	110	2390	850	12000	30000
Middle East	–	51050	2100	52000	130000
Western Europe	9	2710	550	10000	25000
USSR, China P. R., Eastern Europe	–	12700	6300	64000	160000
Antarctic	–	–	–	4000	10000
Total	1505	89140	11600	212000	530000

Source: World Energy Conference 1980: Survey of Energy Resources 1980, Munich, September 1980.

If one considers the distribution of the petroleum over the economic-political blocks (Table 3-13), one can see that the OPEC countries, with about 69% of the *proved recoverable reserves* are far in the lead. They are followed by the Eastern Block, with 14%, and the OECD countries, with about 9%.

Even within the regions or economic-political groups, the *proved recoverable reserves* are very unevenly distributed. Table 3-14 gives the countries with the largest petroleum reserves (12). The key position of Saudi Arabia, with nearly one quarter of the *proved recoverable reserves* for the entire earth, becomes clear from this. The importance of this country in the world politics of the near future cannot be overemphasized. It should also be noted that the USSR has the second largest reserves in the world.

For Western Europe, the petroleum reserve situation is particularly unfavorable compared to the other fossil fuels, coal and natural gas. In the Oil and Gas Journal (Worldwide Issue), L. Auldridge gives the "estimated proved reserves" of Western Europe as $3310 \cdot 10^6$ t (11). By far the largest part (about 88%) of these reserves are offshore, and some lie in areas which are covered with ice (Barents See off Greenland). The tapping of these reserves will be correspondingly difficult and expensive. The North Sea reserves (south of 62° N latitude) are estimated to be about $5000 \cdot 10^6$ t petroleum, and the North Atlantic (north of 62°, the Norwegian part of the Barents Sea), about $6000 \cdot 10^6$ t (16). Fig. 3-5 shows the most important oil and gas fields in the North Sea (68).

Table 3-13: Distribution of the petroleum reserves among the economic-political groups

Group of countries	cumulative produktion up to 1. 1. 1979		proved recoverable reserves on 1. 1. 1979		estimated additional recoverable resources		total amount of oil still recoverable on 1. 1. 1979	
	[10^6 t]	%	[10^6 t]	%	[10^6 t]	%	[10^6 t]	%
OECD	18300	35	7480	9	35000	17	42480	14
Countries with centrally planned economies	7530	14	12700	14	64000	30	76700	26
OPEC	23730	45	61780	69	78000	36	139780	46
Other	3220	6	7180	8	35000	17	42180	14
Total	52800	100	89140	100	212000	100	301140	100

Source: World Energy Conference 1980: Survey of Energy Resources 1980, Munich, September 1980.

Table 3-14: The countries with the largest petroleum reserves

country[1]	cumulative production by 1. 1. 1979 from (12) [10^6 t]	proved recoverable reserves on 1. 1. 1979 from (12) [10^6 t]	%	total amount of oil still recoverable from (2) [10^6 t]	%
1. Saudi Arabia	4500	23000	25.8	43000	14.8
2. USSR	7530	9700	10.9	45500	15.6
3. Kuwait	2600	9100	10.2	10900	3.8
4. Iran	3900	8100	9.1	18000	6.2
5. Irak	1800	4400	4.9	21200	7.3
6. United Arab. Emirates	710	4300	4.8	9100	3.2
7. Mexico	731	4058	4.6	11594[2]	4.0
8. USA	16348	3748	4.2	19720	6.8
9. Libya	1600	3300	3.7	7650	2.6
10. China, P. R.	420	2700	3.0	17000	5.9
11. Venezuela	4928	2621	2.9	n. a.	
12. Nigeria	940	2500	2.8	8660	3.0
13. Great Britain	106	1906	2.1	4175	1.4
14. Indonesia	1085	1400	1.6	8655	3.0
15. Algeria	705	1130	1.3	5550	1.9
16. Canada	1220	729	0.8	10280	3.6
17. Norway	59	550	0.6	8000	2.8
18. Qatar	370	550	0.6	800	0.3
19. Egypt	270	440	0.5	3510	1.2
20. Oman	170	340	0.4	1000	0.3
Italy	35	51		250	
Germany, F. R.	172	42		175	
France	48	8		65	
Japan	29	5		20	

[1] The countries are listed in the order of the size of their proved recoverable reserves.
[2] This figure was taken from (12).

Sources: World Energy Conference 1980: Survey of Energy Resources 1980, Munich, September 1980 (12).
Federal Institute for Geosciences and Natural Resources, Hannover, Federal Republic of Germany, 1976 (2).

The Western European part of the Mediterranean is suspected to cover about $700 \cdot 10^6$ t of petroleum, of which the largest part is expected to be found under the Aegean. The climatic conditions in the Mediterranean are also much more favorable for drilling and recovery than they are in most of the North Sea and North Atlantic. The reserves in Greenland, the North Atlantic Shelf, the Irish Sea and the bordering areas in the Englisth Channel and the Atlantic Shelf off France are judged to be more modest, with less than $300 \cdot 10^6$ t each (2).

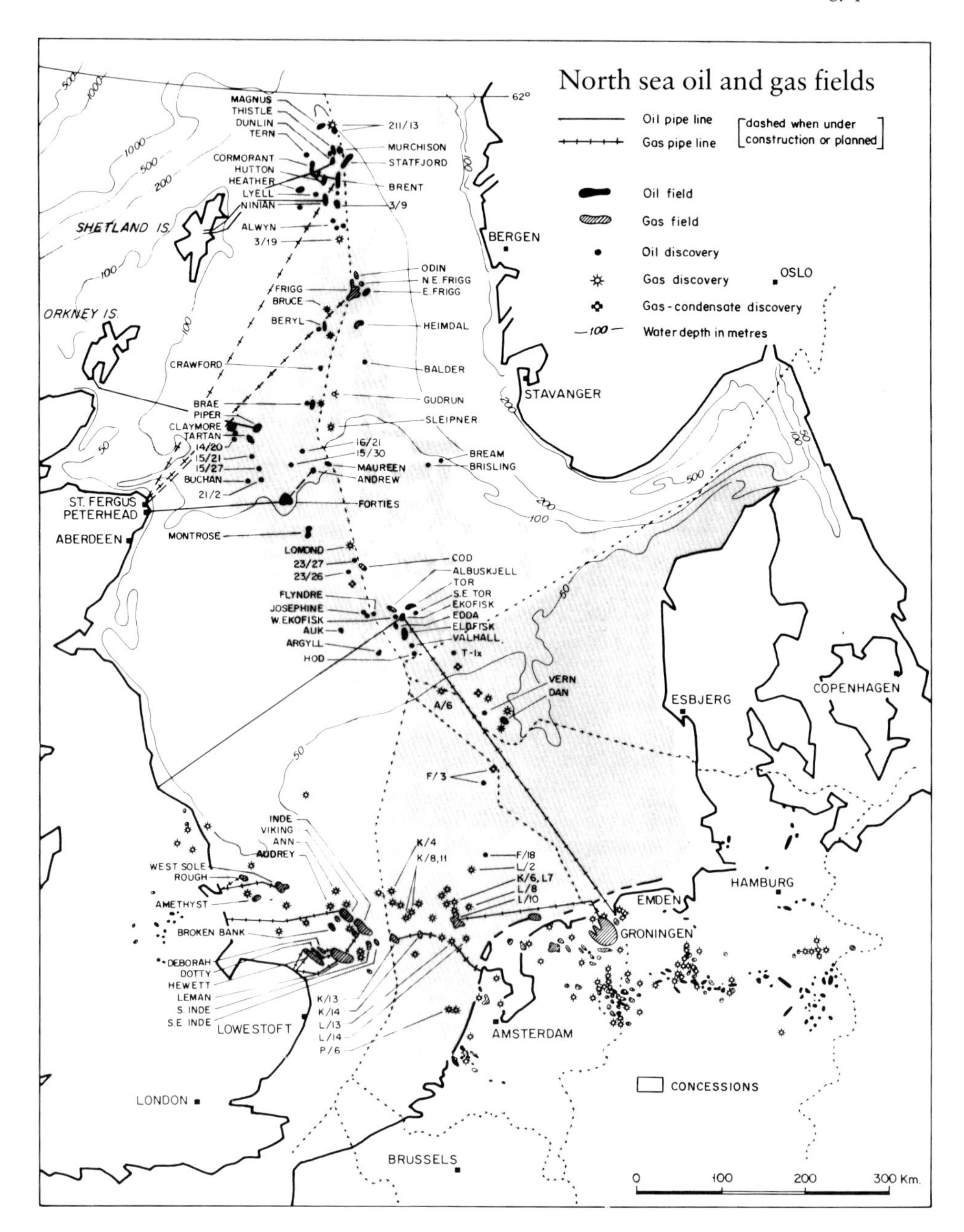

Fig. 3-5: The most important North Sea oil and gas fields

Source: P. A. Ziegler, Geology and Hydrocarbon Provinces of the North Sea, Geo Journal, No 1, 1977.

Although the North Sea is the second largest offshore area in the world, after the Persian Gulf, the exploitation of the North Sea is in a relatively early stage, compared to the Persian Gulf, and this in spite of a 10-year exploration phase. As a result of the OPEC oil policy, the offshore reserves are becoming increasingly important all over the world. About 17% of the petroleum produced in the world at present comes from offshore fields, and it is assumed that in the year 2000, about 50% of the world's production will be from offshore wells.

In addition to conventional oil, there is an enormous potential supply of nonconventional oil (10, 15). This generally includes: deep offshore deposits (water deeper than 2000 m) and polar deposits (see 3.323), Enhanced Oil Recovery (EOR) from old fields (see 3.323), oil shales (see 3.341), oil sands and heavy-oil deposits (see 3.342), synthetics from coal (see 4.64) or biomass (see 4.326.2).

3.322 Centers of petroleum production and consumption

Since petroleum is currently the main fuel, we shall first consider the development of the *proved recoverable reserves* and the world production of petroleum. Table 3-15 shows the development from 1950 to 1980 (69). It can be seen that the *proved recoverable reserves* have increased nine-fold since 1950, while the annual world oil production has increased six-fold. From the "ratio of reserves to production", it can be calculated that, ignoring the uneven geographical distribution, it will be possible to continue the annual world production of oil of about $3000 \cdot 10^6$ t for almost another 30 years.

From 1960 to 1980, the world petroleum consumption quadrupled, although the total consumption of primary energy was only slightly more than doubled (compare Table 2-2). However, if one considers the development of primary energy consumption in the world outside the Eastern block, the change in the comsumption pattern was still more drastic; the fraction contributed by petroleum is even larger here, and amounted to about 50% in 1980 (70). The fraction of the primary energy supplied by petroleum in the largest consumer countries in the world is given in Table 3-16 (70). It is likely that in 1990, more than 40% of the non-East-block countries' energy will be supplied by petroleum (69).

In all the main consumption centers of the western world, the consumption of petroleum far exceeds the production. Fig. 3-6 shows the discrepancy between production and consumption in the individual centers, and Table 3-17 gives the values for a few selected countries (70). In contrast, the Eastern block has a balance of production and consumption of petroleum. The gap between consumption and production within the western industrial nations therefore had to be filled by the countries with large production and low consumption, a pattern which in all probability will not change before the end of this century. The main regions with an excess of production are in the Middle East and Africa. For this reason, the flow of raw petroleum in international trade is highly significant in world economic politics. About every second

Table 3-15: Historical development of the world's proved recoverable reserves of petroleum (conventional oil)

Year	proved recoverable reserves[1)] [10^6 t]	World oil production [10^6 t]	Ratio of Reserves to Production [R/P-ratio]
1950	10600	521	20
1955	21600	771	36
1960	41000	1085	38
1961	42200	1155	37
1962	41800	1251	33
1963	45100	1341	34
1964	46100	1450	32
1965	47600	1547	31
1966	51900	1693	31
1967	55400	1813	31
1968	61900	1976	31
1969	72000	2136	34
1970	83400	2336	36
1971	85200	2463	35
1972	86400	2604	33
1973	87400	2851	31
1974	87600	2873	31
1975	89500	2707	33
1976	87500[2)]	2937	30
1977	87800	3039	29
1978	88900	3056	29
1979	87366	3191	27
1980	88352	3066	29

[1)] Level as of Dec. 31 each year.
[2)] In some countries the figures from previous year were corrected.

Source: Exxon Corporation, Public Affairs and Information Department, Hamburg, 1981.

ton of raw petroleum produced in the world is transported by sea from the producer to the consumer. Fig. 3-7 shows the supply lines for petroleum in 1978 (70), and makes clear the vital importance of oil imports from the Persian Gulf and of oil routes through the Indian Ocean for Western Europe, Japan and, in increasing measure, for the United States. The strategic importance of the sea route around the Cape of Good Hope, the main route for oil supplies for Western Europe and the USA, cannot be overestimated by the western industrial nations.

The further development of petroleum production in individual countries depends on a number of factors, some of which are determined by national interests. A few of the most important are the expected economic growth, possible substitutions for petroleum, especially in the industrial countries with highly petroleum-dependent

Table 3-16: Percentage of primary energy supplied by petroleum in the largest consumers of primary energy

Country[1] / Year	Primary energy consumption 1973 [10^6 tce]	Part supplied by petroleum %	Primary energy consumption 1980 [10^6 tce]	Part supplied by petroleum %
Japan	510	77	521	67
Italy	197	76	209	67
France	259	70	275	58
Germany, F. R.	379	55	390	48
United States	2700	47	2654	43
United Kingdom	324	50	293	40
Canada	275	43	321	39
USSR	1238	37	1680	37
India*	112	27	133[2]	25
China, P. R.	527	15	738	17

[1] The countries are listed in order of the percentage supplied by petroleum in 1980.
[2] This figure is for 1976.

Sources: BP statistical review of the world oil industry 1980, London 1981.
Exxon Corporation, Public Affairs and Information Department, Hamburg 1978*.
United Nations, Statistical Yearbook 1977, New York 1978*.

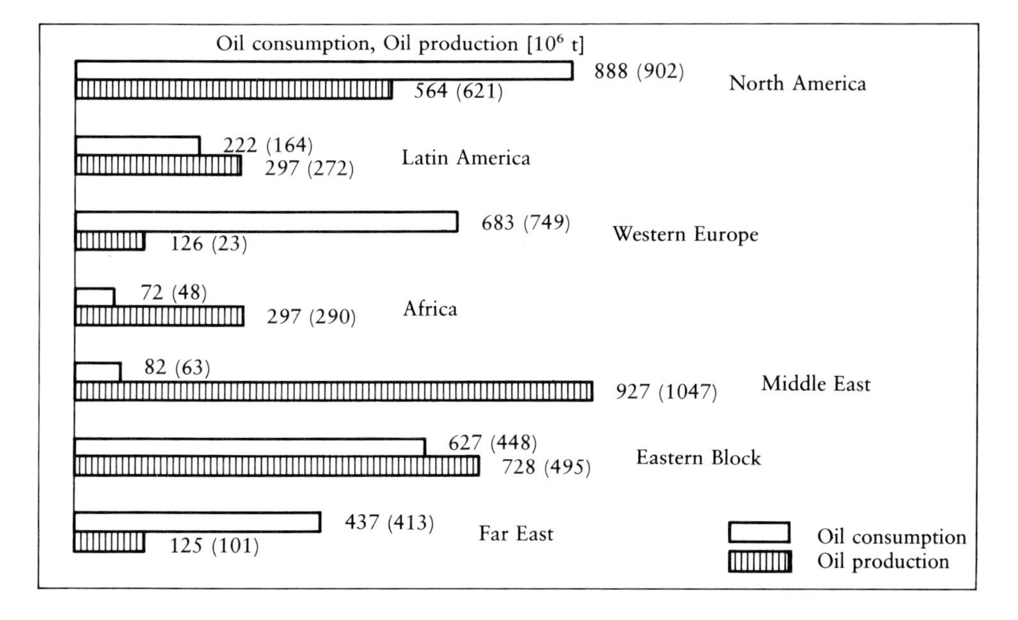

Fig. 3-6: Centers of petroleum production and consumption in 1980 (in 10^6 t)*
(The figures for 1973 are given in parentheses)
*The data are taken from BP statistical review of the world oil industry 1980, London 1981.

Table 3-17: Consumption and production of petroleum in selected countries (in 10^6 t), and static lifetimes in years

Country[1] Year	Consumption 1973 from (69)	1979	1980	Production 1973 from (70)	1979	1980	Static lifetime: proved recoverable reserves[2]	Static lifetime: total amount of oil still recovery[3]
1. USSR	324	426	438	421	586	603	17	78
2. Saudi Arabia	17	19	19	365	470	493	49	91
3. USA	813	861	800	519	481	484	8	41
4. Irak	4	5	5	99	171	130	26	124
5. Venezuela	12	17	17	179	125	116	21	–
6. Mexico	29	44	47	27	81	106	50	143
7. China, P. R.	39	91	92	55	106	106	25	160
8. Nigeria	3	4	4	100	114	102	22	76
9. Libya	1	4	4	105	101	86	32	89
10. Canada	89	91	88	102	84	80	9	122
11. United Kingdom	113	97	83	0.1	78	80	19[4]	42
12. Indonesia	10	19	20	66	79	79	18	110
13. Iran	20	27	25	293	158	74	51	114
14. Kuwait	5	5	5	138	113	71	81	96
15. United Arab. Emirates	n. a.	n. a.	n. a.	63	70	65	61	130
Norway	9	9	8	2	20	26	14[5]	200
Germany, F. R.	150	147	131	7	5	5	8	350
Italy	105	104	103	1	2	2	25	125
France	124	122	113	1	1	1.5	8	65
Japan	281	284	255	0.7	0.5	0.4	10	40

[1] These were the 15 largest producers in 1980.
[2] Based on the production and reserves in 1979 (compare Table 3-14, from (12)).
[3] Based on the production in 1979 and the total amount of oil still recovery (compare Table 3-14, from (2)).
[4] Based on an annual production of $100 \cdot 10^6$ t.
[5] Based on an annual production of $40 \cdot 10^6$ t.

Sources: Exxon Corporation, Public Affairs and Information Department, Hamburg 1981 (69).
BP statistical review of the world oil industry 1980, London 1981 (70).
Author's calculations.

economies, the improvement in the reserve situation brought about by newly discovered reserves, and the development of petroleum prices and their relationship to the prices of other primary sources of energy.

As mentioned above, the Conservation Commission of the World Energy Conference came to the conclusion that the available reserves of petroleum would allow a considerable increase in production (see 2.333 and Fig. 2-13). M. Grenon, Interna-

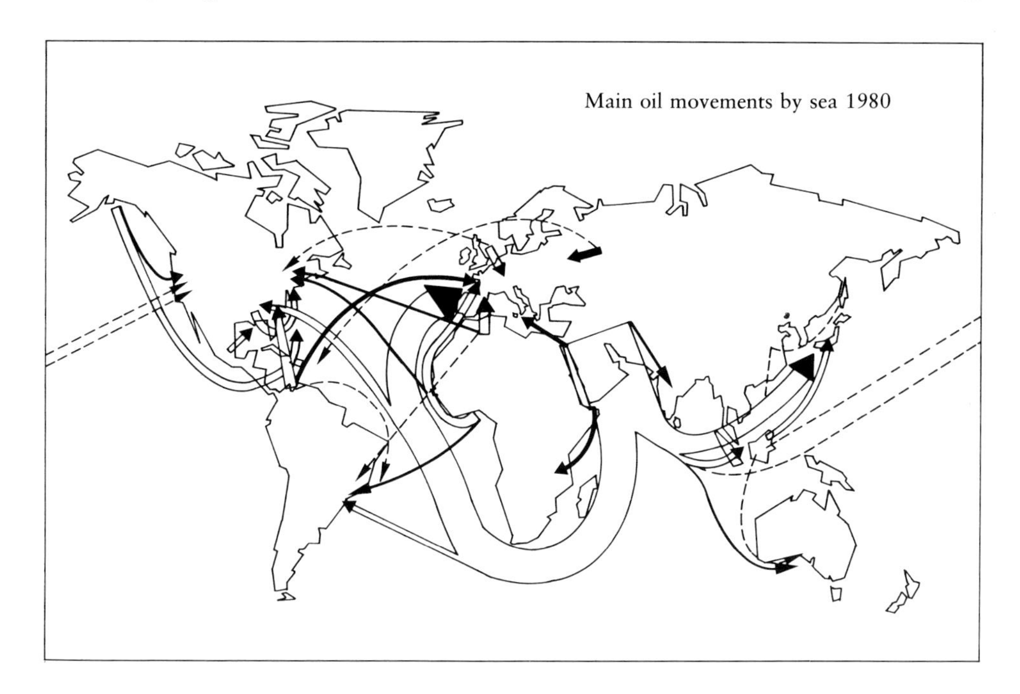

Inter-area total oil movements 1980 (in 10^6 t)

From / To	USA	Canada	Latin America	Western Europa	Africa	South East Asia	Japan	Australasia	Other Eastern Hemisph.	Destination not known	Total Exports
USA	–	5.2	14.8	8.2	0.3	0.6	–	–	–	–	29.1.
Canada	18.6	–	–	1.5	–	0.5	0.5	–	–	1.2	22.3
Latin America	105.4	10.7	13.8	31.1	2.5	–	4.0	0.2	11.7	17.3	196.7
Western Europe	18.1	0.2	–	–	4.2	0.2	–	–	0.5	–	23.2
Middle East	77.2	14.8	80.4	362.3	22.0	86.3	176.1	10.9	34.3	5.0	869.3
North Africa	50.8	0.3	6.7	71.2	0.8	–	2.7	–	7.7	–	140.2
West Africa	46.8	0.2	20.8	49.0	3.0	–	0.7	–	–	–	120.5
East and South Africa	–	–	–	–	–	0.5	–	–	–	–	0.5
South Asia	–	–	–	–	–	–	0.2	–	1.7	–	1.9
South East Asia	20.1	–	–	0.7	1.2	–	52.5	6.9	2.2	–	83.6
Japan	–	–	–	–	–	0.4	–	–	–	–	0.4
Australasia	–	–	–	0.2	–	0.5	0.2	–	–	–	0.9
USSR, E. Europe and China P.R.	–	–	10.2	64.7	–	11.1	8.9	–	4.7	–	99.6
Total Imports	337.0	31.4	146.7	588.9	34.0	100.1	245.8	18.0	62.8	23.5	1588.2

Fig. 3-7: Crude oil supply movements 1980

Source: BP statistical review of the world oil industry 1980, London 1981.

tional Institute for Applied Systems Analysis, Laxenburg/Vienna, has come to the even more optimistic conclusion that it should be possible, given the appropriate efforts and production costs, to increase petroleum production to somewhat more than $7 \cdot 10^9$ tce (about $5 \cdot 10^9$ t oil) by 1990, and thereafter to produce an amount falling to $3.75 \cdot 10^9$ tce (about $2.5 \cdot 10^9$ t oil) annually until 2065. This is based on reserves of $450 \cdot 10^9$ tce (about $300 \cdot 10^9$ t oil, production costs no more than $ 20/bbl, 1976 dollars).

This approach, however, ignores important geographical and political factors. It must be remembered, that the OECD countries can be expected to produce only 15 to 19% of the world's petroleum in the future, while the developing countries will produce between 59% and 63% (compare Table 2-10). In particular, it must be assumed that the OPEC countries will retain their monopoly position among the crude-oil-producing countries. Thus, for example, the production in 1990 of $3.4 \cdot 10^9$ t crude oil (world excluding the Eastern block) will probably be distributed as follows: OPEC countries, 62%; North America, 16%; Western Europe, 8%; others, 14% (69). (Since the Eastern block has a balance between production and consumption, it is realistic to assume that the needs of the rest of the world will be covered outside the Eastern block).

The political explosiveness of these data becomes most clear when one considers the static expected lifetimes of the *proved recoverable reserves* of the 15 largest producers (Table 3-17). The lifetime of the *proved recoverable reserves* in the USA and Venezuela are strikingly short. The static lifetime of the reserves in the USSR, even at a somewhat higher annual production, is twice as long as in the USA. The *proved recoverable reserves* of the Middle East have the longest lifetimes, especially those in Saudi Arabia, Iran, Irak, Lybia and Kuwait. These countries will be able to maintain their current petroleum production several decades longer than other important producing countries.

With regard to the values given for Western Europe, it should be remembered that the petroleum production there (offshore) is still in its infancy. Great Britain expects an annual production of 100 to $120 \cdot 10^6$ t in the early 80's, rising to 150 to $200 \cdot 10^6$ t per year after 1985. Norway expects an annual offshore production of $50 \cdot 10^6$ t in the early 80's and 70 to $90 \cdot 10^6$ t per year after 1985. This production will presumably reduce the dependence of Western Europe on the oil-producing regions (71–77).

Until a few years ago, the search for oil was concentrated for the most part on relatively accessible and cheaply developed deposits on land and in shallow water. Due to the particularly critical situation, the search is moving increasingly to climatically unfavorable areas (e.g. Alaska and the Canadian Arctic), the continental shelves and slopes (78). This has raised the production costs, in some cases considerably. The total costs of production are the sum of investment, operating and capital costs. The investment and operating costs, which together make up the technical costs, are highly influenced by the natural conditions imposed by the site, such as the amount of accessible oil, the depth and structure of the deposit, its distance from other deposits, the distance of the field from the continent, the problems of transport and the environmental factors.

It is generally more expensive to develop a petroleum or natural gas deposit than a coal deposit. A relatively large investment is required to develop the capacity to produce 1 bbl/d ($\triangleq$ 50 t/year) in the North Sea. Table 3-18 shows the investment required to establish a production rate of 1 bbl/d in various areas (79).

By the middle of 1975, the petroleum industry had invested about \$ $15 \cdot 10^9$ in the exploration and development of hydrocarbons in the North Sea. About \$ $3 \cdot 10^9$ was spent on exploration, including the drilling of 550 exploratory wells, which led to the discovery of 60 oil and gas fields. Thus the average cost per exploratory well was about \$ $5.5 \cdot 10^6$, and the chance of striking oil or gas was 1:8. It is estimated that additional investment of more than \$ $40 \cdot 10^9$ (in 1976 dollars), will be required before the end of the 80's to reach the goal of 150 to $300 \cdot 10^6$ t per year (79).

There are operating costs in addition to investment costs, and they also vary widely from region to region. These costs include the price of energy to run the pipeline and the secondary production costs, supplies, insurance premiums, etc. The average operating expense in the Middle East, adjusted to account for inflation, is about 20 US-cts/bbl. In the North Sea it is \$ 1.60–\$ 2.20/bbl (79).

Typical timespans for exploration, development of a field and production are shown in Table 3-19, for a $50 \cdot 10^6$ t/year field. It can be seen that the exploration, development and production cover a span of about 25 years. For larger reserves, this time may be extended by about 10 years.

Due to the large investments required and the relatively long time before production can begin, the fraction of the total represented by capital costs is a major one. In

Table 3-18: Investment costs for oil production

Region	Middle East (onshore) from (79)	Middle East (offshore) from (79)	West Africa (onshore) from (79)	West Africa (offshore) from (79)	North Sea (offshore) from (79)	Oil shales, Oil sands from (15)
Investment costs per barrel per day production rate in 1000 US- \$ (1976 prices)	1.2	2	2	2.5	6–8	15–20

Sources: G. Schürmeyer, The economic aspects of North Sea oil, Marine Technology, Vol. 7, April 1976 (79).
World Energy: looking ahead to 2020. Report by the Conservation Commission of the World Energy Conference, Guildford (UK) and New York: IPC Science and Technology Press 1978 (15).

Table 3-19: Time schedule and investment plan for setting up production in the North Sea (exemplified by a $50 \cdot 10^6$ t field)

	Exploration	Field development	Production	
Operation	Seismic	Planning	Buildup	3 years
	Exploratory drilling	Platform construction	Peak	5 years
	Confirmatory drilling	Construction of pipeline and harbor facilities	Decline	8 years
		Production and injection drilling		
Time	4 years	5 years	16 years	
Investments	$580 \cdot 10^6$	$ $1000 \cdot 10^6$		
Operating expenses			$ $1000 \cdot 10^6$	

Source: G. Schürmeyer, The economic aspects of North Sea oil, Marine Technology, Vol. 7, April 1976.

the past (for example in the Middle East), the capital costs were of secondary importance, because the amount of capital required was relatively small and a project could be financed by the production, which began relatively soon. Now the capital costs are of the same order of magnitude as the technical production costs.

In addition, the cost of petroleum has long been strongly influenced by taxation. The recent sharp increases in the crude oil prices have led the British and Norwegian governments to levy a special tax on the petroleum producers, in addition to the normal corporate taxes. On the whole, however, the price structure of North Sea oil is determined by the technical production costs. In contrast, the technical costs in the Middle East are almost negligible compared to the taxes, and thus have very little effect on the price of this oil (80–82).

The total production costs for undeveloped reserves are, of course, not yet known. However, some idea can be had from the geographical location of the estimated additional recoverable resources of crude oil potential, i.e. reserves in the Middle East can be relatively cheaply developed, while those in the Arctic will be very expensive. It can be expected that about half of the estimated additional recoverable resources of crude oil potential will be relatively inexpensively recovered (83). Table 3-20 shows the distribution of the estimated additional recoverable resources of crude oil potential of the world according to the probable cost of recovery (2). The Shell Oil Company has come to similar conclusions (84).

Table 3-20: Distribution of the estimated additional recoverable resources of crude oil potential of the world (in 10^6 t) according to cost categories

Region	economically favorable I	II	economically unfavorable III	IV	estimated additional recoverable resources crude oil total[1]
Middle East	25680	26000	–	–	51680
Africa	–	13600	16340	3700	33640
North America	–	–	11500	12200	23700
Latin America	–	6200	6200	–	12400
Western Europe	–	–	6360	4300	10660
Far East, Australia	–	4720	4700	–	9420
Western world	25680	50520	45100	20200	141500
Eastern block	–	18800	15000	14000	47800
Antarctica	–	–	–	2700	2700
World (total)	25680	69320	60100	36900	192000

I: Present production costs, Middle East.
II: Present production costs, Africa, Far East, Venezuela.
III: Present average production costs, USA, North Sea.
IV: Significantly more expensive than the highest current production costs.
[1] compare Table 3-12b.

Source: Federal Institute for Geosciences and Natural Resources, Hannover, Federal Republic of Germany, 1976.

3.323 Special aspects of petroleum technology

As mentioned above, about 22% of the $89 \cdot 10^9$ t proved recoverable reserves of petroleum lie offshore, and about 40% of the $212 \cdot 10^9$ t estimated additional recoverable resources of crude oil potential falls into this category. The predictable development of technology will probably make it possible in the next 10 years to tap offshore reserves of petroleum and natural gas at depths of 200–2000 m and deeper, even under unfavorable climatic conditions (85, 86). The offshore technology used in each location will be largely determined by the depth of the water, the volume and area of the deposit, and the weather.

The drilling and pumping equipment used for undersea petroleum (natural gas) can be roughly divided into two categories, those standing on the sea floor, and those which float. Fig. 3-8 shows the classification of offshore drilling and production facilities.

The relatively short time required to drill a well makes it economically unfeasible to use permanent platforms for this phase. In the early 50's, jack-up rigs were de-

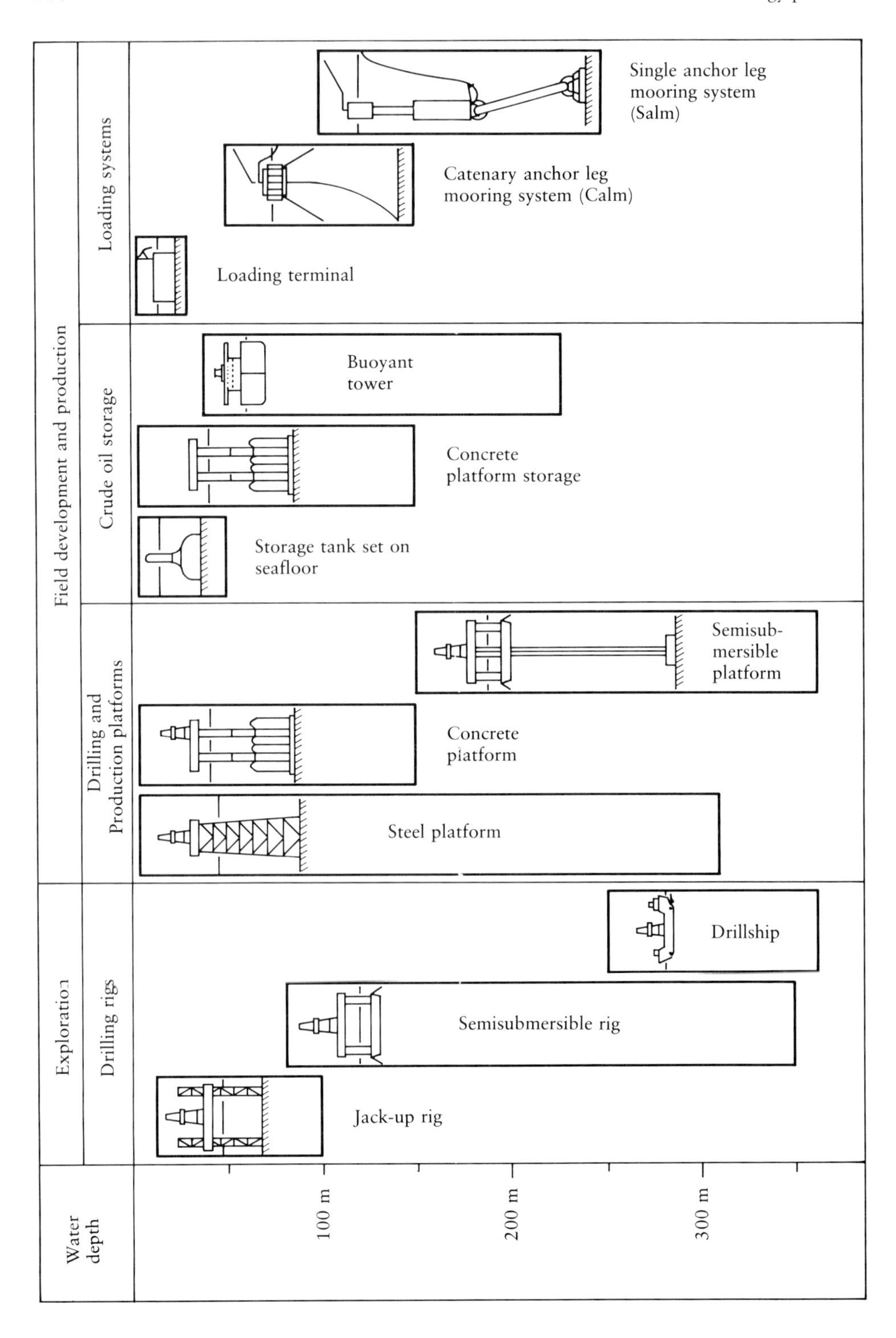
Field development and production
Loading systems
Single anchor leg mooring system (Salm)
Catenary anchor leg mooring system (Calm)
Loading terminal
Crude oil storage
Buoyant tower
Concrete platform storage
Storage tank set on seafloor
Drilling and Production platforms
Semisubmersible platform
Concrete platform
Steel platform
Exploration
Drilling rigs
Drillship
Semisubmersible rig
Jack-up rig
Water depth
100 m
200 m
300 m

veloped for use in water less than 100 m deep. These rigs are platforms which can float, and thus be towed to the site, where they extend supports down to the sea floor. The technology of drilling from a jack-up rig is comparable to drilling on land. In shallow water, the well can be temporarily closed by a well protector, and the drilling platform can leave. In deeper water, the well has to be capped and anchored to the sea floor. It can later be brought into a production system. Relocating it is no particular problem, with modern sonar buoys. In 1978, there were 189 jack-up rigs in operation (85).

If the water is deeper than 100 m, and currents on the surface make it necessary, exploratory and expansion drilling are generally carried out from semisubmersible rigs. In these rigs, the work deck along with the entire drilling apparatus is mounted on supports, about 40 m long, which in turn are mounted on submerged, tubular floats. The floats can be partly filled with water so that they come to equilibrium in calmer water, about 25 m below the surface. The waves do not reach the work deck. The rig is held in place above the drilling site by a number of anchors. Dynamic positioning, in which the rig is maintained over the well by powerful, electronically controlled propellors mounted at right angles to each other, was developed for rigs working in water more than 300 m deep. If the well strikes oil, it is anchored to the sea floor and is later incorporated into the production system. The drilling rig can then be removed. At present, there are 118 semisubmersible rigs, which can be used in water with a maximum depth of 600 m, and for which the record well depth is 9150 m (85).

Drillships were developed for use in still deeper water, and are primarily used at depths greater than 300 m. Like semisubmersible rigs, the drillships are kept in place by anchors or dynamic positioning. Drillships are more mobile than semisubmersible rigs, have more storage capacity for supplies (drilling equipment and pipes) and are therefore essentially independent during the drilling phase. Consequently, they are particularly well suited to work in areas in which it would be difficult to procide supplies. At present, there are 54 drillships, which have a record working depth of 1300 m, off the coast of the Congo. The use of the research ship "Glomar Challenger" for drilling in even deeper water, up to 6240 m, was only for shallow probes to explore the geology of the ocean floor (86–88).

For offshore production systems, platforms rigidly anchored to the sea floor have proved satisfactory. However, the cost of constructing steel or reinforced concrete platforms rises sharply with increasing water depth, which makes it necessary to tap a field, once it has been mapped out by exploratory wells, from a limited number of

---

Fig. 3-8: Classification of offshore drilling and production facilities

Source: H.-G. Goethe, New technologies for the development of energy resources as demonstrated in the production of crude oil and natural gas from shelf areas, VDI-Berichte 338, Düsseldorf: VDI-Verlag 1979.

points. Using turbine drills, it is possible to drill at angles greater than 45° and thus to tap a large area from a single platform. At Brent, in the North Sea, for example, in a 3000 m deep field, it has been possible to drill more than 40 wells from a single platform, thus tapping an area of 10 km^2 (88).

There are many steps in the process of extracting hydrocarbons from a field, and much equipment is required. The oil or gas must be processed. (Oil, for example, must be separated from gas and water pumped up with it.) In the course of production, the pressure of the field drops, and this can be countered by pumping in water or gas. For these reasons, equipment for producing and processing, and for pumping in large amounts of water are planned in most fields. Finally, the crude oil must be transported by pipeline or tanker to shore. Drilling and production platforms which are rigidly mounted on the sea floor have been found most adequate for the operations mentioned above.

The largest steel platform which has been erected to date stands in the Gulf of Mexico, in 310 m of water. The total height of the platform is 335 m, and its total weight is 42000 t (86). Steel platforms are welded together on land and transported to the working site. Some of them are provided with floats so they can be towed to the site, where they are sunk to the sea floor and anchored.

Extreme climatic conditions lead among other things to corrosion of the steel structure. For this reason, drilling and production platforms of reinforced concrete were developed. The lower level of these islands, which may be up to 250 m high, consists of tanks for the temporary storage of the oil. The capacity may be as large as 150000 m^3 (Ekofisk, North Sea).

Deep water and limited reserves are among the factors which may make it impossible or uneconomical to build a rigidly supported platform (compare Fig. 3–8). In this case a semisubmerged platform (underwater production system) must be used.

Production from offshore fields must, if possible, be uninterrupted. The oil can be transported by pipeline (for natural gas, this is at present the only possibility) or, if it is far from the coast or the reserve is small, by tanker. In this case temporary storage in the field is necessary. The crude oil can be stored either in tanks built into the production rig, for example as part of concrete platforms, or in separate containers, which may float or be built on the sea floor (compare Fig. 3-8).

Special systems have been developed for loading crude oil far from the coast. Some examples are the loading island, the loading buoy, a buoy anchored by chains (Cantenary Anchor Leg Mooring System, CALM), and a loading tower (Single Anchor Leg Mooring System, SALM). These systems provide a connection which can vary in length between the sea floor and the surface, and a swinging arm connection between the surface facility and the tanker. In the CALM system, for example, this is simply accomplished by means of flexible plastic pipes (88).

Although drilling techniques are constantly being improved to allow access to deeper and deeper water, production from deeper water is much more difficult and will require the development of new technology (89).

Due to the large increases in cost of producing from offshore reserves, great hope is set in the development of new production technologies which will increase the natural yield (Enhanced Oil Recovery, EOR). It is not possible to pump out all the oil or gas in a reserve (oil or gas in place). When petroleum reserves are reported, only that fraction which can be recovered by the currently technically and economically feasible methods is included. This recoverable fraction of the oil or gas is determined by natural factors such as the type of rock formation, the size of the pores, and the degree to which the oil-bearing formation is fissured. Consequently, the recoverable fraction varies considerably (23, 83, 90).

In 1976, the world average recoverable fraction was about 32% (91), which means that the recoverable reserves at that time were estimated to be $89 \cdot 10^9$ t. In other words, based on this percentage, oil explorers had drilled into formations containing approx. $300 \cdot 10^9$ t crude oil (91). However, with present production methods and at today's prices, only $89 \cdot 10^9$ t can be economically extracted. The amount of oil produced could thus be increased by increasing the recoverable fraction of the reserves, and therefore great efforts are being made all over the world to do just that. An increase of 1% in the recoverable fraction would correspond to more than the current annual world consumption of petroleum (91). The world average recoverable fraction has in fact been increased from about 26% in 1955 to 32% in 1976. (In the early years of the oil industry, as much as 90% of the oil was often left in the wells).

In principle, production methods can be classified as primary, secondary or tertiary. Natural or primary production depends on the pressure of natural gas lying above the oil or dissolved in it, or of underground water to bring the oil to the surface. By choosing the position of wells, one tries to maintain this natural pressure as long as possible. When it is no longer sufficient, pumps are used. Depending on the formation, however, only 10 to 30% of the petroleum actually present can be brought to the surface, and 70 to 90% remains unutilized in the ground.

Secondary production depends on artificially increasing the pressure in the oil formation by pumping natural gas or water back into the underground rocks. These methods allow a recovery of somewhat more than 50%, under the most favorable circumstances.

Tertiary methods can be subdivided into three main types: thermal methods, miscible displacements and chemical flooding (92–94).

The basic idea of the thermal methods is to heat the oil and thereby lower its viscosity so it can be displaced more easily from the reservoir. The oil and reservoir rock can be heated by injecting hot fluids (steam or hot water), or the heat may be generated directly in the reservoir by burning some of the crude oil in place (in-situ combustion). Air is injected to sustain and propagate a burning front (94).

The modern era of thermal processes began in the 1950's when extensive laboratory research and a number of field tests of both hot fluid injection methods and in-situ combustion were initiated. By the early 1960's, the hot fluid injection process was the dominant process because of its success and because of operational problems with in-situ combustion.

Miscible displacement processes using CO_2 and various hydrocarbon gases and liquids have been extensively researched and field tested since the early 1950's. The basic idea is to displace the oil with a solvent which can in turn, be recovered or flushed from reservoir by a cheap fluid, usually a gas. Some fluids are good solvents for crude oil without any modifications; such fluids are miscible at first contact. Liquid hydrocarbons, such as liquefied petroleum gas and products such as butane, propane, and ethane, are examples of fluids that can have first-contact miscibility under appropriate conditions of reservoir temperature and pressure. (Other fluids are not completely miscible with oil on first contact, but can develop miscibility after multiple contacts.) (94).

Chemical flooding methods are surfactant flooding, polymer flooding, and alkaline water flooding. Although a number of polymer floods and field tests of alkaline processes are now in progress, the surfactant flooding process is currently receiving the greatest attention from the oil industry. The term surfactant flooding applies to any process that uses injection of surfactant solutions or dispersions. The composition of the injected mixture is different for the various processes, but they normally include some or all of the following components (in addition to specialized surfactants): water, hydrocarbons, alcohols, polymers, and inorganic salts.

The primary objective of all surfactant processes is to reduce the interfacial tension between the oil and water, so capillary forces no longer trap oil in the reservoir.

Polymer floods employ light molecular weight polymers that are added to the injected water to increase its effective viscosity. This reduces the mobility of the water in the formation and improves the flood's sweep efficiency over that possible with plain water. Polymers now in use are synthetic materials such as polyacrylamides and a biologically produced material which is a specialized polysaccharide. Alkaline water flooding uses chemicals (such as sodium hydroxide, sodium silicate, and sodium carbonate) that can increase oil recovery when added to the injected water (94).

Many enhanced oil recovery processes have been tested all over the world (94). In the United States, thermal recovery methods have been used in nearly two thirds of the enhanced oil recovery projects (excluding conventional water flooding and pressure maintenance by injection of gas or water). Many other thermal projects are being carried out in Venezuela (the Orinoco heavy oil belt) and in Canada (the Peace River and Cold Lake regions). Experiments are also underway in many other countries, including the USSR, the Netherlands, the Federal Republic of Germany, Columbia, Mexico, Trinidad and Indonesia. Other enhanced oil recovery processes, such as miscibles processes, polymer-water floods, and alkaline water flooding, are being tested all over the world.

It is hoped that with the tertiary production methods, most of which are still being developed, it will be possible to attain recoveries of as much as 95%.

In modern production, the three types of method are not always applied in numerical order. Rather, in some fields tertiary methods are applied immediately after primary production. For this reason, all additional measures which help obtain more oil from a deposit than is possible with primary production methods are referred to as "Enhanced Oil Recovery" (EOR) (94).

It is assumed that it will be possible to increase the world average recovery to about 36% by 1985. Starting from the presently *proved recoverable reserves* of approx. $100 \cdot 10^9$ t (recoverable fraction of 32%), the amount of oil which could be recovered, given the following recoverable fractions, can be calculated: at a recoverable fraction of 36%, $112 \cdot 10^9$ t recoverable oil, at 40% recovery, $125 \cdot 10^9$ t, at 44%, $137 \cdot 10^9$ t, at 48%, $150 \cdot 10^9$ t, and at 50%, $156 \cdot 10^9$ t. If, in addition, one takes into consideration the estimated additional recoverable resources of crude oil potential of $212 \cdot 10^9$ t (recovery of 40%), that is, if one bases the calculation of the total amount of oil still recoverable of $301 \cdot 10^9$ t (compare Table 3–13), then the following relationship can be calculated between an increased recovery rate and the total amount of oil still recoverable: At 42% recovery, there are $316 \cdot 10^9$ t oil, at 44%, $331 \cdot 10^9$ t, at 46%, $346 \cdot 10^9$ t, at 48%, $361 \cdot 10^9$ t, and at 50%, $376 \cdot 10^9$ t. Thus a considerable increase in the amount of oil can be achieved by increasing the fraction recovered.

The evaluation of future developments must take account of the fact that most tertiary production methods have only been tested in pilot studies, and their application on a large scale is yet to come. According to the Exxon Corporation, in order to attain a production of $25 \cdot 10^6$ t oil per year by chemical flooding, starting in the middle 80's, it would be necessary to introduce $1.6 \cdot 10^6$ t of chemicals into the oil-bearing layers. Since the tertiary methods are, in addition, very energy-intensive, it would appear in 1978 that they will not become economically feasible until the price of crude oil rises above $ 25/bbl (price of 1978) (95, 96). (The price of Arabian light crude, which is the standard according to which other OPEC countries set their prices, depending on quality and location, was as follows: $ 2.80/bbl on 1/1/1973, $ 10.84/bbl on 1/1/1974, $ 13.34/bbl on 1/1/1979, $ 18/bbl on 1/6/1979, and about $ 35/bbl in the middle of 1981 (96).

As the past has shown, petroleum has decided advantages, both as a fuel and raw material, over other fuels and raw materials. It is also indisputable that petroleum will be irreplaceable as a raw material for a considerably longer time than as a fuel. Even so, it is questionable that all the reserves will ever be recovered, if the costs of recovery continue to rise as sharply as they have in the past few years (97).

3.33 Natural gas

3.331 Geographical distribution of the natural gas reserves

The world consumption of natural gas has increased faster in the past years than that of any other fossil fuel except petroleum, tripling between 1960 and 1979, although the total consumption of primary energy was only slightly more than doubled (compare Table 2-2). The reserves of natural gas should last longer than those of petroleum, given the predicted development of consumption, and furthermore, the geographical distribution of gas reserves is much more favorable for the OECD countries.

A number of authors have attempted to estimate the ultimate recoverable resources of natural gas (conventional gas) on the earth (2, 10, 12, 15), as shown in Table 3-21. Most of the estimates, after subtraction of the cumulative production of

Table 3-21: Estimates of ultimate recoverable resources of natural gas (conventional gas)[1]

Year	Source	$[10^{12}\ m^3]$
1958	Weeks	120–150
1959	Weeks	150
1965	Weeks	180
1965	Hendricks (USGS)	250
1967	Ryman (Exxon)	320
1968	Weeks	175
1969	Hubbert (USGS)	205–320
1971	Weeks	200
1973	Coppack (Shell)	190
1973	Hubbert	320
1973	Linden	275
1975	Moody (Mobil)	200–230
1976	Federal Institute for GNR	235
1977	Institute of Gas Technology (IGT)	260–270
1978	WEC (Conservation Commission)	300[2]
1980	WEC (World Energy Conference 1980)	293

[1] After deduction of the cumulative production of about $27 \cdot 10^{12}\ m^3$ up to 1. 1. 1979.
[2] The production costs on which this is based are equivalent to prices up to $ 20 per barrel of oil (1974 Dollars).

Sources: World Energy Conference 1980: Survey of Energy Resources 1980, Munich, September 1980 (12).
Federal Institute for Geosciences and Natural Resources, Hannover, Federal Republic of Germany, 1976 (2).
M. Grenon, On Fossil Fuel Reserves and Resources, IIASA, RM – 78 – 35, Laxenburg/Vienna 1978 (10).
World Energy: looking ahead to 2020. Report by the Conservation Commission of the World Energy Conference, Guildford (UK) and New York: IPC Science and Technology Press 1978 (15).

about $27 \cdot 10^{12}$ m^3 by 1/1/1979, fall in the range between 200 and $300 \cdot 10^{12}$ m^3. The not yet recovered *proved recoverable reserves* make up about $74 \cdot 10^{12}$ m^3 of this.

The Federal Institute for Geosciences and Natural Resources, Hannover, estimates the *proved recoverable reserves* of natural gas (conventional gas) to be $72 \cdot 10^{12}$ m^3 ($\triangleq$ $96 \cdot 10^9$ tce) (2), while L. Auldridge has published the value of $71 \cdot 10^{12}$ m^3 ($\triangleq$ $94 \cdot 10^9$ tce) (as of 1/1/1978) (11).

Table 3-22 shows the cumulative production, proved recoverable reserves, estimated additional resources and ultimate recovery according to region. The geographical distribution of natural gas throughout the world differs substantially from that of oil (compare Table 3-12a). It seems that the Middle East does not have half such a strong position with natural gas as in the case with oil. But it should be noted that only very little exploration for natural gas has taken place. The largest concentration of natural gas deposits are situated in the region of the USSR, the P.R. of China, and Eastern Europe, where the USSR possesses by far the largest reserves.

The distribution of the reserves over the economic-political groupings is fairly even, compared to the petroleum reserves (compare Table 3-23). Although here, too, the OPEC countries are at the top, with about 40% of the *proved recoverable reserves,* followed by the Eastern block, with about 36%, the OECD countries are still fairly well off, with about 16%.

However, the distribution of natural gas reserves within individual countries is even more concentrated in a small number of nations than are the petroleum reserves (compare Table 3-24). The USSR, with about one third of the *proved recoverable reserves,* has a very pronounced lead. The Netherlands and Great Britain have the largest reserves in Western Europe.

In addition to conventional gas, there is an enormous potential in nonconventional gas sources (10, 15), such as coal beds, shales, tight formations, geopressured gas, coal conversion (see 4.632), and biomass conversion (4.326.2).

Gas (methane) occurs naturally in coal beds, held by adsorption in the coal, in vertical joints and fractures in the coal beds, and in the interfaces with adjacent strata above and below the beds (15). (Approximately 6–8 m^3 of gas is associated with 907 kg (1 short ton) of coal.) The Conservation Commission of the World Energy Conference (WEC) has estimated that there is about 75 to $92 \cdot 10^9$ tce of gas contained in coal beds (coalbed degasification) (15). The amount of gas contained in coal beds in the USA alone is estimated to be between 11 to $30 \cdot 10^9$ tce (10).

Shales are rock formations containing an organic material called "kerogen". Kerogen, in contrast to conventional oil deposits and oil sands, is firmly bound to the non-porous rock of the formation. This rock, on being heated to about 770 K, will release shale oil (a mixture of hydrocarbons similar in composition to petroleum), or, at higher temperatures, a synthetic gas (a mixture of hydrocarbons, CO_2, CO, H_2S, CH_4, O_2 and H_2). The Institute of Gas Technology (IGT) in Chicago

Table 3-22: The cumulative production, reserves, resources and ultimate recovery for natural gas according to region

Region	cumulative production up to 1. 1. 1979		proved recoverable reserves[1] on 1. 1. 1979		estimated additional recoverable resources[2]		ultimate recovery	
	$[10^{12} m^3]$[3]	%	$[10^{12} m^3]$	%	$[10^{12} m^3]$	%	$[10^{12} m^3]$	%
Africa	0.1	0.4	7.3	10.0	26	13.6	33.4	11.4
North America	16.9	63.1	7.5	10.1	42	21.9	66.4	22.7
Latin America	1.8	6.7	4.7	6.3	10	5.2	16.5	5.6
Far East/Pacific	0.2	0.7	3.3	4.6	10	5.2	13.5	4.6
Middle East	1.1	4.1	20.5	27.3	30	15.6	51.6	17.6
Western Europe	1.5	5.6	3.9	5.3	6	3.1	11.4	3.9
USSR, China P.R., Eastern Europe	5.2	19.4	26.9	36.4	64	33.3	96.1	32.8
Antarctic	–	–	–	–	4	2.1	4.0	1.4
Total	26.8	100.0	74.1	100.0	192	100.0	292.9	100.0

1) *Proved reserves* represent the fraction of total resources that has not only been carefully measured but has also been assessed as being exploitable in a particular nation or region under present and expected local economic conditions (or at specified costs) with existing available technology.
Proved recoverable reserves are the fraction of proved reserves in-place that can be recovered (extracted from the earth in raw form) under the above economic and technological limits.

2) *Estimated additional recoverable resources* embrace all resources, in addition to proved reserves, that are of at least foreseeable economic interest. The estimate, provided for additional resources reflect, if not certainty about the existence of the entire quantities reported, at least a reasonable level of confidence. Resources whose existence is entirely speculative are not included.
This system is not quite as detailed as the classification systems used by some countries, but it has the advantage of being internationally combatible.

3) 10^{12} m³ natural gas $\triangleq 39.4 \cdot 10^{18}$ J.

Source: World Energy Conference 1980: Survey of Energy Resources 1980, Munich, September 1980.

Table 3-23: The distribution of the natural gas reserves among the economic-political groups (conventional gas)

Group of countries	cumulative production up to 1. 1. 1979		proved recoverable reserves on 1. 1. 1979		estimated additional recoverable resources		total amount of natural gas still recoverable on 1. 1. 1979	
	[10^9 m^3]	%	[10^9 m^3]	%	[10^9 m^3]	%	[10^9 m^3]	%
OECD	18500	69	11800	16	51000	27	62800	24
Countries with centrally planned economies	5100	19	26900	36	64000	33	90900	34
OPEC	1300	5	29800	40	29000	15	58800	22
Other	1900	7	5600	8	48000	25	53600	20
Total	26800	100	74100	100	192000	100	266100	100

Source: World Energy Conference 1980: Survey of Energy Resources 1980, Munich, September 1980.

Table 3-24: The countries with the largest reserves of natural gas (conventional gas)

Country[1]	proved recoverable reserves on 1. 1. 1979 from (12) [10^9 m³]	%	total amount of natural gas still recoverable from (2) [10^9 m³]	%
1. USSR	21930	29.6	68000	28.9
2. Iran	14000	18.9	24340	10.3
3. USA	5540	7.5	26140	11.1
4. Algeria	5150	7.0	22600	9.6
5. Saudi Arabia	2700	3.6	7160	3.0
6. Venezuela	1951	2.6	5550[2]	2.4
7. Canada	1802	2.4	15460	6.6
8. Mexico	1669	2.3	3230	1.4
9. Netherlands	1650	2.2	2935	1.2
10. Nigeria	1200	1.6	3500	1.5
11. Qatar	1100	1.5	1600	0.7
12. Great Britain	1032	1.4	2360	1.0
13. Kuwait	890	1.2	2400	1.0
14. Malaysia	827	1.1	1430	0.6
15. Iraq	790	1.1	3780	1.6
16. China, P.R.	710	0.9	5000	2.1
Norway	406		3300	
Germany, F.R.	196		615	
Italy	190		495	
France	180		420	
Romania	120		n.a.	
Japan	19		150	

[1] The countries are listed in the order of the size of their proved recoverable reserves.
[2] This figure was taken from (12).

Sources: World Energy Conference 1980: Survey of Energy Resources 1980, Munich, September 1980 (12).
Federal Institute for Geosciences and Natural Resources, Hannover, Federal Republic of Germany, 1976 (2).

has developed a process by which the product can be made either primarily liquid or mostly gaseous by controlling the temperature and degree of hydrogenation. The amount of gas contained in shales in the USA alone is estimated to be between 10 and $22 \cdot 10^9$ tce (10). The world reserves will be considerably greater (15).

The term "tight" formation refers to gas-bearing structures characterized by low permeability, low porosity and lack of structural continuity caused by complex cross-bedding and variations in clay and sand content. The Conservation Commission of the WEC estimates that the gas in tight formations in the USA alone, especially in the Rocky Mountains, amounts to about $22 \cdot 10^9$ tce (15).

Gas in geopressured zones is by far the largest unconventional source of gas. The structures of geopressured zones are characterized by thick sedimentary deposits containing water trapped at higher than normal pressures. The trapped water is estimated to contain quantities of dissolved methane. The following areas may contain large amounts of geopressured gas: the Gulf of Mexico, the Siberian coastal basins, the Indo-China Sea, the Yellow Sea, the Sea of Japan and the North Sea. The Conservation Commission of the WEC has estimated that the Gulf of Mexico alone contains up to about $1700 \cdot 10^9$ tce geopressured gas (15).

In sum, the potential for nonconventional gas sources for the USA alone is estimated as follows: Gas in coal beds, 11 to $30 \cdot 10^9$ tce; gas in shales, 19 to $22 \cdot 10^9$ tce; gas in tight formations, $22 \cdot 10^9$ tce; geopressured gas, $1700 \cdot 10^9$ tce. The total could thus be as high as $1750 \cdot 10^9$ tce, which is several times as much gas as is potentially available from conventional sources. At present, these nonconventional gas sources are producing about $30 \cdot 10^9$ m^3 gas in the USA, which is about 5% of the total production in this country. The ERDA has estimated that some of the geopressured gas can be produced for a price that would be equivalent to \$ 15 to \$ 30/bbl oil (1978 dollars).

3.332 Centers of natural gas production and consumption

As mentioned above, the predicted lifespan of the natural gas reserves, given the expected consumption developments, is somewhat longer than that of petroleum. Let us first consider the development of the "proved recoverable reserves" and world natural gas production. Table 3-25 shows the development from 1960 to 1980, in which time the *proved recoverable reserves* were increased about four-fold, and thus kept pace with the increased annual production.

The important industrial nations have a fairly good balance between natural gas production and consumption (compare Table 3-26). In addition, two West-European countries (Great Britain and The Netherlands) are among the 6 largest producers, and even in the other industrial nations, like the Federal Republic of Germany, Italy and France, the difference between production and consumption is much smaller than is the case with petroleum.

Although it cannot be proved from the past development of natural gas reserves, it does seem likely that the *proved recoverable reserves* will continue to increase in the coming years, especially in light of the fact that exploration for natural gas has only begun in many countries. It has already been mentioned that the Conservation Commission of the World Energy Conference (WEC) came to the conclusion that the available natural gas potential would allow a considerable increase in production (see 2.333 and Fig. 2-13). M. Grenon, International Institute for Applied Systems Analysis, Laxenburg/Vienna, has come to the even more optimistic conclusion that

Table 3-25: Historical development of the world's proved recoverable reserves of natural gas (conventional gas)

Year	proved recoverable reserves[1)] [10^9 m^3]	World natural gas production[2)] [10^9 m^3]	Ratio of Reserves to Production [R/P ratio]
1960	18600	469	40
1961	20400	506	40
1962	22000	551	40
1963	23000	602	38
1964	24600	657	37
1965	25400	704	36
1966	28300	766	37
1967	33600	824	41
1968	37700	890	42
1969	42400	975	43
1970	45000	1074	42
1971	49100	1146	43
1972	53700	1212	44
1973	57900	1281	45
1974	67600	1331	50
1975	63200[3)]	1312	48
1976	65630	1361	48
1977	71760	1418	51
1978	71210[3)]	1482	48
1979	73190	1569	47
1980	74980	1617	46

[1)] Level as of Dec. 31 each year.
[2)] In general, the production figures are for the net production of natural gas and natural gas liquids (gross production minus the gas pumped back into the well or burned off, used on site, and losses).
[3)] In some countries the figures from the previous year were corrected.

Source: Exxon Corporation, Public Affairs and Information Department, Hamburg 1981.

it should be possible, given the appropriate technologies and production costs, to increase the annual production of natural gas to about $5 \cdot 10^9$ tce (about $3.8 \cdot 10^{12}$ m^3) by the turn of the century, and thereafter, to continue production of an amount which would slowly fall to $3.8 \cdot 10^9$ tce (about $2.8 \cdot 10^{12}$ m^3) until the year 2065 (see 3.2).

This outlook, however, ignores important aspects of the problem, such as the geographical distribution of the reserves and political factors. It must be taken into account that the fraction of the world's natural gas produced in OECD countries will presumably fall from more than 50% at present to about 14% by 2020, and the fraction produced by developing countries will rise from less than 20% to 58%

Table 3-26: Consumption and production of natural gas in selected countries (in 10^9 m^3) and static lifetime in years

Country[1)] / Year	Consumption 1973 from (70)	1979	1980	Production 1973 from (69)	1979	1980	Static lifetime: proved recoverable reserves[2)]	total amount of natural gas still recovery[3)]
1. USA	668	538	531	641	580	577	10	45
2. USSR	233	332	354	236	407	433	54	167
3. China, P. R.	15	13	13	27	80	98	9	62
4. Netherlands	38	36	35	72	93	88	18	32
5. Canada	49	54	53	70	76	74	24	203
6. United Kingdom	30	45	44	30	39	35	26	60
7. Norway	–	–	–	–	26	33	17	127
8. Romania	n.a.	n.a.	n.a.	29	34	33	4	–
9. Mexico	n.a.	n.a.	n.a.	15	22	27	76	147
10. Algeria	n.a.	n.a.	n.a.	5	21	23	25	108
Italy	17	25	24	16	13	12	15	38
France	18	25	25	7.5	7.8	7.6	23	54
Germany, F. R.	32	50	49	19	21	19	9	29
Japan	6	21	24	2.6	2.4	2.3	8	63

1) These were the 10 largest producers in 1980.
2) Based on the production and reserves of 1979 (compare Table 3-24, from (12)).
3) Based on the 1979 production and the total amount of natural gas still recovery (compare Table 3-24, from (2)).

Sources: BP statistical review of the world oil industry 1980, London 1981 (70).
Exxon Corporation, Public Affairs and Information Department, Hamburg 1981 (69).

(compare Table 2-11). Therefore the volume of world trade (intercontinental and inter-regional) in natural gas will increase considerably in the coming decades.

Even today, natural gas is produced on all continents and shipped, sometimes over great distances, to consumers in other continents, especially in the countries of Western Europe, the USA and Japan. On the basis of trade agreements which have already been signed, twice as much gas will be shipped from producing to consumer countries in 1985 as in 1978, which will mean that every seventh cubic meter of natural gas produced will be transported from the producing to the consuming country. It is estimated that by the year 2000, the total international trade in natural gas will reach a volume of about $250 \cdot 10^9$ m^3. ($\triangleq 340 \cdot 10^6$ tce) annually (15). This, however, is only about 15% of the present volume of the international petroleum trade. It is expected that in the year 2000, the OECD countries will import about 8%

of their natural gas, but the figure for important industrial nations like Japan and the Federal Republic of Germany will be considerably higher. The Federal Republic of Germany will presumably import 74% of its natural gas in 1985 (in 1978 it was 59%). The Netherlands will supply 36%, Norway, 18%, the USSR, 13%, Iran, 7%, and the Federal Republic itself, 26%. This means that 80% will come from West European sources (Germany itself, The Netherlands and Norway).

A disadvantage of natural gas is that its use depends on the existence of an infrastructure. (In the Federal Republic of Germany, for example, there is a highly branched underground distribution network with a total length of about 114000 km, and the pipelines in the USA have a total length of about 640000 km). Therefore the use of natural gas will be concentrated for the foreseeable future in the industrial countries, such as the United States and Western Europe. (In 1978, Western Europe consumed $209 \cdot 10^9$ m^3 and produced $189 \cdot 10^9$ m^3).

The static lifetimes of the natural gas reserves in individual countries, based on the production and reserves in 1979, are given in Table 3-26. These lifetimes are short in a number of Western states (USA, The Netherlands, the Federal Republic of Germany, Italy and France), but very long for the USSR. For most of the countries of the Middle East and Africa the lifetimes are also very long, due to the fact that large reserves have already been discovered there, but the production has only begun (98).

As mentioned above, about 7% of the total known reserves are found in Western Europe. Great Britain expects an annual production (offshore) of 60 to $80 \cdot 10^9$ m^3, starting in 1980, and 90 to $100 \cdot 10^9$ m^3 from 1985 on. In Norway, the offshore production is expected to reach 20 to $40 \cdot 10^9$ m^3 annually by 1980, and 40 to $50 \cdot 10^9$ m^3 per year by 1985. Thus the supply situation will remain favorable for Western Europe for the near future.

Oil present within a deposit contains a proportion of dissolved lighter hydrocarbons. According to how much the pressure is reduced during the recovery process, part of these light hydrocarbons can on turn into gases in the aboveground installations. Such gases are called "oil gas" or "associated gas", and have a composition similar to natural gas but usually contain less impurities (12). The world availability of associated gas can only be very roughly estimated. The gas-oil-relationship, the volume of gas arising from each volume of oil produced, varies within wide limits. It varies, between a few m^3 gas per m^3 oil to several hundred, and varies not only from one deposit to the next but often from well to well within the same deposit. Assuming a world oil production level as in 1978, the world availability of associated gas is approximately $400 \cdot 10^9$ m^3 to $700 \cdot 10^9$ m^3 per annum. This corresponds to approximately 30–50% of the world gas production for the same year. (Approximately 55% of the world occurence of associated gas is currently flared or reinjected).

Relatively little has been published about the production costs for natural gas. The average wellhead price in the USA in 1974 was still about 30 cts/100 ft^3, which is equivalent to about $ 4/bbl for oil. The corresponding price for North Sea gas was about 60 cts/100 ft^3. The production costs at the new sources to be tapped in Alaska will be at least equivalent to $ 11/bbl oil (99, 100).

3.333 Special aspects of natural gas technology

Natural gas is a mixture of gaseous hydrocarbons, the chief component of which is methane (CH_4). It is frequently found in the same formations as petroleum, which implies that the conditions for the formation of natural gas deposits are similar to those for petroleum.

As mentioned above, natural gas technology is similar in many respects to petroleum technology (compare 3.313 and 3.323). The reason for this is its gaseous state. One of the advantages of natural gas production (conventional natural gas) is that it is usually possible to extract 75% of the gas in place without additional measures, and that some of the remaining 25% can be obtained by using additional measures.

The applications of natural gas as a primary energy source are such that high efficiencies are generally obtained. It is used primarily for heating in the industrial, commercial and household sectors. Natural gas is also unusually non-polluting. These advantages would argue for the increased use of gas (natural or coal gas) for heating, especially in heavily populated areas, and, to a certain extent, as a substitute for heating oil. Furthermore, based on the total efficiency, a gas-driven heat pump used to redistribute stored solar heat is particularly effective (see 4.325).

The overseas transport, storage and distribution of natural gas, on the other hand, are not so advantageous. The creation of value of a product depends on production costs, transportation and taxation. In the case of petroleum the generation of value is distributed about equally over the phases of production, transport and distribution, while for natural gas, 40 to 55% of the final price is due to the cost of transport.

For distances up to a few thousand kilometers, pipeline transport is economical. At present, pipeline transport of natural gas – like petroleum – is the worldwide state of the art (101–106). For example, the United States is planning the construction of a natural gas pipeline more than 7700 km (2500 miles) from Alaska to the Western and Midwestern states. This Alaskan natural gas could replace about 425 000 barrels of oil per day, starting in 1985 (15, 107, 108). About 15% of the gas is used to supply power for pumping the rest over these distances. Because the gas is delivered continuously, while the rate of consumption varies with the seasons, large storage facilities are needed to act as buffers. Depleted natural gas fields or underground salt mines are suitable for this (109).

For transport over still longer distances, it becomes more economical to liquefy the gas (LNG, Liquefied Natural Gas) (110–114). The gas is liquefied at a pressure of $10^5 N/m^2$ and a temperature of 112 K. In this state it has only 1/600 the volume of the gas under normal temperature and pressure, so it is possible for LNG tankers to carry considerable quantities of energy. Before liquefaction, those components which freeze out at 112 K must be removed so that they do not clog up the equipment. These components are mainly water, carbon dioxide and hydrogen sulfide. Special tankers are used for the sea transport of LNG. These are loaded by pumping the liquefied gas from storage tanks, and unloaded by pumping the liquid into storage tanks on land, where it is first stored in liquid form. The gas is allowed to evaporate and is warmed to the temperature for transport in a pipeline. The thermal energy required for liquefaction and re-evaporation amounts to about 25% of the gas produced. The present generation of LNG tankers carry on the order of 130000 m^3 of gas, but the next generation will have about twice this capacity. At present there are about 30 LNG tankers on the high seas, and about 40 more are being built or have been ordered. An estimated \$ $30 \cdot 10^9$ will have been invested in LNG tankers by the year 1990 (15).

In 1978, about $220 \cdot 10^9$ m^3 of natural gas was burned at the heads of oil wells because there was no economical way to use this gas. The amount of energy was about equivalent to the current annual consumption in Western Europe. As the price of gas rises, more and more of this natural gas will presumably be put to use (115).

It has already been mentioned that the potential for geopressured gas (nonconventional gas) is several times as large as the conventional potential. However, geopressured gas cannot be produced by the methods currently in use. One promising method developed in the USA is to break up the gas-bearing formation by hydraulic pressure created by pumping a mixture of water and sand into the well under high pressure (about 675 kg/cm^2). The gas-bearing formation, which lies at depths of 3000 m and more, contains the gas in microscopic pores. The hydraulic pressure creates cracks about 60 cm high in an area extending to about 800 m from the well shaft, and thus provides channels in which the gas can flow to the well. In a large-scale experiment in Fallon Field, for example, almost 4000 t water mixed with 1400 t sand were pumped into the gas-bearing rocks in the first 16 to 24 h. (The Fallon natural gas field lies in the sand and chalk formations of Cotton Valley, which have an area of about 650000 km^2 and reach from central and northeastern Texas to northwest Louisiana and southeast Arkansas.) It is thought that some geopressured gas can be produced for a price equivalent to \$ 15 to \$ 30/bbl oil (1978 dollars).

3.34 Oil shales, oil sands and heavy oils

3.341 Oil production from oil shales

Oil shale is a fine-grained, foliate sedimentary rock which contains an organic material called kerogen. When oil shale is heated to about 770 K, it releases shale oil, a mixture of hydrocarbons with a composition similar to petroleum.

The reserves of oil shale are distributed over the entire earth. The deposits vary in thickness from a few cm to a few hundred meters, and the content of kerogen, i.e. of recoverable oil, also varies widely, from a few liters per ton to 480 l/t in the Marahu Shale in Brazil. (Oil shales with < 40 l shale oil per ton rock (oil content), are at present economically uninteresting and are not considered here (compare Table 3-1).

The literature estimates of oil shale reserves vary widely. This is partly due to the fact that no one has undertaken a classification according to the content of recoverable oil. The total oil content of all the world's oil shale reserves, including those with very low concentrations of oil, has been estimated by the World Energy Conference (1974) to be $75\,000 \cdot 10^9$ tce and by the Colorado School of Mines Research Institute (1975) to be as much as $450\,000 \cdot 10^9$ tce. Although it is unlikely that more than a small fraction can ever be economically recovered, these reserves represent an enormous energy potential for humanity.

Based on formations containing 40–100 l/t, the Exxon Corporation has estimated that the total oil shale reserves have an oil content of $456 \cdot 10^9$ t oil. However, at present it would be economically feasible to recover only about $25 \cdot 10^9$ t of this. G. Bischoff has assumed that at most $28.5 \cdot 10^9$ t of heavy oil could be recovered from oil shales under the present economic conditions: in the USA, $12 \cdot 10^9$ t, in Brazil, $7.5 \cdot 10^9$ t, in Europe, $4.5 \cdot 10^9$ t, in Asia, $3 \cdot 10^9$ t, and in Africa, $1.5 \cdot 10^9$ t (116). The Federal Institute for Geosciences and Natural Resources, Hannover, estimates the total world reserves of shale oil to be $490 \cdot 10^9$ t ($\triangleq 705 \cdot 10^9$ tce), of which $33 \cdot 10^9$ t ($\triangleq 47 \cdot 10^9$ tce) could be economically recovered. In these estimates, only shales containing at least 40 l shale oil/t rock are considered. The geographical distribution of oil shale reserves can be seen in Table 3-27.

At present, shale oil is produced commercially only in the USSR ($37 \cdot 10^6$ t/year) and in the P.R. of China (10^7 t/year). Pilot projects have been started in Brazil and the USA. At Sao Mateus du Sul in Brazil, a few thousand tons of shale oil/year are produced from the Irati shales. There are a number of small pilot plants in which methods for economical and environmentally non-damaging production are being tested.

In principal, there are two ways to obtain oil from oil shales: conventional methods involving underground mining and aboveground retort press and in situ processes. The latter are divided into "true in situ" methods which do not involve mining, and "modified in situ" methods, which do.

Conventional mining of oil shales creates considerable environmental problems. A plant which produces 7000 t shale oil per day must process 80000 t rock in 24 hours, which is about 1 t/second. Since the volume of the extracted shale is increased more than 30% by the milling process, only a part of it can be replaced in the mine, and a depot must be found for the rest of it. In the USA, pilot plants producing up to 1000 t shale oil per day by conventional processes have been built. These processes, however, are at a disadvantage due to the high mining and extraction costs, so that in situ processes appear much more promising.

Table 3-27: Geographical distribution of the oil shale reserves

Country[1]	proved recoverable reserves[2] [10^6 t oil] from (12)[4]	estimated additional recoverable resources[2] [10^6 t oil] from (12)[4]	geological resources[3] [10^6 t oil] from (15)
USA	28 00	236 000	293 000
Morocco	7 400	n.a.	n.a.
USSR	6 820	49 180	n.a.
Thailand	2 015	n.a.	n.a.
Sweden	880	n.a.	n.a.
Jordan	800	n.a.	n.a.
Germany, F.R.	250	n.a.	n.a.
Brazil	84	n.a.	107 000
Spain	12	n.a.	n.a.
Australia	n.a.	490	n.a.
Zaire	n.a.	n.a.	13 000
Canada	n.a.	n.a.	7 000
Italy (Sicily)	n.a.	n.a.	5 000
China, P.R.	n.a.	n.a.	4 000
Other countries	n.a.	n.a.	2 000
Total	46 261	292 670	446 000

[1] The countries are listed in the order of the size of their proved recoverable reserves.
[2] The method of resource classification: compare Table 3-12 a.
[3] Deposits yielding > 10 gal/t.
[4] Results of the enquiry "Survey of Energy Resources 1980".

Sources: World Energy Conference 1980: Survey of Energy Resources 1980, Munich, September 1980 (12).
World Energy: looking ahead to 2020. Report by the Conservation Commission of the World Energy conference, Guildford (UK) and New York: IPC Science and Technology Press 1978 (15).

True in situ processes, i.e. those which involve only the drilling of shafts, depend on breaking up of the shale followed by some form of mobilization of the oil, such as pumping in of hot solutions or gases or in situ combustion. It is difficult to create connections between the shafts in oil shale because the shale expands after it has been ignited, and thus tends to block the channels to the pipes through which the oil is pumped out.

In modified in situ extraction (with additional mining work), underground chambers of certain dimensions are hollowed out in a conventional fashion. The permeability of the rock layers above the hollow is then increased artificially, for example by numerous explosions, in order to create a retort-like formation of very large dimensions. The shale oil is released by in situ combustion of this formation, seeps down into the hollow, and is pumped out.

Such "in situ" technologies are not harmful to the environment, because there are no tailings to dispose of and no air pollution. It can be assumed that these are the only methods which will allow a significant production of shale oil in the future. (There are plans in the USA to use, under certain conditions, small underground nuclear explosions for this purpose. See 3.343.)

The main centers for the public shale oil research in the USA are the Laramie Energy Research Center (LERC) in Laramie, Wyoming, Lawrence Livermore Laboratory (LLL) in Livermore, California, Sandia Laboratories (SL) and Alamos Scientific Laboratory (ASL) in Los Alamos, New Mexico.

The Laramie Energy Research Center supervises the publically sponsored industry programs, for example the studies with the PARAHO retort (up to 350 t/day) in Rifle, Colorado (117). (In 1972, 17 US companies formed the PARAHO group to develop new extraction technologies.) Other US firms active in the development of oil shale technology are The Oil Shale Corporation (TOSCO), Union Oil of California and Occidental Oil Shale.

The world production of shale oil in 1974 was about 10^7 t, and it could rise to about $120 \cdot 10^7$ t per year ($2 \cdot 10^6$ bbl/day) by the end of the century. The future volume of shale oil production will depend both on the development of an adequate technology for production in situ and on the further development of crude oil prices. It has been estimated that the total cost per barrel of shale oil will be about one third lower with in situ technology than with conventional methods. In the past, threshold prices, in dollars/barrel oil, at which the large-scale production of shale oil would be economically feasible, have been repeatedly predicted and repeatedly surpassed without the initiation of a significant shale oil production. The ERDA assumes that oil could be produced from the majority of the shale oil reserves in the USA for $ 25 to $ 35/bbl (July, 1979), and that the production of shale oil is less expensive than liquefaction of coal.

3.342 Oil production from oil sands and heavy oils

Oil sands and heavy oils have been formed by infiltration of petroleum into porous sand near the surface. It has been partially solidified to asphalt by oxidation and loss of the more volatile components.

The estimates of the reserves of oil sands and heavy oils vary; the Exxon Corporation estimates that the world reserves contain about 200 to $300 \cdot 10^9$ t oil, of which about $150 \cdot 10^9$ t are recoverable. The Federal Institute for Geosciences and Natural Resources, Hannover, has estimated a total reserve of about $340 \cdot 10^9$ t oil ($\triangleq 490 \cdot 10^9$ tce), of which $40 \cdot 10^9$ t ($\triangleq 57 \cdot 10^9$ tce) is economically recoverable. The Conservation Commission of the World Energy Conference (WEC) has estimated the total world reserves of oil sands and heavy oils to contain $330 \cdot 10^9$ t oil, of which 15 to $30 \cdot 10^9$ t oil is economically recoverable (15).

The geographical distribution of the reserves of oil sands and heavy oils can be seen in Table 3-28. About 99% of the reserves which at present are economically recoverable lie on the American continents. The largest oil sand reserves on earth are found in Canada, in the province Alberta; the Athabasca, Cold Lake and Peace River oil sands (118, 119). The Athabasca oil sands are the most important, with an area of about 50000 km^2 and an oil content up to 18% by weight. Of the $40 \cdot 10^9$ t of reserves which are at present economically recoverable, about $10 \cdot 10^9$ t can be obtained from surface mines. Although these deposits have been known for about 200 years, industrial mining of them only began in 1967. The plant is run by the GCOS (Great Canadian Oil Sands Ltd.), and produces about 2.5 to $3.5 \cdot 10^6$ t oil per year.

Other oil-sand projects are being built under the direction of Syncrude Canada Ltd., Shell Oil Company and Petrofina Canada (120, 121). According to the Shell Oil Company, the production is to be expanded to 30 to $50 \cdot 10^6$ t/year by 1990, and to $160 \cdot 10^6$ t/year by 2030 ($3 \cdot 10^6$ bbl/d) (15).

There are large reserves of oil sand in Venezuela, north of the Orinoco River (Orinoco Tar Belt). They stretch about 400 km from Guarico to the mouth of the

Table 3-28: Geographical distribution of the reserves of oil sands and heavy oils

Country[1]	proved recoverable reserves[2] [10^6 t oil] from (12)[3]	estimated additional recoverable resources[2] [10^6 t oil] from (12)[3]	geological resources [10^6 t oil] from (2)
Venezuela	20000	50000	104000
Canada	19300	16300	130000
Jordan	700	10000	n.a.
Germany, F.R.	50	n.a.	n.a.
USA	1	n.a.	4000
Columbia	n.a.	n.a.	100000
Madagascar	n.a.	n.a.	270
Total	40051	76300	338270

[1] The countries are listed in the order of the size of their proved recoverable reserves.
[2] The method of resource classification compare Table 3-12 a.
[3] Results of the enquiry "Survey of Energy Resources 1980".

Sources: World Energy Conference 1980: Survey of Energy Resources 1980, Munich, September 1980 (12).
Federal Institute for Geosciences and Natural Resources, Hannover, Federal Republic of Germany, 1976 (2).

Orinoco and are estimated to be 10 to 100 m thick. There are also large reserves in Colombia (in the Llanos area). The oil sand reserves in the USA lie in Utah, California and Kentucky.

The costs of producing oil from oil sands has risen sharply in the last few years, due to high investment costs. Like oil shales, the oil sands cannot be tapped in the same way as normal petroleum, but must be mined and extracted (for example by hot water). With conventional production methods, the total costs for a $6 \cdot 10^6$ t installation in Canada are presently about $ 25/bbl (US, 1976 prices) (10). It is assumed that the production costs can be considerably reduced by the use of in situ technologies. Here, too, the use of underground nuclear explosions is being considered (see 3.343).

3.343 Use of nuclear explosions to mobilize petroleum and natural gas reserves

3.343.1 Physical and political aspects of nuclear explosives[1)]

Beyond any doubt, the discovery of nuclear fission by O. Hahn and F. Straßmann in December 1938 fundamentally changed our world. (The discovery was published in January, 1939, in the journal "Die Naturwissenschaften".) On August 2, 1939, A. Einstein, who was born in Ulm/Donau and had been living in the USA since 1933, wrote his historic letter to President F. D. Roosevelt. In it, Einstein pointed out the possibility that an atomic bomb could be built, and strongly recommended that the USA undertake a program to develop nuclear weapons because of the possibility that the Germans were doing so (122–124).

The Second World War began on September 1, 1939. The USA developed the first atomic bomb under the code name "Manhattan Project". The director of this gigantic project was General Groves. An important part of the work was done in Chicago, and although eventually tens of thousands of people were immediately involved in the project, working in giant installations which had been built overnight, secrecy was maintained.

Because of the wartime atmosphere and the massive destruction of the war, there was no criticism of the project, and public debate of the project was in any case impossible, due to the military secrecy. The Manhattan Project was accomplished in only a few years. The first reactor (CP 1) in Chicago, supervised by E. Fermi, went critical on December 2, 1942, and the first test explosion of an atomic bomb (fission) took place in Alamogordo, in the desert of New Mexico, on July 16, 1945. A few weeks later, on August 6, 1945, after the war was over in Europe, the first bomb (uranium) was dropped on Hiroshima, and on August 9, 1945, a second (plutonium) atomic bomb destroyed Nagasaki (125).

---

1) These aspects will be discussed only to the extent that is necessary for an understanding of the situations treated in 3.343.2 and 5.83.

For several reasons which will not be discussed here, the German research program in nuclear fission was conducted on a very low budget. Near the end of the war, the program was moved to Hechingen (Baden-Würtemberg), and after the Allied occupation, it was found that Germany was still a long way from the realization of a nuclear weapon (126).

After the Second World War, more and more countries developed and exploded nuclear weapons. On August 29, 1949, the USSR detonated its first fission bomb, followed on October 3, 1952 by Great Britain, on February 13, 1960 by France, on October 16, 1964 by the Peoples' Republic of China, and on May 18, 1974, by India, which is so far the last country to do so. (The dates given are those of the first successful test explosion) (127).

The number of nuclear powers has thus been only partially limited by the Treaty on the Non-Proliferation of Nuclear Weapons, or Non-Proliferation Treaty (NPT), which became effective on March 5, 1970. (The purpose of the treaty was to insure that as many non-nuclear powers as possible should refrain from production and purchase of nuclear weapons, and all nuclear powers should refrain from making nuclear weapons available to non-nuclear states) (128). The countries of the world can be divided into two groups, according to whether or not they have signed the treaty, and each group includes both nuclear and non-nuclear powers[1)]. The non-nuclear powers can be subdivided into the so-called "threshhold powers", and the "developing nations." The former are those countries with advanced nuclear technologies which are capable of producing nuclear explosives[2)], and those which could, either independently or with outside help, develop nuclear warheads within the near future[3)] (129–131).

The individual atomic powers have expended enormous effort in the quantitative and qualitative refinement of nuclear weapons. The United States first succeeded in detonating a fusion weapon on November 1, 1952, on the Eniwetok Atoll in the Pacific. Surprisingly, the USSR was able by August 12, 1953, to explode its first hydrogen bomb (127). On May 15, 1957, Great Britain detonated its first fusion bomb, followed on June 17, 1967, by Peoples' Rep. of China, and on August 29, 1968, by France. The extremely high pressure and temperature required to ignite a fusion (thermonuclear) bomb are produced by a fission explosive (see 4.22).

One line of development was intended to produce ever more powerful bombs. An example of this was the three-phase bomb (fission-fusion-fission). A parallel development produced smaller and smaller warheads (tactical nuclear weapons, miniaturized nuclear weapons).

---

1) For example, the parties to the NPT include the USA, USSR, Great Britain, the Federal Republic of Germany and Canada. For example, France, the Peoples' Republic of China, India, Brazil and Israel have not signed the treaty.

2) For example, Israel, Pakistan and the Rep. of South Africa.

3) For example, Argentina and Brazil.

The nuclear deterrent was always an essential component of the various doctrines for the prevention of war after the Second World War. The fact that this system of mutual deterrence has made a decisive contribution to the prevention of a major war for more than three decades does offer a corresponding probability of future effectiveness, but no certainty (132, 133).

Due to the above developments since the discovery of nuclear fission, and especially to the destruction of Hiroshima and Nagasaki, the concept of "nuclear energy" is coupled, in the minds of a large fraction of humanity, with fear and the image of mass destruction. Thermonuclear weapons, however, are a reality which cannot be wished away. Even if a complete, controlled, worldwide disarmament could be achieved, it is to be assumed that the knowledge of the production techniques would be preserved, and thus also the latent danger of their misuse. Humanity must therefore learn to live with "the bomb".

In no other area is the duality of scientific research and technical application so apparent as in nuclear energy, in particular in the release of nuclear energy in nuclear or thermonuclear bombs (see 5.83). However, even this type of nuclear energy is, in itself, amoral. People must decide, whether it is to be used as a blessing or a curse for humanity. For the peaceful use of nuclear energy includes not only controlled reactions in reactors, but also the use of nuclear explosives. The chief problem with this, however, is that a nuclear explosive intended for peaceful purposes cannot be distinguished in function or effect from a nuclear weapon.

This problem is acknowledged by the treaty between the United States and the Soviet Union, signed on May 28, 1976, which regulates and limits the use of underground nuclear explosions for peaceful purposes (Treaty on Underground Nuclear Explosions for Peaceful Purposes or Peaceful Nuclear Explosions Treaty (PNET)) (134). The treaty is a supplement to the 1974 American-Soviet Treaty on the Limitation of Underground Nuclear Weapon Tests, which had not been completed when it went into effect (135). The PNET limits each single peaceful underground nuclear explosion to 150 kt TNT[1]. Under exactly specified conditions- including, for the first time, inspection of the site – group explosions[2] of up to 1.5 Mt TNT are allowed by this PNET.

According to Article III of the PNET, each side may carry out explosions of 150 kt TNT or less not only at any desired site within its sovereign territory or the areas under its control, but also in the territory of foreign states, upon their request. This is

[1] The question of single explosions with more explosive force than 150 kt TNT will be discussed by the parties to the treaty at a later time, which has not yet been agreed on.

[2] *Article II (PNET):* "For the purposes of this Treaty: (a) 'explosion' means any individual or group underground nuclear explosion for peaceful purposes: (b) 'explosive' means any device, mechanism or system for producing an individual explosion: (c) 'group explosion' means two or more individual explosions for which the time interval between successive individual explosions does not exceed five seconds and for which the emplacement points of all explosives can be interconnected by straight line segments, each of which joins two emplacement points and each of which does not exceed 40 kilometres."

an essential prerequisite for the realization of Article V of the NPT (see 7.2), which allows for the provision of international nuclear explosion services (128). These services must be in accordance with the terms of the Treaty Banning Nuclear Weapon Tests in the Atmosphere, in Outer Space and Under Water of 1963, the NPT and the PNET. An additional "agreed statement"[1)] contains a further clarification, namely that test explosions do not count as peaceful even when they are used for the development of nuclear explosives for peaceful purposes; they are therefore subject to the limitations which apply to tests of nuclear weapons (134). The protocol, which is an integral part of the PNET, regulates the explosions down to the smallest details, such as the minimum depth for the placement of the explosives, the requirements for informing the other partner to the treaty, the conditions for the admission of observers from the other treaty partner to the sites of the explosions, the nature of the instruments to be used in observing and testing the explosions, the techniques to be used in determining the strength of explosions, privileges and immunities (136).

In a nuclear detonation, the total energy released is divided in roughly the following proportions among kinetic (pressure) energy (50–60%), heat (30–35%) and radioactivity of various kinds (10–15%). (The proportions vary considerably, depending on the type of bomb and the strength of the explosion.)

The peaceful use of nuclear explosives is only feasible underground, on account of the radioactivity associated with every type of nuclear explosion. If the explosion occurs at sufficient depth under the surface, the fission products are trapped at the site in water-insoluble form (137). The situation would be much more dangerous if nuclear explosions were used for the construction of harbors or canals, for which purpose they would be placed only about 100 m below the surface.

If a fusion device could be built which did not depend on ignition by a fission device, and thus released energy solely from fusion processes, it would not produce radioactive fission products. In such a "clean" explosive, essentially all the radioactivity would be due to the capture of neutrons by the surrounding materials, and since neutrons do not travel far, these would be limited to the immediate surroundings of the explosion. However, at present the conditions required to ignite a fusion reaction can only be obtained in the explosion of a fission device. The fraction of the total energy released as radioactivity therefore depends on the ratio of the amounts

---

[1)] *Agreed statement (PNET):* "The Parties to the Treaty Between the United States of America and the Union of Soviet Socialist Republies on Underground Nuclear Explosions for Peaceful Purposes, hereinafter referred to as the Treaty, agree that under sub-paragraph 2 (c) of Article III of the Treaty: (a) Development testing of nuclear explosives does not constitute a 'peaceful application' and any such development tests shall be carried out only within the boundaries of nuclear weapon test sites specified in accordance with the Treaty between the United States of America and the Union of Soviet Socialist Republics on the Limitation of Underground Nuclear Weapon Tests; (b) Associating test facilities, instrumentation or procedures related only to testing of nuclear weapons or their effects with any explosion carried out in accordance with the Treaty does not constitute a 'peaceful application'."

of energy produced by fusion and fission (138, 139). Since fusion devices not requiring ignition by a fission device would also be of great military interest, both the USA and the USSR are working intensively in this area. Lately there has been great excitement about the possibility that lasers might be made powerful enough to ignite a fusion explosive (140–144). (It is not possible to make an absolutely clean nuclear explosive, which would release no radioactivity when ignited.)

There is an interest in peaceful uses for nuclear explosives. For example, the USSR has not given up the hope of building a canal between the Pechora and the Kama Rivers (145). Some developing countries also hope that by application of the relatively cheap and abundant energy of nuclear explosives, they may alter physical features or open mines (146, 147). This was apparent at the NPT Review Conference, which was provided for in Article VIII, 3 of the NPT (see 7.2) and took place from May 5 to 30, 1975, in Geneva. The purpose of the conference was to review the effectiveness of the NPT. The conference recommended that, according to Article V of the NPT, nuclear explosives for peaceful purposes should also be made available to non-nuclear states under the auspices of the International Atomic Energy Agency (IAEA) (148).

The IAEA is a branch of the United Nations, founded on October 26, 1956 for the purpose of establishing international supervision over atomic research and to make information available on progress in this field. In 1960, the IAEA was made responsible for supervising the delivery of reactors from the USA, Canada and England to other countries, in order to prevent the use of the fissionable material for military purposes. The organization was upgraded considerably when the NPT went into effect, as Article III of the treaty requires all non-nuclear states which agree to it to allow supervision of all their activities in the area of peaceful use of nuclear energy. In addition, as recommended by the NPT Review Conference of 1975, the IAEA has been given the task of making nuclear explosives available to non-nuclear-weapon states for peaceful purposes.

It has been possible, especially since the PNET was signed, to solve important physical, technical and political problems related to underground explosions for peaceful purposes. The first application of a non-nuclear country for a nuclear explosion service has also been made to the IAEA. Egypt wishes to build, using underground nuclear explosions, a 75-mile canal from the Mediterranean to the Qattarah Valley[1)] (149). Quite a number of previously suggested projects ought to be technically feasible, so that the use of underground nuclear explosions may become an important technical method for the benefit of humanity in the future.

[1)] The Qattarah Valley lies south of Al Alamayn and Matruh, and is 137 m below sea level in places. It would be converted by the proposed canal into a large lake. Evaporation from the surface would cause water to flow into it continually, at an estimated rate of about 650 m^3 per second. A hydroelectric power plant utilizing this flow would provide nearly all of the electricity Egypt needs. It is also hoped that the project would improve the climate in the surrounding desert, making it useful for agriculture.

3.343.2 Underground nuclear explosions for the release of hydrocarbons

In 1957, the USA began one of its largest research programs, the PLOWSHARE project, the purpose of which is to investigate possible peaceful applications for nuclear explosions, and to develop appropriate explosives for these uses.

Nuclear explosions are potentially useful because of the enormous amounts of energy released. Both the kinetic energy (pressure) and the thermal energy can potentially be of use in mining natural resources. The use of a nuclear explosive is technically simpler, and thus cheaper, than the use of a corresponding amount of chemical explosive. For example, the nuclear equivalent of 50 kt TNT can be placed in a shaft with a diameter of 25 cm, while the equivalent amount of TNT would require a shaft about 30 m in diameter (147). In addition, nuclear explosives are less expensive – according to the US Atomic Energy Commission, they cost about one tenth as much as an equivalent amount of TNT.

As mentioned above, a large fraction of the known oil reserves are found in the form of oil shales, oil sands and heavy oils (nonconventional oil). Conventional mining of these reserves is both economically unfeasible and associated with considerable environmental problems. In addition, the amount of natural gas in tight formations and geopressured zones (nonconventional gas) is enormous. "In situ" techniques for the production of these huge hydrocarbon reserves are thought to have great promise. On the basis of previous experience with underground nuclear explosions, it is realistic to assume that nuclear explosives can be used in this way.

In order to estimate the possibilities for use of underground nuclear explosions to release hydrocarbons, it is necessary to be aware of a few important facts. The scientific and technical results of a series of underground nuclear explosions have been published (147, 150). The essentials can be summarized as follows: The entire energy is released within one microsecond (10^{-6} s). The temperature in the neighborhood of the explosion center is more than 10^6K, and the pressure is several million atmospheres (10^{11} N/m^2). Depending on the size of the explosive, several cubic meters of the surrounding stone are vaporized, creating a hollow filled with hot, dense gas. The hollow is bounded by a layer of melted rock (melt zone), which contains most of the radioactive material. Within a few thousandths of a second, a shock wave forms and spreads out in all directions, enlarging the hollow by plastic deformation. The melted material flows to the bottom of the hollow and freezes to a glassy substance. The energy of the shock wave, which is still sufficient to shatter the surrounding rock, increases the size of the hollow until the gas pressure, which decreases as the gas expands and cools, is equal to the lithostatic pressure. If the overlying material is strong enough, in spite of the shock wave, to withstand the lithostatic pressure, the hollow is stable. Usually, however, the roof and walls break in, so that a more stable dome is formed. As a rule, the subsequent collapse of the hollow cannot be prevented, due to the thermal tension. The result is a chimney-like area of broken rock, the height of which depends on the porosity of the rock. The tip of the chimney remains as a small hollow.

Experience with about 200 nuclear explosions has shown that about 80% of the explosions released the nominal explosive energy of the charge, ± 20%, and the remaining 20% released within 50% of the calculated energy. (In two cases, a little more than double the expected energy was released) (151). In the past, there have also been efforts to develop suitable simulation techniques in order to learn something about the effects of nuclear explosions, using conventional explosives. However, there are problems of radioactivity in addition to the problems of releasing vast amounts of energy (152).

In completely contained underground nuclear explosions, the melt zone absorbs the thermal and radioactive radiation produced in the first seconds after detonation, so that these are not dangerous. However, the "fallout" resulting from neutron activation of the surrounding rocks creates problems. The neutrons arise both in fission of uranium and in fusion of tritium. The type and amount of radioactivity produced by each type of nuclear explosive are relatively well known (147).

On December 10, 1967, within the framework of the PLOWSHARE program, project "Gasbuggy" was carried out by detonating a 26 kt TNT explosive. The purpose was to determine the feasibility of increasing the productivity of natural gas fields by nuclear explosives. The project was planned and executed by the El Paso Natural Gas Co., the US Bureau of Mines, and the USAEC. The explosion created a fracturing and loosening of the gas-bearing rock formation, so that the gas from an originally less porous area could flow into this zone. (The flow rate is dependent on the extent of the artificially created zone of higher porosity; the larger the zone, the faster the gas can flow to the well.)

The last experiment of the PLOWSHARE program was the Rio Blanco project. On May 17, 1973, a 30 kt TNT explosive was detonated, again for the purpose of gas stimulation, i.e. the loosening of non-porous, natural gas-containing rock layers (147). The gas stimulation experiments were successful, in that they increased gas production by a factor of four to eight.

It is thought that nuclear explosions have also been detonated in the USSR for the purpose of opening petroleum deposits which were difficult to reach. However, the details are not known. (The last two nuclear explosions for peaceful purposes are supposed to have been detonated on August 14, 1974 in Western Siberia, and on August 29, 1974 in the Urals) (147).

As further examples of the possible uses of nuclear explosives, two projects planned by the USA will be described. As mentioned earlier, about 90% of the oil shale and about 99% of the oil sand reserves are found in North and South America (compare Tables 3-27 and 3-28). The USA plans to mine oil from oil shale, which has been pulverized by an underground nuclear explosion, by heating it in situ. The proposed process involves two steps. In the first (nuclear) phase, about 300000 t of oil shale in a spherical cavern about 60 m in diameter would be shattered by detonation of about 10 kt TNT equivalent. This would make the rock porous. In this stage, the kinetic energy (pressure) released by the detonation would be of primary interest, and the thermal energy secondary. The organic components of the shale would be partly decomposed to oil, natural gas and coke, but the thermal energy released by a 10 kt TNT equivalent would, due to the uneven distribution of temperatures, drive at most 2400 liters oil and 300000 m^3 of gas out of the shale.The second (conventional) phase would be the actual production of oil. The heat required to liquefy the oil would be supplied by burning the coke, gas, and if necessary, some of the oil in situ. Five wells would be drilled for this purpose, a central one for pumping the oil and gas to the surface, and four concentrically arranged injection shafts through which air can be supplied. The oil shale in the upper part of the cavern is to be ignited, and by locally controlled air injection, a horizontal burning region established. This will be induced to spread downwards, and it will heat the layers below it to 750 to 950 K, which will liquefy the organic components. The oil flows downward into a resevoir from which it can be pumped into the central shaft and to the surface (137).

The American Atomic Energy Commission, the Canadian government and Richfield Oil Company have cooperatively developed the plans for Project OILSAND with the goal of producing oil from the Canadian oil sand deposits by means of nuclear explosives. The oil sand layers lie from 100 to 500 m under the surface, and are 30 to 80 m thick. At the proposed detonation site, the oil sand lies 350 m below the surface in a 55 m layer. The plans call for placing a 9 kt TNT explosive a few meters below the oil-bearing layer. It is expected to create a cavern about 75 m in diameter which, however, will probably collapse after a short time. This will bring large amounts of the oil sand into the region in which the temperature of the rocks is 370 K and higher, and this heat will decrease the viscosity of the oil to the point that it can be pumped to the surface in conventional fashion (137).

The future development of the use of underground nuclear explosives in the recovery of hydrocarbons will depend on several things. Some important factors are the future demand for hydrocarbons, the future rate of production in relation to the size of the conventionally obtainable reserves, and the development of a "clean" nuclear explosive. The interest in the use of underground nuclear explosives for peaceful purposes is great in both the USA and the USSR; the drafting of the 1976 PNE treaty on the peaceful use of underground nuclear explosives is surely evidence of that interest. This treaty achieved important political prerequisites for the use of underground nuclear explosives.

3.35 Fuels for nuclear fission

3.351 Geographical distribution of uranium and thorium reserves

The first exploitation of nuclear energy was based on fission of the uranium nucleus $^{235}_{92}U$. Natural uranium is composed of 99.274% $^{238}_{92}U$, 0.720% $^{235}_{92}U$ and 0.006% $^{234}_{92}U$. When the nonfissionable $^{238}_{92}U$ absorbs fast neutorns emitted in the fission process, it is converted to the fissionable isotope of plutonium $^{239}_{94}Pu$, which is not naturally present on earth (see 4.213 and 5.822.2).

Although the possibilities for a rapid, large-scale exploitation of nuclear energy to meet energy demands has been overestimated in many industrial nations, everyone assumes that nuclear energy, compared with other sources of primary energy, will have the largest growth rate. Therefore the search for uranium and thorium reserves has been intensified all over the world.

The OECD Nuclear Energy Agency (NEA) and the IAEA have divided the uranium reserves into two cost categories: cost class $<$ \$ 30/lb U_3O_8 ($<$ \$ 80/kg U) and cost class \$ 30–\$ 50/lb U_3O_8 ($\triangleq$ \$ 80–\$ 130/kg U) (\$ 1/lb U_3O_8 $\triangleq$ \$ 2.60/kg U) (153). The low cost reasonable assured resources are defined as reserves.

There is no uniform international classification for the reserves of uranium and thorium. Fig. 3-9a shows the resource categories of the NEA/IAEA and the approximate correlations with terms used in other major resource classification systems. Fig. 3-10 indicates relations between resources and cost categories in a two-dimensional matrix, with the horizontal axis representing the degree of assurance for the existence of the uranium resource, while the economic recoverability of the resource varies along the vertical axis. (This Figure is derived from the McKelvey Diagram, US Geological Survey. The terms illustrated are not strictly comparable, as the criteria used in the various systems are not identical. Nonetheless, based on the principal criterion of geological assurance of existence, the chart presents a reasonable approximation of the comparability of terms.) The resource level of estimated additional resources should be considered as having a potential for later conversion to reasonable assured resources as the possible result of further exploration effort.

The World Energy Conference 1980 estimates the uranium reserves (reasonable assured resources) of the western world (excluding the USSR, Eastern Europe, and the Peoples' Republic of China), in the cost classes up to \$ 50/lb U_3O_8 ($\triangleq$ \$ 130/kg U), to be 2590700 t uranium, and the estimated additional resources to be 2552500 t uranium (12). The sum is 5143200 t uranium ($\triangleq$ $228 \cdot 10^9$ tce). The conversion to tce is based on the efficiency of thermal reactors. Complete utilization is only possible in breeder reactors (the theoretical energy content of 1 t uranium or thorium is about $2.95 \cdot 10^6$ tce); in thermal reactors only about 1.5% of the energy is utilized. The distribution of uranium reserves in the western world is shown in Table 3-29 (12, 15).

Source	Identified		Undiscovered	
US Bureau of Mines & Geological Survey	Demonstrated Measured Indicated	Inferred	Hypothetical (known districts)	Speculative (undiscovered districts)
Canada				Speculative
	Measured Indicated	Inferred	Prognosticated	
U.S. D.o.E.	Reserves	Probable potential		Possible & Speculative
Australia B.M.R.	Demonstrated resources ("in situ")	Inferred resources	Hypothetical resources	Speculative resources
France	Reserves I and II	Perspectives I	Perspectives II	
NEA/ IAEA	Reasonable assured	Estimated additional		(Speculative)*

* the category has the following definition:

"Speculative resources refers to uranium in addition to estimated additional resources, that is thougt to exist mostly on the basis of indirect indications and geological extrapolations in deposits discoverable with existing exploration techniques. The location of deposits envisaged in this category could generally be specified as only being somewhere within a given region or geological trend. As the term implies, the existence and size of such resources are highly speculative."

Fig. 3-9: Approximate correlation of terms used in major resource classification systems which refer to the level of assurance of existence

Source: World Energy Conference 1980: Survey of Energy Resources 1980, Munich, September 1980.

In contrast to the estimates of reserves in the western world, there are very few data on the uranium reserves in the eastern countries. The Federal Institute for Geosciences and Natural Resources, Hannover, estimates the uranium reserves in the eastern countries as follows:

Reasonable assured resources, 150000 to 300000 t uranium, estimated additional resources, 1115000 to 1630000 t uranium[1]. Together this is 1930000 t uranium (upper limit) (2). The uranium reserves of the entire world (reasonable assured resources + estimated additional resources), taking the values given in Table 3-29, can thus be estimated at 7073200 t uranium (upper limit).

[1] These estimates are based on recovery costs up to \$ 8/lb U_3O_8. However, in spite of inflation, they are probably less than \$ 50/lb U_3O_8.

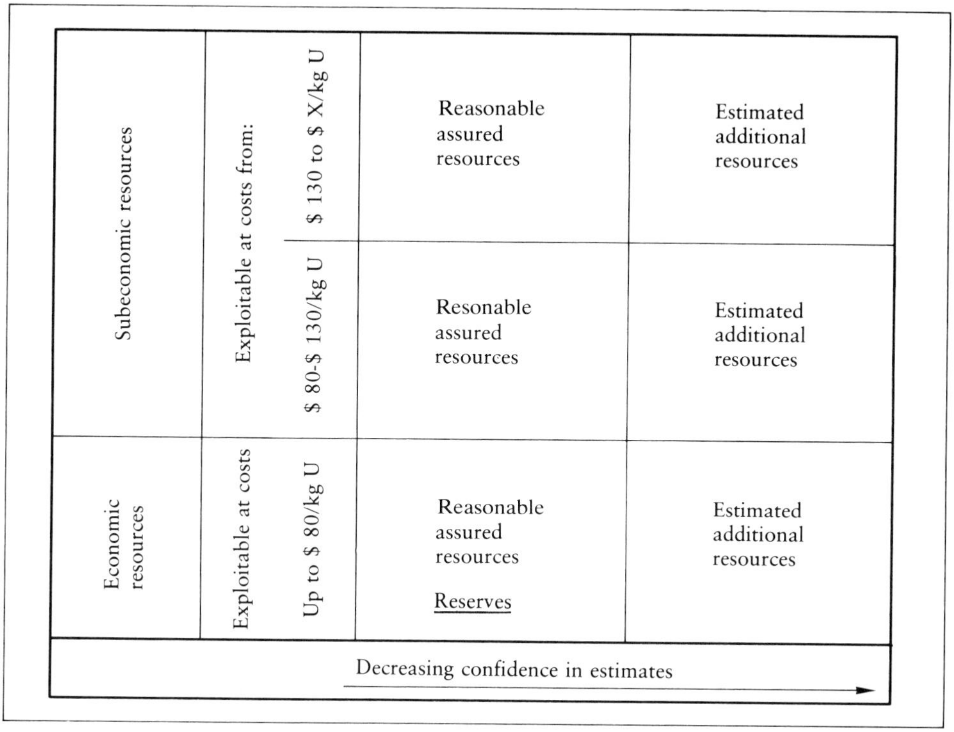

<table>
<tr><td rowspan="2">Subeconomic resources</td><td rowspan="2">Exploitable at costs from:</td><td>$ 130 to $ X/kg U</td><td>Reasonable assured resources</td><td>Estimated additional resources</td></tr>
<tr><td>$ 80-$ 130/kg U</td><td>Resonable assured resources</td><td>Estimated additional resources</td></tr>
<tr><td>Economic resources</td><td>Exploitable at costs</td><td>Up to $ 80/kg U</td><td>Reasonable assured resources
Reserves</td><td>Estimated additional resources</td></tr>
<tr><td colspan="5">Decreasing confidence in estimates →</td></tr>
</table>

Fig. 3-10: NEA/IAEA classification scheme for recoverable Uranium resources

Source: Uranium Resources, Production and Demand, a Joint Report by the OECD/NEA and IAEA, Paris, December 1977 and 1979.

The uranium reserves of the western world are concentrated in a few countries. As can be seen in Table 3-30, the United States has a key position, and the reserves in Western Europe are essentially limited to Sweden, Spain and France. Of the countries of the European Community (EC), France has the largest uranium reserves.

In order to obtain an approximate idea of the uranium reserves which might still be to be discovered in the world, the Federal Institute for Geosciences and Natural Resources, Hannover, undertook a survey on the basis of geological requirements and known deposits. These reserves – in analogy to the classification of ERDA (1976), they are termed *possible and speculative potential* (compare Fig. 3-9) – are estimated as 4550000 to 5300000 t uranium in the western world, and 800000 to 1000000 t uranium[1] in the eastern countries; together, this is 6300000 t uranium (upper limit). The total reserves in the world (reasonable assured resources + esti-

[1] These estimates are based on recovery cost up to $ 30/lb U_3O_8. In spite of inflation, they should be less than $ 50/lb U_3O_8.

Table 3-29: Distribution of the uranium reserves in the western world[1)]

Region	reasonable assured resources[2)] [10^3 t U]	%	estimated add. resources[3)] [10^3 t U]	%	total [10^3 t U]	%
North America	943.0	36.4	1886.0	73.9	2829.0	55.0
Africa (south of the Sahara)	745.4	28.8	264.9	10.4	1010.3	19.6
Western Europe	423.3	16.3	113.6	4.4	536.9	10.4
Australia, Japan	306.7	11.8	53.0	2.1	359.7	7.0
Latin America	109.9	4.2	136.7	5.4	246.6	4.8
Middle East, North Africa	32.3	1.2	74.6	2.9	106.9	2.1
South Asia	29.8	1.2	23.7	0.9	53.5	1.1
East Asia	0.3	0.1	n.a.	–	0.3	–
Western World	2590.7	100.0	2552.5	100.0	5143.2	100.0

1) The figures are based on recovery costs up to 50 \$/lb U_3O_8 (≙ 130 \$/kg U) (US Dollars, 1977).

2) *Reasonably assured resources* refers to uranium that occurs in known mineral deposits of such size, grade and configuration that it could be recovered within the given production cost ranges, with currently proven mining and processing technology. Estimates of tonnage and grade are based on specific sample data and measurements of the deposits and on kowledge of deposit characteristics (see Fig. 3-9, 3-10) (153).

3) *Estimated additional resources* refers to uranium in addition to reasonably assured resources, that is expected to occur, mostly on the basis of direct geological evidence, in: extensions of well-explored deposits, little-explored deposits, and undiscovered deposits believed to exist along a well-defined geological trend with known deposits. Such deposits can be identified, delineated and the uranium subsequently recovered, all within the given costs ranges. Estimated of tonnage and grade are based primarily on knowledge of the deposit characteristics as determined in its best-known parts or in similar deposits. Less reliance can be placed on the estimate in this category than for reasonably assured resources (see Fig. 3-9, 3-10 (153).

Source: World Energy Conference 1980: Survey of Energy Resources 1980, Munich, September 1980.

mated additional resources + possible and speculative potential) are thus estimated to be 13373200 t uranium (upper limit).

If one takes into account the low-grade ores, the total uranium reserves of the world might be on the order of $52 \cdot 10^6$ tons (compare Table 3-2a). Some of the important low-grade ore deposits are the phosphates (100–150 ppm uranium) of Morocco, the USA and Brazil, alum shale (200 ppm uranium) in Sweden, various shales (150–1000 ppm uranium) in France, copper shale (40–60 ppm uranium) in the Federal Republic of Germany, Conway granite (10–30 ppm uranium) in the USA, Illimaussaq-syenite (100–200 ppm uranium) in Greenland (2). Extracting this uranium would cost, depending on the amount in the ore, up to \$ 200/lb U_3O_8.

In addition, the oceans are a nearly inexhaustible source of uranium. However, since the concentration is very low (0.0033 ppm), this source will be of at most limited significance in the foreseeable future. Laboratory experiments in the nuclear research center at Jülich (Federal Republic of Germany) have shown that it is possi-

Table 3-30: The countries of the world with the largest uranium reserves (in 1000 t uranium)

Country[1)]	reasonable assured resources		estimated additional resources		total
	< 80 $/kgU	80–130 $/kgU	< 80 $/kgU	80–130 $/kgU	
1. USA	530.0	178.0	780.0	380.0	1868.0
2. Canada	215.0	20.0	370.0	358.0	963.0
3. USSR	160.0	n.a.	800.0	n.a.	960.0
4. German Dem. Rep.	60.0	n.a.	500.0	n.a.	560.0
5. Rep. of South Africa	247.0	144.0	54.0	85.0	530.0
6. Australia	296.0	9.0	47.0	6.0	352.0
7. Sweden	1.0	300.0	3.0	n.a.	304.0
8. Niger	160.0	n.a.	53.0	n.a.	213.0
9. Namibia	117.0	16.0	30.0	23.0	186.0
10. China, P.R.	166.0	n.a.	n.a.	n.a.	166.0
11. Brazil	74.2	n.a.	90.1	n.a.	164.3
12. Czechoslovakia	25.0	n.a.	120.0	n.a.	145.0
13. France	39.6	15.7	26.2	19.7	101.2
14. Algeria	28.0	n.a.	50.0	5.5	83.5
15. Romania	20.0	n.a.	50.0	n.a.	70.0
16. India	29.8	n.a.	0.9	22.8	53.5
17. Columbia	n.a.	n.a.	51.0	n.a.	51.0
18. Bulgaria	15.0	n.a.	30.0	n.a.	45.0
19. Denmark	n.a.	27.0	n.a.	16.0	43.0
20. Hungary	10.0	n.a.	30.0	n.a.	40.0
21. Gabon	37.0	n.a.	n.a.	n.a.	37.0
22. Argentinia	23.0	5.1	3.8	5.1	37.0
23. Yugoslavia	4.5	2.0	5.0	15.5	27.0
24. Poland	5.0	n.a.	20.0	n.a.	25.0
25. Spain	10.7	n.a.	8.5	n.a.	19.2
Italy	n.a.	1.2	n.a.	2.0	3.2
Germany, F.R.	4.0	0.5	7.0	0.5	12.0
Japan	7.7	n.a.	n.a.	n.a.	7.7

1) The order is based on the total reserves.

Source: World Energy conference 1980: Survey of Energy Resources 1980, Munich, September 1980.

ble to culture one-celled organisms that are able to concentrate uranium from sea water. It has been roughly estimated that the cost of uranium from sea water would be $ 100–$ 300/lb U_3O_8, but this uranium would at least be independent of the influence of foreign countries. It has been estimated that the oceans contain $4 \cdot 10^9$ t uranium (153) (see Table 3–2a). On account of the high cost, however, it is probable that only a few tens of thousands of tons will be extracted from sea water between now and the year 2000 (2).

It has been found that thorium as well as uranium can be used to produce energy in a thorium high temperature gas reactor (THTGR) (see 4.213.3) Like uranium, thorium is found on earth, but its mean abundance is 12 ppm, which is about three times that of uranium. On the other hand, the mean abundance of thorium in sea water, 0.00005 ppm, is much lower than that of uranium (0.0033 ppm) (18). Since the demand for thorium is coupled to the use of advanced reactors, the exploration for thorium has as yet been much less intense than for uranium.

The distribution of the thorium reserves in the western world is given in Table 3-31 (16). The Asian reserves are found primarily in India, those of Western Europe are found primarily in Denmark (Greenland) and Norway, the Latin American reserves are found almost exclusively in Brazil, the North American, about two thirds in the USA and one third in Canada, the African reserves primarily in Egypt. The thorium reserves (reasonable assured resources + estimated additional resources) are thus about 3 893 500 t thorium. By far the largest reserves in the Eastern countries lie in the USSR, which has about 80 000 t. The Federal Institute for Geosciences and Natural Resources, Hannover, estimates the thorium reserves of the entire world to about 8 800 000 t (the low-grade ores are included; without cost estimates). If the areas where deposits might be found (up to 3000 m depth) are included, the figure is $1976 \cdot 10^9$ t thorium (2). (Thorium is often found in significant amounts in deposits of other heavy minerals, for example uranium, tin, zirconium and monazite, so that thorium can be obtained from these minerals as a byproduct).

In the past, the estimated uranium reserves have climbed as time went on (155), partly because the efforts to find them have increased. This has been particularly true

Table 3-31: Distribution of the thorium reserves in the western world[1)]

Region	reasonable assured resources		estimated additional resources		total	
	$[10^3$ t Th]	%	$[10^3$ t Th]	%	$[10^3$ t Th]	%
Latin America	71.4	5.7	1200.0	45.3	1271.4	32.6
Middle East, North Africa	345.0	27.8	720.0	27.2	1065.0	27.4
North America	123.0	9.9	570.0	21.5	693.0	17.8
Western Europe	318.0	25.6	92.0	3.5	410.0	10.5
South Asia	319.0	25.7	30.0	1.1	349.0	8.9
Africa (south of the Sahara)	30.2	2.4	38.8	1.4	69.0	1.8
East Asia	18.3	1.5	n.a.	–	18.3	0.5
Australia	17.6	1.4	0.2	–	17.8	0.5
Western World	1242.5	100.0	2651.0	100.0	3893.5	100.0

[1)] The figures are based on recovery costs up to 75 $/kg Th.

Source: World Energy Conference 1980: Survey of Energy Resources 1980, Munich, September 1980.

in North America. As mentioned above, the mean abundance of uranium in the earth's crust is 4 ppm. It can be assumed, therefore, that large uranium deposits will still be found in those areas which have not been thoroughly explored, since it is very unlikely that about 30% of the world's uranium is deposited in the USA (8% of the land area).

3.352 Centers of production and consumption of uranium and thorium

By comparing the uranium reserves with the expected demand, one can estimate the lifetime of the reserves, or predict when uranium will become scarce. That is, the decisive factor in such a prediction is the extent, for a given amount of uranium reserves, to which nuclear energy will be used in individual countries or regions to meet the total need for energy.

The Conservation Commission of the World Energy Conference (WEC) has assumed, for the time from 1972 to 2020, an annual worldwide rate of increase in consumption of electricity of 4.5% per year. This is broken down for individual regions as follows: OECD, 3.6%; centrally planned economies (CPE), 5.6%; and developing countries (DC), 6.3%. On the assumption that nuclear energy will supply 45% of this by the year 2000, and around 60% by 2020, the development of nuclear energy production could go as follows: 76 GW ($\triangleq 0.1 \cdot 10^9$ tce) in 1975, about 300 GW ($\triangleq 0.6 \cdot 10^9$ tce) in 1985, 1540 GW ($\triangleq 3 \cdot 10^9$ tce) in 2000, and 5000 GW ($\triangleq 10.7 \cdot 10^9$ tce) in 2020 (15). On the further assumption that only thermal converters (light water reactors, LWRs) will be used to generate electricity, the cumulative world uranium demand will be about $3.1 \cdot 10^6$ t up to the year 2000, and $13.9 \cdot 10^6$ t to the year 2020. (Under these conditions, the annual world uranium demand will be 300000 t in 2000, and 880000 t in 2020.) On the assumption that, starting in the 90's, thermal converters (LWRs) will be combined with fast breeder reactors (FBRs), the cumulative world uranium demand to the year 2020 will be $9.5 \cdot 10^6$ t. (The annual world uranium demand under these conditions would be 260000 t in 2000, and 530000 t in 2020) (15).

Some of the other predictions of the fraction of the future energy needs which will be supplied by nuclear energy are much lower. The Energy Program of the Government of the Federal Republic of Germany, for example, predicts nuclear energy production in the entire world of about 1100 GW in 2000. (This corresponds to about $2.4 \cdot 10^9$ tce or 13% of the primary energy demand) (compare Table 2-2). On the assumption that only thermal converters (LWRs) will be used for energy generation, this would lead to a cumulative world uranium demand of about $2.3 \cdot 10^6$ t. (Under these conditions, the annual world uranium demand would be about 200000 t in the year 2000.)

It follows from the above, even assuming a relatively low growth of demand, that by the year 2000 the presently known *reasonable assured resources* of uranium (up to \$ 130/kg U) will be used up. If this trend in demand should continue, the reasonable assured resources + estimated additional resources of uranium (a total of 7073200 t) would be consumed by about 2010, and by 2020, all this uranium, and in addition the possible and speculative potential of a total of $13.4 \cdot 10^6$ t (compare Table 3-2a) would be used. Should the demand rise in this way, it could only be met within the given cost categories if more uranium reserves were discovered by intensive exploration efforts. However, if only the USA is considered, i.e. if one compares the expected consumption to the uranium reserves within the USA, then the supply situation is relatively unproblematic, due to the large reserves (35). If one assumes that a 1000 MW- LWR consumes 4700 t uranium (over its life-span of 30 years), the 713000 t uranium reserves estimated by the Department of Energy, USA, would suffice to supply about 150 1000 MW reactors for 30 years. The *probable resources* of 1160000 t uranium would be enough to run an additional 250 1000 MW reactors for 30 years, and the *speculative resources* of 460000 t uranium could supply another 100 1000 MW reactors for 30 years. It is presently assumed that the nuclear capacity of the USA will be about 185000 MW (35). That means that the United States' uranium reserves are sufficient to meet the expected demand, even if only LWRs are used, until well into the coming century.

The reserves and production of uranium are now concentrated in a few countries, so that for several other countries the future supply situation even of this primary energy source is problematic. The largest uranium producers in the western world are listed in Table 3-32 (153).

It follows from the above that if the future needs of the western world (outside North America) are to be met, it will be necessary to tap additional uranium reserves and to build the associated production facilities in time. Exploration outside North America ought to be especially promising, since it is, as already mentioned, very unlikely that the uranium of the earth has been mostly deposited on the North American continent.

Enriched uranium is becoming more and more significant as a fuel for nuclear power installations, because it is the only fuel which can be used in the light water reactors (LWRs). According to the OECD-NEA, Paris, the total nuclear electric capacity of the western world was 105 GW in 1978, and of this, 90 GW was produced by LWRs. According to this source, the nuclear capacity of the western world will be 1000 GW by the year 2000, and the LWRs will account for 901 GW (compare Table 4-1) (153). Until very recently, the United States Atomic Energy Commission (USAEC) supplied nearly all the enriched uranium for the nuclear generating plants of the western world. In 1975, the European Community (EC) imported 21584 kg ^{235}U. Of this, 19684 kg came from the USA, and 1900 kg from the USSR (156).

However, the situation for the European Community will change when the Eurodif and Urenco enrichment facilities are opened. It is expected that, starting in the early 80's, these facilities will supply more than half of the European needs. (Eurodif is a diffusion facility constructed cooperatively by France, Belgium, Italy

Table 3-32: The largest uranium producers of the western world (in 1000 t uranium)

Country[1)]	Production cumulative end of 1978	Production 1978	Estimated Production attainable 1985	Production capability to 1990
1. USA	260.00	14.00	30.00	37.00
2. Canada	124.70	6.80	14.40	15.50
3. Rep. of South Africa	96.50	4.53	8.10	7.60
4. Namibia	5.69	2.70	5.00	5.00
5. France	27.40	2.18	3.25	3.00
6. Niger	9.78	2.06	10.50	12.00
7. Gabon	9.60	1.00	1.50	1.50
8. Australia	9.03	0.52	12.00	20.00
9. India	0.40	0.20	0.20	0.20
10. Spain	0.85	0.19	1.30	1.27
11. Argentina	0.52	0.11	0.60	0.60
12. Portugal	0.73	0.10	0.30	0.30
13. Germany, F.R.	0.21	0.04	0.20	0.20
14. Brazil	0	0	1.10	1.10
15. Central African Rep.	0	0	1.00	1.00
16. Rep. of China	0	0	0.80	n.a.
17. Sweden	0	0	0.40	n.a.
18. Yugoslavia	0	0	0.40	0.44
19. Mexico	0	0	0.30	0.55
20. Japan	0.06	n.a.	0.30	0.30
Total	545.47	34.43	91.65	107.56

1) The countries are listed in order of their 1978 production or on their estimated production attainable in 1985.

Source: World Energy Conference 1980: Survey of Energy Resources 1980, Munich, September 1980.

and Spain, while the Urenco project, financed by Great Britain, The Netherlands and the Federal Republic of Germany, is a centrifugal facility.)

The production capacity for thorium is estimated to be 1700 t/year. The world production is at present about 730 t Th/year (16). The main producers are Australia, India, Malaysia and Brazil. An increase in the demand for thorium will be primarily due to the development of the high temperature gas reactor (HTGR) which, however, is not expected to come into large-scale use before the end of the century. The thorium reserves are large enough so there should be no difficulties with the supply in the foreseeable future.

This discussion shows that, given the present size of the inexpensive uranium reserves, there may well be difficulties with the supply toward the end of the century unless new uranium deposits are opened. It must be remembered that the uranium reserves are concentrated in a few countries. The first signs of a tight market situation are visible in recent price trends. In 1973, the price of uranium was about \$ 6/lb U_3O_8. By 1977, it was already \$ 40/lb U_3O_8, and in 1980, \$ 50/lb was already being paid. The United Kingdom Atomic Energy Agency (UKAEA) is basing its calculations on uranium prices in the 90's being \$ 100 to \$ 200/lb U_3O_8. However, in this case the production of uranium from many low-grade ores or sea water would become economically feasible.

Although the largest uranium producers, the USA and Canada, are politically and economically stable countries and NATO members, which are guarantees for a high degree of security for the Western European consumers, even here there are certain unmistakable signs of a change in natural resources policy. Canada, for example, to insure its long-term uranium supply, initiated restrictive export limits on September 5, 1974, and the USA passed the Nuclear Non-Proliferation Act of 1978 (157). In Australia, all uranium exports require a government permit.

Although uranium is important only as a primary energy source, and does not, like petroleum, have the double function of raw material and fuel, the supply situation, especially in several Western European countries, is problematic. This must be carefully weighed, along with other, related problems (see 5.8), in the further expansion of nuclear energy based on fission.

3.36 Fuels for nuclear fusion

3.361 Geographical distribution of lithium and deuterium reserves

It was only about four years from the discovery of nuclear fission in December, 1938 by O. Hahn and F. Straßmann, to the time that the first fission reactor, built by E. Fermi, went critical on December 2, 1942 in Chicago, but in spite of immense world-wide efforts in the last decades, no one has yet succeeded in building a fusion reactor FR (controlled thermonuclear reactor CTR) for the production of energy. This is all the more surprising because the underlying processes have been known for decades, and research on extraterrestial plasmas have yielded important information on the processes of energy release in stars (158–160). The first fusion bomb, to be sure, was exploded on November 11, 1952 in the United States, but the extreme temperature and pressure required to detonate it were achieved by the explosion of a fission device.

Although extremely difficult physical and technical problems remain to be solved, confidence is growing that by the end of the century, a pilot fusion power plant will have been built (19). However, the world-wide efforts toward this end will only be justifiable if the contribution of nuclear fusion to the world energy supply is proportionate to the expenditures required to develop it.

Since it is relatively uncertain when controlled nuclear fusion will be achieved, it is difficult to answer questions about the value of a FR. Nevertheless, something can be said about the reserves of fuel, the security of fuel supplies, and the costs of fuels for a FR. However, the developments in the energy sector in the past few years and the observable trends have made precisely these criteria extremely important in the evaluation of a primary energy source.

The trends in research and technology make it seem most probable that energy will be obtained from the D-T process (see 4.22). However, since the T (tritium) does not occur naturally on earth, it has to be produced artificially, which is done using lithium. Thus deuterium and lithium are the fuels for a D-T reactor.

The world reserves of fuels for nuclear fusion are given in Table 3-2b. The lithium reserves in the *measured* category, $1.4 \cdot 10^6$ t, represented an energy potential of $(35 \text{ to } 123) \cdot 10^3$q (1 q $\triangleq 10^{15}$ Btu $\triangleq 3.62 \cdot 10^7$ tce), depending on the type of reactor (compare Table 3-2b). The *indicated + inferred reserves* are $5.2 \cdot 10^6$ t, which correspond to $(128 \text{ to } 455) \cdot 10^3$q. The total resources are estimated to be $1.2 \cdot 10^8$ t $\triangleq$ $(0.29 \text{ to } 1.0) \cdot 10^7$q (19). Since the annual consumption of lithium is small (it was 6345 t in 1972) and the reserves large, there has been no need for lithium prospecting on a large scale. It is therefore likely that intense prospecting would considerably increase the known reserves. (The mean abundance of lithium in the earth's crust is 65 ppm (18)).

The lithium deposits which are currently mined are scattered over the entire land surface of the earth. In 1972, 45.1% was mined in the USA, 28.1% in Rhodesia, 9.9% in the USSR, 7.0% in Canada, and 4.2% in China. Of the total, 85.9% was mined in the western world, and 14.1% in the Eastern block (19).

In addition, sea water contains an average of 0.1ppm lithium, which amounts to a reserve of $1.1 \cdot 10^{11}$ t $\triangleq$ $(2.7 \text{ to } 9.5) \cdot 10^9$q. Under the present market conditions, however, it is financially uninteresting to attempt to recover lithium from sea water, although the required technology would not present any insurmountable problems. Lithium is already mined from the brine of salt lakes, though to be sure, these have higher concentrations than the seas. It is expected that desalinization of sea water will be carried out on a large scale, and lithium could be extracted from the brine resulting from this operation (19). With an average concentration of 0.1 ppm, lithium ought to be much less expensive to extract from sea water than uranium, which has an average concentration of 0.0033 ppm (see 3.351). Therefore, it can be presumed that the necessity of extracting lithium from sea water would not greatly influence the economics of a FR.

In some types of reactors, beryllium is also needed as a breeder material (see 4.22). To date, there has been no need for extensive prospecting for this element, for lack of demand. The average abundance of beryllium in the earth's crust is 6 ppm. (As mentioned above, the abundance of lithium in the earth's crust is 65 ppm, and that of uranium is 4 ppm) (19). The present price of beryllium is about $ 30/kg. It is not possible, for lack of data, to make any clear prediction about the possibility that the energy potential of fusion reactors – at least those requiring beryllium – might be limited by the availability of beryllium. However, the amount of beryllium consumed by both types of reactor being considered is specific, and is less than 1/11 the amount of lithium consumed. The average abundance of beryllium is about 1/11 that of lithium, so it is assumed that even with the two beryllium-requiring reactor types, lithium would be the limiting resource (19).

Deuterium is available in practically unlimited quantities in the oceans, in the form of D_2O or HDO. Natural water contains 16.68 ppm deuterium. The volume of the world's oceans is about $1.37 \cdot 10^9$ km^3, so they contain about $4.6 \cdot 10^{13}$ t deuterium, which corresponds to an energy of about $15 \cdot 10^{12}$q (18).

It can be assumed that lithium as well as deuterium can be extracted from sea water, to which most of the countries of the world have direct access. This is extremely important for the security of the fuel supply, and is one of the reasons that nuclear fusion is an option for an "unlimited" and safe supply of energy for all mankind.

3.362 Fuel costs for a fusion reactor

At the present world-market prices, the lithium for a FR would cost \$ 2.50/kg, the deuterium about \$ 600/kg, and the beryllium, \$ 30/kg. From the fuel costs and the amount of energy which could be released, a D-T fusion reactor could produce heat for about 0.07 cts/GJ (based on liquid lithium as breeder and cooling material), or 0.14 cts GJ based on solid lithium compounds as breeder material (19). Assuming a efficiency of 38% in converting heat to electricity, the cost of fuel is about 0.0006 cts/kWh (liquid lithium reactor) or 0.0013 cts/kWh (solid lithium reactor) (19). Since there is no external fuel circulation for FR (as there is with fission reactors), these are the total fuel costs. By comparison, the fuel cycle costs.for single fission reactors are about two orders of magnitude larger. Since the reserves of lithium and deuterium are unusually favorable, and the geographical distribution likewise, the fuel costs for a fusion reactor can be presumed to remain constant over an unusually long period.

The price of energy is primarily determined by the costs of the fuel and generating plant. No reliable estimate of the cost of building a fusion reactor plant can yet be made, but preliminary estimates are that it will be on the same order of magnitude as a fission power plant (19).

3.37 Solar energy

3.371 Basic data applying to solar energy

The enormous potential for solar energy is hardly exploited today. This is all the more surprising when one considers that $192 \cdot 10^{12}$ tce/a ($\triangleq 178 \cdot 10^{15}$ W) of sunlight falls on the earth's surface annually, about 20000 times the world consumption of primary energy in 1980. If the solar energy falling on 2‰ of the land surface (desert areas) could be used at an efficiency of 20%, it would amount to $15 \cdot 10^9$ tce/year (see Table 3-3). This would be almost twice the world consumption of primary energy in 1976. Let it also be emphasized here that solar energy, compared with other sources, is not detrimental to the environment. The greatest problems in exploiting solar energy, however, are the low power density and the resulting requirement for large surfaces to collect it, and the need for storage and transport forms, since this energy source is highly dependent on time and place (161, 162).

The earth's energy balance is shown in Fig. 3-11, which shows that the natural flow of energy comprises 178000 TW radiant energy from the sun, 32 TW heat escaping from the interior of the earth, and 3 TW in tidal energy (regenerative energy sources) (15). (The present human consumption of primary energy is equivalent to about 9 TW.) This energy flow can be considered constant, at least over a time span comparable to the time human civilization has existed (see 3.1, 3.38, 3.39).

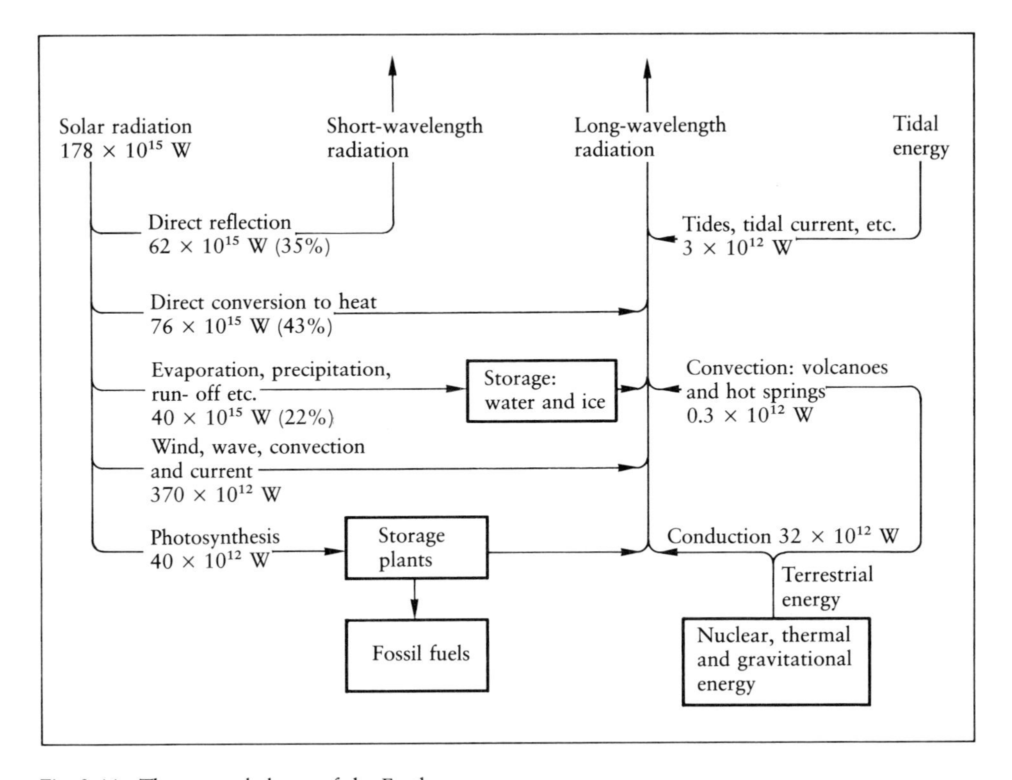

Fig. 3-11: The energy balance of the Earth

Source: World Energy: looking ahead to 2020. Report by the Conservation Commission of the World Energy Conference, Guildford (UK) and New York: IPC Science and Technology Press 1978.

The earth itself is a huge energy reservoir. It contains energy in various forms, including fossil primary energy sources, which were originally formed by transformation of solar energy, over a period of millions of years (non-regenerative energy sources).

The earth's energy balance is dominated by solar radiation, which is converted into other forms of energy by a multitude of natural processes. 35% of the sun's rays are reflected unchanged, for the most part before they reach the earth's surface. About 43% are converted directly to heat and re-radiated at a longer wavelength back into space. About 22% of the energy is consumed by the evaporation of water, and is thus temporarily stored. This portion of the energy thus keeps the water cycle going. All other processes consume vanishingly small parts of the total energy. For example, only 0.2% is converted to wind, wave and ocean current energy, and 0.02% into carbohydrate by photosynthesis (compare Fig. 3-12).

Fig. 3-12 shows schematically the geometry of the earth-sun relationship (it is not drawn to scale) (161). Due to the eccentricity of the earth's orbit, the distance between the earth and sun varies by ± 3%. The solar constant I_o, which is the amount of solar energy falling on a unit surface in space, perpendicular to the sun's rays and at a distance equivalent to the mean earth-sun distance, in a unit time, is 1353 W/m^2

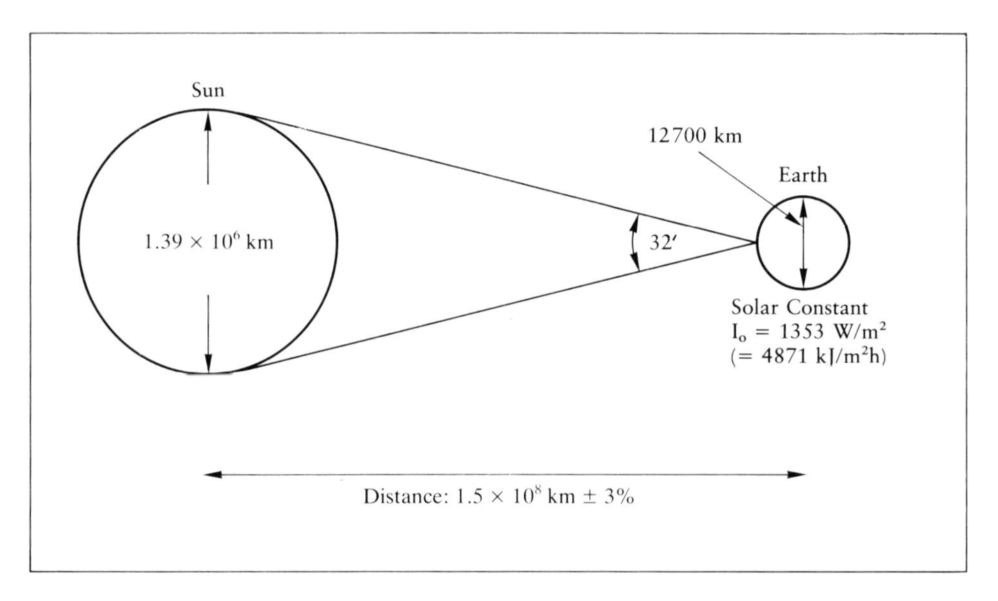

Fig. 3-12: Schematic of sun-earth relationships (not to scale). The angle subtended by the sun at mean earth-sun distance is 32'

Source: J. A. Duffie, W. A. Beckmann, Solar Energy-Thermal Process, New York: John Wiley & Sons 1974.

(161). Scattering and absorption reduce this to 1000 W/m^2 at sea level at the equator, and it can be further reduced by clouds to the range of 100 W/m^2.

It has been possible with satellites to measure the intensity of radiation at the outer edge of the earth's atmosphere directly and exactly. The data indicate that the variation in the total solar emission is less than $\pm$ 1.5%, i.e. the radiation from the sun can be regarded as practically constant. However, the changes in the distance between the earth and sun lead to changes of about $\pm$ 3% in the amount of radiation falling on the earth. Fig. 3-13 shows the variation of the extraterrestial radiation with the season (161).

The spectral distribution of solar energy, in addition to the solar constant, is of interest with reference to utilization of solar energy. Fig. 3-14 shows the spectral distribution at the outer edge of the atmosphere and at sea level (163).

X-rays and other very short-wave radiation are absorbed in the ionosphere, mainly by nitrogen and oxygen, and so is most of the UV light, by ozone. The infrared part of the spectrum ($\lambda > 2500$ nm) has a very low intensity, and is strongly absorbed by CO_2 and H_2O, so that very little of it reaches the earth's surface. It follows that only the wavelengths between 290 and 2500 nm need to be considered as possible sources for terrestial energy use (Fig. 3-14).

Solar radiation in this range penetrates the atmosphere, although there are some scattering and absorption effects. Scattering is caused by air molecules, water drop-

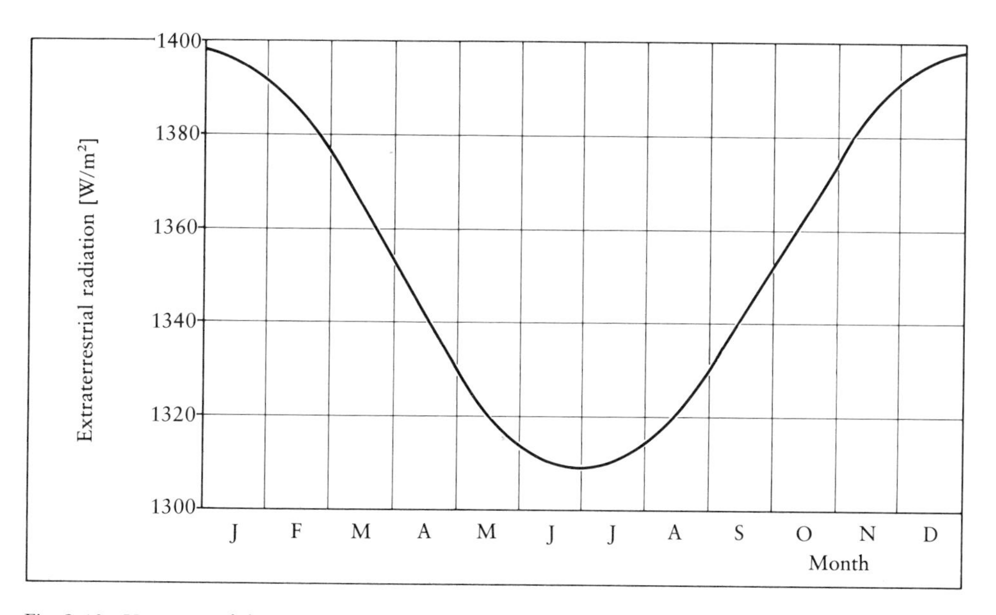

Fig. 3-13: Variation of the extraterrestrial solar radiation with time of year

Source: J. A. Duffie, W. A. Beckmann, Solar Engergy-Thermal Process, New York: John Wiley & Sons 1974.

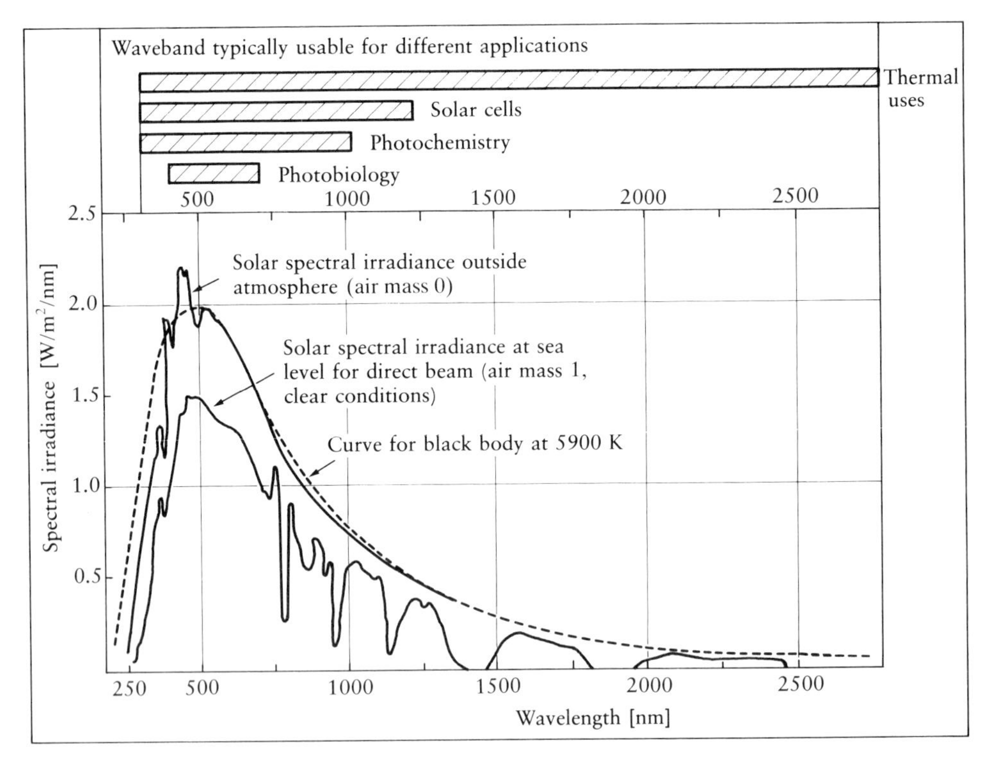

Fig. 3-14: Spectral irradiance curves for direct sunlight extraterrestrially and at sea leval. Wavelengths potentially utilized in different solar energy applications are indicated at the top.

Source: Solar Energy, UK Section of the International Solar Energy Society, London, UK-ISES, 1976.

lets and dust particles, all of which are small compared to the wavelength. Therefore the process is described by the Rayleigh scattering theory, in which the scattering coefficient is proportional to λ^{-4}. This means that short-wavelength light is much more strongly scattered than long-wavelength, and diffuse light has a larger proportion of short wavelength components.

From the above, it can be seen that there are several types of radiation falling on a surface on or near the surface of the earth. Direct solar radiation is that light which reaches the surface without change in its original direction. (In Zürich, for example, the direct radiation received on a sunny day in April is 875 W/m^2, and in December, 775 W/m^2.) Diffuse radiation is that part of the radiation which is scattered toward the earth in the course of penetrating the atmosphere. Diffuse radiation is distributed over the entire lighted half-hemisphere of the sky, and thus comes from all directions rather than from a single source. Nevertheless, diffuse radiation can be utilized. (Even on a cloudy winter day in England, something on the order of 50 W/m^2 is still received.) In terms of technology, the main difference between diffuse and direct radiation is that only the latter can be concentrated by lenses or mirrors. The sum of direct and diffuse radiation is called global radiation.

Measurements of solar radiation usually record global radiation, the intensity of which depends on the place and time. At a given location, the intensity varies in periodic fashion with the time of day and year, and statistically with the weather (compare Fig. 3-15 a and 3-15 b). It is useful, for purposes of direct utilization of solar energy, to know these relationships. Depending on the type of utilization, it may be important to know the dependence of direct as well as global radiation on the area and the time (162).

Radiant energy is transformed into other forms by a multitude of natural processes (compare Fig. 3-11), which means that solar energy can be utilized indirectly (using the energy transformed by natural processes) as well as directly. Direct utilization of solar energy involves either direct conversion of the sunlight into heat (solar thermal conversion) by collectors, direct conversion into electricity (photoelectric conversion), or generation of electricity from solar heat. Indirect utilization of solar

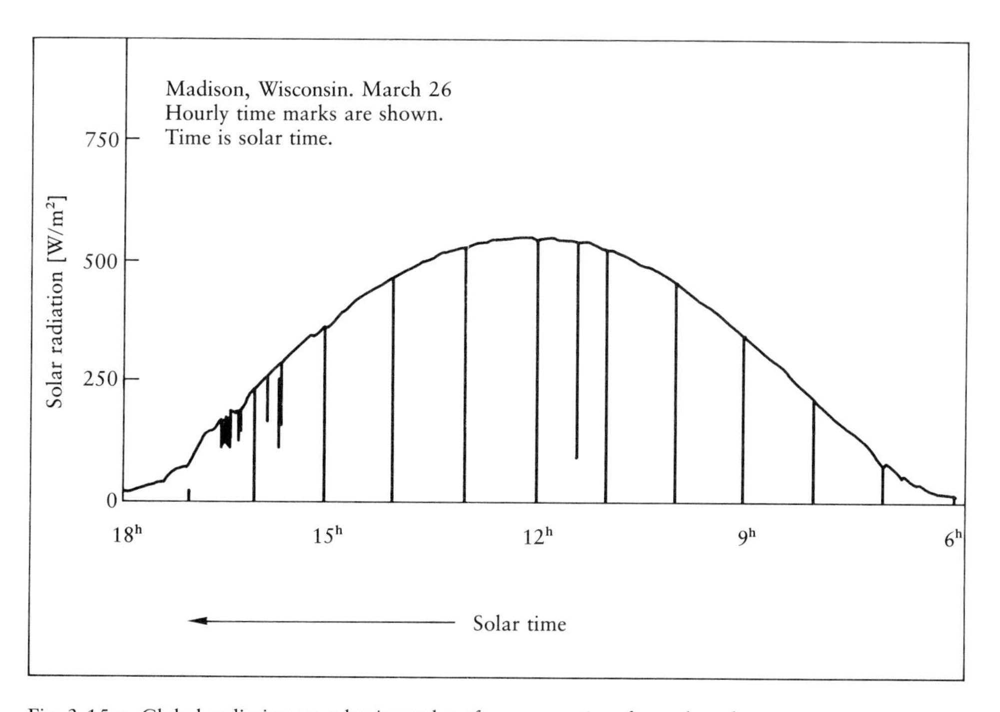

Fig. 3-15 a: Global radiation on a horizontal surface versus time for a clear day

Source: J. A. Duffie, W. A. Beckmann, Solar Energy-Thermal Process, New York: John Wiley & Sons 1974.

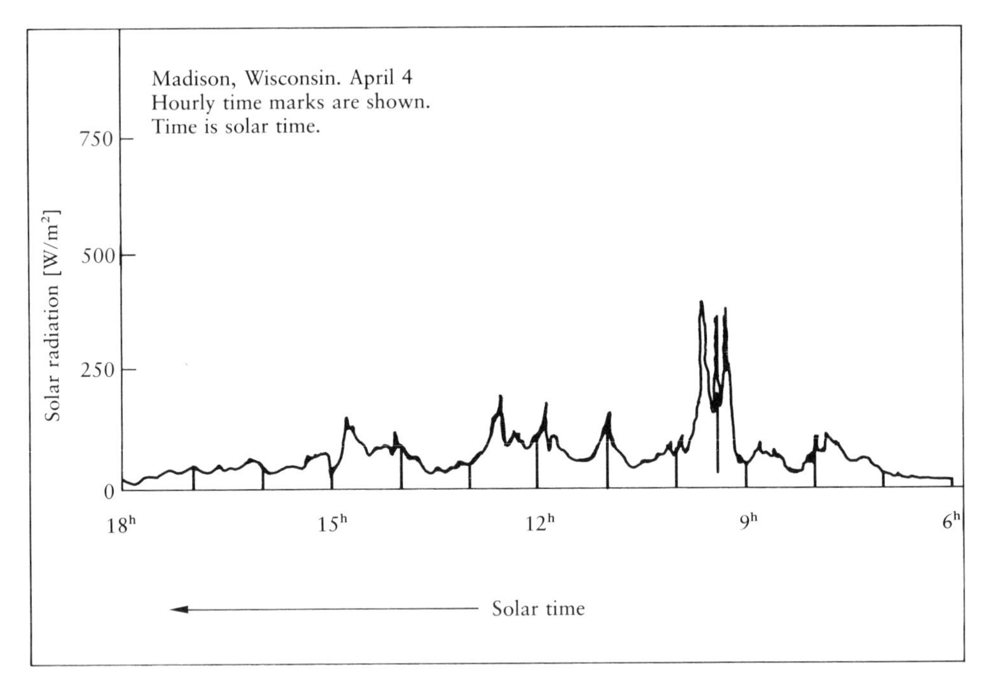

Fig. 3-15b: Global radiation on a horizontal surface versus time for a cloudy day

Source: J. A. Duffie, W. A. Beckmann, Solar Energy-Thermal Process, New York: John Wiley & Sons 1974.

energy is possible through exploitation of water power, wave energy, the heat or currents of the ocean, wind energy, photosynthesis or use of stored heat from the sun by means of heat pumps, to name several examples (compare Fig. 3-1).

In general, information on the solar energy available for a solar installation is not derived from meteorological data, but is obtained by measurements made at or near the site. The pyrheliometer and the pyranometer are two of the instruments which may be used for this purpose. The pyrheliometer consists of a detector with a diaphragm, with which a small part of the sky, including the sun, i.e. the direct solar radiation, can be measured at a perpendicular angle of incidence. The pyranometer is an instrument for measurement of global radiation, i.e. direct and diffuse, usually on a horizontal surface. In order to measure the diffuse radiation, the direct radiation is screened out with a disc (161). The data are given as energy per unit of time and area.

There are also instruments which measure the hours of "bright sunshine". Such an instrument consists of two photocells, one of which is shielded from the direct rays of the sun. If the light is diffuse, the two photocells measure nearly the same intensity of light, but if direct sunlight falls on the unshielded cell, it registers a much higher intensity. The time during which the difference between the intensity measured by the two cells exceeds a certain value is the duration of "bright sunshine."

When data on solar radiation are recorded, they should include the time or time span over which the measurements were made, the position of the receptor surface (horizontal, vertical or a specific angle), whether the values were instantaneous or integrated over a period of time, e.g. a day or an hour, and whether direct, diffuse or global radiation was measured. If an average is reported, the corresponding time span must be given (e.g. the monthly average of the daily global radiation) (161). Table 3-33 shows, as an example, the monthly means of the daily global radiation falling on a horizontal surface for various locations. (This is the form in which radiation data are most frequently reported.)

Average values of solar radiation are often drawn onto maps. Fig. 3-16a and 3-16b show the monthly mean values of daily global radiation on a horizontal surface in Europe for the months of June and December (161). These maps are useful for making rough estimates of the areas in which exploitation of solar energy is likely to be rewarding. However, such large-scale maps should be used with caution for areas in which the solar radiation changes abruptly, such as the edges of mountainous areas or urban centers.

If no data on the solar radiation are available, other meteorological data related to solar energy may be used to estimate the potential for utilizable solar energy. In many countries, for example, there are numerous meteorological measurement stations from which one can obtain data on the actual hours of sunshine, the possible hours of sunshine and the degree of cloudiness. Although radiation measurements at a given site yield the most exact information for evaluation of solar potential, the lacking data can be estimated from the actual and possible hours of sunshine and the cloudiness, using empirical formulas.

3.372 The utilization of solar energy: possibilities and limitations

It can be seen in Fig. 3-16a and 3-16b that the monthly mean value of the global radiation falling on a horizontal surface in Europe in the month of June varies between 4.5 kWh/m^2d in northern England to 8.5 kWh/m^2d in southwestern Spain. Even in December, the corresponding averages lie in the range of 0.5 kWh/m^2d to 2.5 kWh/m^2d (Additional data can be taken from Table 3-33.) For a comparison, 2.2 to 2.5 kg coal has to be burned in a conventional power plant to generate 1 kWh.

There are various methods of exploiting solar energy. In the following, only basic problems which apply to direct utilization, i.e. direct conversion to heat (with collectors) or direct conversion to electricity (with solar cells or via sun-generated heat) will be discussed. The methods for indirect utilization of solar energy (compare Fig. 3-1) have either been in use for years (e.g. hydroelectric generators) or are now technically feasible (e.g. utilization of stored solar heat with heat pumps) or they will

Table 3-33: Global radiation data, on horizontal surface, averaged by months [kW/m² d].

Station	Latitude φ	Alt. [m]	Annual	Radiation [kWh/m²d] Jan	Feb	Mar	Apr	May	Jun	Jul	Aug	Sept	Oct	Nov	Dec
Madison	43° N	270	3.92	1.67	2.42	3.51	4.28	5.9	6.25	6.41	5.37	4.63	3.22	1.95	1.38
Cancaneia (Brazil)	25° S	10	4.09	6.38	5.83	9.3	3.41	2.8	2.37	2.64	3.13	2.93	3.86	5.9	5.59
Calcutta	22° N	0	7.97	6.19	7.18	8.15	9.08	9.12	9.5	9.49	9.3	7.5	7.25	6.48	5.83
Tokyo	36° N	0	3.04	2.21	2.69	3.19	3.63	3.99	3.52	3.91	3.93	2.95	2.35	2.15	1.97
Yangambi (Congo)	1° N	140	4.77	4.76	5.23	5.33	5.19	5.09	4.61	4.09	4.2	4.75	4.77	4.92	4.34
Dakar	15° N	0	6.07	5.35	6.26	7.36	7.29	7.2	6.75	5.95	5.3	5.4	5.22	5.26	5.47
Pretoria, S.A.	26° S	1418	5.52	7.09	6.05	5.7	4.77	4.19	3.95	4.19	5.00	5.82	6.16	6.63	6.75
Canberra	34° S	177	4.93	7.20	6.35	5.15	4.00	3.07	2.38	2.65	3.52	4.95	5.87	6.91	7.41
Athens	38° N	0	4.48	2.16	3.07	3.97	5.34	5.55	6.86	7.26	6.44	5.44	3.34	2.33	1.93
Lisbon	39° N	0	5.18	2.37	3.37	4.63	6.26	7.05	8.12	8.36	7.48	5.71	3.97	2.69	2.16
Brussels	51° N	0	2.77	0.65	1.26	2.4	4.02	4.72	5.13	4.72	4.12	2.92	1.84	0.88	0.55
Stockholm	59° N	0	2.80	0.34	0.91	2.34	3.58	5.43	6.01	5.82	4.56	2.83	1.30	0.37	0.21

Source: J. A. Duffie, W. A. Beckmann, Solar Energy-Thermal Processes.
German translation: G. Bräunlich, Munich: Pfriemer 1976.

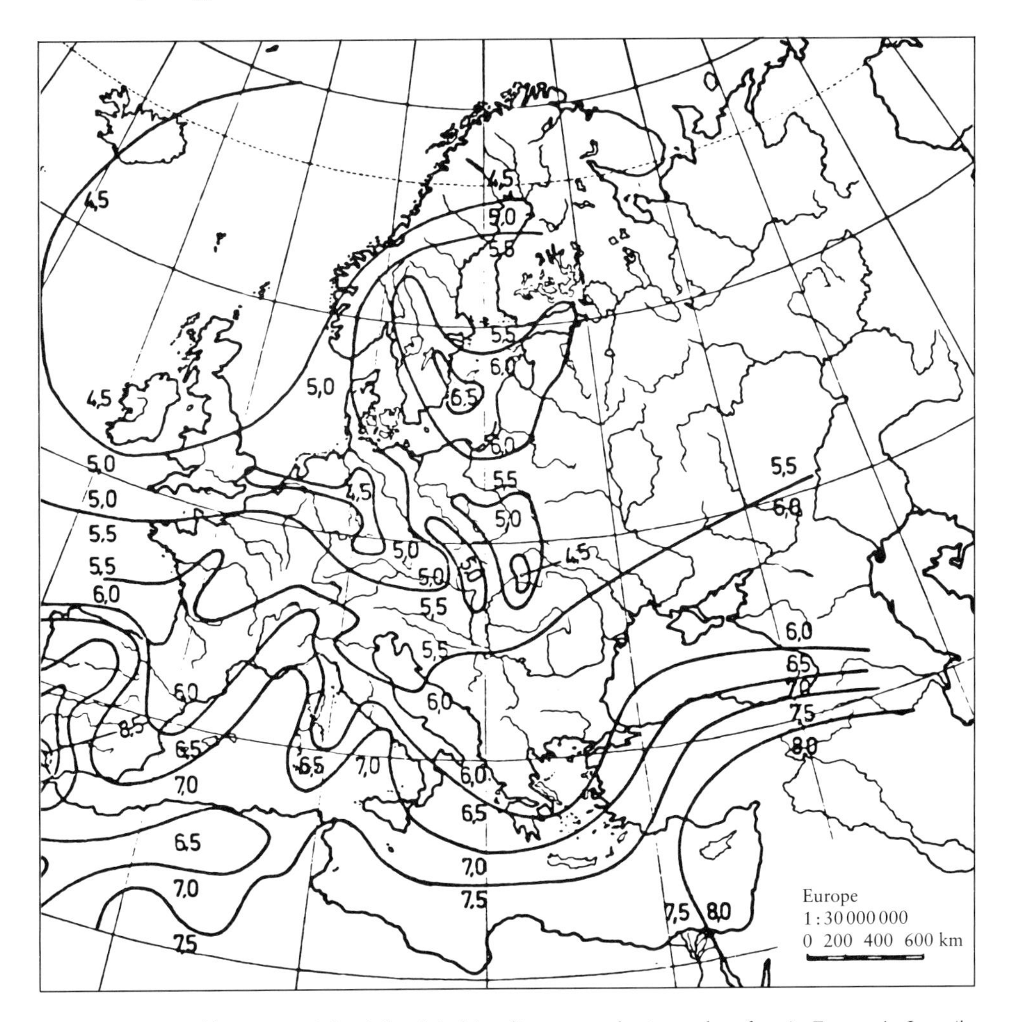

Fig. 3-16a: Monthly means of the daily global irradiance on a horizontal surface in Europe in June (in kWh/m^2d)

Source: J. A. Duffie, W. A. Beckmann, Solar Energy-Thermal Process. German translation: G. Bräunlich, Munich: Pfriemer 1976.

not play any significant role – in terms of the world economy – in the foreseeable future (see 4.3).

No matter which method is chosen, there are problems with direct utilization of solar energy which result from the low power density and the local and temporal variability of the radiation. In general, the low power density results in a requirement for a relatively large area. For example, if one exploits the solar radiation falling on an area of 34 km^2 with an efficiency of 15% to generate electricity (photovol-

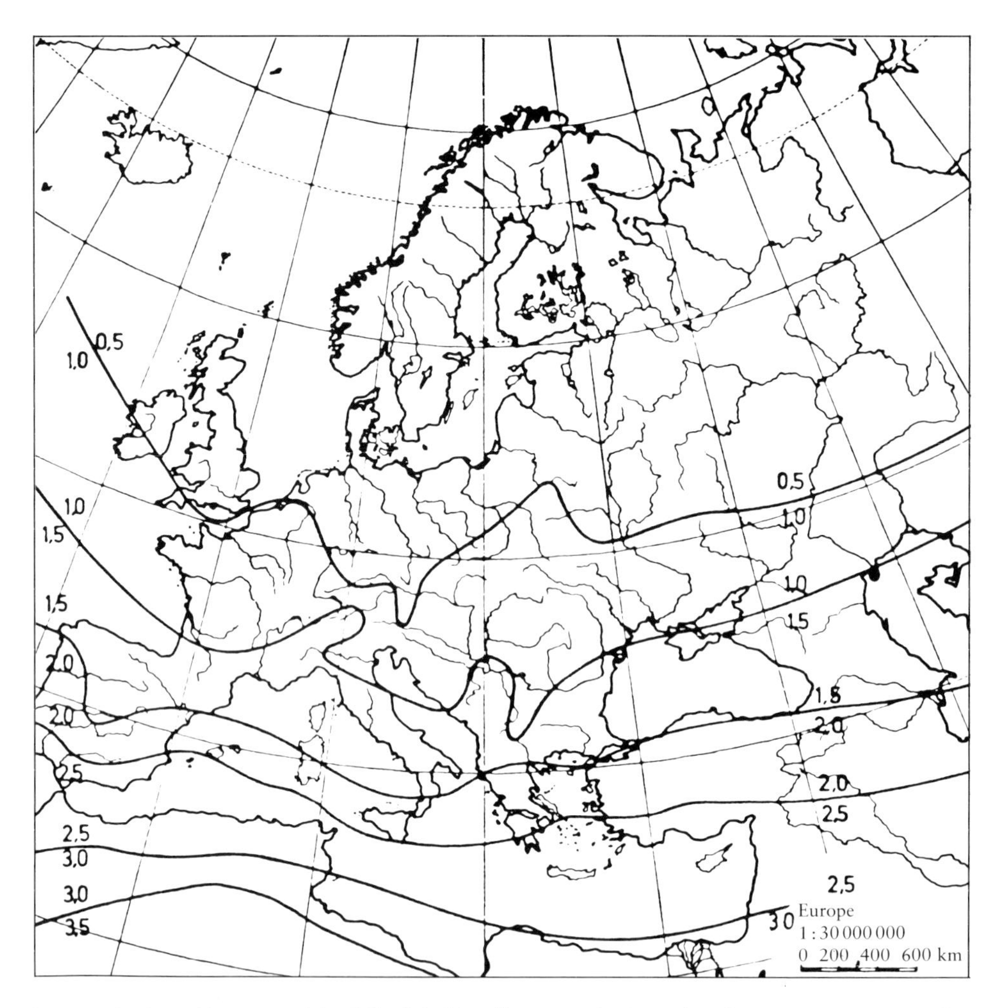

Fig. 3-16 b: Monthly means of the daily global irradiance on a horizontal surface in Europe in December (in kWh/m²d)

Source: J. A. Duffie, W. A. Beckmann, Solar Energy-Thermal Process. German translation: G. Bräunlich, Munich: Pfriemer 1976.

taic energy conversion – this is a typical value for silicon solar cells), assuming the average value of 200 W/m², which is the amount, averaged over the year, received in much of the USA and southwestern Europe (compare Fig. 3-17 a, b) one can generate an average of 1000 MW. (The peak power is several GW). Assuming an efficiency of 15% and the average radiation of 130 W/m² which is received in middle and western Europe, a surface of about 50 km² is required for an average power output of 1000 MW.

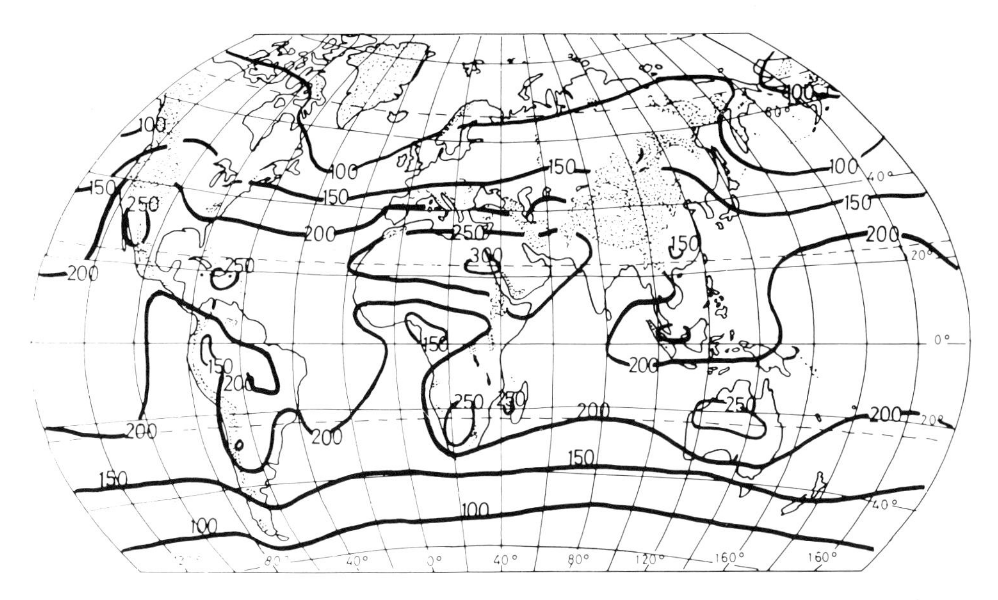

Fig. 3-17a: Annual mean global irradiance on a horizontal plane at the surface of the earth (W/m^2 averaged over 24 hours)

Source: Solar Energy, UK Section of the International Solar Energy Society, London, UK-ISES, 1976.

Because the sun's rays are not steadily available, one of the main problems in utilizing them is that of energy storage. Solar radiation is also not evenly distributed over the surface of the earth, and the countries with the most intense sunshine have, in many cases, the lowest energy requirements (compare Fig. 2-9a and 3-17a, b). Therefore, in the long run it could become important to convert solar energy to some form in which it could be transported from countries with intense solar radiation to the countries with large energy consumption. This means that the problem of energy transport would be added to that of storage (162, 164). The transformation of radiant energy into chemically bound energy could solve both problems at once, and hydrogen might be the substance of choice, especially in view of the possible development of a hydrogen economy.

The direct conversion of solar radiation into electricity, without moving parts, chemicals or high temperatures can be regarded as technically feasible, for example with silicon solar cells.

In principle, solar cells can be made from any number of materials or combinations of materials. Single crystals of silicon have proved especially suitable (165), however. These solar cells have been used for years to provide electricity for weather and news satellites, where they cannot be serviced. They have an efficiency of 13% to 16%, and it appears possible that an efficiency of 20% will be achieved in the

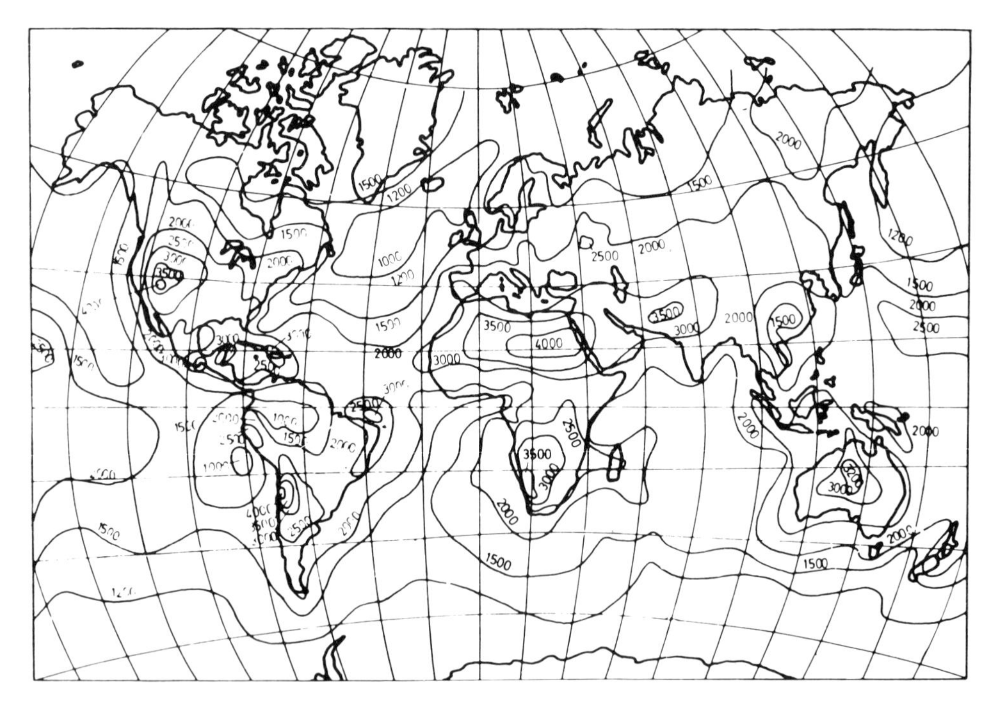

Fig. 3-17b: World map of sunshine hours/year

Source: World Energy Conference 1980: Survey of Energy Resources 1980, Munich, September 1980.

near future. However, the cells used in space are so expensive that large-scale terrestrial use of them is impossible for economic reasons. (The investment required for electricity generated from solar cells is at present around \$ 10000 to \$ 25000/kW, which is about 30 times as expensive as a conventional power plant.) For this reason there are research programs all over the world, but especially in the USA and the European Community, which have the goal of producing cheaper solar cells (166, 167). The production of solar cells from polycrystalline or amorphous silicon seems promising. There have also been experiments with CdS, CdTe and GaAs cells (168, 169). The US Department of Energy calculates that by the middle of the 80's, solar cells will be available for \$ 500/kW (peak power), and by the 90's, for \$ 100 to \$ 300/kW (peak power) (all in terms of constant dollars) (170, 171). Since solar cells have a favorable efficiency even when the sky is cloudy, their use in middle and northern Europe would be advantageous.

At present, the use of solar cells on earth is only sensible when electric power on the order of 10 W to 1 kW is required, in places where it is not available centrally. Whether larger power stations using solar cells will be built in the future will depend critically on the further developments in the price of solar cells.

Other promising methods for utilization of solar energy are the direct conversion of solar radiation to low-temperature heat and solar thermal-electric conversion.

Radiant energy can be relatively simply converted to heat using surface collectors (see 4.311). The temperatures which can be reached, up to about 100°C, depend on the design and mode of operation, especially on the flow of a heat carrier. Usually this is a liquid, ordinarily water. As discussed in 2.331, most industrial states convert a large fraction of their primary energy into low-temperature heat for space heating, due to the climatic conditions. This means that heat collected with surface collectors could make a significant contribution to the energy economies of these countries (mostly for space heating, but also for hot water). In principle, solar energy could be substituted for a large part of the petroleum which is now used for these purposes.

The following is an estimate of the energy which can be made available by systems based on surface collectors. Assuming a mean radiation level of 130 W/m^2 over the entire year (this applies to much of western and middle Europe), and an average system efficiency of 25%, it would require a surface of about 3000 km^2 to collect $100 \cdot 10^6$ tce/a. (This corresponds to 1.3% of the total surface area of the Federal Republic of Germany, or 0.6% of the area of France). Assuming a system efficiency of 25% and the mean radiation of 200 W/m^2, which is the annual average in much of the USA, it would require 2000 km^2 to produce $100 \cdot 10^6$ tce/a.

In the past few years, systems have been developed primarily for heating water. Solar hot water heaters appear, even in the temperate zones of middle Europe, to be on the threshhold of economic feasibility. (When solar energy is used to heat hot water, the problem of heat storage for short times is essentially solved.) However, space heating is of greater importance in the energy economy of most industrial countries. The main problem here is that to use solar energy for space heating, it must be stored for months (medium and long-term storage). In principle, there are already solutions to this problem, but they are in general not yet economically feasible.

The important factors in the total cost efficiency of solar installations are the costs of collectors and storage units, the lifespan of the system, the cost of conventional heat and the intensity of solar radiation at the site.

In many parts of the earth, including much of the USA, air conditioning is almost as important as heating is in the temperate zones. The fact that the need for cooling is greatest at the times when solar radiation is most plentiful is a great advantage for this application of solar energy.

Electricity can be generated from heat obtained from the sun either at low or at high temperature. Low-temperature collectors (surface collectors) do not require concentrating elements, but must have a low-boiling liquid to drive the turbines. Concentrating collectors do make it possible to use conventional steam generators, but are most applicable in regions with relatively low proportions of diffuse radiation. A few considerations of the principles applicable to these methods are presented below.

The radiation M emitted by a black body is described by the Stefan-Boltzmann law

$$M = \sigma T^4 \tag{1}$$

with M in Wm^{-2}, T in K and $\sigma = 5.7 \cdot 10^{-8} Wm^{-2}K^{-4}$. The emission of bodies which are not black can be described approximately by introducing the emissivity e and absorption capacity a. The energy balance of a stationary surface collector (ignoring losses due to convection and conduction) is given by

$$a\,S - Q_N = e\sigma T_c^{\,4} \tag{2}$$

S is the radiation intensity in Wm^{-2}, Q_N is the heat which can be removed by a heat-conducting medium, and T_c is the collector temperature. For a given S and T_c,

$$Q_N = a\,S - e\sigma T_c^{\,4} \tag{3}$$

The collector temperature T_c is given by

$$T_c = \left[\frac{a\,S - Q_N}{e\,\sigma}\right]^{\frac{1}{4}} \tag{4}$$

From these equations can be seen the difficulty of generating electricity from heat collected by solar collectors. On the one hand, the heat which can be extracted with a heat-collecting medium decreases with the fourth power of the collector temperature, but on the other, a high collector temperature is needed for a reasonable thermodynamic efficiency, which is given by

$$\eta_{th} = \frac{T_c - T_m}{T_c} \tag{5}$$

(T_m is the temperature of the coolant in the turbogenerator.) In short, a higher thermodynamic efficiency can only be had at the price of smaller amounts of heat energy being extracted from the collectors.

As an example of this sort of playoff, take the situation at 50° north latitude (e.g. Prague or Frankfurt), where S = 500 W/m² (cloudy sky). For a good collector, let a = 0.90 and e = 0.10. The turbogenerator is run by solar-heated steam at a temperature of 450 K, with cooling water at 290 K. The thermodynamic efficiency at these two working temperatures in 35%. The efficiency of conversion of solar to electrical energy is 0.1 (170). Increasing the collector temperature would not improve the situation. Increasing the thermodynamic efficiency from 35% (T_c = 450 K) to 55% (T_c = 650 K) would require, according to equation (2), a radiant intensity of S =

1570 W/m^2, which is higher than the solar constant. Therefore, it would be necessary to focus the radiation (e.g. with mirrors) to obtain the desired temperature. However, this would increase the amount of capital required and would lead to a disproportionate increase in the complexity of the system (171) (see 4.31).

Solar thermal generation of electricity is only at the beginning of its development. At present, there are plans for plants with power outputs up to 100 MW, the projected cost of which is between $ 1000 and $ 5000/kW at the end of the 80's (15).

Economic questions aside, a critical factor in the utilization of solar energy on a large scale is the question whether enough surface can be set aside for this purpose. As has already been shown, for a system with 15% conversion efficiency (photovoltaic conversion) and 200 W/m^2 average radiant energy, the surface required to generate 1000 MW is 34 km^2. In southwest Europe (the Iberian Peninsula), it might not be impossible to set aside this much area, although not as an uninterrupted unit. Fig. 3-17a shows that in northern Africa, the Arabian Peninsula and the western part of the USA, the surface required for 1000 MW is smaller. With a system efficiency of 15% and an annual average of 250 W/m^2, about 27 km^2 is needed. In these areas, the use of solar energy is especially promising. The USA in particular is the only country on earth with both the technical know how and economic strength to utilize solar energy, and the necessary amount of radiant energy (172, 173).

The conditions for exploitation of solar energy are not so favorable in the thickly populated regions of middle and western Europe and Japan, because the average annual solar radiation in these regions is only about 130 W/m^2. As already mentioned, under these conditions, and with an efficiency of 15% for photovoltaic energy conversion, a surface of about 50 km^2 is needed to provide 1000 MW of electricity. That is to say, the amount of area needed is very large, but it ought to be less than prohibitive, in some areas of western Europe, since a continuous surface is not required, provided that the generation of electricity from solar energy can be technologically and economically achieved. However, due to the low power density, and to the current trends in technological development, solar electricity is not likely to become competitive with conventional power in the temperate climatic zones in the near future.

The situation is more favorable in the case of generating low-temperature heat with flat collectors. As shown earlier, the area required to provide $100 \cdot 10^6$ tce/a, assuming a system efficiency of 25% and an annual average radiation of 200 W/m^2, is 2000 km^2. These figures apply to large areas of the USA. In part of middle and western Europe and Japan, where the annual average radiation is 130 W/m^2, a surface of 3000 km^2 would be required. It should be remembered that in highly developed industrial states, about 50% of the terminal energy demand is for heat in the temperature range between about 30° and 100°C (space heating, hot water and processing heat in the low temperature range) (see 2.331). A large fraction of the petroleum is used for space heating and hot water, and could thus be replaced by solar energy.

Since the USA consumes a relatively large amount of petroleum, a forced development of solar energy utilization in this country would considerably relieve the pressure on the world petroleum market (see 3.322).

In addition to utilizing the solar energy falling within their borders, the economically strong countries of western Europe have the option of cooperating with northern Africa. Aside from geographic proximity, there is the fact that the countries of the Sahara are among the poorest in the world, and lack the technical know how and economic strength to utilize this natural energy source. Comparison of Fig. 3-17a, b with Fig. 2-9a shows that many of the poorest countries of the world lie in the regions with the most intense solar radiation. In view of the area needed, the uninhabited regions of the Sahara would provide ideal conditions for the large-scale technical utilization of solar energy.

Such a cooperation between western Europe and northern Africa would naturally bring with it political problems, which the Europeans should include in their political calculations from the start. The secondary energy obtained from solar energy would have to be transported over long distances, and possibly stored. (This is why electricity is not necessarily the best form of secondary energy for solar energy. From all appearances, hydrogen would be better, since it is easier to transport and store.) From the above, it can be seen that in all probability, even after large-scale utilization of solar energy has been realized, western Europe will remain dependent on the countries of northern Africa and the Middle East for its energy.

It is understandable that the USA, Japan and the countries of the European Community (EC) are making serious efforts to utilize solar energy. Even the North Atlantic Treaty Organization (NATO) has initiated a program, within the framework of the Committee on the Challenges of Modern Society (CCMS), to encourage the use of solar energy. This use, however, is only just beginning. The total output of all the solar collectors presently installed in Europe is only about 1 MW, an amount which is negligible in comparison to the total energy balance.

3.38 Tidal Energy

Tides are due to the fact that the gravitational attraction of the sun, moon or planets is greater on the side of the earth closer to the attracting body, and therefore, as the earth rotates, the force exerted on a given location varies periodically (173, 174). The local changes in gravitational force causes the rise and fall of the tides, and the masses of water moved by them contain the corresponding energy of motion.

The total potential energy of the tides has been estimated to be $3.2 \cdot 10^9$ tce/a ($\triangleq 3 \cdot 10^{12}$ W), and the technically exploitable tidal potential, only about $60 \cdot 10^6$ tce/a (compare Table 3-3) (15). The economically feasible exploitation of tidal energy in a power plant requires a difference between high and low tides of about 5 m. The height of the tides varies widely. A spring tide in the Bay of Fundy (Canada) may be as much as 21 m. The average tidal difference on the coasts of the Atlantic, Indian and Pacific Oceans is 6–8 m; on the Baltic Sea it is only a few dm, and on the Mediterranean, only about 10 cm.

Because the economic utilization of the tides depends on the physical geography, such as the width and depth of bays or estuaries, in addition to the average difference in water level, there are relatively few sites where tidal power plants would be economically feasible. The economically utilizable tidal potential has been estimated to be 6 to 15 GW (electric power) (15). (Other estimates have been several times larger (174)). However, it follows that tidal energy can probably be only of local significance, and, on the global scale, it will probably never account for a significant fraction of the energy supply.

3.39 Geothermal energy

Based on geological and geophysical data, it is thought that the core of the earth has a temperature between 3400 and 10000 K. The core is surrounded by the mantle and crust, and the temperature decreases from the core outward. (In the boundary region between the mantle and the crust, which is between 10 and 60 km deep, depending on the thickness of the crust, the temperature is assumed to be between 600 and 1300 K.) This temperature gradient results in a flow of heat from the interior to the surface of the earth. In addition, the decay of radioactive minerals in the crust releases heat. The total heat flow amounts in most places to only about 0.06 W/m^2, which is very small in comparison to the solar radiation, which may be as high as 1000 W/m^2 on a sunny day. Therefore a direct utilization of the heat flow from the interior is not feasible (162). However, in the regions of geothermal anomalies, the heat flow may be orders of magnitude larger.

Those anomalies which could be exploited, given the technical and economical processes to do so, are called geothermal mines. There are basically four different forms of geothermal mines: hot steam (vapor-dominated hydrothermal), hot and warm water (liquid-dominated hydrothermal), hot dry rocks, and geopressured reservoirs (175).

Geothermal energy is already being exploited, for example in the USA, Mexico, Iceland, Italy, New Zealand, Japan and the USSR. The energy is extracted essentially as follows:

– Dry, superheated steam, which is pumped or flows out of geothermal reservoirs, can be used directly to turn a turbine. This type of source is found, for example, in Larderello, Italy, and in California.

– Water reservoirs which are under pressure yield a mixture of steam and water at a temperature of 450–640 K. The steam can be separated and used to generate electricity or for industrial processes. The hot water can be used for space heating. (Such installations are in operation in Wairakei, New Zealand, Japan and Ireland).

– Hot-water springs produce water at atmospheric pressure and a temperature of about 320–350 K. This can only be used for generation of electricity via heat exchangers and a low-boiling liquid, e.g. freon or isobutane. Because the ground water does not come into contact with it, the turbine does not have to be made of corrosion-resistant material. The hot water can also be used for space heating. There are hot-water springs in several of the states in the western part of the USA.

The conventional uses of geothermal energy depend on the presence of steam or water in the source; the source is tapped by drilling and removing the heat-transporting medium. However, like tidal energy, this form of energy can be at best of local significance, since there is too little of it to make a major contribution to the world's energy supply.

The situation changes, however, when one takes into consideration the widespread "hot dry rocks" found up to several thousand meters depth. The extraction of energy from these rocks lies within the reach of present drilling technology (hot dry rock technology) (176). The temperature of these rocks increases about 3 K per 100 m down from the surface (geothermic temperature gradient). The estimates of the available energy depend on the depth which can be reached by present technology and the assumed cooling of the rocks. If one were to cool all the rocks under the land surface of the earth to a depth of 5000 m to a temperature of 373 K, one could extract about $8 \cdot 10^{14}$ tce, which is about 10^5 times the present world demand for primary energy. The heat content to a depth of 10000 m is about $90 \cdot 10^{14}$ tce (see Table 3-3) (2). (The total heat content of the earth is about $4 \cdot 10^{20}$ tce).

It can be safety assumed that such amounts of heat will never be quantitatively extracted, not even to a few thousand meters depth. The hot dry rock technology has yet to be developed, and it is not known whether it will ever be economically feasible. To extract heat from hot dry rocks, water will be pumped into them through shafts from the surface, heated, collected as steam at a shallower shaft, and returned to the surface.

With a few exceptions, however, the hot dry rocks are compact and impenetrable. They must therefore be shattered, so that the water can penetrate them and be heated by exposure to as large a surface area as possible. There are plans to use both conventional and nuclear explosives for this purpose (see 3.343). More recently, it has been proposed to open the hot dry rocks by pumping water into them under high pressure. The rigid rock is expected to shatter, and the thermal tension created by the cold water pumped into the shaft to be heated will expand the shattered zone. The realization of these technologies to utilize hot rock formations will be crucial to the future role of geothermal energy, because the potential energy of hot dry rocks is far greater than that of any other geothermal reservoirs.

In the course of drilling for oil in the Gulf of Mexico, geopressured reservoirs of hot water, at a temperature of 380 K, have been discovered at 3300 m depth. The water trapped in this sediment basin is under high pressure, is very salty, and contains large amounts of methane. The utilization of this geopressured reservoir is likely to be extraordinarily difficult, due to corrosion, the necessity of removing the methane, and the enormous pressure on the wellhead.

The fraction of the world's energy supplied by geothermal sources is at present very slight, and will in all probability remain an insignificant part of the total through the year 2000. A geothermal electric capacity of 1000 MW is expected in 1980, and 100000 MW in 2000 (2, 20). However, geothermal energy may increase to provide a significant portion of the energy consumed in several countries, such as Italy, the United States, Mexico, Ireland, New Zealand, the Soviet Union and Japan. The non-electrical use for heating and hot water at present amounts to about 5500 MW throughout the world.

4. Energy supply systems

4.1 *The role of secondary energy carriers*

Most primary energy carriers, such as nuclear fuels or solar energy, cannot be directly utilized, but, with the exception of natural gas, must be converted to other forms of energy, such as electricity or heat. This explains the importance of conversion technologies in energy supply systems. Conversion is never achieved without loss. The immediately usable, transportable and storable forms of energy are called secondary energy carriers (see Fig. 3-1).

Since about 1950, the trend in secondary energy carriers has been from solid to fluid forms, especially in the industrial countries, and it can be assumed that this trend will continue. However, in the long run, the application of liquid secondary energy carriers will probably be limited to the transportation sector.

For economic rassons, secondary energy carriers are frequently produced in large amounts and usually at some distance from the consumer. They must therefore be transported and distributed. Electricity, for example, is usually generated in major power plants, but the transport of electricity is the most expensive form of energy transport, per unit of energy. However, if the cost of transporting electricity is divided by 3, which is a realistic compensation for the average 35% efficiency of conversion of fossil fuels into electricity, then it is only slightly more expensive than the conventional railway transport of coal. This, however, is about 70% more expensive than the pipeline transport of gas. (It can be assumed that pipeline transport of hydrogen, a potential secondary energy carrier, would be roughly as economical.) The pipeline transport of gas is about three times as expensive as pipeline transport of oil (1, 2). Thus electricity is very expensive to transport, but it is very versatile, and has almost no environmental impact at the consumption stage.

Storage of secondary energy carriers has become more and more important in the last few years, partly because the installations which convert primary energy carriers (e.g. coal, oil or uranium) into secondary forms (e.g. electricity or heat) have become so expensive that it is important to use them to maximum capacity. However, since the demand fluctuates, it is logical to try to find storage forms to even out the load. It is relatively easy to store liquid or gaseous secondary energy carriers in tanks or underground caverns. In contrast, it is extremely difficult to store electricity, so that the peaks in demand can only be met by building additional complicated facilities into the network. One example of this type of installation is a system for pumping water uphill into a reservoir when the demand for electricity is low, and letting it run through turbines at times of peak demand.

There are other criteria besides the ease of transport, storage and convenience which are becoming ever more important in the evaluation of secondary energy carriers or energy supply systems, including economy and environmental impact. The above criteria are the most important, but not the only ones which must be applied to the individual secondary energy carriers, and they are also not independent of one another. Measures to reduce the environmental impact, such as generation of electricity or conversion of coal to gas, generally increase the total cost. Furthermore, the relative importance of the various criteria differs from region to region. The environmental impact is generally less important in thinly settled regions than in thickly populated industrial countries. Also, in many cases primary and secondary energy carriers cannot be evaluated separately. For example, it is conceivable that coal can be more economically refined using high-temperature nuclear reactors than fossil fuels.

This chapter deals with the conversion of nuclear, solar, tidal and geothermal energy into secondary forms. Only the basic principles and problems are indicated; special references to the literature sources of detailed information are made. Finally, a few promising secondary energy carriers or energy supply systems are discussed, including electricity, the uses of waste heat, the products of petroleum and coal refining, latent heat, and the possible use of hydrogen as a secondary energy carrier.

4.2 *Secondary energy from nuclear sources*

4.21 Energy from nuclear fission

4.211 Basic reactor physics

Heavy atomic nuclei (uranium, plutonium) can be split by absorbed neutrons, a process which can be represented as follows:

$$^{235}_{92}U + ^{1}_{0}n \rightarrow ^{236}_{92}U \rightarrow ^{90}_{38}Sr + ^{144}_{54}Xe + 2^{1}_{0}n + 195\ \text{MeV} \quad (1)$$

$$^{235}_{92}U + ^{1}_{0}n \rightarrow ^{236}_{92}U \rightarrow ^{89}_{36}Kr + ^{144}_{56}Ba + 3^{1}_{0}n + 195\ \text{MeV} \quad (2)$$

This means that the reaction of a uranium nucleus $^{235}_{92}U$[1] with a neutron n produces first an "intermediate nucleus" of $^{236}_{92}U$, which is unstable and splits into two medium-sized fragments. Depending on the fission pathway, an average of 2 or 3 neutrons and about 195 MeV of energy are also released.

The pathways indicated in (1) and (2) are only examples; about 300 possible fission products are known. This is due to the fact that the uranium nucleus can fission in various ways, producing a variety of primary fission products, and also, these products are themselves radioactive and produce a series of secondary nuclei. Fig. 4-1 shows the yields of fission products of ^{235}U in percent, as a function of the mass. (^{235}U fission induced by thermal neutrons). The curve has two maxima, i.e. the most probable fission patterns yield product nuclei with mass ratios of 2:3. (The fission of ^{239}Pu and ^{233}U by thermal neutrons has a similar product curve) (3, 4).

The neutrons released by every fission pathway can be classified as "prompt" or "delayed", depending on whether they are emitted directly after fission (within 10^{-14} second) or after a delay of 12 to 80 seconds. When fission is induced by thermal neutrons, about 0.75% of the emitted neutrons are delayed, and these are the decisive factor in the control of energy release in a reactor.

If two neutrons are released by a fission event (1st generation), each of them can cause another uranium nucleus to split, thus releasing a total of 4 neutrons (2nd generation). After the tenth generation, there are thus 1024 neutrons (chain reaction). (In the detonation of a nuclear fission bomb, this chain reaction occurs in "uncontrolled" fashion, in about a millionth of a second.)

Of the 195 MeV released by the fission, about 162 MeV is in the form of kinetic energy of the fission products, which fly apart at high velocity. They are slowed by interaction with the surrounding material, the kinetic energy being converted to heat. The neutrons arising from the fission carry off about 6 MeV, which means that the neutrons fly apart with an average energy of about 2 MeV apiece. This corresponds to a velocity of about 10000 km/h (fast neutrons). The remainder of the energy, about 27 MeV, is emitted as β and γ radiation, or as neutrinos. Due to their lack of charge and to the fact that their mass is extremely small compared to other particles, neutrinos scarcely interact with matter, and thus do not need to be considered here.

[1] The nuclear charge Z is equal to the number of protons in the atomic nucleus, and thus equal to the number of the electrons in the shell of an electrically neutral atom. Z determines the chemical properties of the atom. The mass number A of a nucleus is the sum of the number of protons (Z) and the number of neutrons (N), i.e. A = Z + N. A is the mass of the nucleus.

Isotopes are nuclei with the same value of Z (the corresponding atoms have the same chemical properties), but different neutron numbers A, and thus mass numbers N.

The following convention is used to designate an isotope: $^{A}_{Z}$chemical symbol. A few examples are $^{1}_{1}H$, hydrogen nucleus (1 proton); $^{2}_{1}H$ = D, deuterium nucleus (1 proton + 1 neutron); $^{3}_{1}H$ = T, tritium nucleus (1 proton + 2 neutrons); $^{3}_{2}He$, helium (2 protons + 1 neutron); $^{4}_{2}He$ (2 protons + 2 neutrons); $^{235}_{92}U$, uranium nucleus (92 protons + 143 neutrons); $^{238}_{92}U$, uranium nucleus (92 protons + 146 neutrons).

Naturally occuring elements are usually mixtures of isotopes. For example, naturally occuring uranium consists of 99.274% $^{238}_{92}U$, 0.720% $^{235}_{92}U$ and 0.006% $^{234}_{92}U$.

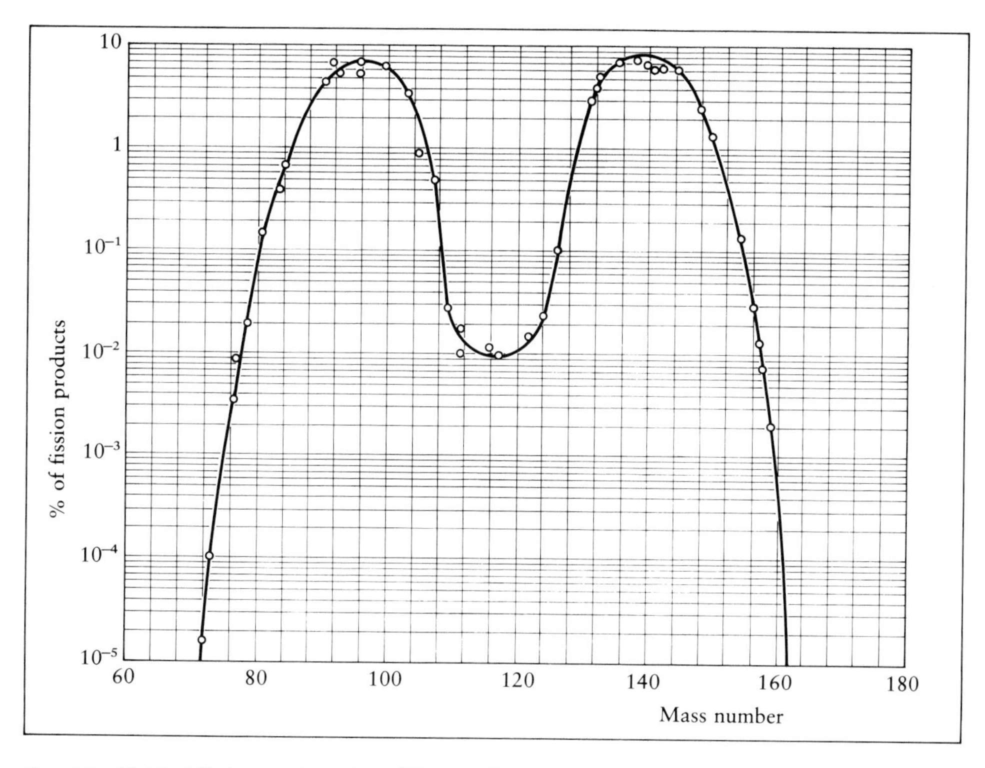

Fig. 4-1: Yield of fission products from ^{235}U as a function of mass

Source: E. W. Schpolski, Atomphysik, Teil II, Berlin: VEB Deutscher Verlag der Wissenschaften 1969.

Before discussing the most important reactor types, we shall consider the essential elements of a nuclear reactor, the fuel elements, the moderators, the control mechanism, the coolant and the radiation shielding. Fig. 4-2 shows the basic plan of a reactor.

The fuel elements consist of a number of fuel rods, which contain the reactor fuel. For example, the core of the nuclear reactor Biblis A, Federal Republic of Germany, has 193 fuel elements, each composed of 236 fuel rods. The reactor fuel includes not only the fissionable material proper, i.e. ^{235}U, ^{239}Pu and ^{233}U, but also, depending on the type of reactor, the breeder material ("fertile materials"), such as ^{238}U or ^{232}Th, and other materials which improve certain properties of the fuel, such as its heat conductivity. The fuel is contained within a sheath (cladding) (for example of a zirconium alloy) which prevents the fuel or the fission products from contaminating the coolant. The fissionable material has the property that its nuclei can be induced by neutron capture to split in an energy-yielding process, and, under certain conditions, the fission process is self-sustaining. The ^{235}U nucleus is split by capture of a slow

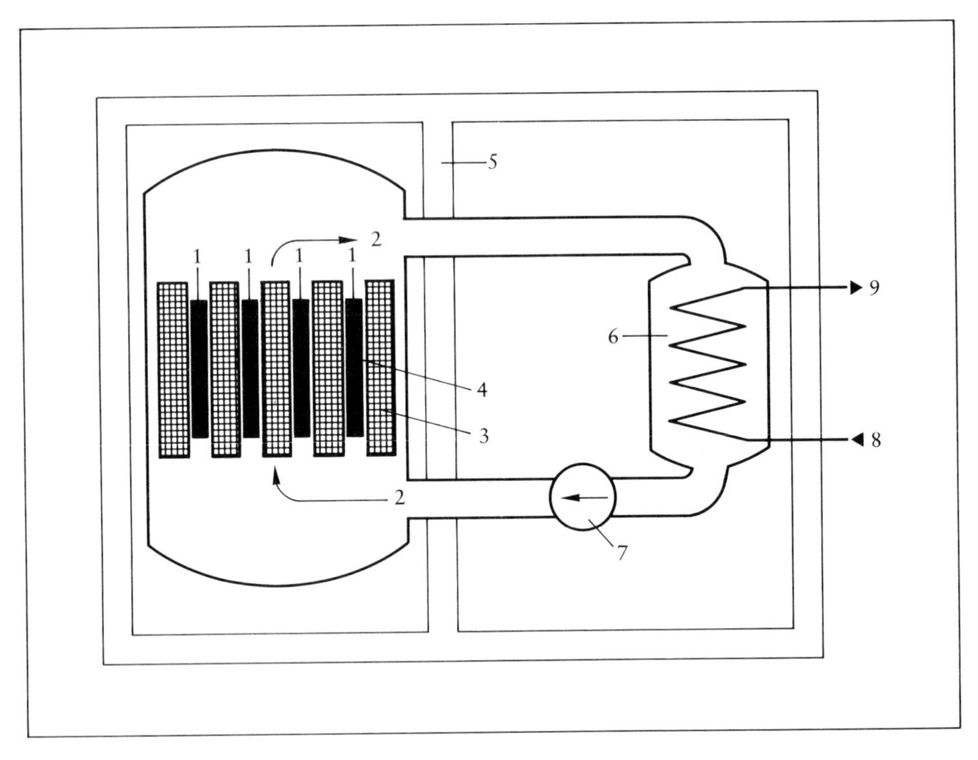

Fig. 4-2: Basic scheme of a reactor
1 = Control rods; 2 = Coolant; 3 = Moderator; 4 = Fuel elements; 5 = Radiation shielding; 6 = Heat exchanger; 7 = Pump or bellows; 8= Feed water; 9 = Steam.

Source: K. J. Euler, A. Schramm (Eds.), Energy Supply of the Future, Munich: Karl Thiemig 1977.

(thermal) neutron. Since natural uranium contains only 0.720% ^{235}U, however, it must be enriched before it can be used. (^{238}U, which makes up 99.274% of natural uranium, can be made to fission by fast neutrons, but its reaction cross section is orders of magnitude smaller than that of ^{235}U. For this reason, although it does fission, the occurrence is so rare that it can be ignored. ^{234}U makes up only 0.006% of natural uranium, and is thus insignificant.) Besides ^{235}U, there are plutonium $^{239}_{94}Pu$ and ^{233}U, which are generated from ^{238}U and thorium $^{232}_{90}Th$, respectively, and can also be used as fissionable fuels. The man-made nuclei ^{233}U and ^{239}Pu can, like ^{235}U, be fissioned by thermal neutrons[1], but their cross sections are different (see 4.212

[1] Depending on the type of neutrons primarily responsible for inducing fission, reactors can be classified as thermal or fast. The former use neutrons with energy less than 1 eV and require a moderator (see 4.212 and 4.213). Most of the reactors which have been built to date are thermal. In fast reactors, the neutrons are not slowed down by a moderator, but are used at their original energy (> 10^5 eV) (see 4.214 and 4.215).

and 4.214). In addition, it is possible to run a reactor on natural uranium, although this requires the use of heavy water as moderator, and it is expensive (heavy water reactor, HWR) (5). Aside from its material composition, the fuel is characterized by its *burn-up,* the proportion of its fissionable material which has already split, or the amount of energy released. The complete fission of 1 g of ^{235}U releases 0.913 MWd of heat, and other fuels release a similar amount of energy. A simple conversion gives 1% burn-up $\triangleq 9.13 \cdot 10^3$ MWd/t (4).

There are three basic types of fuel material: metallic, ceramic and dispersion fuel elements. The metallic group includes uranium, plutonium and thorium (as a fertile material together with a fissionable material) and their alloys. Uranium metal is used, in reactors together with natural uranium. Pure plutonium, however, is not used in reactors. One reason is that in the appropriate geometry, its critical mass can be 10 kg or less. The main ceramic fuels are the oxides and carbides of uranium, thorium and plutonium, which are used in sintered form in the fuel elements. Modern reactors use almost entirely ceramic fuels. If highly enriched fuels are used, and no breeding or conversion is planned, the fuel can be used in diluted form in a metallic carrier and built into a matrix. Metallic ceramic compounds are one important form of dispersion fuel material. To date, however, dispersion fuel elements have not been used in power reactors (4).

Fuel rod claddings prevent losses of the fuel and fission products into the coolant, and also protect the fuel from the coolant. There are four main types of cladding material, aluminum, magnesium, steel and zirconium. (In high temperature reactors, which are a special case, the technique of coated particles is used. The particles of fuel, for example uranium carbide or uranium oxide, are coated with graphite, oxides or carbides.) Aluminum and magnesium have relatively small absorption cross sections for thermal neutrons, and are therefore used mainly in natural uranium reactors. The magnesium alloy *magnox* is the most frequently used in reactors, especially in CO_2-cooled, graphite-moderated reactors (GGR), which are therefore also called magnox reactors (6). Steel cladding materials, such as chromium-nickel steel or molybdenum-nickel steel, are strong and resistant to high temperatures and corrosion, but they have relatively large cross-sections for thermal neutrons. Zirconium has a relatively low neutron absorption cross section, and has good mechanical, thermal and corrosion resistance. Zirconium, especially in the alloys zircaloy-2 and zircaloy-4 (zirconium with Sn, Fe, Cr and Ni), is used preferentially in water-cooled reactors.

A moderator is required in a nuclear reactor to slow the neutrons down to the point that they can interact with the fissionable nuclei. The moderator should therefore absorb as few neutrons as possible, but each collision between a neutron and a moderator atom should transfer as much energy as possible. This is the case when the masses of the colliding particles are similar. Depending on the type of reactor, the moderator can be light or heavy water, an organic liquid (C_xH_y) or a solid like graphite or beryllium (as BeO). The moderator can sometimes be the same material that is used as coolant, for example, when the latter is light or heavy water. Graphite is used as moderator in nearly all gas-cooled reactors because it is very resistant to stresses induced by high and changing temperatures. (CO_2 is often used as the coolant in a graphite-moderated reactor.) If the fuel and moderator are separate entities, the reactor is called heterogeneous, and when they are a single unit, it is called homogeneous. In the early stages of reactor development, efforts were made to build a homogeneous reactor using a solution of uranium salts in water. However, the corrosion problems were so serious that this line of development was abandoned. Modern power reactors are heterogeneous. When light water is used as moderator, a relatively large percentage of the neutrons generated by fission is lost, due to the very large absorption cross section of hydrogen nuclei for neutrons. Therefore, light water can only be used if a sufficient excess of neutrons is produced, for example when enriched uranium is used as fuel. When natural uranium is used as fuel, the absorption of neutrons by light water is so great that the chain reaction is not maintained. For this reason, heavy water, carbon (graphite) or beryllium must be used as moderator with natural uranium.

Another important component of a reactor is the mechanism by which the chain reaction can be regulated. Regulation is achieved by rods of a material with a very large absorption cross section for neutrons, for example cadmium or boron steel, which can be inserted to various depths into the core of the reactor. These control rods are magnetically controlled; in practice a number of the fuel elements contain control rods. The multiplication factor K (the number of neutrons from each fission event which successfully initiate further fission events) is equal to 1 in steady operation; the parameter $\varrho = (K-1)/K$, which is called the reactivity, is then equal to zero, which means that the number of neutrons per unit volume does not change with time. (The delayed neutrons are included in the equilibrium neutron balance.) If $K = 1$, the neutron losses due to absorption or escape from the system are exactly large enough that on the average, one of the neutrons from each fission event initiates another fission. (Neutrons are absorbed by the coolant, the structural elements, materials inserted for the purpose of breeding new fuel, and control rods.) Under these conditions, the number of fissions per unit of time and the power output of the reactor is constant; the reactor is *critical* or *on*. If, starting from the steady state $K = 1$, the number of neutrons per volume unit is decreased ($K < 1$) by increasing the number absorbed by the control rods (as they are shoved farther into the core), the power produced by the reactor diminishes, and it is *subcritical* or *off*. If, starting from $K = 1$, the number of neutrons per unit volume increases because fewer are absorbed (the rods are withdrawn from the core), the power output increases. The control mechanism is designed so that K is always less than 1.0075, i.e. the multiplication factor never exceeds 1 by more than 0.75%. Under these circumstances, there are never enough *prompt* neutrons released immediately by fission to make $K > 1$. The delayed neutrons are required to shift the balance. Therefore, with $1.0075 > K > 1$, the power output of the reactor can only begin to increase moderately fast after a delay of about 12 seconds, due to the 0.75% of the neutrons which are delayed. (This condition is called *supercritical.*) The reactor is thus designed so that if the production of neutrons should increase, the full effect will be delayed by about 12 s, which is enough time for the automatic or mechanical control mechanisms to be turned on and to take effect. At $K > 1.0075$, the power output would increase very rapidly. The reactor would become *prompt supercritical,* because the chain reaction could be sustained by the prompt neutrons alone, without the delayed neutrons (which would also be present, however). In this condition, the reactor would quickly produce so much heat that the cooling system would be inadequate, and damage to the fuel rods, primary cooling system and structural elements could ensue. The control system of the reactor must therefore be designed in such a way that this condition cannot occur (7). In an emergency shutdown *(scram),* all the control rods must fall from whatever position they are in all the way into the center of the core, as quickly as possible. This is accomplished in different ways in different types of reactor. In a pressurized water reactor (PWR), for example, the lack of current through the electromagnets which control the control rods would result in the rods falling into the core by gravity. In the boiling water reactor (BWR), by contrast, the control rods move into the core from below, so in the event of a scram they must be moved by some other source of power. In this design, there is a hydraulic system to push the control rods up into place. The basic concept of gas-cooled, graphite-moderated reactors (GGR) makes large core dimensions necessary, so that their control rods have to be 6 to 8 m long. In these reactors, the control rods fall into the core by gravity in an emergency shut-down. It should also be mentioned that the temperature of the core has, within certain limits, a self-regulating effect on the power output. As the temperature rises, so does the average neutron energy, which results in a higher average absorption. This decreases K (negative temperature coefficient).

A coolant is required in every reactor to remove the heat of fission. Liquids, such as light or heavy water or organic fluids, gases, e.g. carbon dioxide or helium, or liquid metals like sodium or potassium are used (8).

The radiation protection measures about a nuclear reactor must be particularly strict because nuclear fission produces intense radiation of every kind (α, β and γ rays, neutrons and fission products). Therefore the reactor, and especially the core, must be shielded, for example with special concrete. In addition, neutrons and γ rays generate activation products when they are absorbed by nuclei. Essentially all the materials used in the structure and operation of the actual reactor become radioactive by exposure to radioactivity. In addition, gaseous, liquid and solid radioactive wastes are generated in the operation of the reactor, and these must be carefully monitored. Finally, the disposal of about 300 different radioactive isotopes with widely varying half-lives is an extraordinarily difficult problem. As yet there is no satisfactory solution for the final disposal of highly active radioactive waste, but the results of intense research and development in several countries give reason to hope that the problem will be solved satisfactorily (see 5.822).

The first controlled chain reaction in an arrangement of metallic natural uranium (fission material) and graphite (moderator) was started by E. Fermi in Chicago on December 2, 1942. Since then, many possible methods of using nuclear energy for peaceful purposes have been developed by varying the combination of components, particularly the basic fuel, the moderator and the coolant.

In a relatively short time, a few types of reactor evolved for commercial power production, namely water-cooled and water-moderated reactors, and gas-cooled, graphite-moderated reactors. The water-cooled and water-moderated types include heavy water (HWR) and light water (LWR) reactors, which can be functionally classified as boiling water (BWR) or pressurized water reactors (PWR) (compare Table 4-1). The gas-cooled, graphite-moderated reactors include the gas graphite reactors (GGR), advanced gas reactors (AGR) and the high temperature gas reactors (HTGR or HTR), which are still in the developmental stage. The fast breeder reactors comprise another line of development. In these, the fission is induced by fast neutrons, in contrast to the reaction in thermal reactors. This means that fast breeders contain no moderator. For years, many countries have been investing large sums in the development of the liquid-metal-cooled fast breeder reactors (LMFBR); in the last few years, the USA and the Federal Republic of Germany have been attempting to build a gas-cooled fast breeder reactor. Nearly all of the currently operational and planned power plants fall into one of these categories, with the LWRs dominating (compare Table 4-1) (8, 9).

Table 4-1: Total reactor distribution of the western world (present trend)[1] in GW electric power

Year	LWR[2]	HWR	AGR	GGR	HTGR	FBR	Total
1978	90	5	3	6	–	1	105
1980	126	7	6	6	–	1	146
1985	251	13	6	6	1	1	278
1990	460	28	6	6	2	2	504
1995	683	50	6	4	3	4	750
2000	901	75	6	2	6	10	1000

[1] The "present trend" estimate takes cognizance of current patterns of energy utilization supply as well as present delays in the construction of new reactors, and generally assumes a continuation of these trends. Nuclear share of electrical capacity: 1985 16%; 1990 23%; 2000 32%. (Reactor lifetime: 30 years).

[2] Reactor abbreviations: LWR: Light Water Reactor, HWR: Heavy Water Reactor, AGR: Advanced Gas Reactor, GGR: Gas Graphite Reactor, HTGR or HTR: High Temperature Gas Reactor, FBR: Fast Breeder Reactor.

Source: Uranium Resources, Production and Demand, A Joint Report by the OECD Nuclear Energy Agency and the International Atomic Energy Agency, Paris, December 1977.

It is customary to classify power reactors according to the primary coolant used in them. (Sometimes, for more exact specification, the moderator is also named.) The primary coolant determines not only the design of the primary circulation system, but affects the secondary system as well, for example, by limiting the attainable temperature. Therefore, the primary coolant essentially determines the type of installation, and also the economic feasibility of the power plant.

4.212 Water-cooled and water-moderated reactors

4.212.1 Heavy water reactors

From the beginning of reactor development, heavy water has been considered as coolant and moderator. Heavy water (D_2O) differs from light water (H_2O) in having a very much smaller neutron absorption and a somewhat lower braking action than the latter. However, D_2O is considerably more expensive than H_2O (see 3.362).

The lower neutron absorption of D_2O makes it possible to use natural uranium as a fuel in heavy-water-moderated reactors, instead of the enriched uranium (about 3% ^{235}U) needed in light water-moderated reactors. It is an advantage to dispense with the uranium enrichment, but much more D_2O than H_2O is needed to brake the neutrons. This means that heavy-water-moderated reactors are always much larger than light-water reactors of comparable power output.

Since less D_2O is needed as coolant than as moderator, the individual fuel rods are not distributed evenly throughout the reactor core, as they are in LWRs with adjacent fuel rods, but are bundled together and arranged around coolant channels (5). This functional separation of moderator D_2O and coolant makes it possible to use another substance as coolant, for example CO_2, H_2O steam, or an organic liquid. However, most D_2O-moderated reactors are also D_2O-cooled, i.e. it is heavy-water-cooled and heavy-water-moderated reactors (HWR) which are fueled with natural UO_2. It is also possible to replace the coolant channels with high-pressure pipes (*pressure tube* reactors), and thus to dispense with the pressure chamber.

The heavy-water-cooled and heavy-water-moderated reactor has been developed to the point of commercial use, primarily in Canada and a few other countries, including the Federal Republic of Germany and the United Kingdom (5, 9).

In Canada, the experimental reactor NPD (Nuclear Power Demonstration, Ontario), with a capacity of 22 MW, went critical in 1962. Since then the following CANDU (Canadian Deuterium Uranium) power reactors have gone into operation: Douglas Point, Ontario, 208 MW (1967); Pickering A, Ontario, 514 MW (1971); Gentilly 1, Quebec, 250 MW (1971); KANUPP Karachi Nuclear Power Project), Pakistan, 125 MW (1971); RAPP 1 and 2 (Rajastan Atomic Power Project), India, both 203 MW (1972); Bruce A, Ontario, 745 MW (1976); Gentilly 2, Quebec,

600 MW (1979). Other CANDU reactors are being planned or are under construction, including Point Lepreau, New Brunswick, 600 MW; Cordoba (Argentina), 600 W; Pickering B, Ontario, 514 MW; Wolsung 1 (Korea), 600 MW; Bruce B, Ontario, 750 MW; Darlington, Ontario, 800 MW (5).

In the Federal Republic of Germany, only limited funds have been available for the development of heavy-water-cooled and heavy-water-moderated reactors, primarily for export. Because these reactors, unlike LWRs, can use natural uranium for fuel, and thus do not require an isotope enrichment facility, they allow for greater independence with regard to fuel supply. An example of these reactors is ATUCHA, Argentina, which was delivered by the Federal Republic of Germany and went into operation in 1974.

In Great Britain, the 100 MW SGHWR (Steam Generating Heavy Water Reactor) went into operation in 1968 in Winfrith, Dorset. On the basis of the good performance of this reactor, it was decided to build more of this type of PTR (Pressure Tube Reactor) with capacities of 660 MW (10–12). The PTRs use boiling H_2O as a coolant and D_2O (outside the pressure tubes) as moderator. Due to the H_2O cooling, these reactors use enriched uranium, with about 2.1% ^{235}U content, which makes them more like a boiling water reactor (BWR) than a heavy-water reactor (see 4.212.2).

Due to the more efficient neutron economy of a D_2O-cooled and moderated reactor (HWR), it has a higher conversion rate than an LWR (see 4.213.3), i.e. an HWR produces more plutonium than an LWR. Using natural uranium as fuel (0.720% ^{235}U), an HWR produces 607 kg total plutonium $^{238-242}Pu$, of which 450 kg is fissionable (^{239}Pu and ^{241}Pu), for each GWa of electric energy and 21 GWd/t burn-up. By comparison, an LWR using enriched uranium (about 2.3% ^{235}U) produces 285 kg total $^{238-242}Pu$, of which 194 kg is fissionable ^{239}Pu and ^{241}Pu, for each GWa electricity and burn-up of 33 GWd/t (see 5.83).

There is no significant difference in the fuel costs for HWRs and LWRs, because the lower cost per kg of natural uranium is compensated by the lower energy production per kg of fuel in an HWR. The specific installation costs for an HWR, however, are considerably higher than for an LWR, in part due to the D_2O inventory and the lower power density.

In the USA and the European continent the HWRs have not been widely adopted, partly for the reasons mentioned above. However, due to their independence from enrichment facilities, the HWRs are a significant development. As mentioned, there are HWRs in India, Pakistan and Argentina, which have not yet become members of the Non-Proliferation Treaty (NPT) (see 5.83).

4.212.2 Light Water Reactors

The majority of the 620 nuclear reactors in the world (in operation, under construction or on order) were of the light water type (compare also Table 4-1). Table 4-2

Table 4-2: The nuclear power plants of the world

Country[1)]	Operating reactors		Reactors under Construction		Planned reactors	
	Number	Electric power (net) in MW	Number	Electric power (net) in MW	Number	Electric power (net) in MW
USA	70	51550	85	93319	42	46935
USSR	33	12616	15	13680	12	12600
United Kingdom	33	6980	10	6840	5	3694
Japan	24	14994	8	6745	10	8622
France	23	15409	29	30230	12	13270
Germany, F.R.	14	8606	10	10636	11	13117
Canada	11	5494	14	9751	–	–
Sweden	8	5515	4	3931	2	1960
German Dem. Rep.	5	1694	4	1644	4	1632
Switzerland	4	1940	1	942	–	–
Italy	4	1382	3	1966	2	1900
India	4	809	4	880	–	–
Finland	3	1740	1	420	1	1000
Belgium	3	1664	4	3807	–	–
Spain	3	1073	7	6258	8	7734
China, P. R.	2	1208	4	3716	–	–
Bulgaria	2	816	2	828	–	–
Czechoslovakia	2	800	6	2520	7	2940
Netherlands	2	498	–	–	–	–
Rep. of Korea	1	564	6	4953	2	1860
Argentina	1	335	1	600	1	560
Pakistan	1	125	–	–	–	–
Iran	–	–	4	4182	4	4800
Brazil	–	–	3	3116	–	–
Rep. of South Africa	–	–	2	1842	–	–
Mexico	–	–	2	1308	–	–
Hungary	–	–	2	816	2	816
Austria	–	–	1	692	–	–
Yugoslavia	–	–	1	632	–	–
Philippines	–	–	1	621	–	–
Cuba	–	–	1	408	1	408
Poland	–	–	–	–	2	816
Luxembourg*)	–	–	–	–	1	1247
Egypt*)	–	–	–	–	1	622
Rumania	–	–	–	–	1	408
Bangladesh*7	–	–	–	–	1	200
Total	253	135812	235	217283	132	127141

1) The countries are listed according to the number of operating reactors.

Sources: International Atomic Energy Agency (IAEA), Vienna 1979 and 1980.
*) Atomwirtschaft – Atomtechnik, Vol. 24, November 1979.

lists the nuclear power reactors of the world (13, 14). The total capacity is 135 812 MW, which is about 6% of the total electric generating capacity of the world.

According to the prognosis of the OECD/NEA, Paris, the LWRs will retain their dominant position in the foreseeable future (compare Table 4-1). Their success relative to other types of reactor is due primarily to their relatively simple construction and their relatively low construction costs, advantages which result from the favorable neutron absorption and thermodynamic properties of light water. The IAEA, Vienna, estimates that the nuclear power plant market in the developing countries up to 1990 will be divided as follows: 140 plants with 150 to 400 MW capacity (total of 38 000 MW), 86 plants with capacities of 500 to 600 MW (total of 50 000 MW) and 129 plants with more than 600 MW capacity (total of 133 000 MW). The needs of a few African countries, which will have an estimated demand for 7100 MW nuclear power by 1990, are further broken down as follows: Algeria, 450 MW; Egypt, 5000 MW; Ghana, 300 MW; Morocco, 400 MW; Nigeria, 500 MW; and Tunesia, Zambia and Uganda, 150 MW each. That is, with the execption of Egypt, most of the demand will be for small and medium-sized reactors. It is likely that most of these will be LWRs (15–17).

Light water reactors (light-water-cooled and moderated reactors) can be classified as pressurized water reactors (PWRs) or boiling water reactors (BWRs). The PWR has become the more frequent of the two types. LWRs require enriched uranium with about 3% ^{235}U as fuel. The construction of a PWR is shown in Fig. 4-3 (18). The heat generated by fission in the reactor core is transferred by means of a sealed primary coolant circulation system to the steam generators, where feed water is evaporated (secondary circulation). The separation of the primary and secondary coolant circulation systems insures that no radioactive materials leak out of the reactor into the feedwater steam circulation, which is an advantage over the BWR (see below). The steam generated in the steam generator drives the turbogenerator. The circulation of water in the primary coolant system is driven by the main coolant pump. A pressure regulator in the primary circulation system maintains sufficient pressure (about 155 bar) to keep the water from boiling, in spite of the relatively high temperature (about 590 K) at which it leaves the reactor. Saturated steam is generated in the steam generator (about 54 bar and 543 K). (Depending on the size of a reactor, there are two to eight primary circulation loops with independent pumps and heat exchangers.)

The reactor core consists of a large number of fuel elements, which themselves are composed of individual fuel rods. The entire arrangement is surrounded by a core container, and this in turn by the reactor pressure vessel. As an example, the core of the Biblis A reactor, Federal Republic of Germany, (3517 MW thermal power, 1150 MW net electric power output, net overall efficiency of 32.7%) has a total of 193 fuel elements, each composed of 236 fuel rods (9, 16). Some of the fuel elements include control rods, which are rods of material with an especially large cross section

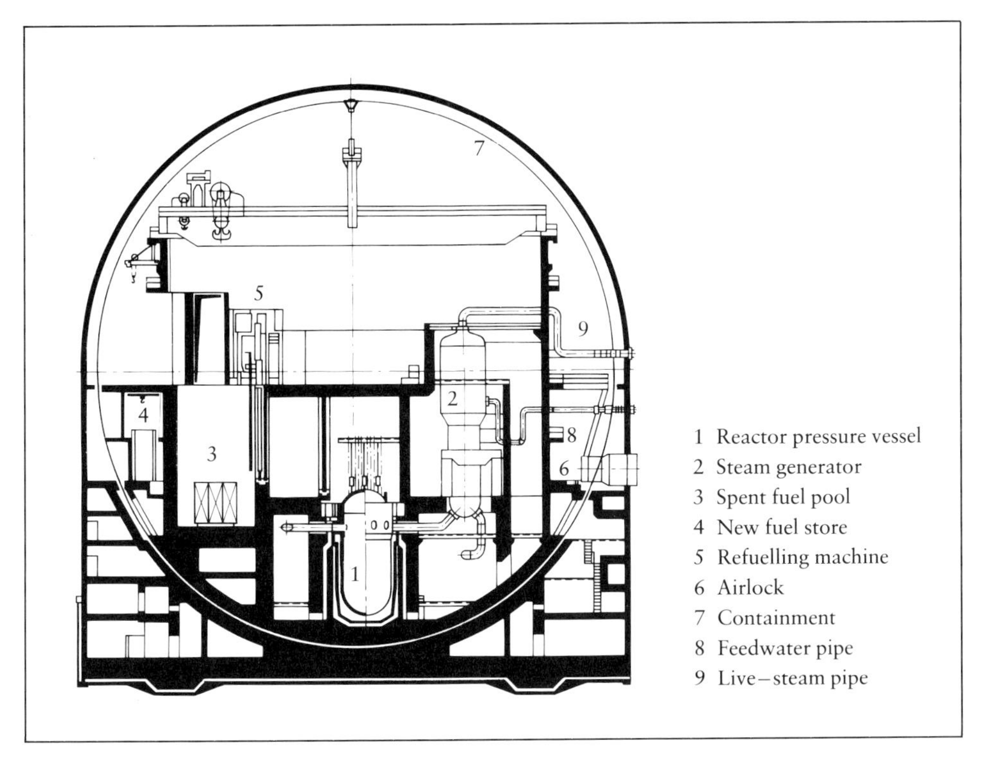

Fig. 4-3: Nuclear power plant with pressurized water reactor (reactor building)

Sources: H. H. Fewer, W. Mattick, Economic and Safety Advantages of Standardization, 10th World Energy Conference, Istanbul, September 1977.

for neutron absorption. Further data on this reactor are: uranium in the primary core, 102.7 t; inner diameter of the reactor pressure container, 5000 mm; total weight of the reactor pressure container, 530 t; 4 main coolant circulation systems; 72 000 t/h coolant flow; coolant pressure at the reactor exit, 155 bar; median coolant temperature at the reactor exit, 590 K.

The fuel rods contain sintered wafers of enriched UO_2, with about 3% ^{235}U content. (Natural uranium, which contains 0.720% ^{235}U, cannot be used in an H_2O-moderated reactor because of the neutron absorption of H_2O.) The cladding material for the fuel rods, chosen for its relatively low neutron absorption, is a zirconium alloy. The heat generated in the fuel rods is removed by water flowing from bottom to top through the fuel elements. The water serves both as coolant and moderator; it slows the fast neutrons emitted by fission down to thermal velocities. (Thermal neutrons are required for fission of ^{235}U.)

The specific heat production of modern PWRs is 30 to 40 kW per kg uranium, and the mean thermal power density is about 100 MW per m^3 core volume. Thus about

30 t enriched uranium and about 10 m^3 reactor core volume is needed per 1000 MW thermal output. The attainable mean burn-up is about 33000 MWd per ton uranium, and the mean use time is about 1000 full power days. (In practice, one third of the fuel is renewed once a year, and the fuel elements which remain in the core are moved about to achieve an even power density distribution.)

The power output of the reactor is regulated with neutron absorbers. Rapid changes in reactivity are controlled by the control rods as discussed previously; slower changes in reactivity are controlled with boric acid dissolved in water, the concentration of which can be varied within wide limits by means of auxiliary systems.

All of the pressurized components of the primary circulation system are located within a cylindrical or spherical reactor building, which serves as a containment vessel. The containment is arranged in such a way that even in a loss of coolant accident (LOCA), the safety devices would still work, and the environment of the reactor would not be endangered (19–21) (see 5.842).

The control scheme of modern PWR reactors makes relatively rapid changes in load possible, within certain limits. For example, the reactor Biblis A can change load at a rate of 130 MW/minute, in the range of 60 to 100% of full capacity. This corresponds to a rate of change of 10% per minute.

The second type of LWR is the boiling water reactor (BWR). It differs from the PWR primarily with respect to the method of removing heat from the reactor core (22). As mentioned above, the heat from a PWR is transferred to the steam generator by a sealed, primary coolant system. In the BWR, however, the steam generated in the reactor itself is used directly to drive the turbines (direct circulation). The steam is heated to about 560 K, and is under about 70 bar pressure, which means that the reactor operating pressure is considerably lower in a BWR than a PWR, at the same operating temperature.

Each type of reactor has certain advantages and disadvantages. On the one hand, the method of heat removal from the reactor in a BWR simplifies the circulation, as the heat exchanger, and the pressure regulator are superfluous, and the pumps are smaller than in the PWR. On the other hand, there is always the possibility with direct circulation that a damaged fuel element could leak radioactive materials into the coolant and thus contaminate the turbines. Therefore, the turbines and circulation must be shielded against radiation and placed within the containment vessel. (The entire condensate has to be decontaminated in a special installation.) There are also certain differences in the reactors themselves.

The BWR has a lower power density than the PWR, but the fuel rods for the BWR can have a larger diameter than in the PWR, which makes them less expensive to produce. In addition, there has to be room for steam baffles in the BWR, which makes its pressure vessel larger and more expensive than that of a PWR.

An LWR using 3.2% ^{235}U produces 285 kg total plutonium $^{238-242}Pu$, of which 194 kg is fissionable (^{239}Pu and ^{241}Pu) per GWa of electric energy, at a burn-up of

33 GW/t. The plutonium can be used as fuel either in a fast breeder (see 4.214) or in an LWR, after recovery from the spent fuel rods. The recycled plutonium can be used in existing LWRs (uranium-plutonium fuel cycle) at about the same rate as it is produced by them (see 5.822.1). (Commercial recycling of plutonium was begun in the nuclear reactor at Obrigheim, Federal Republic of Germany, in 1973.)

Although LWRs have been adopted throughout the world, they do not heat the steam to sufficiently high temperatures to obtain a good thermodynamic efficiency. For this reason, there have been experimental attempts to cycle the saturated steam from a BWR back through fuel elements of the same core, and thus to superheat it to about 820 K. The reactor is still a thermal one, due to the presence of water, so it is called a thermal superheat reactor or hot steam reactor. (Since water vapor is superheated to about 820 K, this is a special form of gas cooling) (see 4.213).

The superheat reactor has undoubted advantages, due to its higher thermodynamic efficiency. However, since zircaloy cannot be used to clad fuel elements exposed to this temperature, they become complicated and expensive. For this reason, the superheat reactor has not yet been adopted.

The light water reactors (LWRs), the BWRs and PWRs, form the basis for the world-wide commercial utilization of nuclear energy, both now and in the near future. They have become competitive for the production of electricity, the PWR becoming the dominant type. Light water reactors have higher power density, simpler design and lower installation costs than other types of reactor (see 4.212.1 and 4.213), but their relatively low operating temperature and the resultant low efficiency of 33% (compared to about 40% in modern oil and coal plants) conversion of heat to electricity are disadvantages.

4.213 Gas-cooled and graphite-moderated reactors

4.213.1 Gas graphite reactors

Like D_2O, graphite brakes neutrons efficiently, but has a low absorption cross section for them. In principle, therefore, graphite-moderated reactors can be run on natural uranium. However, they require a larger volume of moderator than D_2O reactors, and therefore have a larger core volume than D_2O reactors of the same capacity, and a much larger core volume than H_2O-moderated reactors (8, 9).

The oldest and largest group of gas-cooled, graphite-moderated reactors (GGR) are cooled by CO_2. These reactors use metalic natural uranium as fuel, and the fuel elements are sheathed with a magnesium-aluminum alloy called *magnox*. The reactors are therefore called magnox reactors; an example of the type is the power plant which went into operation in 1956 in Calder Hall, Great Britain (26).

Due to the relatively low maximum allowable temperature for the magnox cladding (about 720 K), and to the large moderator volume, the average power density in the core of these reactors is only about 0.87 MW/m^3. This value is about 100-fold smaller than that for a PWR, and the construction costs are correspondingly relatively high. The net efficiency of the power plant, in spite of the higher exit temperature of the coolant (about 680K), is lower than that of a PWR, because of the higher consumption of energy to run the plant (for example for pumping the gas). Since the burn-up is also about a factor of 100 less than in a PWR, the relatively low cost of the fuel elements cannot compensate for the disadvantages.

Between 1956 and 1970, about 30 magnox reactors with capacities up to 590 MW were built; a total of 5000 MW has been installed. Most of these plants are in Great Britain and France. (The cladding material for the fuel elements in the French reactors is a magnesium-zirconium alloy.)

The magnox reactors were not competitive with the LWRs developed in the USA, and are no longer being built, but they can be regarded as the precursors of the advanced gas reactor (AGR) and the high temperature gas reactor (HTGR).

4.213.2 Advanced gas reactors

Advanced gas reactors (AGRs) are a British improvement on the magnox reactor. The CO_2 coolant and the graphite moderator were retained. To increase the power density, the exit temperature of the coolant, and the burn-up, the fuel elements are bundled rods of enriched UO_2 in chromium-nickel steel. This cladding allows a higher core temperature, but their absorption of thermal neutrons is greater than that of magnox, so the AGRs require enriched uranium with about 1.5 to 2% ^{235}U content (27, 28).

As mentioned above, CO_2-cooled, graphite-moderated reactors have a relatively large core volume. The construction of steel reactor pressure vessels for the AGRs caused difficulties, so here, for the first time, the pressure vessel was built of reinforced concrete, and has the dual task of containing the coolant and shielding against radioactivity.

An example of this type of plant is the English Dungeness B power plant, which went into operation in 1974, and has two AGRs, each with 1595 MW thermal capacity. The total electric power output is 1200 MW. The volume of the reactor core is 390 m^3. The fuel rods are filled with UO_2 pellets containing 1.5% ^{235}U. The median pressure of the coolant gas is about 20 bar, and the median power density in the core is 9.6 MW per m^3 of core volume. The coolant exit temperature is about 920 K, which makes possible a plant efficiency of 41.5% (29).

So far, AGRs have been built or ordered only in Great Britain. When all the plants presently on order have been completed, Great Britain will have 5 double plants with 10 AGRs (a total of 6234 MW electric capacity): Windscale, 34 MW; Dungeness B, twice 600 MW; Hinkley Point B, twice 625 MW; Hunterston B, twice 625 MW; Hartlepool, twice 625 MW; Heysham, twice 625 MW (11).

The AGRs have some disadvantages compared to the LWRs, including the fact that the burn-up is much lower, and the power density is a factor of 30 smaller. Therefore the AGRs, in spite of their high efficiency and their superiority to the magnox reactors, are not economically competitive with the LWRs.

4.213.3 High temperature gas reactors

The term high temperature gas reactor (HTGR, or HTR for short) is generally applied to reactors with coolant exit temperatures above 1000 K. Metal cladding and CO_2 cooling cannot be used at such temperatures, as the reduction of the gas by the graphite moderator becomes significant around 920 K. Therefore helium is used as coolant in HTRs. Some of the advantages of helium cooling are the single-phase operation (there is no temperature-dependent phase change in the coolant), the low neutron absorption of the He (it remains practically non-radioactive), and the chemical inertness of the gas.

Analyses of the energy market, particularly in industrial countries, show that the production of heat is very important (see 2.331). The HTGR is especially suited to this task, because it can deliver high temperature process heat, which also means that a relatively high conversion efficiency to electricity – about 40% – is possible. Another important advantage of the HTR is that this type of reactor can use the large reserves of thorium to generate energy, and thus can expand considerably the primary energy base (see 3.35).

Natural thorium, unlike natural uranium, does not contain an isotope which can be fissioned by thermal neutrons. However, upon absorption of thermal neutrons, ^{232}Th is converted to ^{233}U which, like ^{235}U, can be split by thermal neutrons. The formation of fissionable nuclei (for example, ^{233}U) from a non-fissionable (by thermal neutrons) breeding material is called "breeding" or "conversion". The conversion proceeds as follows:

$$^{232}_{90}\mathrm{Th}\ (n, \gamma)\ ^{233}_{90}\mathrm{Th} \xrightarrow[T\,=\,22.4\ \mathrm{Min.}]{} {}^{233}_{91}\mathrm{Pa} + e^- + \bar{\nu}_e \xrightarrow[T\,=\,27.4\mathrm{d}]{} {}^{233}_{92}\mathrm{U} + e^- + \bar{\nu}_e \qquad (3)$$

The intermediate nucleus formed by neutron capture, ^{233}Th, is converted by spontaneous emission of two electrons e- and two neutrions $\bar{\nu}_e$ to ^{233}U. (The times noted under the arrows are the half-lives of the corresponding β decays.) Because of its relatively short half-life of $1.62 \cdot 10^5$ years, ^{233}U is not found in the earth's crust, but for practical purposes it is stable, and because of its high η value, it is a valuable fission material (see below).

Because each conversion process begins with the capture of a neutron, the yield of new nuclear fuel is directly dependent on the neutron balance. Of the average η neutrons released by fission of a nucleus, one is needed to maintain the chain reaction. (η is called the yield of fission neutrons.) The neutrons absorbed by structural materials and coolant, or which escape from the system without reacting, account for the loss V. (For example, V is about 0.55 in a PWR.) The fraction remaining for the conversion of breeder material is thus

$$A = \eta - 1 - V \tag{4}$$

For a unit of volume and time,

$$A = \frac{\text{Number of newly generated nuclei}}{\text{Number of nuclei consumed}} \tag{5}$$

Nuclei can be consumed either through fission or by simple neutron capture. If $A > 1$, more than one fissionable nucleus is produced per consumed nucleus, and the reactor is breeding. For this to occur, η must be > 2. Typical values of η are $\eta = 2.23$ (^{233}U, thermal neutrons) or $\eta = 2.93$ (^{239}Pu, fast neutrons). $A = B$ is called the breeding rate (or breeding ratio). If $1 > A = C > 0$, less than one fissionable nucleus is produced per fissionable nucleus consumed. $A = C$ is called the conversion rate (converter reactor or converter).

The values of η for various nuclear fuels are such that breeding is possible with ^{233}U, ^{235}U and thermal neutrons (thermal breeder) or with ^{239}Pu and fast neutrons (fast breeder), i.e. with an arrangement of ^{235}U as fuel and ^{232}Th as fertile material (thermal breeder) or with ^{239}Pu as fuel and ^{238}U as fertile material (fast breeder) (see 4.214) (9). In principle, higher breeding rates are possible in fast breeders than in thermal breeders; it has been calculated that $A = 1.25$ to 1.40 for fast breeders, and $A = 0.9$ to 1.1 for thermal breeders (30, 31).

The conversion rate in light water reactors is about 0.5 to 0.6, which means that an LWR started with $^{235}U/^{238}U$ generates and burns an increasing amount of fissionable plutonium (^{239}Pu and ^{241}Pu). In a thorium high temperature gas reactor (THTGR or THTR), the conversion rate is 0.6 to 0.8, which means that a THTR started with ^{235}U and ^{232}Th generates and burns ^{233}U, and has a more favorable conversion rate than an LWR burning $^{235}U/^{238}U$. (To a large degree, ^{233}U can be used instead of the naturally occuring ^{235}U, and thus reduce the amount of natural uranium needed for the HTR.) Special "high converters", which are designed to achieve high conversion rates, can approach a rate of one. In fast breeder reactors (FBR), the typical values of breeding rates are 1.14 to 1.28. For example, the breeding rate obtained in actual operation of the French prototype plant Phénix (electrical output 250 MW) was 1.16.

Like the LWR, the THTR uses thermal (slow) neutrons to maintain the chain reaction. The fast neutrons released by the fission of uranium nuclei are slowed down by the moderator. The heat generated in the core is removed with a helium coolant. Both the inner core construction and the fuel rods are made of ceramic materials, so operating temperatures of 1300 K can be realized. (The thermal stability of the graphite is sufficient to guarantee an intact core structure, in the event of malfunction, up to a temperature of about 3700 K (32)).

The HTR is flexible in its fuel requirements. It can be run on any of the following: 1. a pure uranium-plutonium cycle, i.e. with slightly enriched uranium (< 10% ^{235}U content); 2. a uranium-thorium cycle with medium uranium enrichment (about 20% ^{235}U content); 3. a uranium-thorium cycle with highly enriched uranium (about 93% ^{235}U content).

With the first type of fuel, which in contrast to the second and third uses only enriched uranium (8.53% ^{235}U content), the conversion rate is 0.58 and the mean burn-up is 101.5 MWd/kg. However, since the THTR can make use of the world's large thorium reserves when operated with the second and third cycles (uranium-thorium cycles), there is particular interest in these cycles. The following values are typical for the second fuel variant (medium uranium enrichment): 19.8% ^{235}U content, a conversion rate of 0.58, and mean burn-up of 99.8 MWd/kg. For the third variant, typical values are 93% ^{235}U content, a conversion rate of 0.74 and mean burn-up, 69 MWd/kg.

In light of the discussion of non-proliferative cycles, the first two variants are interesting, especially the second (see 5.83). Because of the relatively small amount of ^{238}U in the fuel elements, the amount of plutonium produced in a THTR fueled with the second variant is only 23% of the amount produced with the first; and only 10% of the amount produced by a LWR (32). Due to the low ^{238}U content of the third fuel variant, the amount of plutonium produced in this cycle is "negligible"; the main problem with this fuel mixture is that highly enriched uranium 235 required.

The experimental HTRs Dragon (Great Britain), Peach Bottom (USA), and AVR-Jülich (Federal Rep. of Germany) have provided data on which further HTR development can be based. In the reactors at Dragon and Peach Bottom (both have since been taken out of operation), coolant gas temperatures of 1120 K were maintained over long periods. The AVR reactor (16 MW electric power) has been in operation since 1967. In 1974, its coolant temperature was increased from 1120 to 1220 K, and the plant runs smoothly (33).

Construction on the first HTR prototype at Fort St. Vrain, Colorado (USA) was begun in 1968. This plant has an electric capacity of 330 MW. Construction was begun in 1971 on a 300 MW THTR (thorium high temperature reactor) in Uentrop (Federal Republic of Germany). The plans for this plant reflect the experience gained with the AVR plant. It is expected to be completed in the early 1980's (34).

Fig. 4-4 shows the construction of a THTR (35). The fuel for this reactor consists of particles of 0.8 mm diameter which are mixed crystals of UO_2 and ThO_2. (A mixed carbide of UC_2 and ThC_2 is also possible). To contain the fission products, these particles are coated with up to three layers of pyrocarbon. They are then embedded in a graphite matrix and formed at high pressure into fuel grains of 50 mm diameter, and finally these are coated with a fuel-free graphite sheath several millimeters thick. The final product is a fuel particle (fuel element) with a diameter of

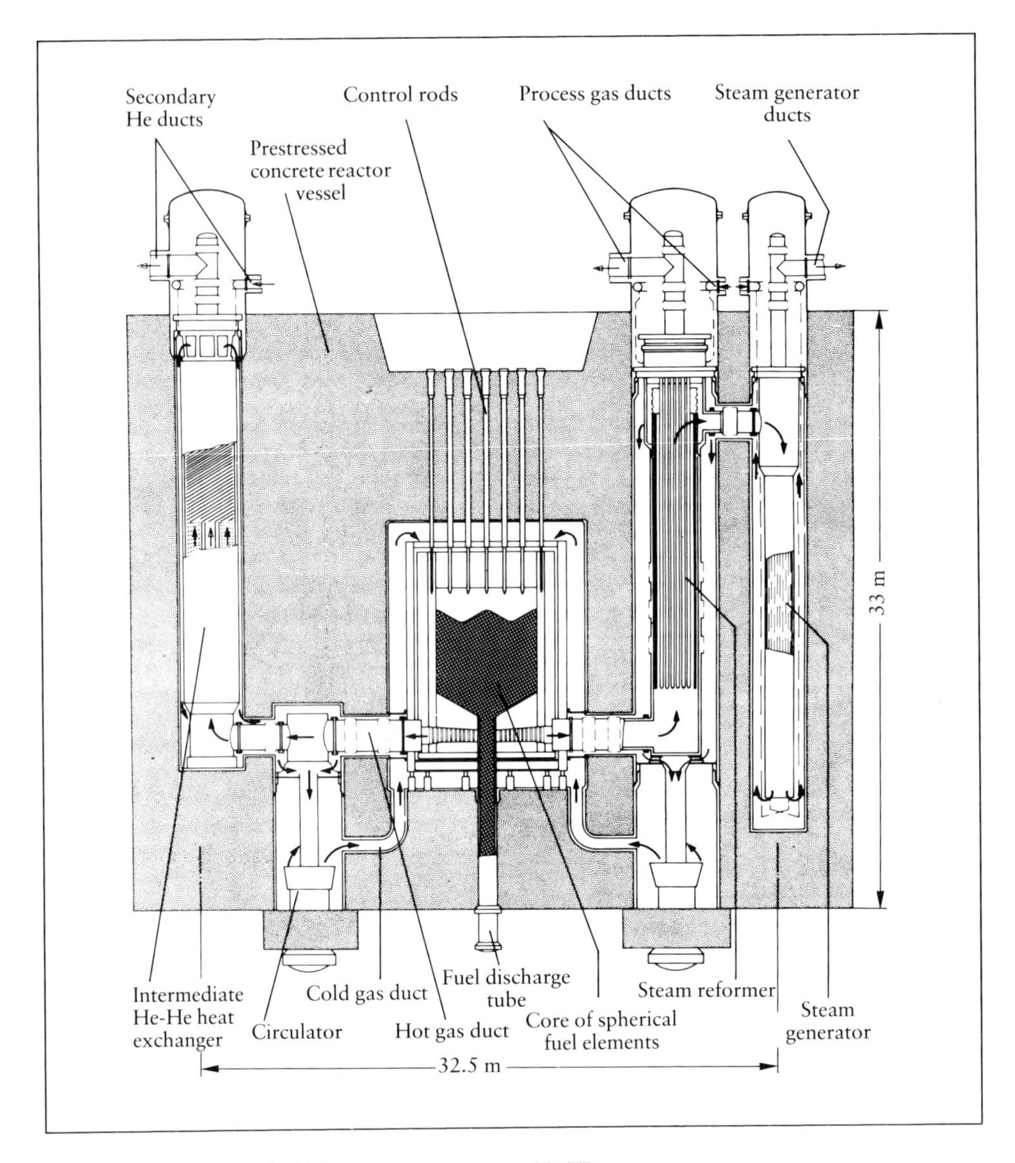

Fig. 4-4: Components of a high temperature reactor (THTR)

Source: GHT-Gesellschaft für Hochtemperaturreaktor-Technik, Köln 1980.

about 60 mm. The highly enriched uranium originally consists of 93% ^{235}U as fissionable material. ^{232}Th is converted to ^{233}U by capture of thermal neutrons, and as the burn-up proceeds, the ^{233}U is responsible for an increasing fraction of the fission. Each fuel element contains 190 g carbon, 0.96 g ^{235}U (in 1.03 g U) and 9.62 g ^{232}Th

(9). The core consists of a pile of about 675 000 fuel particles which are contained in a graphite vessel about 5.6 m in diameter and about 6 m high. The average power density in the core is about 6 MW per m^3 of core volume, which is about 6% of the power density in a pressurized water reactor and about 2% of that in a liquid-metal-cooled fast breeder reactor (LMFBR). This means, on the one hand, that cooling a THTR is relatively simple, but, on the other hand, that a relatively large volume must be enclosed by the reactor pressure vessel.

In contrast to LWRs or LMFBRs, the THTR has a gas (helium) instead of a liquid as coolant. (The operating pressure is about 40 bar). The waste heat from a THTR is exhausted over a dry cooling tower, which allows more flexibility in the choice of a site for the reactor, compared to an LWR, and puts less stress on the environment. Furthermore, the waste heat is vented at a fairly high temperature, and if a slightly lower efficiency in generating electricity is accepted, it can be used for a district heating network.

HTR plants are typified by the presence of a stressed concrete container. Because the core volume is relatively large, it was not practical to build the reactor pressure vessel of steel, as in an LWR. Therefore stressed concrete was chosen for the pressure vessels of the two prototype reactors Fort St. Vrain and THTR-Uentrop.

Both these constructions (Fort St. Vrain and THTR-Uentrop) contain uranium as the fissionable material and thorium as the fertile material, graphite as moderator and reflector, and helium as coolant. The two differ essentially only in the type of fuel element used. The THTR at Uentrop contains the fuel, as described earlier, in the form of graphite spheres, i.e. the fuel elements are spherical. The Fort St. Vrain reactor, however, has its fuel in the form of prismatic graphite blocks. The fuel elements have a hexagonal cross section and vertical conduits for the flow of coolant gas. The fuel, which consists of *coated particles* containing a mixed carbide of UC_2/ThC_2, occupies separate shafts. The core of the reactor contains 247 fuel element columns, each of which contains 108 coolant channels.

The use of spherical fuel elements (THTR system) makes it possible to load and unload the reactor continuously during operation, so that it does not need to be shut down to exchange fuel elements. There are two available concepts for loading the reactor, the MEDUL and the OTTO (36). In the MEDUL (Mehrfach Durchlauf, "several passes") system, the fuel elements are withdrawn continuously during operation of the reactor, and examined to determine the degree of burn-up (and to check for damage). Undamaged spheres which have not yet reached maximum burn-up (about 100 MWd/kg) are pneumatically returned to various positions on the surface of the core. (As a rule, the fuel elements attain maximum burn-up after three to six passes through the core). With this system it is possible to regulate the power output distribution within the core. In the OTTO system (Once Through Then Out) the fuel elements are also removed continuously, without measurement of the degree of burn-up, and replaced by fresh ones. The rate at which the spheres are removed is adjusted so that they have essentially reached maximum burn-up when they leave the core. The spheres have been shown by years of tests at the AVR reactor in Jülich to be mechanically very stable, so that an OTTO loading system is planned for the THTR-Uentrop.

Three types of HTR plant are presently being tested, the double circulation model, the single-circulation model, and the process heat model (37). The double-circulation HTR has been realized in the two prototype plants, Fort St. Vrain and THTR-Uentrop, in which helium heated in the primary circulation passes through a heat exchanger where its heat is transferred to a secondary steam circulation, which corresponds to a conventional steam plant. The primary and secondary circulations are thus separated, as for example is also the case in a pressurized water reactor (PWR) (see 4.212.2).

There is a high potential for further development of the HTR, because there are plans to eliminate the heat exchanger and to direct the hot helium into a helium turbine (singlecirculation HTR). Only when this step has been technically achieved will the HTR attain its optimal economy. As mentioned above, operating temperatures of about 1300 K can be achieved, because both the inner core construction and the fuel are ceramic materials. The temperatures of steam power plants are limited by metallurgical and economic considerations to about 815 K, so that a gas exit temperature above 970 K cannot affect the steam circulation and thus improve thermodynamic efficiency. By contrast, in a single-circulation HTR with a gas turbine, the working temperature and thus the overall efficiency in generating electricity can be increased, because the limiting temperature for the metals involved is higher. An efficiency of somewhat over 40% could be achieved, so that the specific advantages of HTRs are especially pronounced for the single-circulation plants. These advantages are the possibilities for dry cooling (see 5.2) and the use of the waste heat (temperatures up to 440 K) for district heating (see 4.62). In addition, the process is simpler and the construction of the plant is correspondingly more compact. For example, in a double-circulation plant only the boiler is built inside the stressed concrete vessel, but in a single-circulation HTR, the gas turbine is also contained within it.

The high temperature of the helium coolant makes the HTR suitable for production of high-temperature process heat in addition to electricity. The increase in gas exit temperature to 1220 K which was successfully demonstrated at the AVR-Jülich makes possible a number of applications, including the use of the heat for the following processes (preparation of secondary energy sources): petroleum refining (e.g. thermal cracking of heavy oils, see 4.63), refining of coal (e.g. coal gasification, see 4.642), production of methanol (see 4.643), nuclear long distance energy (see 4.65), and nuclear splitting of water (see 4.661). (There are many processes for which heat is required at temperatures of 1000 K to 1200 K.) The possibility of using waste heat for district heating of buildings has been mentioned (see 4.62). In particular, the high-temperature heat could make it possible to replace petroleum products and natural gas by secondary energy forms made from coal. It could also be used for the direct reduction of iron ore. In addition, there are many industrial processes in which heat is used in the form of hot steam, for instance in the chemical and paper industries, so that steam from a power plant would be an ideal carrier for this heat.

Less experience has been accumulated with the HTR plants with respect to problems of safety and recycling or disposal of fuel (see 5.822.2) than with the commercially operating LWRs (38). It is clear that the high efficiency in electric generating and the possibilities for utilization of waste heat will lead to a relatively low heat load on the environment. The safety advantages of the HTR are due to the low power density and the large heat capacity of the graphite used in the core, the high temperature resistance of the fuel elements cladding, the chemical inertness of the coolant, and the resistance of the stressed concrete to pressure. As a result of these features, HTR plants can be built which are inherently safe. In addition, in light of the concern over fuel cycles which do not encourage proliferation of nuclear weapons, the HTR could contribute to international security (see 5.83).

On the basis of its past and predictable development, the HTR has a good chance of being adopted as the second generation commercial reactor, especially as it will make nuclear energy available not only for electric generation, but also for the heat market, which accounts for the largest fraction of the final energy demand.

4.214 Liquid-metal-cooled fast breeder reactor

The reason for developing fast breeder reactors (FBRs) is to utilize the primary uranium fuel much more efficiently than is possible in a thermal reactor. The goal is to develop reactors which produce both commercially utilizable energy and more fissionable material than is consumed in the reactor, i.e. to breed fissionable material. (Like HTRs, FBRs are considered second generation reactors.)

^{239}Pu is generated as follows from ^{238}U by absorption of fast neutrons from nuclear fission:

$$ {}^{238}_{92}\mathrm{U}\ (\mathrm{n}, \gamma)\ {}^{239}_{92}\mathrm{U} \xrightarrow[T\,=\,23.5\ \mathrm{Min.}]{} {}^{239}_{93}\mathrm{Np} + e^- + \bar{\nu}_e \xrightarrow[T\,=\,2.34\mathrm{d}]{} {}^{239}_{94}\mathrm{Pu} + e^- + \bar{\nu}_e \qquad (6) $$

The intermediate ^{239}U nucleus which is formed by neutron capture decays via two electron emissions e^-, each associated with the production of an antineutrino $\bar{\nu}_e$, to ^{239}Pu. (The times shown under the arrows are the half-lives of the corresponding β decays) (32).

As mentioned in 4.213.3 the formation of fissionable nuclei such as ^{239}Pu from nuclei such as ^{238}U which do not undergo fission on neutron capture is called *breeding* or *conversion*. The process goes on in every reactor in which uranium is used as fuel, including LWRs. However, at the same time, the fissionable ^{235}U nuclei are consumed. As mentioned above, the conversion rate is $1 > A = C > 0$. In LWRs, the conversion rate is between 0.5 and 0.6, which means that more fissionable nuclei are consumed than are produced.

By using ^{239}Pu instead of ^{235}U as the fissionable material, and fast instead of thermal (slow) neutrons, the conversion of ^{238}U as fertile material can be considerably improved. This is accomplished in a fast breeder reactor (FBR). In this reactor, $A>1$, which means that more than one fissionable nucleus is produced for each nucleus that is consumed. In order to attain a breeding rate $A > 1$, the yield of fission neutrons η must be greater than 2. With neutron energies $E_n > 50$ keV (fast neutrons), η is large enough to breed more fissionable material than is consumed in the course of the chain reaction. On the average, one neutron is used to maintain the chain reaction, a second one produces a new ^{239}Pu nucleus to replace the one which was split, and the remainder $(\eta-2)$ can cover the neutron losses and breed additional ^{239}Pu nuclei. The typical values of η for ^{239}Pu are in the range of 2.5 (for $E_n = 50$ keV) to 2.9 (for $E_n = 10^3$ keV). For FBRs, values of $A = 1.25$ to 1.40 have been calculated (30, 31). The typically attainable values lie between 1.14 and 1.28. In the French prototype power plant Phénix (250 MW electrical output), for example, a breeding rate of 1.16 has been realized in operation.

Thus the operation of FBRs generates from ^{238}U more fissionable material in the form of ^{239}Pu than is consumed by fission in the reactor. This means that breeder reactors can generate fuel for themselves and for other nuclear reactors, which is the major advantage of fast breeders over thermal reactors (39).

Another important characteristic of a breeder, in addition to the breeding rate, is the doubling time. This is the time required to double the amount of fissionable material. (The excess of fissionable material is the difference between production and consumption of fissionable nuclei.) It can be shown that the doubling time is inversely proportional to $(A-1)$, and to the specific power output (thermal output per gram fissionable material). For example, if $A = 1.3$ and the specific power output is 550 MW/t, the doubling time is about 17 years (9).

In a reactor which uses uranium as fuel, the conversion process produces a mixture of isotopes, rather than pure ^{239}Pu. For example, the mixture from an LWR has the following composition: 55–60% ^{239}Pu, 20–25% ^{240}Pu, 10–15% ^{241}Pu and 5–10% ^{242}Pu. (At present, 15 plutonium isotopes with mass numbers of 232 to 246 are known.) Some of the ^{239}Pu is converted by neutron capture to ^{240}Pu which, like ^{238}U, is not fissionable by thermal neutrons. However, it is a breeding material, which is converted by capture of a further neutron into ^{241}Pu, which is as easily fissionable as ^{239}Pu. A part of the ^{241}Pu captures another neutron, however, and is converted to the non-fissionable ^{242}Pu, which has no practical significance (quantitatively) as a breeding material. This mixture of plutonium can be used as fuel either in a fast breeder or in a thermal reactor.

Because the ^{238}U is also used to produce energy in an FBR, this technology makes possible a 60 to 70-fold better utilization of the energy content of uranium than is possible with a thermal reactor. For example, LWRs "burn" ^{235}U, which makes up

0.720% of natural uranium, by fission with thermal neutrons. At the same time, fissionable ^{239}Pu and a small amount of ^{241}Pu are produced by conversion processes, and they contribute proportionately to the energy production in a LWR. This means that in an LWR, about 1% of the energy content of natural uranium is used. About a third of the uranium requirement can be saved if the fuel is recycled and the plutonium is put back into the reactor. Essentially the same is true for the other thermal reactors, water-cooled and water-moderated, or gas-cooled, graphite-moderated. This means that the recycling of fuel for thermal reactors can improve the utilization of uranium. There is no essential difference whether ^{238}U or ^{232}Th is used as conversion material (40).

Since FBRs use fast neutrons, they do not contain moderators, in contrast to thermal reactors. Their main elements are thus the fuel and breeder elements, the regulatory mechanism, the coolant and the radiation shielding.

The fuel for FBRs is generally a mixed oxide UO_2/PuO_2 packed in stainless steel sheaths. The fissionable material is ^{239}Pu and the breeding material is ^{238}U. The first FBRs, most of which were built in the USA, had metallic fuel elements. However, for metallurgical reasons, these permitted only a small degree of burn-up, so UO_2/PuO_2 mixed oxides are now used instead. Recently monocarbides (UC/PuC) have been tested as fuel, because the physics of neutron absorption are more favorable in them. Two O atoms moderate more strongly than one C atom, so that the breeding properties can be improved by going to the monocarbides.

On the international scene, the only type of breeder reactor which is at an advanced stage of development is the sodium-cooled fast breeder reactor (SFBR) (41–43). Table 4-3 gives a summary of the development and construction of FBRs in the world (44–47).

The lines of development followed in various countries are similar in their essential characteristics. This is especially true for the fuel and breeding materials and the coolants. As an example, the core of the sodium-cooled fast breeder reactor SNR-300[1)] consists of a central fission zone (205 fuel elements, each containing 166 fuel rods with UO_2/PuO_2 fuel, 6 mm in diameter and about 1 m long) and a concentric outer breeding zone (96 fuel elements, each containing 61 individual rods with UO_2 breeding material 11.6 mm in diameter). The individual fuel rods are organized in fuel elements with hexagonal cross sections. The inner fuel elements have a lower Pu content (in the inner fission zone, about 14% ^{239}Pu) than the outer fuel elements (outer fission zone, about 18% ^{239}Pu). Steel reflector elements are arranged concentrically around the outside to protect the wall of the tank from radiation. The central fission zone is also flanked in the axial direction, above and below, by a breeding

[1)] The SNR-300 is being built by the INB at Kalkar, on the lower Rhine, in the Federal Republic of Germany. INB (Internationale Natrium-Brutreaktor-Bau GmbH) is a joint subsidiary of the companies INTERATOM (Federal Republic of Germany) BELGONUCLEAIRE (Belgium), and NERATOOM (Netherlands).

Table 4-3: Development and construction of fast breeder reactors in the world

	Name	Country	Thermal power in MW	Electric power in MW	Coolant	Coolant Temp. at core outlet in K	Date of Operation	Fuel
First Generation Fast Breeder Reactors	CLEMENTINE	USA	0.25	0	Hg	393	1949[1]	Pu metal
	EBR-I	USA	1.2	0.2	NaK	593	1951[1]	U metal
	EBR-II	USA	62.5	20	Na	743	1965	U metal
	EFFBR	USA	200	66	Na	703	1966	U metal
	BR-1	USSR	0	0	n. a.	n. a.	1955[1]	Pu metal
	BR-2	USSR	0.1	0	Hg	333	1956[1]	Pu metal
	BR-5	USSR	5	0	Na	723	1959	PuO_2
	DFR	UK	72	15	NaK	623	1963	U metal
Second Generation Experimental Fast Breeder Reactors	SEFOR	USA	20	0	Na	703	1969	UO_2/PuO_2
	FFTF	USA	400	0	Na	753	1978	UO_2/PuO_2
	BOR-60	USSR	60	12	Na	873	1969	UO_2/PuO_2
	RAPSODIE	France	40	0	Na	803	1970	UO_2/PuO_2
	KNK-II	FRG	58	20	Na	823	1977	UO_2/PuO_2
	PEC	Italy	130	0	Na	798	n. a.	UO_2
	JOYO	Japan	100	0	Na	773	1977	UO_2/PuO_2
Fast Breeder Prototype and Demonstration Reactors	PFR	UK	600	250	Na	835	1975	UO_2/PuO_2
	CFR-1	UK	2900	1250	Na	835	1978[2]	UO_2/PuO_2
	PHENIX	France	563	250	Na	833	1973	UO_2/PuO_2
	SUPER-PHENIX	France	2910	1200	Na	808	1976[2]	UO_2/PuO_2
	SNR-300	FRG	736	312	Na	819	1973[2]	UO_2/PuO_2
	BN-350	USSR	1000	350	Na	773	1973	UO_2
	BN-600	USSR	1480	600	Na	823	1980[3]	UO_2/PuO_2
	CRBR	USA	950	360	Na	813	1977[2]	UO_2/PuO_2
	MONJU	Japan	714	300	Na	813	1978[2]	UO_2/PuO_2

[1] Shut-down. [2] Start of Construction. [3] Operation began on April 22, 1980.

Source: W. Häfele, J. P. Holdren, G. Kessler, G. L. Kulcinski, Fusion and Fast Breeder Reactors, International Institute for Applied Systems Analysis, Laxenburg/Vienna, Austria RR-77-8, July 1977.

zone, so that the entire fuel zone has the shape of a squat cylinder. Above the upper breeding material zone in each fuel rod there is a hollow space in which the gases formed by fission can collect (44). Some of the fuel elements have control rods in the interior, i.e. a bundle of absorber rods filled with boron carbide. The lower ends of the fuel elements rest on a base plate.

Because the fission reactions in a FBR are initiated by fast neutrons, only substances which do not moderate, such as sodium (and possibly helium, see 4.215) can be considered for use as coolants. Water cannot be used as coolant for this reason. Because of the characteristics of nuclear fission with fast neutrons, there is a relatively high power density in the core. In a 1000 MW SFBR, the power density in the core is about 340 MW per m^3 of core volume, which is about 3.5 times as high as in a PWR. This means that the cooling must be especially intensive. Sodium, which melts at 371 K and has good heat transfer properties, is usually used as coolant. It flows through the core from bottom to top.

Because sodium reacts violently with water and becomes radioactive in the core, special safety measures are required with this type of reactor. The primary circulation is not directly coupled to the steam circulation; there is a secondary sodium coolant system in between. This is intended to prevent damage to the core in the event that a pipe breaks and the sodium comes in contact with the water. Fig. 4-5 shows schematically the construction of a sodium-cooled fast breeder reactor (prototype fast reactor, UKAEA) (48). This heat transfer system is rather elaborate and is one reason for the high construction costs (compared to LWRs) (49, 50). Only time will tell whether this disadvantage will be compensated by the lower fuel costs.

There are two basically different arrangements for the reactor coolant system, the pool and the loop. In the pool arrangement, all the components of the primary circulation (reactor jacket, intermediate heat exchanger, primary sodium pump) are enclosed in a large, sodium-filled container, so that loss of coolant, in the event of a broken pipe, cannot occur. (The French reactor Phénix and the British prototype fast reactor (see Fig. 4-5) are the pool type.) In the loop arrangement, the components of the primary circulation are physically separated, but connected by pipes. The arrangement is such that even in the case of a leak, the sodium level would remain high enough to cover the core. The SNR-300 and the Clinch River Breeder Reactor are of the loop type (51, 52). The pool type SFBR plants contain about 1300 m^3 of sodium per GW thermal power, while the loop type contain about 330 m^3 (32). In general, the reactor exit temperature in a SFBR reactor is 830 K, which makes an efficiency of 40% possible.

The development of FBRs goes back nearly as far as reactor technology. The first electricity generated in a nuclear plant, in 1951, was from an FBR (EBR-I) in the USA (see Table 4-3). The first FBR in the USSR went into operation in 1955. The FBRs of the 1950's used metallic fuel (mostly ^{235}U). In the 1960's, fuel in the form of oxides was adopted. In these two decades, the broad basic development was of primary concern, and most of the reactors were experimental (first generation fast

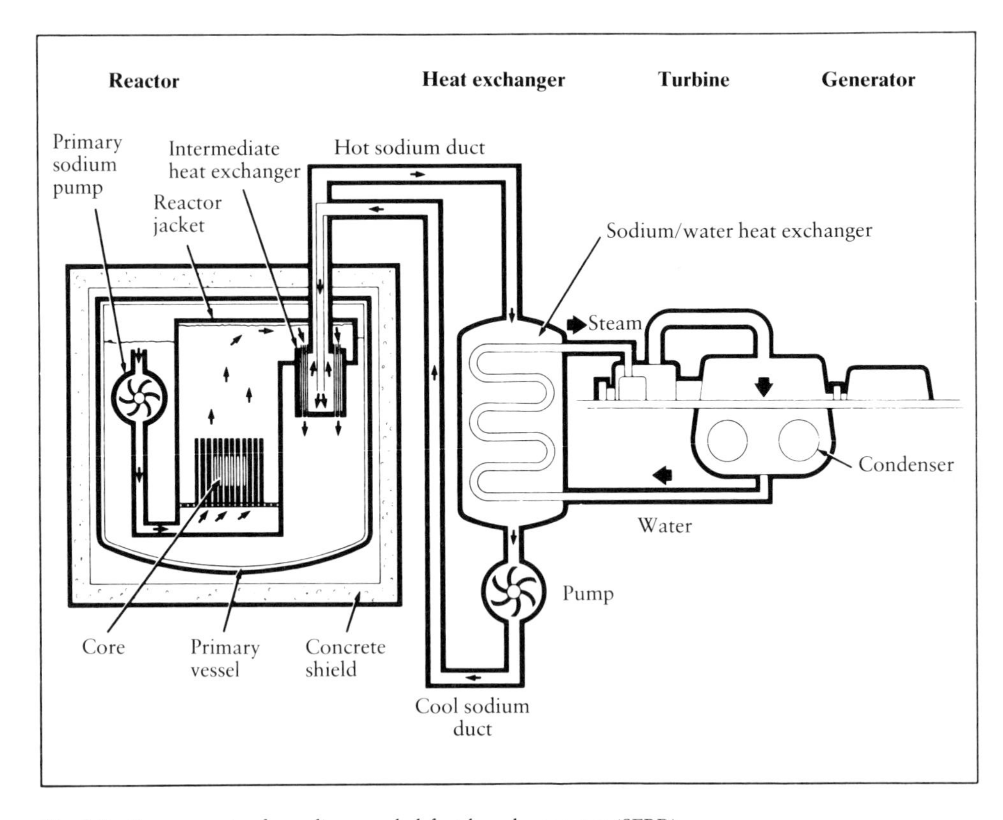

Fig. 4-5: Components of a sodium-cooled fast breeder reactor (SFBR)

Source: F. R. Farmer, How safe is the fast reactor?, Nature, Vol. 278, 12 April 1979.

breeder reactors). In the 1970's, the construction of larger power plants with fast breeders was begun in many countries (fast breeder prototype and demonstration reactors). Most of these plants have capacities of about 300 MW. Some examples are BN-350 (USSR), Phénix (France), and the Prototype Fast Reactor (United Kingdom), all in operation. A number of comparable plants are under construction: SNR-300 (Federal Republic of Germany, in cooperation with Belgium and The Netherlands), Monjou (Japan), and the Clinch River Breeder Reactor (USA).

This size of reactor, however, is not economically optimal. Further steps on the way to economic fast breeders are the BN-600 (USSR) with 600 MW electric capacity, which went into operation on April 22, 1980, and the 1200 MW, Super-Phénix (France), on which construction was begun in 1976. The following data are typical for the Super-Phénix: 3000 MW thermal power, 1200 MW electric power, 40% efficient conversion to electricity, 364 fuel elements, each containing 173 fuel rods with UO_2/PuO_2 fuel, 4.6 t PuO_2 and 26 t UO_2 in the core, 14.5% ^{239}Pu content in the inner fission zone, 18.5% ^{239}Pu in the outer fission zone, 68.3 t UO_2 breeding material, projected mean burn-up in the first core about 70000 MWd/t, breeding rate about 1.25, pool-type reactor, coolant exit temperature 833 K, fresh steam temperature 760 K, fresh steam pressure about 175 bar (53, 54).

FBRs, like thermal reactors, are controlled by the fraction of delayed neutrons (see 4.211). Not all the neutrons released by the fission of an atomic nucleus are *prompt*; some are released from the fission products after a delay. These *delayed* neutrons are important for the control of FBRs, just as in thermal reactors. The control system is designed so that the *prompt* neutrons released by fission do not suffice to make the reactor go critical; the *delayed* neutrons are needed as well.

With respect to safety and environmental aspects, the essential differences between SFBRs and other reactors are the fact that they are cooled with liquid sodium metal, and that they use plutonium as fuel. Some of the important safety features of the SFBRs are:

1. The low coolant pressure, which makes possible a nearly pressure-less heat removal, and the high boiling point (1156 K) of sodium. The likelihood of boiling is reduced by the difference of more than 400 K between the operating temperature and the boiling point of the sodium (32). This means that in the event of leaks and breaks, the core is unlikely to go dry (55).

2. For this reason, the emergency cooling system does not need to pump in replacement coolant, but can be limited in its function to removing excess heat from the coolant under conditions similar to normal operation.

3. In SFBRs of the size under consideration, 300 to 1300 MW electric capacity, there is a positive sodium temperature coefficient and a positive bubble coefficient of reactivity in the central core (55). This means that if the sodium boils, due to a serious accident, its reactivity will be increased by the movement of large bubbles of gaseous sodium through the core. This increase in reactivity is called the *void reactivity*.

4. Assuming a massive failure of safety systems, the core of a light water reactor could melt. Due to the facts outlined in 3., it is possible in an SFBR to have a reactivity excursion (power excursion or *Bethe-Tait excursion*) which would lead not only to melting of the core, but also to a release of mechanical energy. In such a "hypothetical failure", the core would be destroyed, i.e. it would be a core-disruptive accident (CDA) (55).

In order to prevent disruption of the core, SFBRs are equipped with diversified and redundant fast shut-down systems. Redundance means that there are more systems for each safety function than are needed, so that if one should fail, the second or third could perform the function. The systems are diversified in the sense that the safety systems operate on different principles, and as far as possible, are triggered by different physical parameters. For example, the reactor can be shut down either by absorber rods or by liquid boric acid. The redundant and diverse systems are also, as far as possible, physically separated. For example, the SNR-300 has two completely separate fast shutdown systems which work on different principles. The neutron absorbing elements of the first system are suspended above the core; in case of need they would fall into the core by gravity. If this system should fail, the second fast-shut-down system is activated. Its neutron absorber elements are flexible, so that even if their shafts are deformed, they can still be driven into the core.

In addition, the containment vessel is designed to absorb the mechanical energy which might be released and it can cool the core melt, unlike that of the LWRs (55). This means that the safety systems can prevent accidents, or at least, they can limit the effects of a system failure.

5. Since the sodium in the core becomes radioactive and reacts violently with air and water, additional technical precautions are required. As mentioned above, this is the reason for the secondary sodium circulation between the primary circulation and the steam circulation. In addition, in the SNR-300, the entire area about the radioactive primary circulation is enclosed in an inert gas atmosphere, so that combustion cannot occur (51). It should be mentioned here that much experience in sodium technology has been gained in non-nuclear applications.

The greatest potential danger from any nuclear reactor, including the SFBRs, comes from the radioactive fission products which are formed during operation of the reactor (56). Therefore the safety systems must be designed to prevent the escape of radioactive substances into the environment. The most important safety systems of the SFBR are a) the containment of the primary coolant circulation; b) the core and the reactor containment vessels; c) the systems which protect the reactor; d) the emergency cooling system and e) the containment vessel (see 5.842). All the other systems are similar in function and construction to the corresponding systems in LWRs (55).

The introduction of breeder reactors makes necessary the handling of plutonium on a technical scale. Plutonium is both radiotoxic and a chemically poisonous heavy metal (see 5.822.2), which means that strict security measures are required in handling it. Furthermore, the critical mass of ^{239}Pu is only about 5 kg, so its misappropriation for the purpose of building nuclear explosives must be prevented (see 5.83).

There are important points of similarity in the fuel cycles (uranium-plutonium) of the fast breeder and light water reactors. Both use fuel in the form of sintered oxide discs which are enclosed in sheaths to form a fuel rod; a number of these fuel rods fastened in a matrix comprise the fuel element. Past experience with the preparation of LWR fuel elements will contribute to the technology of preparing fast breeder fuel elements. There are also similarities in the recycling of spent fuel elements. Breeder fuel, like LWR fuel elements, is recycled by the modified PUREX process (see 5.822.2). As a rule, LWR fuel contains about 1% plutonium, and LWR fuel with recycled plutonium contains about 3%, while FBR fuel contains about 18%. In addition, the burn-up of FBR fuel (about 70000 MWd/t) is much higher than that of LWR fuel (about 33000 MWd/t). Due to the basic similarities in the fuel cycles of LWRs and FBRs, the transition to FBR fuel will require further developments, some of them extensive, but not a basically new technology (32).

The decisive advantage of fast breeder reactors is that they make it possible to utilize the energy content of uranium 60 to 70-fold better than is possible with any other known reactor system. One result of this is that with the FBR technology, poorer uranium ores (with higher extraction costs) can be economically exploited than with LWR technology (57). Therefore, the use of breeder reactors in countries which do not have enough energy raw materials to produce electricity would make an important contribution to the security of their national energy supplies.

4.215 Gas-cooled fast breeder reactors

In parallel to the efforts to develop an LMFBR, General Atomic in the USA and Kraftwerk Union in the Federal Republic of Germany are working on the development of a gas-cooled fast breeder reactor (GFBR) with helium as the coolant (58–60). This is basically a further development of the high temperature reactor (HTR). In contrast to the sodium-cooled fast breeder reactor, the helium-cooled reactor has a much higher gas pressure (coolant pressure 80 to 120 bar), due to the general properties of gas cooling. The coolant exit temperature is 830 K.

A helium-cooled FBR has the following advantages: since helium is less dense than sodium, it has less of a moderation effect on fast neutrons, and this makes a higher breeding rate (about 1.45) possible (57). Furthermore, it is impossible for bubbles to form in the helium.

The arrangement of the components of a helium-cooled FBR is essentially the same as in a HTR. The reactor and all the components of the primary circulation are integrated into a stressed concrete container. However, the core does not contain a moderator and is built similarly to that of a sodium FBR, with fuel rods containing oxide fuel, or later, carbide fuel (57, 59).

In terms of providing adequate safety features, the helium-cooled FBR is at a disadvantage with respect to the HTR because of the much smaller heat capacity of its core, and the correspondingly smaller time constants. However, emergency cooling of a helium-cooled FBR can be achieved (9). There are not yet any definite time schedules for the achievement of the various steps in the development of these reactors (33).

4.22 Energy from nuclear fusion

4.221 Physical basis

As mentioned earlier, the achievement of controlled nuclear fusion in a controlled thermonuclear reactor (CTR) would provide a practically unlimited supply of energy (see 3.36 and Table 3-2b). It is therefore understandable that enormous efforts are being made all over the world to solve the extraordinarily difficult physical and technological problems which stand in the way of fusion energy.

Nuclear fusion is fundamentally different from nuclear fission, in which a heavy nucleus is induced to break apart when it absorbs a neutron. In fusion, light nuclei are bound together into a heavier nucleus.

The possibility for obtaining energy from fusion is based on the following reactions:

$$D + D \rightarrow T + p + 4.04\ \text{MeV} \qquad (7)$$
$$D + D \rightarrow {}^{3}He + n + 3.27\ \text{MeV} \qquad (8)$$
$$D + {}^{3}He \rightarrow {}^{4}He + p + 18.34\ \text{MeV} \qquad (9)$$
$$D + T \rightarrow {}^{4}He + n + 17.58\ \text{MeV} \qquad (10)$$
$${}^{6}Li + n \rightarrow {}^{4}He + T + 4.78\ \text{MeV} \qquad (11)$$
$${}^{7}Li + n \rightarrow {}^{4}He + T + n' - 2.47\ \text{MeV} \qquad (12)$$

Reactions (7) to (10) are the actual fusion reactions, while (11) and (12) are the breeding reactions. If deuterium is used as fuel, reactions (7) and (8) are equally probable. The products, ^{3}He and tritium (T), react further according to (9) and (10). For technological reasons, however, the D-T reaction (10) by itself will be easier, and therefore earlier achieved, than reactions (7) and (8). There is almost no naturally occurring tritium on earth, however, so it has to be produced artifically. The lithium-neutron process is a good candidate for this purpose. ^{7}Li is important because when it captures a neutron, it emits a neutron on fissioning to ^{4}He and T. This neutron can then react with ^{6}Li (breeding process). To achieve this, the reaction space is surrounded with a lithium "blanket", so that the neutrons released in reaction (10) can react according to (11) and (12). In other words, deuterium and lithium are the fuels for a D-T reactor (61, 62).

The possibilities for production of tritium are essentially these:

- The use of liquid lithium as both fertile material and coolant. (For example, the proposal UWMAK-I) (63).
- The use of a eutectic mixture of the salts LiF and BeF_2 as the fertile material and blanket cooling with helium. The mixture, called "flible", consists of 46.9% LiF and 53.1% BeF_2. (For example, the proposal of the Princeton Plasma Physics Laboratory, PPPL) (64).
- The use of a solid lithium compound, $LiAl0_2$, as fertile material, Be as neutron multiplier and helium as coolant. (For example the proposal UWMAK-II) (65).

There is also the possibility of changing the isotope composition of natural lithium (7.42% ^{6}Li and 92.58% ^{7}Li) by enrichment of one part or the other. This could be important because the effective cross sections of the reactions (11) and (12) depend on the neutron energy in very different ways, i.e. the tritium yield of the "blanket" and thus the fuel consumption per unit of energy released depends very strongly on the nature of the fertile and structural materials (66).

The following is a discussion of some of the physical conditions for the occurrence of the fusion process. In order to fuse, two nuclei must be brought close enough together, against the coulombic repulsion, so that the short-range nuclear forces can become effective. The effective cross section of the reaction depends on the relative energy of the reaction partners, and the D-T process has both a larger reaction cross section and a more advantageous relationship between energy and cross section than the D-D process (57, 66–68).

It must be taken into account that the fusion cross section is generally orders of magnitude smaller than the coulomb cross section for elastic scattering, i.e. collisions which lead to fusion are much less frequent than those which lead to elastic scattering. Therefore, conditions must be achieved under which the number of fusion collisions becomes large enough to measure. For this, the particles must have very high kinetic energies, or, in other words, the gas mixture must be heated to extremely high temperatures (about $100 \cdot 10^6$ K). At these temperatures, the D and T atoms are completely ionized, so one speaks of a fusion plasma, or a plasma.

To attain a positive energy balance, it is necessary that the energy produced by fusion is greater than the energy lost by the plasma to the environment, for example by heat conduction, diffusion, convection, or radiation. This means that the particles must have sufficient opportunity to undergo fusion. The probability of fusion reactions depends not only on the length of time the plasma is enclosed τ_E, but also on the ion density n_i, i.e. the number of particles per cm^3. Theory shows that it is the product $n_i\tau_E$, which, in addition to the ion temperature T_i, is of critical importance.

It can be calculated that for the D-T reaction, the minimum value for $n_i\ \tau_E$ is $n_i\tau_E \geqq 10^{14}\ cm^{-3}$ s (Lawson criterion), and the lowest temperature, also called the ignition temperature, is about 10 keV (about $100 \cdot 10^6$ K; 1 keV $\triangleq 11.6 \cdot 10^6$K) (69). To achieve a positive balance, one must go farther into the "Lawson area", as the above figures are minima, or threshold values (70).

These parameters indicate the most important conditions for the achievement of a fusion reactor. Efforts are now being made to approach the necessary conditions step by step, i.e. to generate plasmas which are hot and dense enough, and which exist long enough.

4.222 The developments toward fusion reactors

A D-T fusion power plant, based on reactions (10), (11) and (12) would use the fuel cycle shown in Fig. 4-6 (63). In the fuel system (upper section), deuterium and tritium from their respective storage vessels would be mixed to give the proper starting amount of tritium and conducted into the reaction space. Here a few per cent of the fuel mixture would react according to (10). The remainder, with the ^{4}He generated in the reaction, would be removed by the vacuum system to the gas separation unit. The helium is removed and the remainder of the fuel is put back into the mixer. In the tritium extraction process, (lower section) the blanket is originally charged

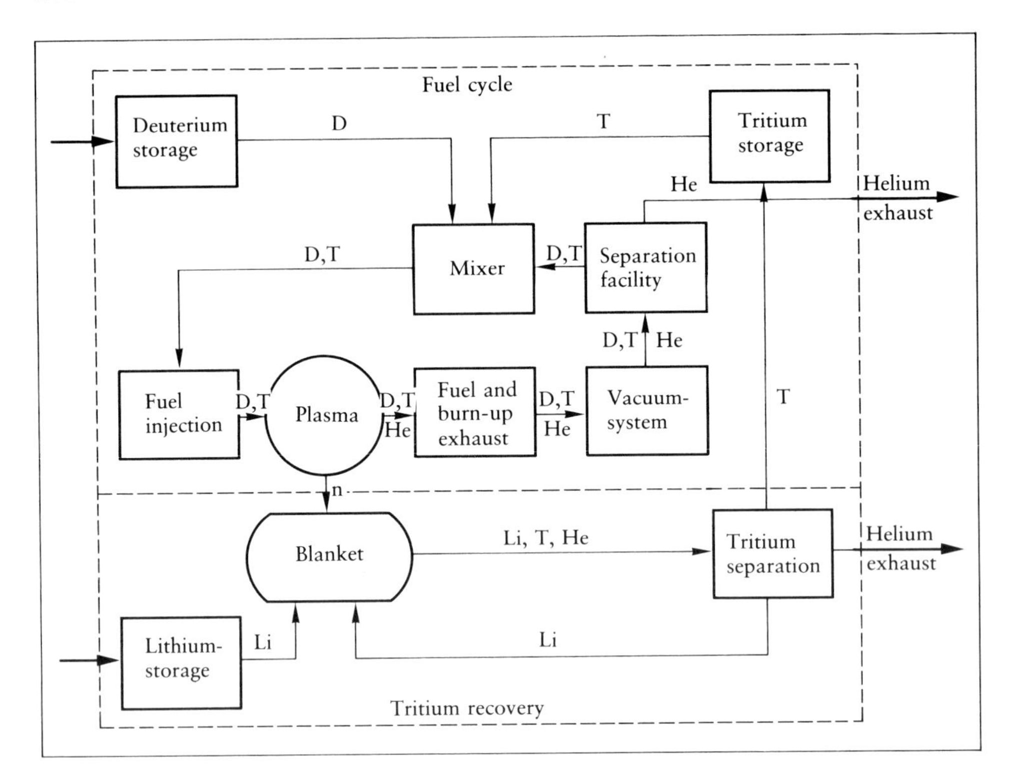

Fig. 4-6: Fuel cycle of a D-T-fusion power plant

Source: Max-Planck-Institut für Plasmaphysik, Garching/Munich, ASA-ZE/09/78, 1978.

with a certain amount of lithium, in which neutrons leaving the reaction space generate tritium according to equations (11) and (12). This tritium, along with the ^4He produced, is removed from the lithium. The tritium goes to the tritium storage vessel, the helium is removed, and the lithium is returned to the blanket. The amount consumed in the reactions of (11) and (12) is replaced from the lithium storage. Thus once the starting charge of tritium has been added, the only further additions to the reactor from the outside are deuterium and lithium (66).

Given the high temperatures needed for fusion, it is understandable that plasmas cannot be contained by material walls. In principle there are two different ways to confine a plasma, inertially and magnetically (61, 67, 70).

The principle of inertial confinement has only been pursued since the end of the sixties. In this technique, small spheres of solid (frozen) D-T mixture, with a mass of 10^{-4} to 10^{-3} g and a radius of about 1 mm, are subjected to a pulse of light from high-energy lasers. This pulse, which is distributed as evenly as possible over the surface of the pellet, imparts so much energy that fusion conditions are realized (10^5 to 10^6 J, in less than 10^{-9} s). The light generates a hot plasma corona, which expands rapidly, and a high pressure shock wave. This shock wave generates a spherically convergent compression wave which travels inward in the pellet. When it reaches the center, where the matter is highly compressed by it, the shock wave induces an outward-moving shock wave, and the conditions for fusion are achieved in the center of the pellet. (Material compression up to 10^4 times the normal density are possible) (71–74). Due to the extreme density, a relatively short confinement time is sufficient, so that the inertia of the matter allows fusion conditions to be realized. The energy is released as a result of microexplosions.

The processes involved in inertial confinement have been intensively investigated in the past few years, especially in the USA (LASL, Los Alamos; UCRL, Livermore; KMS Fusion, Ann Arbor), Russia (Lebedev Institute, Moscow), France (Limeil) and Japan (Osaka). Three types of high power lasers have been most often used, neodymium-gas, CO_2 gas and iodine lasers. The main problems with laser-induced fusion are related to the relatively low efficiency of the lasers and the relatively poor transfer of energy to the pellet material. For these reasons, there have been experiments lately on the use of high-energy electrons or ion beams to compress and heat the pellets (75–77).

Although important steps have been made toward the realization of controlled fusion with inertial confinement (78, 79), it would appear that the use of magnetic confinement is more promising. This second approach to fusion reactors has come farther, partly because some of the important experiments were already being done in the fifties (80).

Hot plasmas are good conductors of electricity, and can therefore interact with magnetic fields. For instance, the pressure from an external confining magnetic field of 1 Tesla is enough to maintain equilibrium with the gas kinetic pressure of a fusion plasma ($T_i \approx 15$ keV; $n_i \approx 10^{14}$ cm^{-3}).

A number of arrangements for magnetic confinement have been tested, including mirror machines, pinch arrangements (toroidal and linear pinches), stellarators and tokamaks[1)] (81–83).

The state of the art of magnetic confinement can most simply be expressed by the parameter $n_i\tau_E$ and the ion temperature T_i. Fig. 4-7 shows the experimentally achieved values of $n_i\tau_E$ and the corresponding ion temperatures T_i and the Lawson limit curves (70). The tokamaks are in the lead with respect to the $n_i\tau_E$ values, especially the ALCATOR-A (84, 85). The highest ion temperatures T_i can be obtained with mirror machines, but their $n_i\tau_E$ values are relatively low. From this one can see that the various systems have special properties and different problems. At the present state of development, it is still difficult to quantify essential aspects of controlled nuclear fusion. Therefore the only realistic possibility for evaluating experiments at

[1)] The ratio β of plasma pressure p to the pressure of the external magnetic fiels ($\sim B^2$) is an important parameter for magnetic confinement arrangements. If $\beta \approx 1$, one refers to high-β plasmas. This group includes toroidal pinches. If $\beta < 1$, one refers to low-β plasmas. This group includes tokamaks and stellarators.

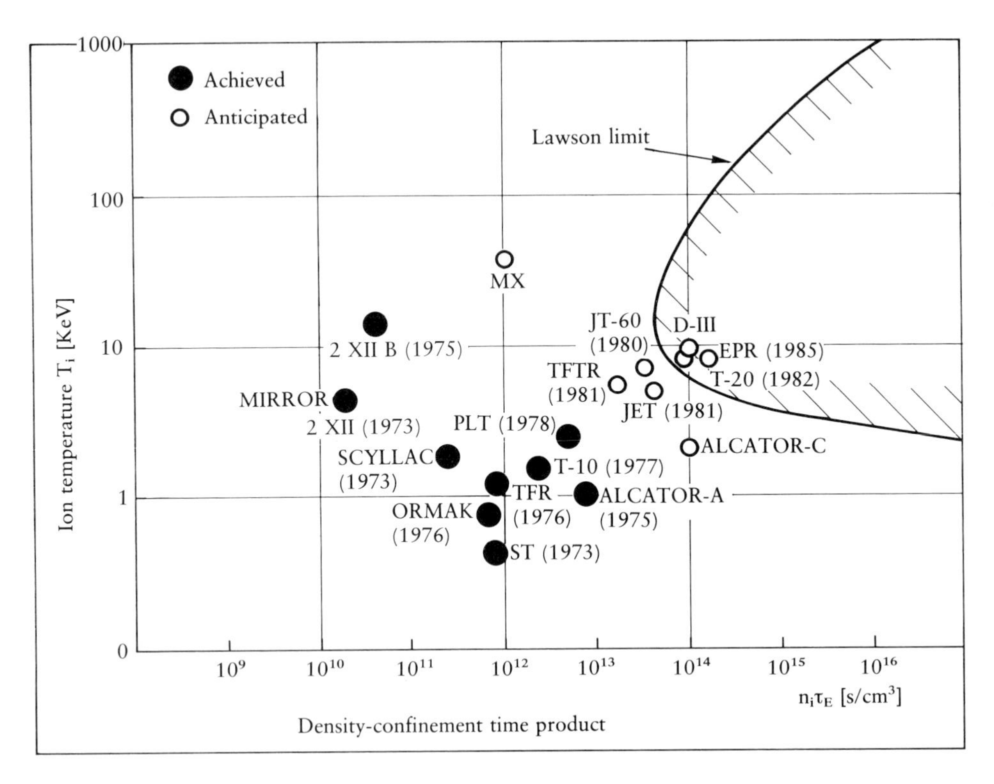

Fig. 4-7: Research progress in D-T-fusion power*
*The following abbreviations are used in Fig. 4-7: ALCATOR, high-magnetic-field tokamak (Massachusetts Institute of Technology); D-III, Doublet-3, tokamak of non-circular cross-section (General Atomic Company); EPR, experimental power reactor (General Atomic Company); JET, Joint European Torus (Euratom project, Culham, Great Britain); JET-60, Japanese tokamak (Japan); ORMAK, Oak Ridge tokomak (Oak Ridge National Laboratory); PLT, Princeton large tokamak (Princeton University); SCYLLAC, toroidal theta pinch experiment (Los Alamos Scientific Laboratory); ST, C-Stellarator, rebuilt in tokamak geometry (Princeton University); T-10, Tokamak (Kurchatov Institute, Moscow); T-20, demonstration thermonuclear tokamak reactor (Kurchatov Institute, Moscow); TFR, tokamak fusion reactor (France); TFTR, tokamak fusion test reactor (Princeton University); 2 XII, 2 XII B, MX, mirror experiments (Lawrence Livermore Laboratory).

Sources: W. Häfele, J. P. Holdren, G. Kessler, G. L. Kulcinski, Fusion and Fast Breeder Reactors, IIASA, Laxenburg/Vienna, RR-77-8, July 1977.
World Energy: looking ahead to 2020. Report by the Conservation Commission of the World Energy Conference, Guildford (UK) and New York: IPC Science and Technology Press 1978.

present is in terms of $n_i\tau_E$ and T_i. On this basis, the tokamak principle is the international favorite in fusion research (86–89).

Fig. 4-8a shows the basic arrangement of a tokamak (94). The main (toroidal) magnetic field is generated by ring-shaped magnet windings which enclose the torus. (The field lines run in the direction of the large torus diameter.) The plasma is generated inside the toroidal vacuum chamber which lies within the ring-shaped magnet

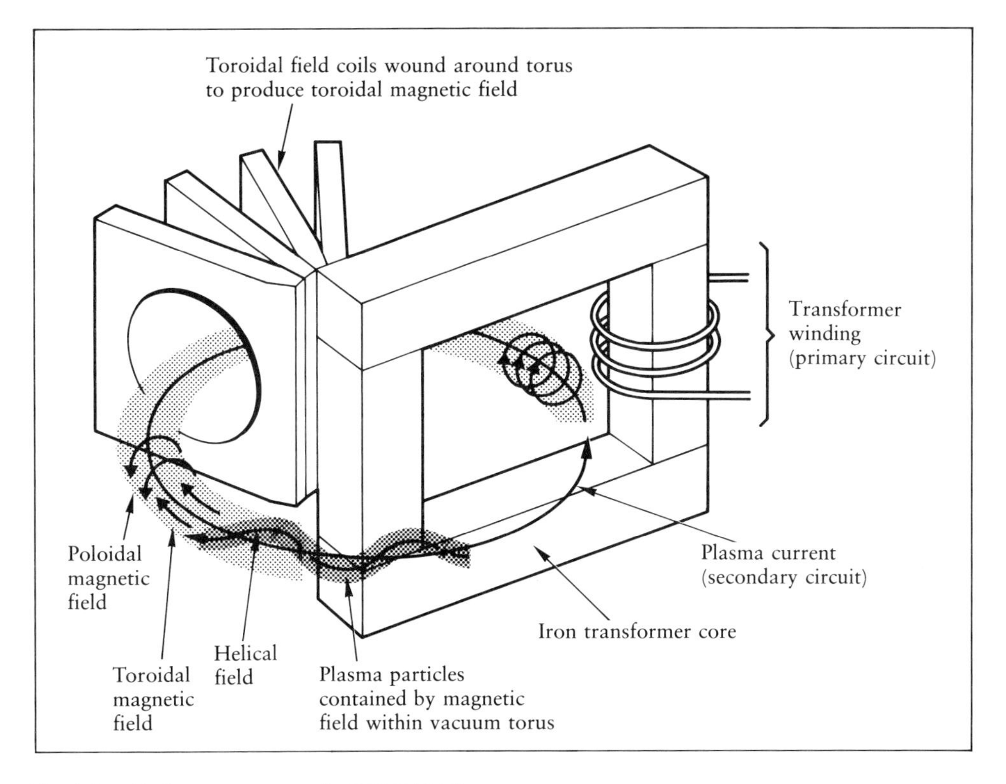

Fig. 4-8 a: Basic arrangement of a TOKAMAK

Source: A. Gibson, The JET project – a step towards the production of power by nuclear fusion, Die Naturwissenschaften, Vol. 66, No 10, October 1979.

windings. The poloidal magnetic field is generated in a tokamak by induction (transformer principle) of an electric current in the closed ring of plasma (impulse current). That is, a transformer is used to induce a strong electric current parallel to the axis, with the torus being used as the secondary winding. The poloidal internal magnetic field is additive to the external toroidal field, with the resultant lines of force wound in helical fashion around the torus.

Because of the induced current, a tokamak cannot be run continuously. The induction principle requires a pulsed mode of operation. After ignition, the reactor "burnes" for a definite time (100 to 6000 s); the length of time depends partly on the plasma. The cycle time, $t_c = t_b + t_o$, where t_b is the burn time and t_o is the dead time. (For a burn time of 5400 s, the dead time is reported to be 450 s) (66). The fuel which is consumed or lost by diffusion is constantly replaced, for example, by injection of pellets of solid deuterium and tritium. At the end of a pulse, the reaction volume has to be pumped out and filled with new fuel.

The induced current serves not only to achieve the magnetic field structure needed to confine the plasma, but also heats it (ohmic heating). (The electric conductivity of a plasma at a few million K is comparable to that of metallic conductors. The passage of current generates heat and thus heats the plasma.) However, since the electric resistance of the plasma decreases with increasing temperature ($\sim T^{-3/2}$), but the Bremsstrahlung losses increase with rising temperature ($\sim T^{1/2}$), the current heating will presumably not suffice to bring the plasma to an ion temperature $T_i \approx 8$ keV at which self-heating through the formation of ^4He is predicted. There are several possibilities for supplemental heating, including irradiation with high-frequency energy or a beam of high-energy hydrogen nuclei. Both methods have been successfully tested, i.e. significant increases in temperature were achieved. The injection of high-energy hydrogen nuclei is at present being preferentially tested (ORMAK, TFR) (see Fig. 4-7). It may also serve to replenish the particles in the reactor plasma and to regulate the reaction. As an example, the method of particle injection was used to achieve a two-fold increase in the ion temperature (66).

At present, the tokamak appears to have the greatest chance of being further developed to a fusion reactor, or in other words, the successful experiments in the past few years are grounds to proceed to fusion plasmas in tokamaks, as they will occur in a fusion reactor (94, 95).

In the seventies, a number of proposals for reactor design were prepared. All of them were based on the D-T process, and most were based on the tokamak principle, which reflects the present lead in the tokamak confinement technique. The mode of operation of a tokamak D-T reactor can be seen in the fuel-flow scheme in Fig. 4-6. Most of the energy released in the core of the reactor is carried off by neutrons and radiation and is absorbed in the lithium blanket. Of the 17.58 MeV released by reaction (10), 3.52 MeV is carried off by the ^4He, and 14.06 MeV by the neutron. The heat can be transferred from the blanket to a conventional heat power plant and transformed into electricity, or possibly, it could be used directly as process heat. Some of the energy would have to be stored to be fed back into the operation of the reactor, and some to even out the power impulses resulting from the discontinuous nature of the fusion part of the cycle. The tritium bred in the lithium will be separated and fed back into the reactor, along with deuterium, as fuel.

The schematic concept for a D-T fusion reactor in Fig. 4-8b is derived from the fuel-flow scheme in Fig. 4-6 (66). The plasma is surrounded by the so-called first wall, which is often a vacuum wall, and the lithium blanket, which is typically 1 to 2 m thick. Here tritium is generated, according to equations (11) and (12), and the kinetic energy of the neutrons is converted to heat. Outside the blanket is the shielding which protects the magnetic coils from neutrons and γ-rays. As mentioned above, liquid lithium, eutectic mixtures of lithium salts or high-pressure helium have been proposed as coolants. Either normal or superconducting magnetic windings have been proposed, depending on the concept and the length of the reactor pulse.

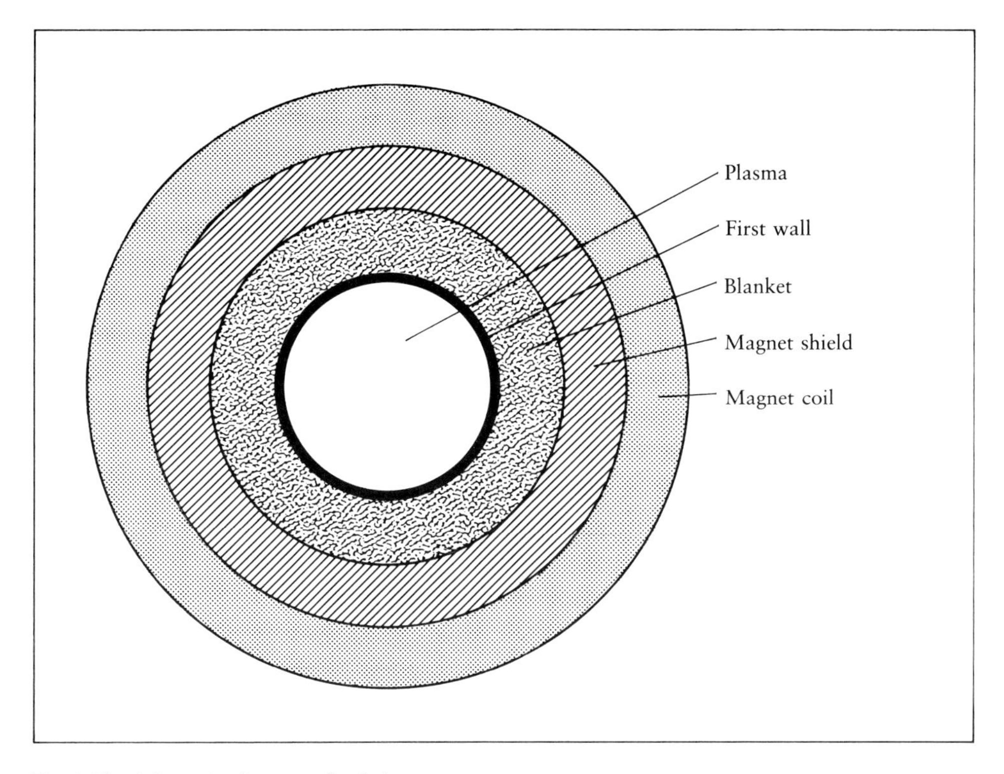

Fig. 4-8 b: Schematic diagram of a fusion reactor

Source: Max-Planck-Institut für Plasmaphysik, Garching/Munich, ASA-ZE/09/78, 1978.

The most important proposals have been conceived at Oak Ridge (ORNL proposal) (96), University of Wisconsin (UWMAK-I, UWMAK-II, UWMAK-III) (63, 65, 95), Princeton University (PPPL proposal) (97), Culham (MK-I, MK-II, JET) (94), Japanese Atomic Energy Research Institute (JAERI proposal), and the Kurchatov Institute (Moscow) (66).

These plans include a large range of physical and technical parameters, but they have important characteristics in common, such as the use of strong magnetic fields, large geometric dimensions and high thermal outputs. The following are a few parameters which are typical of these tokamak proposals: plasma radius, 2 to 2.75 m; large torus radius, 8.1 to 16.7 m; plasma volumes, 1000 to 6000 m^3; ion density, $0.5 \cdot 10^{14}$ to $2 \cdot 10^{14}$ cm^{-3}; ion temperature, 10 to 30 keV; maximum magnetic field, 8 to 16 Tesla; fusion power density in the plasma, 0.46 to 4.75 MW/m^3; thermal power 1 to 1.54 GW; burn time, 100 to 6000 s; thermal heat power, 10 to 100 MW; neutron stress on the walls, 0.83 to 5.7 MW/m^2; ratio β of plasma pressure to the pressure of the external magnetic vield, 1.45 to 9.3% (63). Fig. 4-9 is a typical proposal for a tokamak reactor (UWMAK-III) (95).

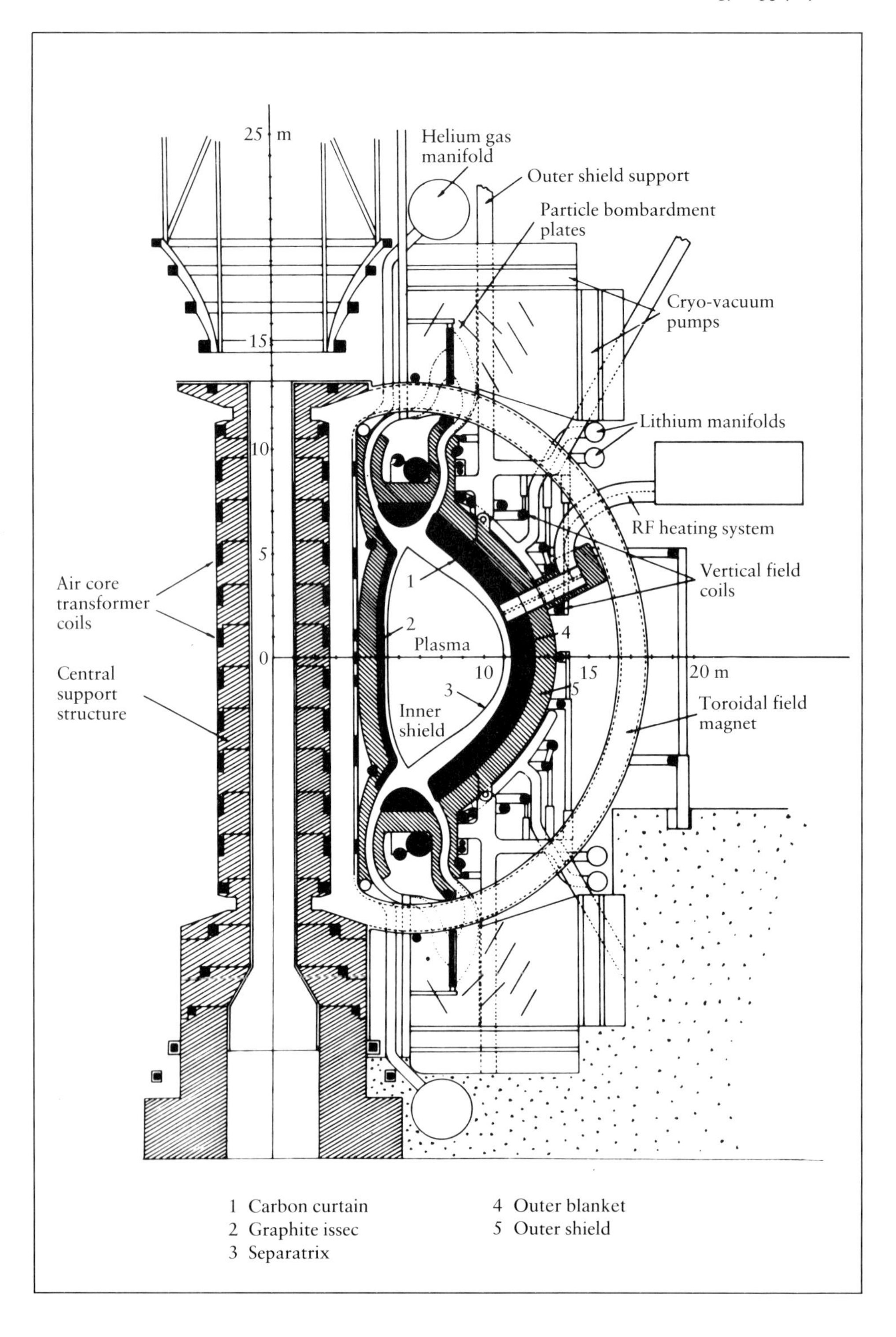
25 m
15
10
5
0
10
15
20 m
Helium gas manifold
Outer shield support
Particle bombardment plates
Cryo-vacuum pumps
Lithium manifolds
RF heating system
Vertical field coils
Toroidal field magnet
Air core transformer coils
Central support structure
Plasma
Inner shield
1
2
3
4
5
1 Carbon curtain
2 Graphite issec
3 Separatrix
4 Outer blanket
5 Outer shield

In addition to the physical problems of fusion research, there are still many technological problems to solve. Fusion research, system studies and reactor proposals all serve to uncover important technological, economic and ecological problem areas, some of which will be discussed below.

One of the key problems is the question of materials, especially for the first wall between the plasma and the blanket. On one side it interacts with the plasma, and on the other, with the lithium (compare Fig. 4-8b). It is exposed not only to the bombardment of ions, electrons and atoms from the plasma, but also to the intense neutron flux of about 10^{14} $cm^{-2}s^{-1}$ 14-MeV particles. Conversely, the effect of the wall on the plasma is very significant. The bombardment with plasma particles results in a high rate of erosion of the wall material. Any impurities in the plasma, especially heavy metals, affect the physical processes in it, such as the burn time and the energy balance. A few countermeasures are being tested experimentally, for example diverters (systems to suck off unburned fuel, products and impurities) (compare Fig. 4-6), and low atomic weight solids to coat the wall (98, 99). In addition to these considerations, there are several operational parameters and technical conditions for the wall material, including high temperatures, resistance to corrosion by lithium or lithium salts, vacuum tightness, and impermeability to tritium and helium, which make the problem all the more difficult. For these reasons, the designers are considering regular replacement of construction elements, particularly of the first wall (66). (The problems mentioned in connection with the first wall also apply to the material for the supporting structures.)

A further problem area is the development of large superconducting magnet windings, which will be subjected to extreme conditions in a fusion reactor, such as intense neutron and gamma irradiation, strong mechanical forces, cooling to very low temperatures with liquid helium, and large changes in the magnetic field. A great deal of development work will be needed before reliably operating magnet systems can be built for fusion reactors (100, 101).

At the present level of research and development on nuclear fusion, it is extremely difficult to make even somewhat reliable estimates of the absolute costs of fusion power plants (for example capital costs for the fusion-specific components and the operating costs of a plant) or of the absolute cost of fusion-generated electricity. The estimates of the absolute costs for electricity given with the various reactor proposals have varied widely. One reason for this is that they have proposed to use different forms of heat transfer and energy conversion, which lead to different overall plant efficiencies. The plant efficiency for UWMAK-I (reactor cooled with lithium) was es-

---

Fig. 4-9: Cross-section view of the right half of the Wisconsin Tokamak fusion reactor (UWMAK-III) (The system is toroidal and therefore symmetric about the vertical axis)

Source: W. Häfele, J. P. Holdren, G. Kessler, G. L. Kulcinski, Fusion and Fast Breeder Reactors, IIASA, Laxenburg/Vienna, RR-77-8, July 1977.

timated to be η (net)=32% (63), and that for the PPPL proposal (helium cooled reactor) was with η (net)=38% (97).

It is more useful to compare the cost structures of possible fusion power plants with each other, than to estimate absolute costs. Detailed analyses of cost structures (the proportion of the total costs accounted for by each component) make it possible to estimate the effect of changes in the costs of individual components on the cost of generating electricity. The main components of a fusion power plant include the reactor, the turbines and the electrical equipment. These main components can be further subdivided. As an example, the reactor includes the reactor proper (magnets, cooling system, blanket, shielding, fuel supply system, heating, etc.), the heat transfer system, the fuel circulation, and other equipment (66). On the basis of the relationship between system parameters and the cost of each component, one can estimate the influence of these parameters on the cost of the electricity. An example of a system parameter is β, the ratio of plasma pressure to the pressure of the external magnetic field. Since the fusion power density is proportional to β, a smaller value of β will result in higher capital costs for the components of the reactor and correspondingly costs for the electricity.

In light of the importance of the heat market in the energy economy (see 2.331), experiments on the use of fusion reactors not only for electric generators, but also for process heat sources or for both are very significant (102, 103). In this reactor concept, which is also based on a tokamak arrangement, part of the blanket is used for the generation of high-temperature process heat. It has been estimated that about 75% of the blanket volume is sufficient for tritium generation (breeding blanket), and the other 25% can be used to generate process heat (non-breeding blanket). The heat generated in the breeding blanket is converted to electricity, with an efficiency of 36%; the heat in the non-breeding blanket can be used, for example, to superheat steam (1420 K, 50 bar) (103). This high-temperature heat can be used in many processes for the production of secondary energy carriers (see 4.63, 4.642, 4.643.1, 4.65, 4.661). One advantage of fusion reactors which produce process heat could also be a continuous full-load operation of the whole plant. Because of the possibility of storing a secondary carrier such as hydrogen or substitute natural gas (SNG), the plant could produce these during slack times, provided sufficient intermediate storage capacity were available.

The studies which have been made on the hazards which would be unique to a fusion reactor have come essentially to these conclusions: No long-lived radioactive isotopes arise in the fuel cycle of a fusion reactor. The radioactivity induced in structural materials can be reduced by the right choice of material. This should make the closing down of a reactor easier than that of a fission reactor. The tritium inventory of a fusion reactor would be the main problem, but it seems to be manageable, since a "nuclear excursion" is physically impossible for a fusion reactor. Finally, in contrast to fission reactors, fusion reactors would not produce any materials which could be used to build nuclear weapons (see 5.83) (104).

Clear progress has been made towards the development of controlled nuclear fusion. Further development towards an economically feasible fusion power plant might be roughly divided into the following stages: 1) Proof of the physical possibility of controlled nuclear fusion (achievement of ignition conditions, not necessarily with a net release of energy); 2) Achievement of a D-T machine with a positive energy balance; 3) Construction of a demonstration power plant with a fusion reactor; 4. Construction of an economically operating fusion power plant. The estimates of when, if ever, these stages will be reached vary, in some cases widely (57, 66, 91, 104). The time horizon will probably depend on the amount of research and development invested in the project. It is generally expected that if fusion power can be realized, it will be about half a century before commercial power plants can be built, so that the development lead of HTRs or FBRs over the fusion power plants is 4 to 5 decades (104, 105).

The following is a discussion of fusion-fission hybrid reactors, which are essentially schemes to use the intense neutron flux generated in a fusion core. Three possible uses of fusion-fission systems are being discussed: 1) For the generation of electricity (106), 2) For the production of fissionable material for (separate) fission reactors, and 3) For the transformation on long-lived, heavy radioactive isotopes into relatively short-lived, light isotopes.

In principle, a fusion-fission hybrid reactor consists of a fusion core which is surrounded by a fission blanket. The main function of the latter is to increase the energy yield from the 14 MeV neutrons. If the energy yield per neutron could be increased, it might be possible to obtain energy from nuclear fusion before an economically feasible "pure" fusion reactor is available (107, 108).

Fusion-fission systems for production of fissionable material would produce fuel which could be used in separate fission reactors. They would breed ^{239}Pu from the non-fissionable ^{238}U (see 4.214) or ^{233}U from ^{232}Th (see 4.213.3). It has been estimated that a fusion-fission system could provide fuel for 5 to 10 LWRs (or HTGRs). Unlike a "pure" fusion reactor, this type would not need to achieve a net release of energy.

It has also been suggested that fusion-fission systems could be arranged in such a way that they would transform dangerous long-lived radioactive isotopes from fission reactors into relatively short-lived light isotopes (transmutation of long-lived radioactive wastes.) For example, when strontium-90 and cesium-137 absorb neutrons, they are converted to isotopes with shorter half-lives which are also less radiotoxic. The neutron flux required for this is extremely high, however, and the irradiation time very long, so that this method would produce still more radioactivity problems. However, the irradiation of long-lived actinides with fusion neutrons would lead to their fission. The products would be highly radioactive, to be sure, but their half-lives are much shorter (57), so this method seems more promising. In any case, further progress toward controlled fusion will be required for the achievement of hybrid fusion-fission systems.

Based on the fuel reserves (see 3.36), and factors of economy, ecological stress and safety (see 5.8), it would appear from the present standpoint that the utilization of fusion energy in a reactor, if it can be acheived, would be an attractive alternative to fission reactors for the long-term energy supply. Therefore it is quite understandable that the achievement of a fusion reactor is one of the main goals of scientific and technological research and development all over the world.

4.3 Secondary energy from solar energy

4.31 Direct methods[1]

4.311 Solar thermal conversion

4.311.1 Components of a solar thermal system

The essential components of a solar thermal system are the solar collectors (flat or concentrating), heat carrier and heat storage. The radiant energy from the sun is converted into heat in the solar collectors, and is transported by the heat carrier to the consumer. Due to the variation in the solar radiation available, heat storage is needed to provide for the times when there is little or no sunlight. There are different types and combinations of each component of a solar thermal installation, which will be discussed in the following.

One promising technique for the commercial utilization of solar energy, especially in the temperate zones, is the direct conversion of solar energy into low-temperature heat using flat (low temperature) collectors which absorb direct and diffuse radiation (see 3.371 and 3.373). The basic technologies required for this have already been developed. However, it is still necessary both to improve the individual components (e.g. increasing the efficiency and lifespan of the collectors while decreasing their cost) and to optimize the entire solar system, both technically and with respect to costs.

Fig. 4-10 is a schematic cross section of a flat collector. The transparent cover presents a problem, because it should not noticeably decrease the amount of light, either by reflection or by absorption, but it must at the same time prevent the re-radiation of infrared from the heated absorber plate. The absorber plate converts the incident light into heat, which is carried to the consumer by a transport medium, e.g. water or a glycol with a high boiling point. To attain a high efficiency, a flat collector must

[1] In this section, the discussion in 3.372 will be illustrated with a number of concrete examples.

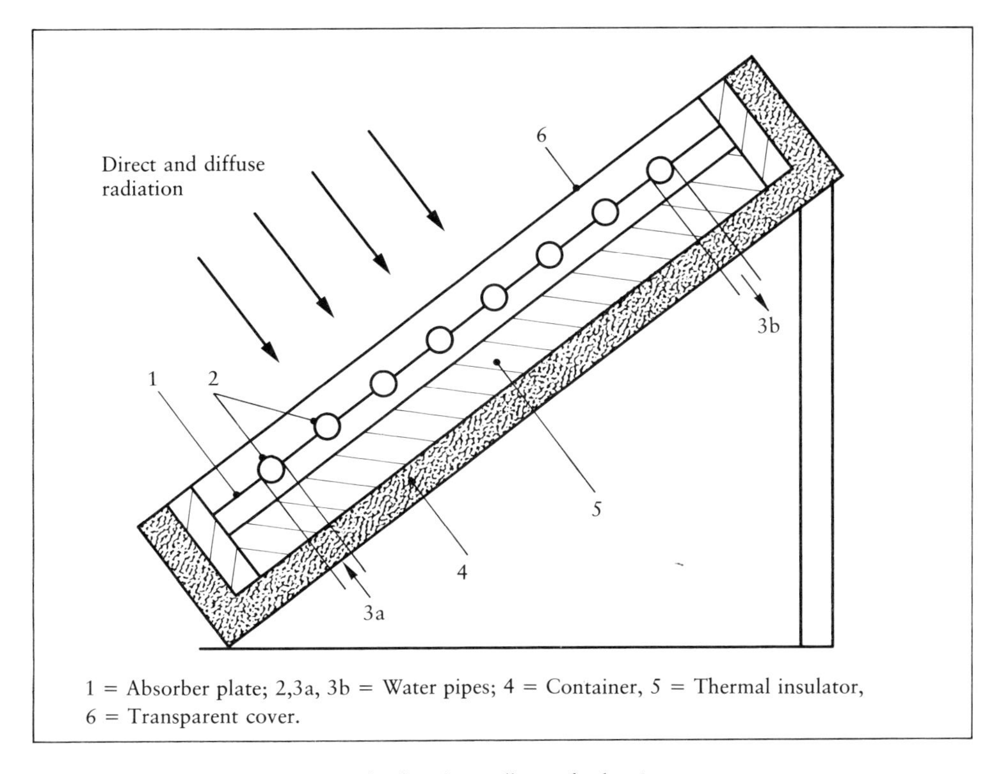

1 = Absorber plate; 2,3a, 3b = Water pipes; 4 = Container, 5 = Thermal insulator, 6 = Transparent cover.

Fig. 4-10: Schematic cross-section of a flat plate collector for heating water

Source: E. Justi, Stand und Aussichten der Sonnenenergie, in: Heizen mit Sonne, German Solar Energy Society, Proceedings, Göttingen 1976.

absorb most of the incident light and also have low losses through convection, heat conduction and radiation. Even at relatively low working temperatures (30–100°C), the radiation losses (which are proportional to T^4) from a flat collector are several-fold greater than other losses (see 3.372) (109, 110). As a consequence, flat collectors are generally suitable only for low temperatures (< 100°C), unless their absorber plates have selective layers which absorb well in the visible and emit longer-wavelength infrared poorly (111). Collectors are quantitatively described by means of characteristic curves. Fig. 4-11 shows as an example the characteristic curves of a flat plate collector for heating water. On the inset abcissa, the temperature difference for the special case I = 500 W/m² is plotted. The collector efficiency η is defined as the ratio (useful power per unit area of the collector): (solar power per unit area). The useful power produced by the collector is the difference between the absorbed power and the thermal losses. It can be seen from Fig. 4-11 that the efficiency η of the collector is relatively good at low operating temperatures (112). The

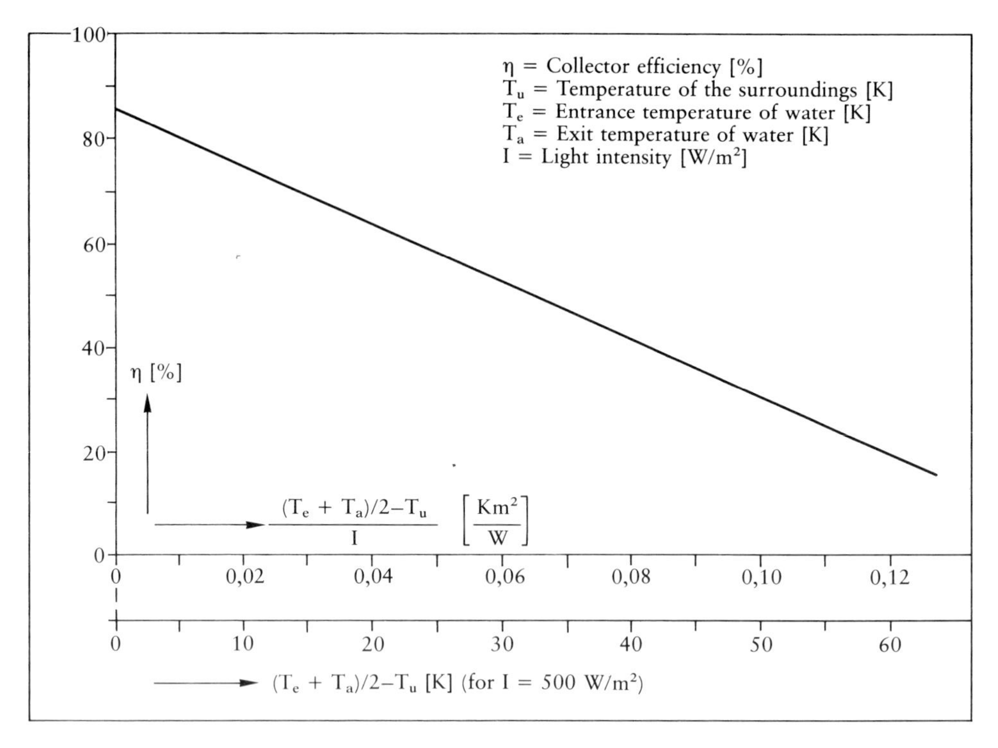

Fig. 4-11: Example of a conversion characteristic of a water-operated flat plate collector

Source: D. Orth, Heat Pump Operation to utilize regenerative Energy Resources, 1st German Solar Energy Forum, Proceedings, Vol. I, Göttingen 1977.

efficiency of collectors averaged over the year in the temperate zones is about $\eta = 50\%$. If one then takes into account the system losses, e.g. in pipes transporting the heat carrier, in storage and from non-optimal control of the system, the overall efficiency of systems equipped with collectors might lie around 25%. The daily and yearly changes in the height of the sun actually requires a constant "tracking" by the collector, because the sun's rays should fall on it as close to perpendicularly as possible in order to minimize reflection losses and maximize the efficiency of absorption. However, for the uses to which the collectors are put, the machinery for following the sun is too expensive. They are therefore mounted (for example on a roof) so that the normal of the collector points south, and the surface is at an angle of 40° to 45° from horizontal.

To obtain higher temperatures than are possible with flat collectors, it is necessary to amplify the intensity of the solar radiation falling on the absorber surface with suitable concentrating optical systems, such as lenses or mirrors. That is, concentrating solar collectors[1] (high-temperature collectors) employ a reflecting (or refracting) material to direct the sunlight to the receiver. In this way the absorber surface can be kept small, and thermal losses are correspondingly reduced (see 3.372, equation (3)). However, two other types of loss become important with these collectors: loss of diffuse radiation, since concentrating systems can only work with direct rays, and optical losses. This means that diffuse radiation, which may account for a large fraction of the energy collected by flat collectors, is practically unused, and there are optical losses associated with the use of concentrating devices. Furthermore, concentrating collectors (lenses or mirrors) must follow the apparent path of the sun. The machinery required for tracking represent an investment which is not required for flat collectors (113). The degree of concentration, i. e. the ratio of the reflector surface to the absorber surface, can be as high as 10000. The concentrating collectors which are currently in use have a ratio of about 10 between the surface of the parabolic cylinder of the reflector and that of the absorber tube which lies along the "focal line" of the mirror.

Depending on the degree of concentration, concentrating collectors can achieve temperatures well above 100°C. Even with low degrees of concentration, temperatures of 250°C to 550°C are possible, so that electricity can be generated in steam turbines. Systems with high degrees of concentration are referred to as "solar furnaces". The solar furnace at Odeillo, France, reaches a temperature of about 3800°C (compare Fig. 4-14) (114). Theoretically, using extraterrestrial sunlight, one could reach the temperature of the surface of the sun (about 5600°C), but in reality, this temperature could not quite be reached (115). (A higher temperature than that of the solar surface is forbidden by the second law of thermodynamics.) For practical applications, temperatures of a few hundred °C are sufficient. To obtain the desired amount of energy, a number of individual collectors are usually linked in batteries. Because it is possible with concentrating collectors (high-temperature collectors) to obtain heat at considerably higher temperatures than with flat collectors (low-temperature collectors), the former have correspondingly higher efficiencies in the further conversion of the heat into mechanical or electrical energy.

The heat absorbed in the collectors must be transported to the consumer, either directly or via a heat storage stage. The most suitable heat carriers are liquids, such as water or glycols, or gases, such as air. Up to temperatures of 100°C, water is most often used, due to its high heat capacity and noncorrosiveness. At higher temperatures, the higher pressure would make it necessary to invest more in the collectors, pipes and storage system. To prevent freezing, antifreeze is added to the water, and corrosion is also reduced by additives. This is not very severe, because a dual circula-

[1] The term "solar collector" generally refers to the entire system of receiver and concentration elements. The receiver, including the absorber, insulation, etc., is the element which absorbs the radiation and transforms it to another form of energy. The concentration system (optical system) is that part of the system which directs the sunlight into the receiver.

tion (primary and secondary circulations) is used. Air is rarely used as a heat-transport material, partly because of its low heat capacity. Oil may be used as the heat carrier in concentrating collectors.

One of the main problems in utilizing solar energy is that of energy storage, since the amount of energy available varies greatly, both diurnally and seasonally, and the times when the most energy is needed are not the times when the most is available. In a solar power plant, energy can be stored as sensible heat, by raising the temperature of a fluid or solid material; as latent heat, in a substance in which addition or loss of heat leads to changes in the aggregate state or structure, but not to changes in temperature; or as chemical energy in the product of a reversible chemical reaction (heterogeneous evaporation storage) (110, 116).

As mentioned above, the use of solar energy by direct conversion to low-temperature heat ($T < 100°C$) in flat collectors has promise in the areas of water and space heating. For this application, sensible heat storage is suitable. In general, for the amount of heat ΔQ added to a sensible heat store:

$$\Delta Q = mc\,(T_2 - T_1) \tag{13}$$

where m is the mass of the storage material, c is its specific heat capacity, and $T_2 - T_1$ is the temperature difference between the "loaded" and "empty" storage material. For a given storage material, with a specific heat capacity c and quantity of heat ΔQ, the larger $T_2 - T_1$ is, the smaller the mass m (and hence the volume) of the storage material can be. If the use dictates a high minimum value for T_2 in the "loaded" condition, the heat losses from the reservoir to the environment are increased (assuming no change in the amount of insulation), and the same holds for increasing the storage time. However, the storage material can only be loaded when the temperature of the heat carrier is higher than that of the storage material. This means that as T_2 is increased, the efficiency of the collector decreases (compare Fig. 4-11), or the collector circulation can only be run when the radiation intensity is above a threshold value, and some of the solar energy is thus unutilized (116). Since, according to equation (13) a large storage mass, with a temperature only a little higher than that of the environment (this is favorable for preventing losses), can store the same amount of heat as a smaller storage mass at higher temperature, it is possible to use heat pumps as a means of increasing efficiency (see 4.325). In this way, the temperature of the storage mass can be independent of the temperature at which the heat is used, but it does increase the investment cost for the solar installation. From the above it is clear that the collector, heat carrier and storage components must be mutually optimized, and in addition, the changes in weather conditions and consumer habits must be kept in mind. At present, water is the most frequently used storage medium. In installations for solar heating of swimming pools, for example, the water in the pool itself is the storage medium. It has a very large heat capacity, and solar heat is essentially added only when the sun is shining. In solar hot-water and solar space-heating installations, too, sensible heat storage in water is suitable, although in the latter case, large volumes and heat pumps are generally required. It is possible, for example, with sufficient insulation, to store warm (80° C) water in steel containers for 6 to 7 months, with a temperature drop of only about 2°C.

In latent heat storage, a substance takes up or releases heat as it undergoes a change in physical state without a change in temperature. Examples of such changes are freezing and thawing, evaporation and condensation, and sublimation and crystallization. The heat stored in these processes is latent, because no temperature change is involved. Such reversible processes can in principle be used to store heat (latent heat storage) (116). For practical purposes, the volume changes associated with the changes in state should be as small as possible, since the material is kept in closed containers. Therefore the most likely candidates are transitions between the liquid and solid state. A substance to be used for latent heat storage should have the following characteristics: a large specific heat of melting, a melting point in the temperature region of interest, and a low vapor pressure of the liquid, since it is kept in a closed container. It should be chemically stable and not corrode its container, the volume changes associated with the change in state should be small, it should have good thermal conductivity in both the liquid and solid states, it should be safe for the environment, that is, non-poisonous, non-explosive and non-flammable, and it should be manufactured cheaply and on a large scale (116). For storage of solar energy, conversions at low temperatures ($T < 100°C$) are of particular interest, and there are a number of hydrated salts which melt in this range. One is $Na_2SO_4 \cdot 10\ H_2O$, with a melting point of 32°C, density of 1458 kg/m^3 and specific heat of melting 367 MJ/m^3. Some other materials which might be used for latent heat storage at low temperatures are $Na_2CO_3 \cdot 10\ H_2O$, $Na_2CO_3 \cdot 12\ H_2O$, $FeCl_3 \cdot 6\ H_2O$, $Ca(NO_3)_2 \cdot 4\ H_2O$, $Na_2S_2O_3 \cdot 5\ H_2O$ (116, 117). However, these salts do not adequately meet the criteria listed above, so that only sensible heat storage with water as storage medium has so far been used.

Heterogeneous evaporation storage is based on the thermal decomposition of some chemical compound AB (liquid or solid) into a gaseous and a liquid or solid component under the proper conditions of temperature and pressure. If the gaseous component B is added back to the condensed component A, the reaction goes in the opposite direction and generates heat. In sum, the reaction is

$$\text{AB (solid or liquid)} \underset{\text{heat emitted}}{\overset{\text{heat absorbed}}{\rightleftharpoons}} \text{A (solid or liquid)} + \text{B (gaseous)} \quad (14)$$

A number of salts, e.g. $FeCl_2$, $CaCl_2$, $ZnCl_2$ and $CaBr_2$, can bind large amounts of NH_3. A heat storage unit based on this principle consists of two storage areas, one for the condensed and one for the gaseous component of the reaction. Since the components are stored separately, and at environmental temperature, they can in principle be stored for long periods. A disadvantage of heterogeneous evaporation storage is that the reactions are not completely reversible – there are some hysteresis effects (116).

Metal hydride storage is also a form of heterogeneous evaporation storage. With these compounds, H_2 is released when heat is added, and absorption of H_2 releases heat (see equation (14)). Solar energy could be used to release the hydrogen from the metal hydrides (an endothermal process). The hydrogen could be stored in a separate container and returned to the hydride storage as heat is needed; the heat could be used either directly or indirectly, via heat pumps.

Progress could undoubtedly still be made on the problem of energy storage, which is one of the main problems in utilization of solar energy. The technology is in principle already available, but the methods, especially those for use with solar heating installations (for longterm storage) are in general not yet economically feasible.

4.311.2 Solar thermal installations

Techniques for converting solar energy into heat, especially low-temperature heat, are already in practical use, for example for heating swimming pools, water heaters, space heating, air conditioning, solar furnaces, purification of drinking water and salt production (118, 119).

It has been found in practice that it is relatively easy to build an economically favorable solar heating installation for heating a swimming pool, because the water in the pool itself has a large capacity for heat storage and needs only to be heated during the hours when the sun is shining. A collector surface equal to only about 50% of the water surface is able to heat the water several degrees above the environmental temperature during the entire swimming season. There are a number of public swimming pools which have solar-heated water.

The commercial use of solar thermal conversion to heat water is somewhat more difficult than heating swimming pools, because hot water temperatures up to about 50°C are needed, and not only during the hours of sunshine, which leads to storage problems (120). There are a number of different types of solar hot water heaters. They all consist essentially of three components, a collector, a storage unit for warm water with a built-in heat exchanger, and a system of pipes and regulating elements (connecting pipes, pumps, thermostat and a supplementary conventional heat source (compare Fig. 4-12). The water heated in the collector is circulated through a heat exchanger, in which the water to be consumed is heated (dual circulation system). The separation of the heat carrier from the hot water is useful, because it is then possible to use special liquids in the collector circulation. The heat carrier is circulated by a pump which is regulated by temperature, i.e. the pump is turned on when the difference between the temperatures in the collector and the storage tank exceeds a threshold value. The carrier medium is then pumped through the heat exchanger and the water in the storage tank is heated, starting with the coolest region at the bottom of the tank, where fresh water is drawn in to replace the hot water removed for use.

If the system is designed to supply a single family for a few days when there is little sun, it will typically have a storage volume of 300–500 liters of water and, depending on the climate, 4 to 6 m^2 collector surface. Because of the irregularity of the supply of sunshine, especially in the temperate zones, a solar water heating system must be combined with a conventional heater. Technically, this is relatively easy to do, but it adds to the cost of the system. For example, the water can be preheated by the solar system and then, whenever this is inadequate, brought to the desired temperature inside (or outside) the storage tank. In Europe, especially in apartment buildings, the hot water for the building is often heated by the furnace. However, when an oil furnace is used only to heat water during the summer, its efficiency is only 20%, compared to 70% when it is being used to heat the building as well. This makes solar water heating especially attractive for the summer months.

Solar space heating is of particular interest both because space heating accounts for a large fraction of the energy economy, and politically, because of the large amount of petroleum used in this way (see 2.331). For these reasons, a large number of projects have already been built with solar heating. However, solar energy can also be utilized by appropriate architectural design: windows facing south, and with good thermal insulation (double glass); building mass which can store the heat; and good insulation. (In general, good insulation alone can reduce the energy required for space heating by about 50%, regardless of the type of heating (109). Fig. 4-12 shows schematically the plan for a solar space and water heating system[1]. Collectors for space heating work best at low temperatures, where their efficiency is greatest, (compare Fig. 4-11) i.e. producing heat at 35°C to 45°C for a large-surface heating system, such as subflooring warm-water heating. For utilization of solar energy for space heating, storage becomes an extremely difficult problem, because the supply and demand peak at opposite times of year. In a temperate climate like that of middle Europe, more than 60% of the demand for heat comes in the period from November to February, during which time only 12% of the annual solar energy supply is available (compare Fig. 3-16a and b). However, in much of the USA, the winter solar radiation is twice as high as in middle Europe. As yet, there is no technical solution to the problem of long-term storage which would be economically feasible as well. Therefore, for the present, space heating with solar collectors is limited to direct utilization during the transition months[2]. In the main heating period, except in favorable climatic areas, it will continue to be necessary to use conventional heating systems for space heating for a shorter or longer period of the year (multiple

[1] Collectors installed for space heating are generally designed to heat water as well, so that the large amount of solar energy available in summer can be utilized. A space and water heating system differs from a pure hot water system only in having a larger collector surface, a larger storage volume for hot water, and a heat exchanger with which the heat can be taken from the hot water reservoir for the purpose of space heating. This system provides hot water during the summer months and solar space heating during the spring and fall.

[2] Solar heating can be used for a longer period if the solar collector is supplemented by a heat pump.

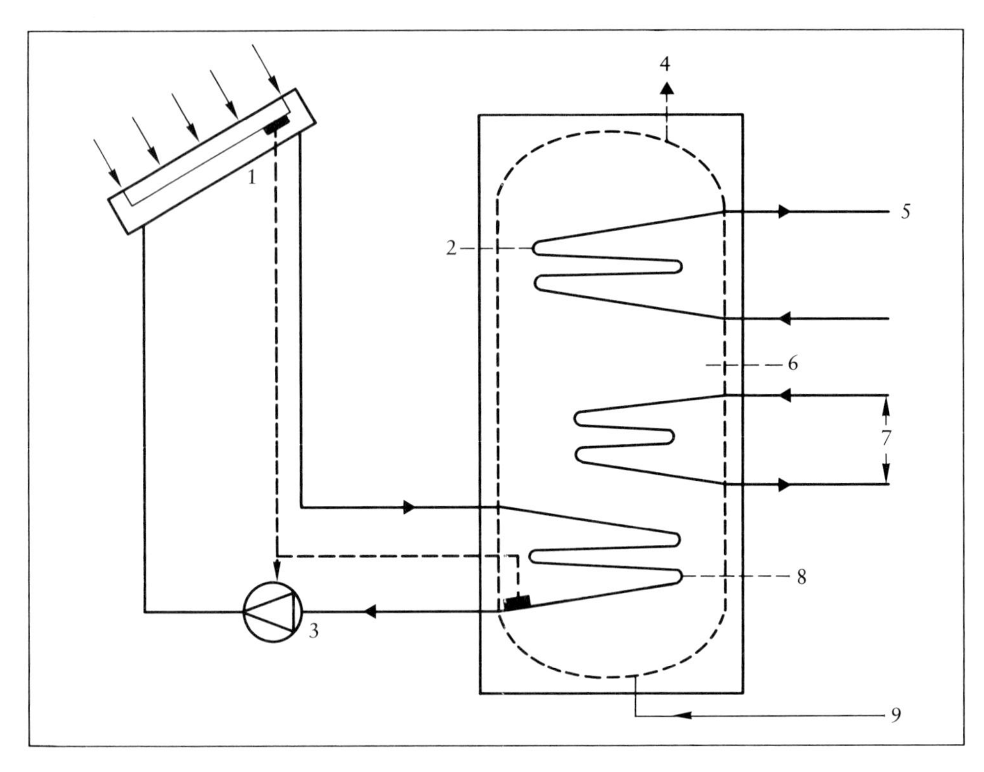

Fig. 4-12: Diagram of a solar installation for space and water heating
1 = Collector; 2 = Heat exchanger; 3 = Temperatur-regulated pump; 4 = Hot water; 5 = Space heating circulation; 6 = Hot water tank; 7 = Supplementary conventional heat source; 8 = Heat exanger; 9 = Cold water.

Source: This Fig. partly corresponds to the Fig. of G. Wagner, Brauchwasser-Speicher für die Nutzung von Sonnenenergie; in: Heizen mit Sonne, German Solar Energy Society, Proceedings, Göttingen 1977.

heating systems). In this case, the additional investment costs must be balanced against the savings in energy costs (for example, the cost of oil). In general, these savings do not (yet) justify installation of solar heating systems for purely economic reasons. For example, take a normal single-family house which uses about 5000 liters heating oil per year. Under the relatively unfavorable climatic conditions in western and middle Europe, it is possible with the currently available solar systems, with about 30 m² collector surface, to save about 40 to 50% of the heating oil costs, but the necessary investment in solar heaters is on the order of 20000 DM (1980 price) (109).

Because solar radiation during the winter months is not intense in the temperate zones, collectors alone are not sufficient to provide heating over the entire year. There is therefore a stronger trend to use solar energy indirectly, by means of heat pumps, which draw on the solar heat stored in the environment (the air, the earth or ground water) (see 4.325).

The question of economic feasibility of solar hot-water and space-heating systems, or combinations of the two, in the various regions depends on a number of factors, including climatic conditions, the price and availability of conventional primary energy carriers, and the prices of the solar systems. For example, in the USA, a solar hot-water system for a single-family house costs $ 2000 to $ 4000, and a combined space-heat and hot-water installation, between $ 3000 and $ 10000; collectors are sold as 32 ft^2 panels for $ 375 (1981 price, not including installation), which comes to about $ 115/m^2. For a pure hot-water system, a pay-back time of about 5 to 10 years would be required on the basis of actual cost, but due to federal and state tax incentives, the pay-back time in practice is only 3 to 5 years. For the combined systems, the pay-back time based on actual cost is 10 to 20 years, which means that they are not so economically advantageous as the pure hot-water systems. There is a tendency to install solar collectors (active heating) for hot water only, although "passive" heating (architectural designs to maximize utilization of solar energy) is becoming widespread. The improvement of individual components and optimization of their interaction with other components, as well as other factors such as mass production and the increasing costs of other primary energy sources, can be expected to contribute to a steadily increasing degree of economic feasibility for solar installations, even in temperate climates.

A number of experimental solar houses, in which a larger or smaller portion of the thermal energy demand is supplied by solar energy, have been built, for example the Philips Experimental House in Aachen, Federal Rep. of Germany. In some of them electricity is also supplied by solar cells, for example the Solar One House, Delaware, Md., USA (109). Most solar houses were built for experimental purposes; in them, the individual system components for utilization of solar energy are tested.

Solar houses have relative large areas of flat collectors, usually on the roof and oriented toward the south. They also have large reservoirs, often in the basement, and especially effective thermal insulation of the outer walls and windows. Fig. 4-13 is a schematic drawing of the Philips Minimum Energy House at Aachen. There are 20 m^2 of solar collectors on the roof. The house has three water reservoirs, one for long-term storage (annual storage reservoir, 42 m^3, 5°C–95°C), a hot water tank (4 m^3, 45°C–55°C) and a waste water tank (1 m^3). The house also has systems for recovering heat from waste water and air which is removed in ventilation, and heat pumps for utilization of the heat stored in the earth. The living area of the house is 120 m^2. The heat loss through the walls is $k = 0.14$ W/m^2K (a typical value is 1.23 W/m^2K) and the loss through the windows is $k = 1.5$ W/m^2K (a typical value is 5.8 W/m^2). The Philips house has a computer in the attic which simulates the energy consumption of an average family and records and analyses the data. The first results show that all the measures together can reduce the annual consumption of heating oil to about 10% (500 liters) of the average.

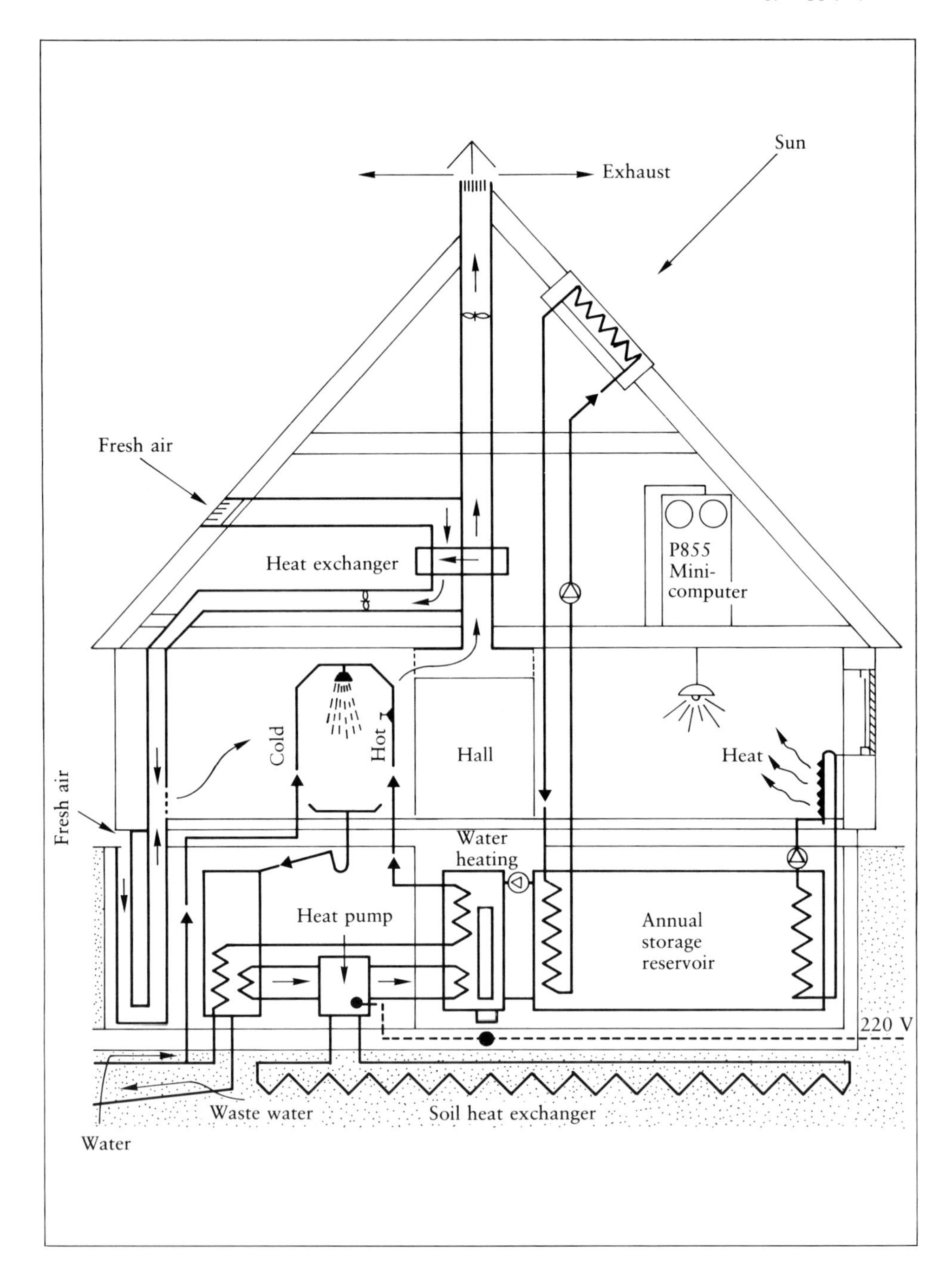

Fig. 4-13: The Philips Minimum Energy House

Source: Solar Energy, UK Section of the International Solar Energy Society, London, UK-ISES, 1976.

Solar air conditioning can be of great practical significance, because in countries with large amounts of solar radiation, the USA for example, the consumption of energy for air conditioning is also large. One advantage of this use of solar energy is that the supply of energy and the demand rise and fall at the same times of the year. This applies not only to air conditioning, but also to industrial cooling, for example for food storage. There are a number of different physical principles on which solar refrigeration can be based. One is the compression cooler, and another is the absorption refrigerator, which has been thoroughly tested (116). (Small absorption coolers have been used for years in refrigerators). In conventional absorption refrigerators, the cooling circulation is maintained by a basic form of energy, such as electric current. When solar energy is used for refrigeration, however, the heat from a collector is used to run the absorption cycle. The combination of coolant and absorption medium can be ammonia and water, for example. The principle of operation is as follows: in the evaporator, heat is withdrawn from the medium to be cooled (e.g. the air) by the spontaneous evaporation of the refrigerant (e.g. ammonia). The refrigerant vapor is pumped into the absorber, where it is absorbed by the absorption medium (e.g. water). The resulting solution is then pumped into the separator, where the heat from a solar collector is used to separate the refrigerant from the absorption medium. (The operating temperature of about 90°C makes flat collectors suitable.) The absorption medium then returns to the absorber and the refrigerant vapor passes into the condensor where it loses heat and returns to the liquid state, thus closing the cycle. (It is also possible to use the waste heat from a refrigerator, in which case it is working as a heat pump (see 4.325)). Because the need for air conditioning is greatest at the times when solar energy is most abundant, the outlook for solar cooling is more promising than for solar heating.

As mentioned in 4.311.1, concentrating collectors can be used to generate heat at temperatures well above 100°C, depending on the degree of concentration. If this is very high, up to about 10000, one refers to solar furnaces. Fig. 4-14 shows the solar furnace at Odeillo, France (122). It has 63 moving flat mirrors (heliostats), each with an area of 42 m^2, and built from 180 individual mirrors. These reflect the sun's rays onto a 2000 m^2 parabolic mirror which has a smelting furnace at its focal point. Temperatures up to about 3800°C are reached with it (114). The installation, which is used among other things to prepare special alloys melted in vacuum, can be operated about 2000 hours per year (116).

Solar energy has been used for years to distill drinking water from polluted fresh or sea water. The technique is very simple: the water is channeled into low troughs which are made of a material with a high absorption capacity. They are covered with slanted glass or plastic "roofs" (the sunlight should pass through them with as little absorption as possible). The water is evaporated by the solar heat and condenses on the covers, from which it runs down into gutters, and from there into a collection tank. Since there are many coastal areas in which there is intense sunlight, but no drinking water, such installations are of great interest. On the Greek island Patmos,

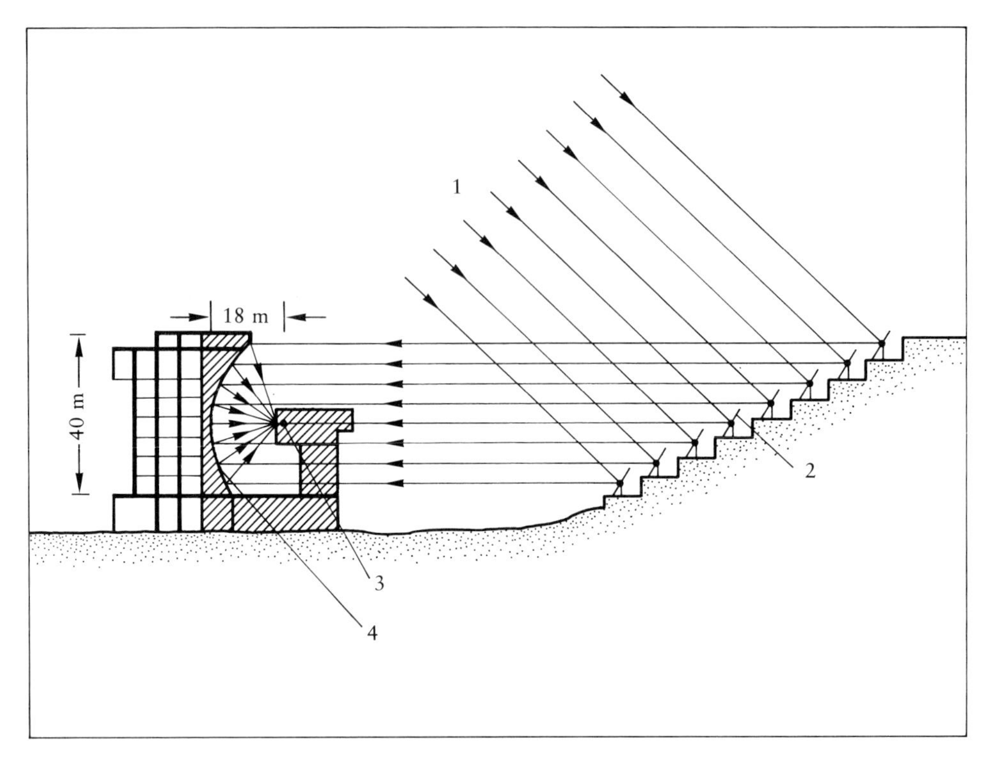

Fig. 4-14: The solar furnace at Odeillo
1 = Sunlight; 2 = 63 movable mirrors; 3 = Focus; 4 = Parabolic mirror.

Source: M. Clemot, B. Dessus, C. Etievant, F. Pharabod, Solar Power Plants: French Realisations and Projects, 10th World Energy Conference, Istanbul, September 1977.

an installation for desalinization of drinking water has been built. It has an evaporation area of 8670 m^2 and delivers 26 m^3 of drinking water per day. Furthermore, the brine remaining after evaporation is an important intermediate in the preparation of salt from sea water (116).

4.312 Solar thermal power plants

Solar-generated heat can be used for generating electric current as well as for direct heating. In principle, either low-temperature flat collectors or high-temperature concentrating collectors can be used. The conversion of heat to electricity is accomplished with conventional generators, i.e. the only difference between solar-thermal and conventional power plants is the way in which heat is generated. Development work on solar thermal power plants has been underway for several years in Italy (Genoa), France (Odeillo), and the USA (Atlanta, Albuquerque) (114, 123).

When low-temperature collectors ($T < 100°C$) are used, the working medium for the evaporation cycle cannot be water, but must be a liquid with a lower boiling point, such as freon. However, due to the low total efficiency (electric power per unit of radiant energy received) of only a few per cent, this technique can only be used for outputs of a few kW.

The total efficiency can be considerably increased by using concentrating collectors. If the degree of concentration is about 10, a working temperature of 140°C can be attained, and, if freon is the working medium, an efficiency of about 10% is achieved. With a 400-fold concentration, which can be achieved with cylindrical parabolic collectors, temperatures of 500°C can be reached, and the electricity can then be generated with conventional steam generators. There are two different concepts for solar thermal power plants with concentrating collectors, the solar tower and the solar farm. In January 1980, nine member countries[1)] of the International Energy Agency (IEA), Paris, began construction of a pilot plant of each type in Almeria, in southern Spain. They are intended to compare the two concepts in plants of the same size (500 kW electric power) as well as to test existing components in a complete power system. The electricity to be generated will be fed into the local Spanish net. In the following, the two pilot installations will be discussed.

In the solar tower concept (see Fig. 4-15), a mirror field reflects the sunlight onto a central receiver, which is mounted at the top of a tower. The mirror field consists of 160 heliostats, each built up of 12 individual mirror segments mounted on a carrier structure. The total reflecting surface is about 4000 m^2, and the degree of concentration is 450. The heliostats follow the sun under the control of a computer. The radiation absorber is an insulated hollow space mounted on a 43 m tower. (This concept is also called a "central receiver system", CRS, because the radiation absorber is located in the center of a large mirror field.) The solar energy penetrates the exterior of the hollow space and is absorbed and converted to heat in a spiral tube containing liquid sodium. The hot sodium (525°C) is pumped into a heat reservoir, where it is either stored or pumped through a heat exchanger where steam is generated. The steam is heated to 510°C at 100 bar pressure. The entire infrastructure of the solar power plants covers a surface of 300 m × 300 m, most of this area being occupied by the mirror field. Taking into account the use of the energy reservoir, a net electric output of 500 kW can be produced in about 2000 hours of operation. The plant is supposed to attain the following efficiencies: electric energy (net)/thermal energy $\approx$ 23%; electric energy (net)/radiation received $\approx$ 14% (122, 123).

In the solar farm concept, a large number of linear parabolic collectors are distributed over the area of the power plant. In the focal line of each collector (the degree of concentration is 40) there is an absorber pipe in which a heat carrier (heat-resistant oil) flows and transports the heat to a thermal power plant or a reservoir. The exit temperature from the collector is 295°C. The collectors are mounted so they can

[1)] Belgium, Federal Republic of Germany, Greece, Italy, Austria, Sweden, Switzerland, Spain, USA.

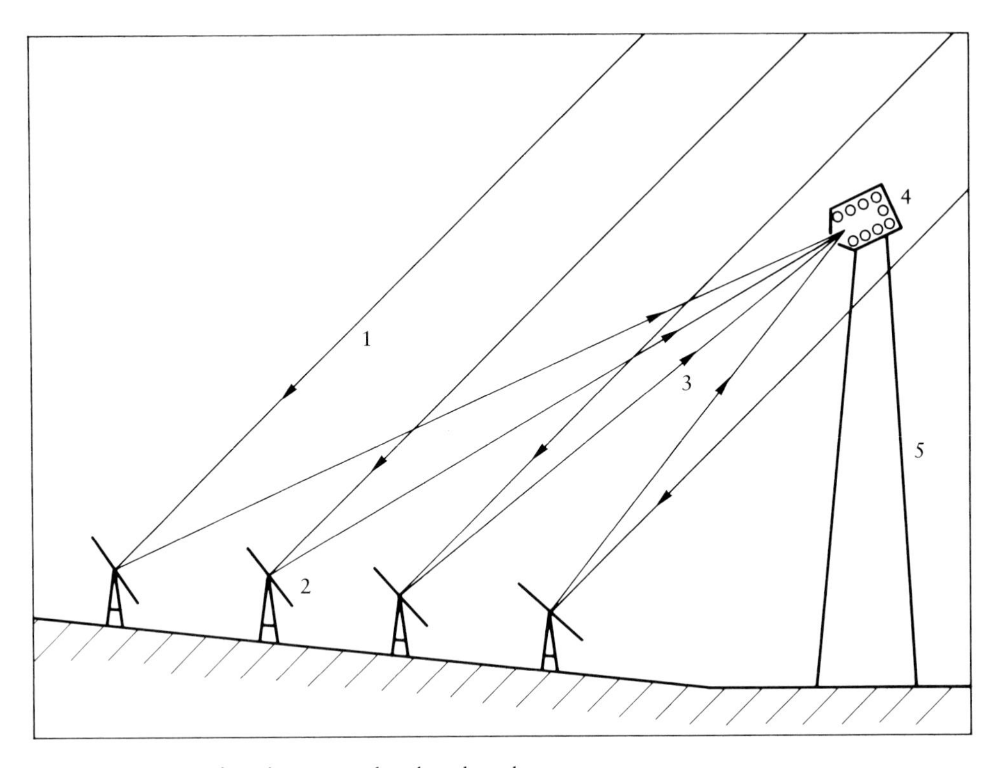

Fig. 4-15: Scheme of a solar power plant based on the tower concept
1 = Incident Insolation; 2 = Heliostats; 3 = Reflected Insolation; 4 = Receiver; 5 = Tower with heat machine and generator.

Source: M. Clemot, B. Dessus, C. Etievant, F. Pharabod, Solar Power Plants: French Realisations and Projects, 10th World Energy Conference, Istanbul, September 1977.

turn and have motors which direct them toward the sun. The entire power plant occupies a surface area of 250 m × 300 m; and the net electric power output is 500 kW. The following efficiencies are expected: electric energy (net)/thermal energy ≈ 19%; electric energy (net)/radiation received ≈ 10% (123).

Solar thermal power plants with concentrating collectors consist essentially of mirrors with drive systems, absorbers, pipes, reservoirs, heat exchangers and generators, i.e. of components which have already been developed. However, their application in a new type of system does bring specific problems. The low energy density of the solar radiation makes large-area concentrating collectors necessary, and these are subject to environmental stresses such as earthquakes, windstorms and hail. In addition, mirrors with up to 40 m^2 of surface must be kept pointed directly at the sun, with an accuracy of a few minutes of arc.

Development of solar thermal electric generation is only beginning. There are already plans for plants with capacities up to 100 MW, however. The costs of these plants are estimated (for the end of the eighties) to be \$ 1000 to \$ 5000/kW (114). However, reliable estimates of the cost of energy from solar thermal power plants, based on empirical data from operating plants, will not be available before the middle of the eighties. Still, something can be said about the cost structure to be expected. For a fossil-fuel power plant, the investment costs represent only about 10% of the final energy cost, while the cost of operation and fuel account for about 90%. With a solar or nuclear plant, the relationship is exactly reversed: the investment costs account for 90% of the power costs, while only about 10% is due to fuel and operational costs.

There are many countries in which there are favorable locations for small power plants. This is especially true for regions with high average solar intensity. With small plants, it would be possible to supply small communities with electricity without building a far-flung and expensive distribution system (decentralized energy supply). In India, for example, there are still several hundred thousand villages without electricity. A decentralized energy supply system could also help to prevent further encroachment of the desert in many regions of the earth where trees and shrubs are constantly being cut to supply energy.

4.313 Photoelectric conversion

4.313.1 Solar cells

Direct conversion of solar energy into electricity is possible, without the roundabout conversion to heat and mechanical energy, and has been used for years in satellites. However, the investment costs for generating electricity with solar cells are still a factor of about 30 higher than for conventional power plants (see 3.372) (114).

Direct conversion of solar radiation into electric energy occurs when photons interact with the crystal lattice of a semiconducting solid and release charges. The generation of a voltage as a result of photon absorption is called the photovoltaic effect. According to the quantum theory, any kind of radiation can be considered to consist of elementary quanta (photons). A photon has an amount of energy $E = h\nu$ (h is Planck's constant and ν is the frequency of the wave; the relationship between the wavelength λ and the frequency ν of electromagnetic radiation is $\nu\lambda = c$, where c is the speed of light.) That is, the higher the frequency (or the shorter the wavelength), the more energy is carried by the photons of the electromagnetic radiation. According to the photon model, a radiation intensity of 1 kW/m^2 corresponds to a stream of photons containing about $6 \cdot 10^{21}$ photons per m^2 and second.

In the following, the principle of energy conversion in solar cells will be discussed. If sufficiently energetic photons collide with a semiconductor, they can knock electrons out of their positions in the crystal. In the band model of a semiconductor, this process means that by absorbing a photon with an energy $h\nu > \Delta E$, an electron is excited from the valence band E_V into the conduction band E_C. ($\Delta E = E_C - E_V$ is the difference in energy between the two bands in the particular semiconductor). Both the negatively charged electron in the conduction band and the hole in the valence band, which acts as a positive charge, are relatively free to move through the material (109). If there is an internal electric field in the material, the positive and negative charge carriers move in opposite directions, and can be induced to flow through a resistance load R by appropriate connections to the semiconductor. In this way the photocurrent I and the photovoltage U can be measured (compare Fig. 4-16). (The electric power is the product of the current and the voltage.) The electric field which drives the photovoltaic current in the solar cell arises from a potential drop at the contact surface between two oppositely doped semiconductor materials, or between a semiconductor material and a metal. (The cause of the potential drop is the difference in the work an electron must do to leave the material.) In solar cells, this potential drop can be achieved in one of three ways: 1) by a homogeneous p-n contact between a p- and an n-doped region of the same semiconductor material (for example, a silicon solar cell, Fig. 4-16); 2) by the contact between a semiconductor and an appropriate metal, a so-called Schottky metal-semiconductor contact (e.g. between an n-doped gallium arsenide (GaAs) and gold (Au); or 3) by a heterogeneous p-n contact between two different semiconductor materials, e.g. n-doped cadmium sulfide (CdS) and p-doped copper sulfide (Cu_2S), as in the cadmium sulfide/copper sulfide cell.

The semiconductor layers can be either single-crystal discs or thin polycrystalline layers. If an electron-hole pair generated by absorption of photons in the semicon-

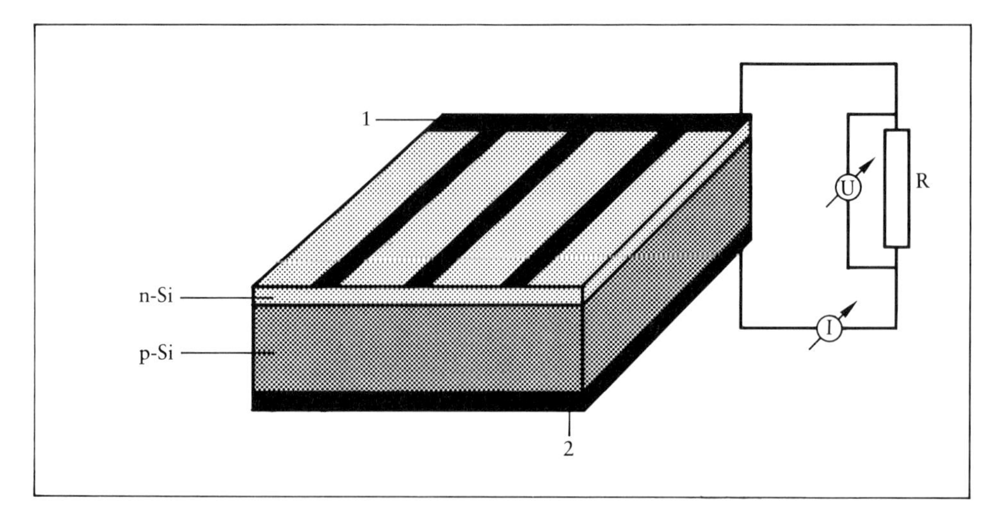

Fig. 4-16: Diagram of a silicon solar cell
1,2 = Metal electrode.

Source: M. Selders, D. Bonnet, Solarzellen, Physik in unserer Zeit, Vol. 10, No 1, 1979.

ductor material (the holes in n-regions, and the electrons in p-regions) diffuses into a boundary region in which there is an electric field, the electron will be accelerated into the n-region, and the hole into the p-region. This causes the n-region to accumulate a negative charge, and the p-region, a positive charge, which results in a photovoltage U. If there is a closed external circuit, a photocurrent I and the photovoltage U can be measured by the external resistance R (124, 125). The efficiency η of a solar cell is defined as the ratio of the maximal electric current from the cell to the radiant energy falling on the cell. (Even when the sky is covered with clouds, solar cells have relatively favorable efficiencies.)

For a maximal photocurrent I, the number of electron-hole pairs generated by solar photons and separated by the electric field must be maximized, since only those charge carriers which are separated contribute to the photocurrent. A photon cannot generate more than one electron-hole pair, and for each semiconductor there is a minimum energy ΔE, such that only photons with $h\nu > \Delta E$ can knock electrons out of the bound state. For the semiconductors used in solar cells, the values of ΔE lie between 1 and 2.5 eV. The semiconductor layer must be thick enough to absorb essentially all the sunlight, but as most of the materials used have relatively high absorption coefficients in the visible, a few μm suffice. GaAs and CdS, for example, are suitable for thin-layer cells. Silicon, on the other hand, has a low absorbance, so that layers of about 300 μm are needed of single-crystal silicon. The following values are typical for a silicon cell: $\Delta E = 1.1$ eV, photocurrent density under air mass 1 conditions (see 3.371) (this corresponds to about $6 \cdot 10^{21}$ photons per m^2 and s) about 35 mA/cm^2. The maximal power output of a solar cell depends on the type of cell, but is on the order of 100 mW.

Reflection losses at the surface of the solar cell are strongly reduced by layers which eliminate reflection. In order to minimize interference with light absorption, and to keep ohmic losses small, the electrodes on the surface are shaped like combs or meshes.

Silicon solar cells (single crystals, doped with boron or phosphorous) are most often used, because Si is the most common semiconductor material and the technology for its production is most advanced. The efficiencies of these cells are between 15% and 16% under air mass 0 conditions (see 3.371). The cells have dimensions of about 2 cm × 2 cm. The technology for their production is complicated, elaborate and expensive. The raw material obtained by chemical purification still contains impurities which must be removed by other techniques, e.g. zone melting techniques. The material produced in this way is polycrystalline and is the starting material for the production of single crystals. There are a number of techniques for obtaining single crystals, such as the Czochralski method and the floating zone technique (126).

Another material with good semiconductor properties is gallium arsenide (GaAs single crystals). Because of its high absorption coefficient, a layer about 2 μm thick is enough to absorb the sunlight (thin-layer cells). Because the charge carriers are generated in the immediate neighborhood of the surface, there is a large component of surface recombination, and thus a low photocurrent. This surface recombination can be nearly eliminated by a further GaAlAs layer, about 1 μm thick, which is prepared by the liquid phase epitaxic technique. Such GaAlAs-GaAs solar cells have the highest efficiency of photovoltaic conversion of sunlight that has so far been achieved, but they are extremely expensive to produce, so that for the present they can only be justified in space vehicles or for use with concentrated sunlight, where much smaller surfaces are needed. A particular advantage of these cells over Si cells is that they have good efficiencies even at high temperatures. For example, with a degree of concentration of 300, they still have 19% efficiency, and at a temperature of 200°C, an efficiency of 14% was obtained (124).

Other types of single-crystal solar cells are CdS-InP (14% efficiency); and CdS-$InSe_2$ (12% efficiency). Single-crystal cells are expensive, because their preparation is elaborate, and their use for terrestrial applications is therefore only feasible in exceptional cases. It is easier, and therefore cheaper, to make polycrystalline silicon solar cells, but the efficiency here is only 10 to 12.5% (124). For practical purposes, one could for example mount 32 5-cm by 5-cm polycrystalline Si cells as a module with a power output of about 8 W. (AEG Telefunken, Federal Republic of Germany, has been able to produce polycrystalline Si solar cells which are 10 cm by 10 cm and have an efficiency of 10%.)

Since the middle of the seventies, the possibility of using amorphous silicon (a-Si) as the semiconductor material for solar cells has been studied. For one thing, amorphous silicon has an absorption coefficient which is one or two orders of magnitude larger than that of monocrystalline silicon. Thus thin layers (1 μm thick), which are easy to produce, can be used. The efficiencies which have been obtained are about 5.5%. The main problems, however, are in the reproducibility of the production conditions.

For very cheap solar cells, polycrystalline thin-layer cells are of interest. The best known example is the Cu_2S-CdS cell, which has efficiencies about 9%. The advantage of polycrystalline Cu_2S-CdS thin-layer solar cells is the possibility that they might be cheaply produced in large-area arrays. Two other examples of polycrystalline thin-layer cells are the CdTe-CdS solar cell with 5% efficiency and the CdS-$CuInSe_2$ cell with 7% efficiency (109).

4.313.2 Possible uses

Solar cells have been used mainly in space vehicles. For the most part, Si cells have been used, to provide current for satellites, for instruments and transmitters. Solar cell generators with outputs of 10 to 20 kW can be made today. The cells are usually 2 cm × 2 cm in area and a few tenths of a mm thick. To obtain the desired voltages and currents, a large number of individual elements are connected, with groups of cells connected either in series or in parallel. A solar cell generator with 1 kW output requires a surface of about 10 m^2. (Such solar cell surfaces and their connecting "wiring" are glued to a carrier material and transported in folded form into space, where the carrier is unfolded.) An example of this type is a 9 kW solar generator developed by AEG Telefunken (Federal Republic of Germany). It consists of two wings with a total span of 22 m and width of 2.8 m, and has about 92 000 solar cells. When rolled up, the generator fits into a case 286 cm × 90 cm × 45 cm. When solar cells are used in space, they are subject to radiation damage by electrons and high-energy particles, so that in the course of time their efficiency declines by about 30%. (The lifetime of generators in space is 10 to 15 years.)

Radiation damage would also be a disadvantage for a project proposed by P. E. Glaser, in which a solar generator consisting of solar cells would be placed in geostationary orbit – i.e. about 36 000 km from the earth – and used to generate electricity. Glaser suggests that his satellite solar power station would have an output of 10 GW, and the energy would be transmitted to earth by microwaves (128, 129). According to his estimates, a solar cell field with an area of about 60 km^2 would be needed, and the weight of this field, along with the supporting structure, would be about 40 000 t. On the assumption that a space shuttle would have a transport capacity of about 100 t, about 400 flights would be needed to build such a station. The expense would be so high, that the plan could not be realized in the foreseeable future.

Due to the high cost of solar cells, terrestrial applications of solar cell generators are limited to those cases in which there is no available electric supply net (decentralized energy supply) (130, 131), such as relay stations and weather stations. (Nickel-cadmium batteries are generally used here to store the energy.) As an example, to operate a television set about 4h a day, a solar generator with a cell surface of about 1 m^2 is needed (132).

The use of solar generators to run water pumps, especially in sunny and dry developing countries, is very promising. For instance, if the daily irradiation is 6 kWh/m^2, a typical value for sunny countries, a 4 m^2 solar generator with a 5% efficiency can provide 1.2 kWh daily. If the motor and pump have an efficiency of 50%, they can bring 10 m^3 water from a depth of 10 m per day (126). Many such projects have already been built, for example in the Sahel zone in Africa.

As mentioned in 3.372, given a system efficiency of 15% and an annual average irradiation of 200 W/m^2 (this applies in much of the USA and southwest Europe), a surface of 34 km^2 is needed to supply 1000 MW electric power. For the same system efficiency and an annual average irradiation of 130 W/m^2 (this applies to much of middle and Western Europe), a surface of 50 km^2 is required for 1000 MW. The present cost of solar cells is so high that their terrestrial application for commercial power production is not (yet) economically feasible. The investment required for solar cell generators is presently about \$ 10000 to \$ 25000/kW, and is thus about 30 times as expensive as conventional power plants (114). In many countries, especially in the USA, the work on the development of more economical solar cells has been intense. The US Department of Energy, Wasthington D.C. estimates that by the middle of the eighties, solar cells will be available for \$ 500/kW (peak load), and by the beginning of the nineties, the price will drop to \$ 100 to \$ 300/kW (peak load), all in terms of constant dollars (133, 134).

The degree to which solar cell generators will contribute to future energy production will depend essentially on the success of efforts to reduce the cost of solar cells by mass production and thus to make the electricity produced by them competitive with that produced by other methods. It seems likely that for small plants (decentralized energy supply) in the kW range, solar cell generators will often prove a favorable solution.

4.32 Indirect methods

4.321 Water power

Water power is one of humanity's oldest energy sources. Because it is a regenerative energy source, kept in motion by solar energy, efforts have been made all over the world to increase the utilization of the hydraulic potential beyond the present level. Hydroelectric plants often change the natural water flow, but the resulting changes in the environment are not always negative; sometimes the climate is positively affected.

The technically utilizable hydropower potential of the world is presently estimated to be about $2200 \cdot 10^3$ MW generating capacity. In 1978, the installed electric capacity was only $372 \cdot 10^3$ MW (see Table 3-3) (114). However, the technically utilizable potential is much less than the theoretically available total, because the upper and lower parts of rivers cannot be used (the latter due to the fact that the alluvial plains are nearly flat. The technically utilizable potential, which amounts to about 50% of the total, is geographically distributed as follows: Asia, 27%; Africa, 20%; Latin America, 19%; Europe, including the USSR, 19%; North America, 13%; and Australasia, 2% (135).

The economically attractive fraction of the total potential is still less than the technically utilizable. However, it cannot be exactly stated, because it depends on many factors. The changes in the world energy market in the last few years have increased the number of economically attractive sites for hydropower. An important advantage of hydroelectric plants over thermal power plants is that the former make much more efficient use of the primary energy, the potential energy of the water. (The overall efficiency of a hydroelectric plant is 80 to 90%.)

In a comparison of the generating costs for hydroelectric vs. thermal-electric power, criteria other than the investment cost per kW output must be considered. For one thing, hydroelectric plants age less rapidly than others, and operating costs are lower. Although in the Federal Republic of Germany, for example, the investment in a pump-storage-reservoir power plant costs 600 to 700 DM/kW, while a gas-turbine plant costs only 350 DM/kW, the generating costs per kW are comparable. Furthermore, pump-reservoir plants can be used to supply power only during the periods of peak demand, thus allowing a more economical use of large thermal plants (especially nuclear plants) which are most efficient when operating at full capacity, and also, because of their low environmental impact, the expenditures to preserve environmental quality are very low for hydroelectric plants (136).

Hydroelectric power contributed about 5% of the primary energy consumed in the world in 1980, or 8% of the consumption in the world excluding the East Block.

Table 4-4: Fraction of electricity generated from individual energy carriers in several countries

	Net electric output[1)] in GWh in 1978				
Country	Geothermal energy	Hydro-electric	Nuclear energy	Conventional thermal	Total
Austria	–	24608	–	12321	36929
Belgium	–	496	11872	35988	48356
Canada	–	234190	29435	72712	336337
Denmark	–	23	–	19500	19523
Finland	–	9744	3106	21133	33983
France	–	68537	28999	199716	217252
Germany, F. R.	–	18204	33856	280500	332560
Greece	–	2978	–	16766	19744
Ireland	–	1013	–	8416	9429
Italy	2384	47138	4159	113733	167414
Japan	1000	74187	56131	406022	537340
Luxembourg	–	311	–	1007	1318
Netherlands	–	–	3811	55151	58962
Norway	–	80101	–	129	80230
Portugal	–	10532	–	3048	13580
Spain	–	41007	7302	46927	95236
Sweden	–	57074	22718	10532	90324
Switzerland	–	32510	7995	1845	42350
Turkey	–	9185	–	11277	20462
United States	2900	284001	276403	1722111	2285415
United Kingdom	–	5194	32462	231148	268804
USSR	–	168000	41500	926000	1135500
EC 9	2384	140916	115159	865159	1123618
World	6500	1510000	583000	5030000	7130000

1) i.e. after subtraction of the plant's own consumption.

Source: Statistical Office of the European Communities 1980, Office for Official Publications of the European Communities, Brussels 1980.

Table 4-5: The largest hydroelectric facilities of the world

Name of dam		Country	MW present	MW ultimate	Initial operation
Itaipu		Brazil/Paraguay	–	12600	U.C.[1]
Grand Coulee		U.S.A.	3563	9780	1941
Paulo Afonso		Brazil	1299	6774	1955
Guri		Venezuela	524	6500	1967
Tucurui		Brazil	–	6480	U.C.
Sayanskaya		U.S.S.R.	–	6400	U.C.
Krasnoyarsk		U.S.S.R.	6096	6096	1968
La Grande		Canada	–	5416	U.C. 1979
Churchill Falls		Canada	5225	5225	1971
Bratsk		U.S.S.R.	4100	4600	1964
Sukhovo		U.S.S.R.	–	4500	U.C.
Ust-Ilimsk		U.S.S.R.	720	4320	1974
Ilha Solteira		Brazil	3200	4100	1973
Cabora Bassa		Mozambique	2000	4000	1975
Inga		Zaire	350	3700	U.C.
Rogunsky		U.S.S.R.	–	3600	U.C.
John Bay		U.S.A.	2160	2700	1968
Nurek		U.S.S.R.	–	2700	U.C.
Sao Simao		Brazil	–	2680	U.C.
Volgograd – 22nd Congress		U.S.S.R.	2560	2560	1958
Chicoasen		Mexico	–	2400	U.C. 1978
Volga V. I. Lenin (Kuibishef)		U.S.S.R.	2300	2300	1955
W. A. C. Bennett		Canada	1816	2270	1969
Foz do Areia		Brazil	–	2250	U.C.
High Aswan (Saad el Aali)		Egypt	2100	2100	1967
Iron Gate		Rumania/Yugoslavia	2100	2100	1970
Tarbela		Pakistan	700	2100	1975
Bath County	P.S.[2]	U.S.A.	–	2100	U.C.
Itumbiara		Brazil	–	2100	U.C.
Chief Joseph		U.S.A.	1024	2069	1956
Salto Santiago		Brazil	–	2000	U.C.
Robert Moses-Niagara		U.S.A.	1950	1950	1961
Salto Grande		Argentina/Uruguay	–	1890	U.C. 1979
Dinorwic	P.S.	Great Britain	–	1880	U.C. 1980
Ludington	P.S.	U.S.A.	1872	1872	1973

[1] U.C. = Under Construction (Date indicates estimated date of initial operation).
[2] P.S. = Pump Storage Installation.

Source: T. W. Mermel, Contributions of Dams to the Solution of Energy Problems, 10th World Energy Conference, Istanbul, September 1977.

(In some countries, however, the fraction is larger (137)). It is estimated that the magnitude of this fraction will not change much up to the year 2000 (compare Table 2-2, Fig. 2-3). The fraction of electricity generated hydroelectrically in 1978, however, was about 23% (see also Table 4-4)(138). It is predicted that this fraction will have declined by 2000 to 16% (114).

In the industrialized countries, the hydraulic potential has already been largely put to use, so that the possibilities for further development of hydroelectric power are mostly limited to parts of Asia, Africa and Latin America. Table 4-5 shows the largest hydroelectric plants in the world (135).

For the sake of completeness, it should be mentioned that even the use of meltwater from glaciers might possibly be used for generating electricity. The conditions for this exist in Greenland and Antarctica, but at present it is not possible to make any concrete statements about an eventual utilization of this energy source.

4.322 Wave energy

The ocean waves which might be used as sources of energy arise from the effects of wind on the surface of the water. The energy of a wave can be calculated from the potential energy released by the orbital motion from the peak of the wave to the bottom of the trough. In violent storms, waves can rise to a height of 34 m from peak to trough, and can have a period of 16.5 s. The average power of ocean waves is about 77 kW/m length of wavefront (139–141).

The following values are typical for waves at the North Sea coast of Germany: average height, 1.52 m; average period, 6.42 s; and average power, 14.4 kW/m. Thus for a 250 km wavefront, the power is 3.6 GW and the potential energy is 64 TWh/a (142). According to the Marine Science and Technology Center, Japan, the supply of wave energy on the Japanese coast is between 17 kW/m and 100 kW/m for 3000 hours of the year.

Wave generators, which produce from 70 to 500 W, are already in use to provide power for isolated buoys, lighthouses, etc. The problem in large-scale utilization of wave energy is to develop a mechanism with which to convert the diffuse, changing forces in the waves into concentrated, direct forces which would work reliably in both low and high waves. Two suggestions have been made, a paddlewheel or a pressure hose arrangement near a coast. The paddlewheel, which would be half submerged, would utilize the orbital motion of the waves, with an efficiency of 70% and an output of kilowatts to megawatts. Fig. 4-17 shows one of the components for utilization of the orbital motion of waves (paddlewheel) (143). In a pressure hose generator, the changes in pressure below the water surface are transferred to a working fluid, which drives a turbogenerator. The attainable outputs should also be in the megawatt range. Systems with higher outputs could be built by connecting such generator units in parallel or in series.

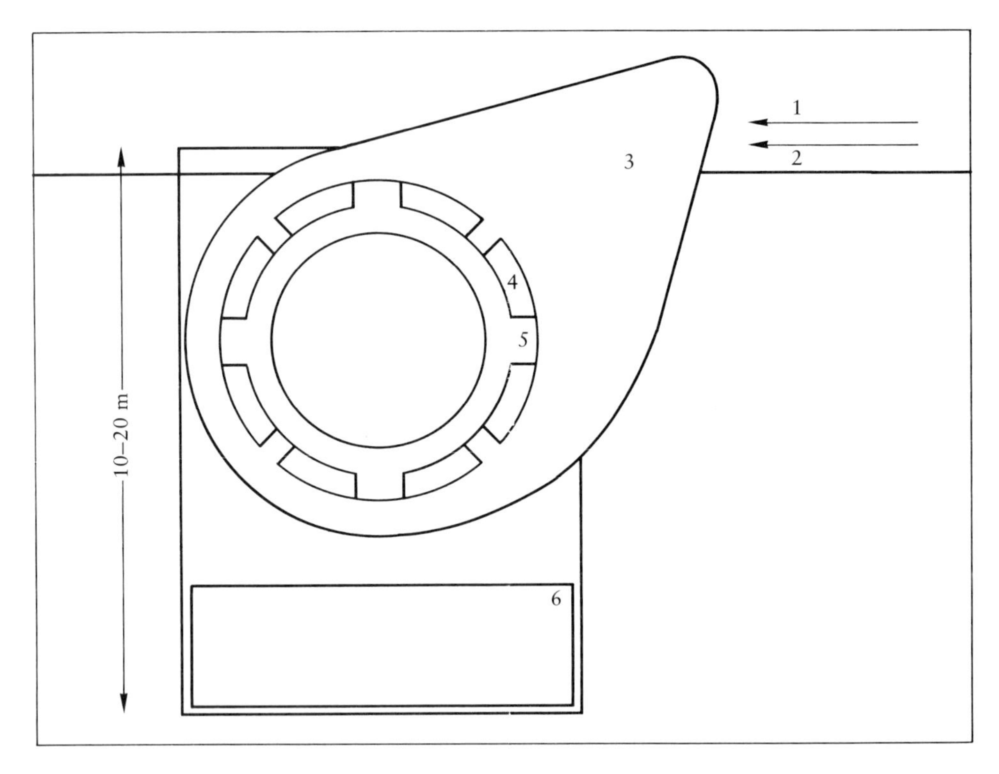

Fig. 4-17: Wave-energy converter
1 = Direction of wave propagation; 2 = Water surface; 3 = Rotating outer member; 4 = High pressure chamber; 5 = Inner cylinder; 6 = Support structure.

Source: J. Fricke, W. L. Borst, Energie aus dem Meer, Physik in unserer Zeit, Vol. 10, No 3, 1979.

It is in principle possible to operate wave energy converters on the open sea. The problem of energy storage could be solved by using the current to electrolyse water and transporting the resulting hydrogen to land. This type of power plant should not cause significant environmental problems, although the operation of wave power plants near the coast could have negative effects on the local biological equilibrium. The reduction in the orbital motion could reduce the transfer of oxygen and plankton between the surface and lower layers. It is not possible at present to estimate the cost of current generated by large-scale wave generating plants, if they should be realized.

4.323 Ocean heat and currents

Most of the solar energy absorbed by the earth is stored in the oceans, primarily near the surface. It has been estimated that the energy stored in the tropical oceans would suffice to supply about $6 \cdot 10^9$ people with as much energy per capita as was consumed in the United States, per capita, in 1970 (145). The main problem with utilization of this stored energy is the small (vertical) temperature difference to 20 to 30 K, which would preclude efficient energy conversion. For example, an ocean heat power plant working between 300 K and 280 K would have a Carnot cycle efficiency of about 7%. After allowing for all the losses which would occur, the effective efficiency would not be more than about 3% (Ocean Thermal Energy Conversion, OTEC) (116). What is more, a very large amount of sea water would have to be pumped through the heat exchanger of such a system.

It has been suggested in the USA (Offshore Technology Conference, Houston, 1975) that floating power plants (100 to 1000 MW) could be used to generate electricity from the temperature gradients of the ocean (compare Fig. 4-18) (144–147). The evaporator, condensors, turbines, generators and other parts of a power plant

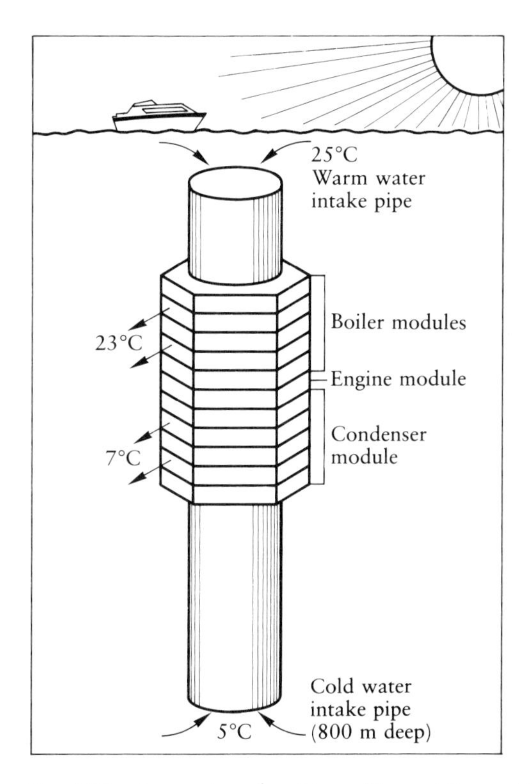

Fig. 4-18: Conception of an Ocean Thermal Energy Conversion (OTEC) Plant

Source: Project Interdependence: US and World Energy Outlook through 1990. A Report printed by the Congressional Research Service, U.S. Government Printing Office, Washington D.C., November 1977.

would be mounted on a platform, about 100 m in diameter and about 120 m high. A cold-water pipe about 15 m in diameter, made of fiber-glass-reinforced plastic, would extend several hundred meters down into the water and bring cold water (about 275 K) up to the platform. Warm surface water would be used to evaporate ammonia, which would drive the turbines. The gas would be condensed by heat exchange with the cold water. There are still technical problems to be solved before such a plant could be built, and corrosion by sea water and microbial growth would cause difficulties.

Utilization of the energy of ocean currents seems very unlikely. The currents arise because of regional differences in the amount of sunlight falling on the sea. Some of the important ones are the Gulf Stream, the Guinea, the Brazil, the Mozambique and the Australian Currents. The Gulf Stream has the highest velocity, more than 2 m/s. The power of the core of the Gulf Stream (average width 50 km and depth 120 m) is 24 GW. The currents could be used to drive turbines arranged along their course. If on assumes that for ecological reasons only 20% of the Gulf Stream could be utilized, then assuming an efficiency of 40%, about 2 GW electric power could be extracted from it (143). This estimate shows that utilization of the kinetic energy of ocean currents cannot contribute a significant fraction of the world energy supply.

4.324 Wind energy

Like water power, wind energy was one of the first energy sources to be tapped. The reason is that wind can be converted relatively easily to mechanically useful energy. It has been estimated that in the middle of the 19th centruy, 14% of the energy supply in the USA was provided by wind. However, cheap and abundant fuels steadily replaced wind power, because it is a variable source of energy. It is also difficult to use because of the low energy densities. (This also applies to the other methods of indirect utilization of solar energy, such as wave, ocean heat and ocean current energy.) In addition, the power which can be obtained from a wind turbine increases with the cube of the wind velocity (148).

The exact potential for usable wind energy can only be determined from measurements of wind velocity. Cost analyses for a large-scale utilization of wind energy (feeding into the electric network) in the Federal Republic of Germany have shown that it is feasible only in areas with an annual average wind speed of 4 m/s or more (the standard height for these measurements is 10 meters above the ground). In the USA there are areas in which the wind speed is 13 m/s or higher for 4000 to 5000 hours per year; in Europe, however, there are only a few areas (e.g. the northern coasts) where the wind velocity is greater than 4 m/s. The technically usable wind energy potential (with the efficiency of a 3 MW installation) have been estimated as follows: world total, $2.9 \cdot 10^5$ TWh/a; EC countries, 4590 TWh/a; Federal Republic of Germany, 220 TWh/a (142).

Due to the new situation in the world energy markets and the large potential for wind energy, there is an increasing interest in the potential commercial utilization of wind energy in many countries. Wind energy converters are at a high level technologically. At present, two types seem suitable for commercial utilization of wind energy: the horizontal-axle turbines with simple rotors and two or three rotor blades, of which many models have been tested, and the vertical-axle wind turbines based on the Darrieus principle (149). Fig. 4-19 shows examples of a large wind turbine generator. The wind turbines based on the Darrieus principle have the general advantage that they do not have to turn to catch the wind, but the disadvantage in comparison to horizontal-axle turbines that they develop only about 75% of the power of an optimally constructed horizontal axle turbine of the same size.

In a number of countries, pilot installations with wind turbines in the 100 kW range, and prototype installations with several MW power are now being built. However, wind generators cost about two to four times as much as conventional generators with the same output (114).

The area required for large-scale utilization of wind energy could be problematic. If 100 single 3 MW turbines are linked in a 300 MW power plant, the total land area

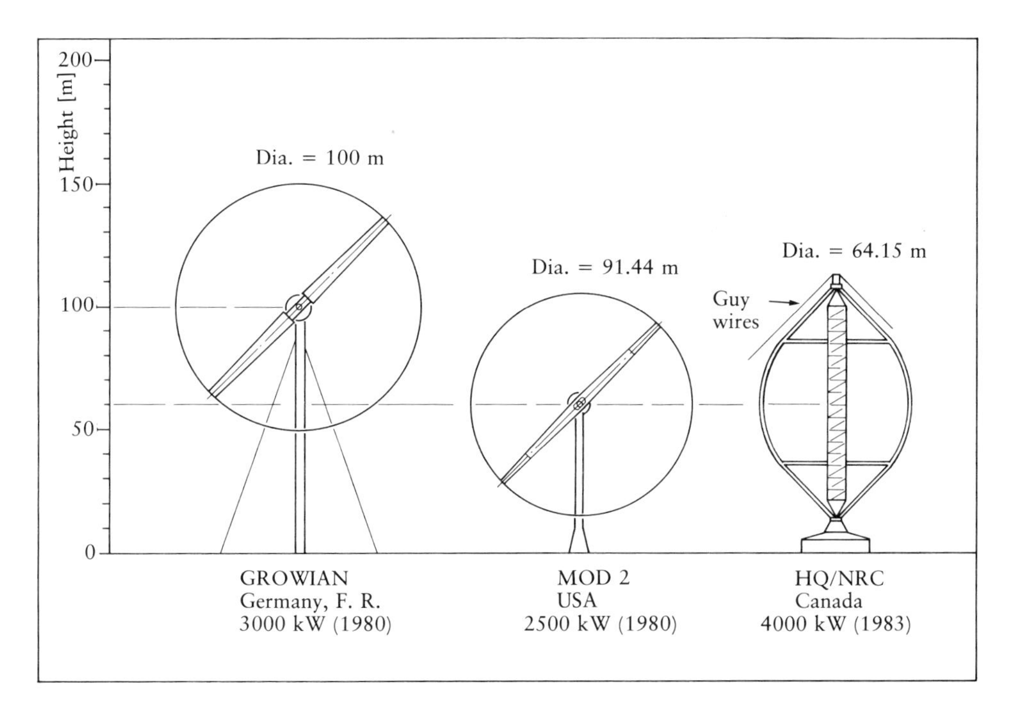

Fig. 4-19: Examples of large wind turbine generators

Source: S. Quraeshi, Large wind turbine generators for electrical utility application, 11th World Energy Conference, Vol. 1A, Munich, September 1980.

required is 10 to 12 km^2 (142). However, only a small part of this area (0.2 km^2) is required for the converter system itself. An intolerable noise load is not expected (150).

Wind energy cannot be expected to make a significant contribution to the total energy supply until either large-scale energy storage systems are available, or the wind-derived energy can be integrated into an existing energy supply system. It can be assumed that some countries will be interested in wind energy converters with outputs up to 100 kW for supplying power in remote areas (decentralized energy supply) (151). Larger installations with 100 kW to several MW could be integrated into existing energy supply networks or used to run pump-reservoir systems.

4.325 Utilization of stored solar heat with heat pumps

In principle, heat pumps can be used to supply heat for space and hot water heating. Heat pumps can utilize low-temperature heat, which is "pumped" to a higher temperature at the expense of mechanical energy, and is thus available for space or water heating. The source of heat can be the solar heat (environmental heat) stored in water, air or the soil; or it can be solar heat from flat collectors; or waste heat from households or industry can be used (heat recycling).

The process of heat pumping is a thermodynamic cycle. Fig. 4-20 shows the basic arrangement of a compression heat pump, which is the most advanced type (152, 153). It consists essentially of an evaporator, a compressor, a condenser (liquefier) and the expansion valve. The working fluid is a "coolant", which is evaporated at the temperature T_u in the evaporator. (The heat Q_z is transferred from the heat source to the working fluid in this step.) The working fluid is pumped out of the evaporator by the compressor, which then expends mechanical energy W to re-compress the fluid. The fluid is warmed by compression and is pumped under higher pressure into the condensor. There it condenses at the temperature T_k, the boiling point of the liquid at the pressure maintained in the condensor, and releases the heat of condensation Q_a to a heat carrier. Finally, the pressure is reduced by opening the expansion valve. The liquid working medium is returned to the evaporator, where the cycle begins again.

The efficiency ε of a heat pump is generally defined as the ratio of the heat delivered, Q_a, to the mechanical energy W expended in the compression step. For a Carnot-cycle heat pump process,

$$\varepsilon_c = \frac{Q_a}{W} = \frac{T_k}{T_k - T_u} \tag{15}$$

T_k is the condensation temperature and the temperature at which the heat is used, and T_u is the temperature of the heat source. Both are in Kelvin. The efficiency of an actual heat pump ε_p is always less than the theoretical value ε_c, due to losses occur-

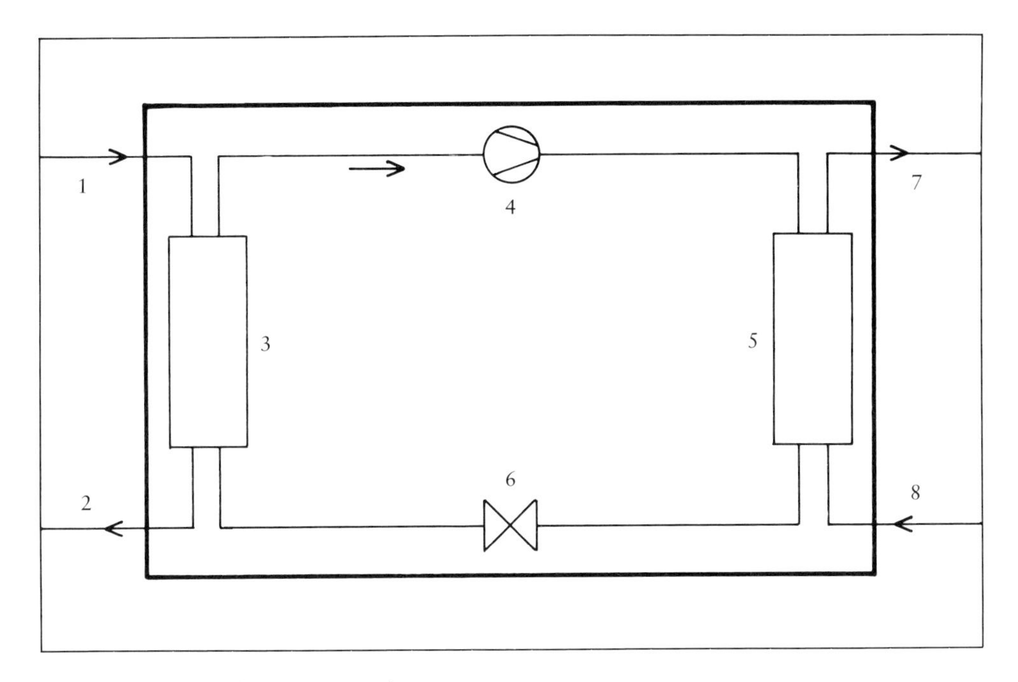

Fig. 4-20: Diagram of a compression heat pump
1 = From the heat source; 2 = To the heat source; 3 = Evaporator; 4 = Compressor; 5 = Condensor; 6 = Expansion valve; 7 = To heating system; 8 = From heating system.

Source: H. Michel, J. Pottier, J. Jaéglé, Compression and Adsorption Heat Pumps. Fields of Use and Future, 10th World Energy Conference, Istanbul, September 1977.

ring in the process ($\varepsilon_p \approx 0.5\ \varepsilon_c$). The practically attainable values of ε_p depend on the climate and the heat source, but the annual average is generally between 2 and 3.5.

The efficiency is a measure of the quality of the heat pump, and it should be as high as possible, so that the work W done to pump a given amount of heat Q_a can be as small as possible. Equation (15) shows that the efficiency depends on the temperature difference ($T_k - T_u$). At low environmental temperatures the efficiency is low, which leads to a high requirement for mechanical energy W. For this reason, a bivalent operation, in which the heat pump is supplemented by a second heating system, is suggested. When the temperature of the heat source falls below a certain limit, the second heater either takes over or supplements the output of the heat pump. In monovalent systems, the heat pump supplies the heat alone.

The choice of working fluid and the tuning of the individual components is important for the effective use of heat pumps. The maximal condensation temperature can be as high as 60° C, so that in principle hot water can also be supplied by a heat pump. As with plate collector systems, the heating surface should be large, so that the heat carrier temperature can be as low as possible, as in sub-floor systems, for

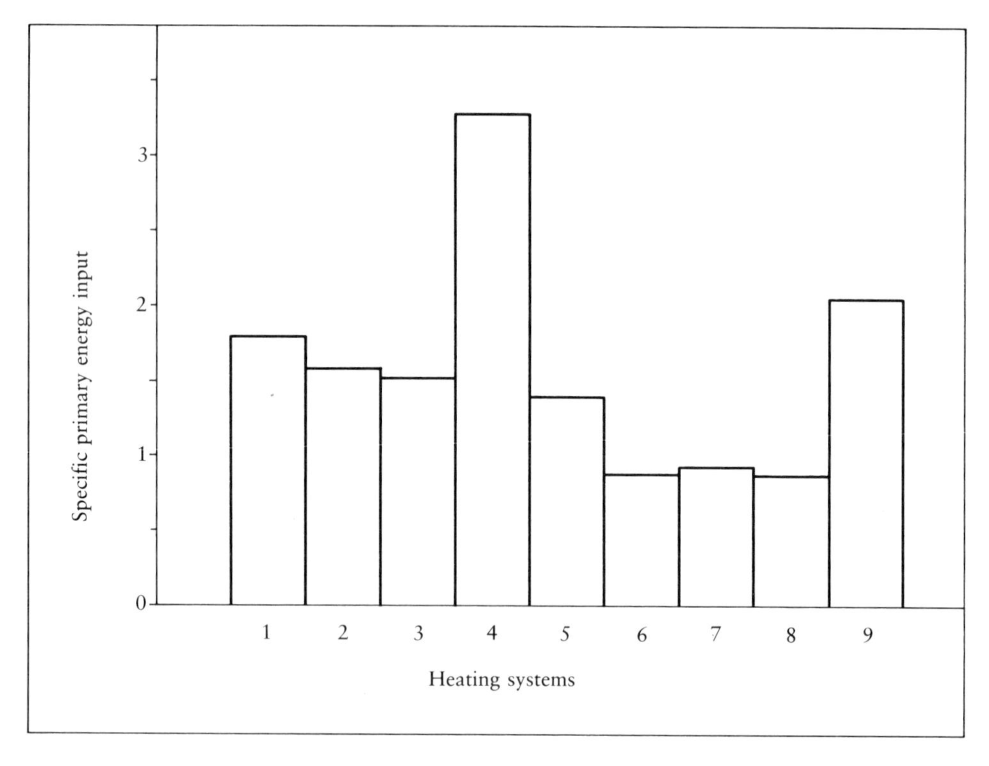

Fig. 4-21: Primary energy input required to produce one unit of usable heat from various heating systems 1 = District heating; 2 = Oil furnace; 3 = Gas furnace; 4 = Electric night storage unit; 5 = Bivalent electric heat pump; 6 = Monovalent diesel heat pump; 7 = Bivalent diesel heat pump; 8 = Solar + oil furnace; 9 = Motor-generator with supplementary furnace.

Source: D. Oesterwind, Aspects of the soft-hard discussion – Centralized and dezentralized energy systems as a common option, Atomenergie-Kerntechnik, Vol. 34, No 3, 1979.

example. Ammonia or fluoro-chloro-hydrocarbons (e.g. freon) is used as working fluid. The condensor may work either with pistons or a turbine.

Heat pumps are usually run with electric motors, but they can be run with combustion engines which burn natural gas or diesel oil, in which case the waste heat from the engine is also utilized. In short, the use of heat pumps to utilize stored solar energy always requires a "high value" (exergy-rich[1]) source of energy (153, 154).

In terms of the amount of primary energy used, a heat pump driven by a combustion engine is more favorable than one driven by an electric motor. For one thing, the heat generated by the combustion engine can be used for space heat, and for another, the electricity is generated at a relatively low efficiency of about 33%. Fig. 4-21

[1] Exergy is energy which can be completely converted to work. Anergy is energy which cannot be converted to work. Energy = exergy + anergy.

shows the amount of primary energy used to produce one unit of consumable energy (specific primary energy consumption) (155).

As mentioned in 4.311.2, the use of a heat pump in combination with solar collectors extends the season in which, in a temperate climate, solar heat alone is adequate to provide for space heating. However, in such multiple systems (heat pump/solar collector and conventional furnace), the savings in fuel are not (yet) sufficiently large to justify the investment on economic grounds alone. Since the irradiance in the temperate zones is low during the heating season, a collector system which can be installed for a reasonable cost is not adequate, by itself, to supply space heating for the entire year. A new and interesting solution is the combination of an energy roof or facade with a heat pump (156). In this system, a solar absorber (the roof or facade) takes on the function of solar collector. Its surface temperature is kept low by a flow of cold liquid through it; the liquid is cooled by the heat pump, which removes the heat from it. The advantage is that the solar absorber makes use not only of radiant solar energy, but also of environmental heat, which can be provided by atmospheric moisture (rain) or the air. In other words, the energy roof or facade functions even at night or when the sky is completely cloudy, so that storage problems are less serious. However, for extremely cold days, supplemental heat is needed.

In evaluating the future of heating systems, one must consider not only the amount of primary energy requried to produce a unit of consumable energy, but other factors as well, such as economics and politics. All costs which arise from installation and use of a heating system must be considered, along with the differences in the amount of environmental pollution resulting from use of the various systems (155). It should also be kept in mind that, as discussed in 2.331, the industrialized countries are presently using a large fraction of their petroleum to heat water and buildings (low-temperature heat), and that some of these countries import most or all of their petroleum. Therefore, it could be entirely reasonable in these countries to use electrically driven heat pumps to deliver low-temperature heat, to substitute for petroleum, and thus to reduce their dependence on imports.

4.326 Photochemical conversion

4.326.1 Photosynthesis

The techniques for utilization of solar energy which have been discussed so far have been based either on use of heat or natural processes resulting from the sun's radiation, or on direct conversion of solar radiation to electricity (photo-voltaic conversion). Another type of method is based on solar chemical processes, i.e. photochemical or photoelectrochemical processes which are induced by light.

Photosynthesis by green plants is nothing other than a photochemical reaction in which CO_2 and H_2O are converted to carbohydrates and oxygen (157, 158). The process can be described by the following equation for the production of glucose:

$$6\,CO_2 + 6\,H_2O \underset{\text{Respiration}}{\overset{\text{(light)}}{\overset{\text{Photosynthesis}}{\rightleftharpoons}}} C_6H_{12}O_6 + 6\,O_2 \tag{16}$$

Glucose is the starting material from which starch, cellulose and protein are biosynthesized. The latter substances are used as food by bacteria, animals and people, i.e. they are either transformed into body substance or they are broken down and combined with oxygen to provide energy for the life processes. A human being, for example, produces about 1 kg CO_2 per 24 hours. (CO_2 is also produced in the process of rotting of organic matter by bacteria and fungi.)

The formation of photosynthesis products, i.e. carbohydrates (biomass), by green plants is thus associated with the uptake of carbon dioxide and the release of oxygen. Therefore, the process of photosynthesis could not develop until carbon dioxide was present in the atmosphere, which was about $3.5 \cdot 10^9$ years ago. That is to say, free oxygen has been accumulating for about $3.5 \cdot 10^9$ years, released partly by photosynthesis (originally in blue-green algae), but also by direct decomposition of water by ultraviolet radiation in the upper atmosphere. The higher land plants appeared about $400 \cdot 10^6$ years ago, and since then the accumulation of oxygen has proceeded at a relatively rapid rate. The atmosphere has had its present composition for the last 2 to 3 million years: 78.08% nitrogen (N_2), 20.95% oxygen (O_2), a locally variable amount of CO_2, which depends of life processes and human activities (the average is 0.03%), and traces of noble gases and other compounds. (The volume percentages are for dried air – the amount of water vapor varies greatly.)

All life on earth depends on photosynthesis, and as a result of this process in all land and water plants, a biomass of about $1.7 \cdot 10^{11}$ t is synthesized each year.

Plants also produce free oxygen (see equation (16)). The annual assimilation of CO_2 is about $27.5 \cdot 10^{10}$ t, or about 5% of the amount present in the atmosphere, and the production of O_2 is about $2 \cdot 10^{11}$ t. About $2.3 \cdot 10^{11}$ t water is needed for the annual O_2 production. (The earth has about $1.5 \cdot 10^{18}$ t water, of which $1.37 \cdot 10^{18}$ t is in the oceans (159)).

As discussed earlier, the fossil fuels, which at present supply about 90% of the world's energy, are none other than solar energy which was stored for millions of years (see 3.1). They have been produced by complicated transformations of biomass, which was originally produced by photosynthesis. Coal, for example, was produced by geochemical effects such as high pressure and temperature. The process, which leads to increasing enrichment in carbon, is called carbonization.

The free oxygen produced by plants is consumed not only by respiration, but also by technological combustion. However, large amounts of CO_2 are produced by burning of fossil fuels, and the CO_2 content of the air has been rising as a result, because photosynthesis and other natural processes have not been able to assimilate the excess CO_2 (see 5.32).

Although photosynthesis is the largest global production process, it is very complicated, and has not been completely worked out in detail (109, 160). In essence, however, the chlorophyll in green plants absorbs a relatively wide range of sunlight with wavelengths between 400 and 700 nm. Through a rather complicated series of biophysical reactions, the absorbed light quanta increase the electronegativity of certain compounds to the point that they are able to "electrolyse" water. The oxygen is released as the gas O_2, but the hydrogen atoms are transferred to CO_2, thus reducing it to carbohydrate (160). The primary process in photosynthesis is thus the light-induced decomposition of water (photolysis):

$$H_2O \xrightarrow{h\nu} H_2 + \frac{1}{2} O_2 \quad (17)$$

Water, however, can only be directly photolysed by ultraviolet light with wavelengths shorter than 200 nm, and only a vanishingly small fraction of the sunlight which reaches the earth's surface has wavelengths that short (compare Fig. 3-14). (Ultraviolet light splits water into H-and OH-radicals.) In order to make use of visible light, which accounts for about 45% of the sunlight reaching the earth's surface[1)], plants depend on a multiple-quantum process catalysed by chlorophyll. Chlorophyll is embedded in the membranes of chloroplasts, which are small bodies within plant cells[2)] (161). It is associated with a number of other pigments, including a complex called the electron transport chain. When light is absorbed by chlorophyll, one of the electrons in the molecule is excited to a higher energy state (as in the photovoltaic effect in metals or semiconductors). By a process which is not understood, the energy of this electron is transferred to other electrons within the electron transport chain, which are then "pumped" in a two-step process to a sufficiently high energy to be transferred to the oxygen of water. This releases highly reactive hydrogen atoms, which are used by the cell to make energy-rich carbohydrates.

4.326.2 Biomass

Plants are the only organisms which are able to utilize a substantial part of the available solar energy. There are plans to increase the efficiency of conversion of solar energy into biomass by appropriate plant breeding. The efficiency of light conversion is calculated on the basis of the dry mass synthesized by photosynthesis, on the assumption that it is all polymeric carbohydrates. As a first approximation this is adequate, because plant material contains 90% or more C, H and O. (5–10% is nitrogen and minerals such as phosphorus, sulfur and potassium). 1 g of dry biomass

---

1) The peak intensity lies at 550 nm, which is green light.

2) Chloroplasts are about 3 × 5 μm in size. They contain densely packed membranes in which the chlorophyll is embedded. The chlorophyll is responsible for the green color of leaves.

corresponds to 1.7 kJ. (This is only the net gain in substance, because the photosynthesizing plant uses some of the energy itself, for instance for transpiration.) Higher plants (cultivars) need 300 to 800 kg water for each kg dry biomass they synthesize. Most of the water taken up by the roots is evaporated through the leaves, and heat (sunlight) is needed for this. About $1.7 \cdot 10^{11}$ t biomass is synthesized per year by all the land and water plants on earth. The energy content of this biomass is about $3 \cdot 10^{21}$ J, which is about 10 times the present annual world energy consumption. Therefore it is understandable that many countries are investigating the extent to which photosynthesis can contribute to the energy supply (161, 162).

On the world-wide scale, the efficiency of photosynthesis of biomass is not high. One reason is that the CO_2 content of the atmosphere, 0.03%, is considerably less than the optimum for photosynthesis. On land, only about 0.3% of the irradiance is converted to chemically bound energy in the form of biomass. In the oceans, it is only 0.07%. For the entire earth, the efficiency is 0.12% (161).

Different ecosystems have very different capacities for synthesis of plant mass. A few typical values for annual production of dry plant matter follow: tropical forest, 2.0 kg/m^2 (with a total surface of $20 \cdot 10^6$ km^2, this makes $40 \cdot 10^9$ t/year); temperate zone forest, 1.3 kg/m^2 (with a surface of $18 \cdot 10^6$ km^2, this makes $23.4 \cdot 10^9$ t/year); temperate zone grasslands 0.5 kg/m^2 (with a surface of $9 \cdot 10^6$ km^2, this makes $4.5 \cdot 10^9$ t/year); ocean, 0.13 kg/m^2 (with the water surface of $332 \cdot 10^6$ km^2, this makes $43 \cdot 10^9$ t/year); average for the entire surface of the earth, 0.32 kg/m^2 (total surface is $510 \cdot 10^6$ km^2, which makes $168 \cdot 10^9$ t per year) (162). There is no technical production process on earth which even approaches a comparable order of magnitude.

The effectiveness of biomass production can be increased by using suitable cultivars and modern agricultural techniques, such as irrigation and fertilization. In temperate climates, for example, the following yields can be obtained per square meter and day: for grain, about 20 g, for sugar beets, 28 to 31 g, for maize (in the USA), 50 g. Maize converts 1.3% of the radiant energy to chemical energy (in the form of biomass), which means that the efficiency here is 1.3%. However, intensive agriculture is technically elaborate and energy intensive. In Western agricultural countries, 1 J energy input is required to raise biomass equivalent to 6 to 9 J (163, 164). The energy input is compounded from the energy expenditure for the production of fertilizer and agricultural machinery (to replace manpower) and for diesel fuel consumed during cultivation. In the USA, for example, the ratio of energy expended to biomass energy harvested has risen in the case of maize from about 1:4 in 1940 to 1:2.8 in 1970. A similar trend can be seen with other crops, for example soy beans and potatoes (162).

The rapidly growing species of trees (such as poplars) and sugar cane are among the plants with the highest efficiency of conversion of solar energy to biomass. There are plans to raise such plants on a large scale (energy farming), and then to use the biomass as a raw material for the production of fuel or plastics. One example of such plans is the cultivation of sugar cane in Brazil. In general, the production of biomass for energy only has not yet been economically feasible (165). The greatest disadvantage of renewable biomass compared to fossil fuels is the low specific energy content (due to the relatively large amounts of water and oxygen in it). Another serious disadvantage is the large amount of land required to raise the biomass (due to the relatively low yield per unit of area). This leads to a large requirement of energy for collecting and transporting the biomass.

The American Nobel Prize winner M. Calvin (Berkeley) began experiments on "agricultural" oil production from plants of the Euphorbia family. These are found in arid regions, and their sap consists of a mixture of water and hydrocarbons similar to those found in petroleum. It would be possible to make gasoline or other oil products from them (166, 167). There has also been some success in raising one-celled algae (Chlorella, Scendesmus) instead of the usual cultivars.

None of these methods has yet proved economical. It can also be objected that the same area is needed for the production of food (or algae for the production of food or animal feed) more urgently than for production of fuels (see 2.32), especially since intensive farming for the production of biomass is energy intensive and requires elaborate technology. Furthermore, the use of agriculturally produced fuels on a large-scale would cause environmental problems similar to those caused by the use of fossil fuels (see 5.3).

The world production of vegetable food is about $1.4 \cdot 10^9$ t dry matter per year, which is about 1.3% of the total annual photosynthetic production on the land area. A human being needs about $4 \cdot 10^9$J per year as food, which corresponds to 250 kg carbohydrate. Purely from the standpoint of energy, the present world harvest would be enough for $5.6 \cdot 10^9$ people, but in spite of this, the amount of food available in many parts of the earth is insufficient (see 2.32). This is partly due to the fact that almost no diet is strictly vegetarian, and about 9/10 of the energy is lost when the plant food is fed to animals. (The world nutrition problem is also partly a matter of distribution, because in several areas of the world, including the USA, Canada and Western Europe, there is sometimes an overproduction of food.) At present, it is assumed that an area of 4000 m^2 of agricultural land is needed to feed a single person. (In the Federal Republic of Germany, for example, the area of agricultural land is 2000 m^2 per capita.) A total of 21% of the surface of the earth might be suitable for various forms of agriculture; 11% is already so used. It follows that the potential agricultural land would be enough to feed at least $8 \cdot 10^9$ people, given the present agricultural capacities (see 2.31) (162).

The planting of special energy farms would appear, especially for densely populated industrial countries, to be unpromising, for the reasons mentioned. The generation of energy from biomass could be useful, however, wherever large amounts of biomass are generated for other reasons, for example in garbage disposal in industrial areas or for the utilization of agricultural and forest wastes. The latter seems especially promising in developing countries which lack an energy infrastructure (reads, power networks, etc.), i.e. where decentralized energy supplies are needed.

Thermal, chemical and biological techniques can be used to convert biomass to energy. Of the thermal techniques, the burning of wood or straw is the oldest known to man. For some time, in the industrialized countries garbage-burning plants have been built in the population centers, and the energy so released can be used. (Agricultural operations sometimes include straw burners which are used to produce energy.) Large-scale burning of biomass for energy production is only practical when there is a continuous demand for energy or a continuous supply of biomass. If these conditions do not apply, it is sometimes desirable to convert biomass into fuel which can be stored. In this case, the biomass can be converted by pyrolysis to oil or combustible gas.

In the USA, a chemical process (Eco-Fuel II process) has been developed to convert biomass into a dry, finely granular (particle size < 0.8 mm) material by use of chemicals. This fuel can be stored and burned in coal-dust burners.

The two most important biological techniques are production of biogas and production of ethanol from biomass (biological waste materials). Biogas consists of 50–70% methane, and is formed by anaerobic rotting of biomass. The biological wastes which can be utilized in this way include human and animal excrement, or plant wastes such as grass or straw; water is also required for the decomposition process. For optimal anaerobic decomposition, the ratio of carbon to nitrogen in the waste material should be about 25 : 1. (Excrements have relatively high nitrogen contents, while plant matter has a relatively high carbon content.) The time required for decomposition is 15 to 25 days at temperatures between 30°C and 40°C. The decomposition in biogas plants is carried out by microorganisms. The heat content of the gas is about 20 MJ/m^3. Because of the temperatures needed, the conditions for the use of biogas plants are most favorable in developing countries like Pakistan, India, Bangladesh, Indonesia and the P.R. of China, which have relatively high average temperatures and, due to their lack of infrastructure, need decentralized energy supplies (168).

Because they are liquids, alcohols have come under increasing consideration as fuels. At present, the chemical synthesis of alcohols for industrial purposes is most economical when petroleum is used as the starting material. Nevertheless, the further development of biological alcohol production (ethanol) from biomass has recently been seriously considered. The starting materials are plant products which contain sugars, starch or cellulose. In alcoholic fermentation, sugar-containing solutions are fermented by certain yeasts at temperatures of 30°C–40°C. However, only sugar-containing substrates can be directly fermented; the others must first be "released" (see 4.671). In Brazil, for example, sugar cane is being used for ethanol production. The ethanol is then used in a mixture of 20% ethanol and 80% gasoline as a fuel for automobiles, which is possible without major changes in the engines (169).

4.326.3 Photolysis of water

As mentioned above, there seems to be little point in planting special energy farms, except in a few countries, for the purpose of large-scale energy extraction from biomass, because this would compete with other land uses which are more valuable (food production). Therefore, attempts are being made to generate free hydrogen by photolysis of water, i.e. to use the electron transport system without reducing CO_2. In other words, the hope is to use a synthetic membrane system to produce hydrogen without forming biomass (170).

As mentioned in 4.326.1, water itself does not absorb visible light, but only ultraviolet with a wavelength $\lambda < 200$ nm, which splits it into H- and OH-radicals (171). If suitable absorbers (photocatalysts) are mixed with the water, it is possible to obtain photolysis with visible light. The photocatalyst absorbs the sunlight and transfers the energy to the water, with as little loss as possible. In this way, photo-

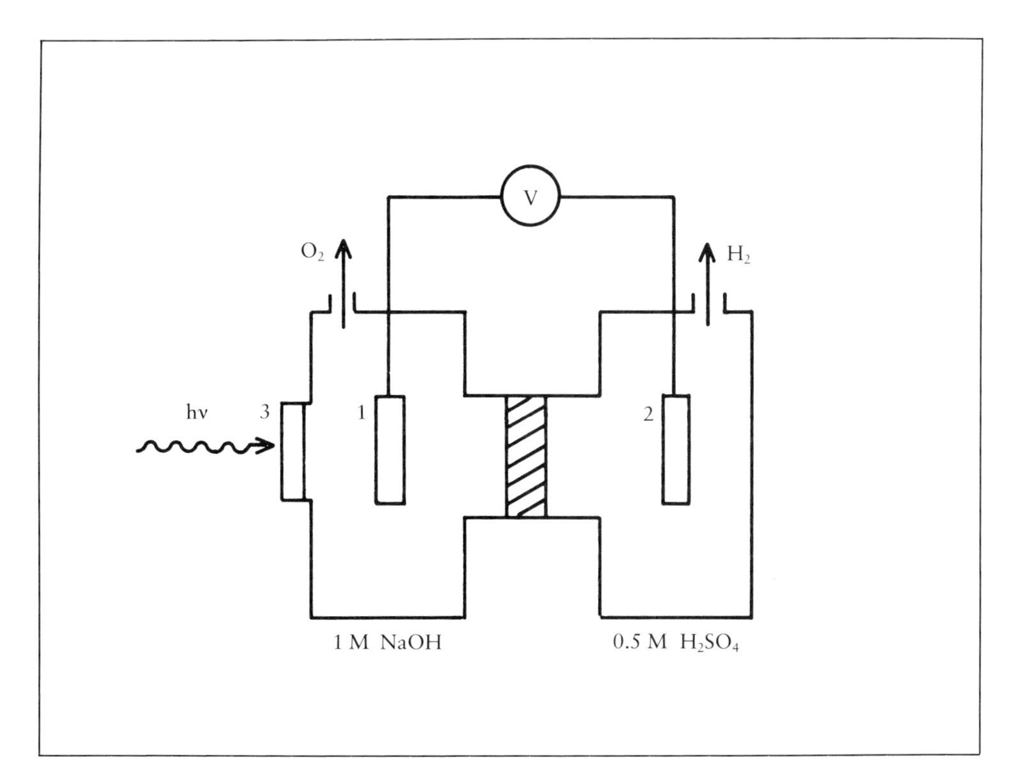

Fig. 4-22: Diagram of the Fujishima-Honda cell
1 = Rutile crystal; 2 = Pt electrode; 3 = Quartz window.

Source: N. Gethoff, S. Solar, M. Gohn, Solar Energy Utilization, Die Naturwissenschaften, Vol. 67, No 1, 1980.

lysis of water according to equation (17) is possible (172). In green plants, of course, the photocatalyst is chlorophyll.

Certain soluble salts, including some cerium and iron compounds, and some semiconductors, e.g. a rutile crystal, n-TiO_2, are photocatalysts. In solution, their ions are present in different oxidation states. Absorption of light changes the oxidation state of the ions, which can then either donate electrons to water to produce hydrogen, or extract electrons to generate oxygen (109, 116).

If the appropriate semiconductors are exposed to light, electrons are excited from the valence to the conduction band. (The semiconductor itself is the photocatalyst in which the multiple-quantum process takes place.) The electron-hole pairs generated in the semiconductor effect the decomposition of the water. The semiconductor (e.g. n-TiO_2) is used as the light-absorbing electrode (anode), and is submersed in an electrolyte (e.g. 1M NaOH); the cathode is platinum in 0.5M H_2SO_4.

With this cell, its inventors (A. Fujishima and K. Honda) were the first to observe photolysis of water on the surface of a n-TiO_2 semiconductor. The oxidation reaction (formation of O_2) occurs at the anode, while the reduction (H_2 formation) occurs at the cathode. Fig. 4-22 is a schematic representation of the Fujishima-Honda cell (172). The original cell had a total efficiency (energy of the H_2/light energy falling on the cell) of a few tenths of a per cent (173), but by doping the n-TiO_2 with strontium, the efficiency was increased to a few percent.

This discussion shows that it is possible in principle to utilize solar energy in the form of photolytically generated H_2. Hydrogen is in many respects a versatile and non-polluting secondary energy carrier (see 4.66 and 4.672). However it will undoubtedly be necessary to make a serious effort, both in applied and basic research, before hydrogen can be generated by photolysis for a reasonable cost.

4.4 Secondary energy from tides

As discussed in 3.38, economic utilization of tidal energy depends not only on the average difference between high and low tides, but also on geographical conditions like the extent and depth of bays or estuaries. There are relatively few sites where the necessary conditions occur (114).

The first large tidal power plant was built at the mouth of the Rance near St. Malo in northern France in 1966. It has an output of 240 MW, and an expansion to 320 MW is being considered. The Soviet Union also has a tidal power plant at Kislogubsk (on the Arctic Sea), with 0.8 MW capacity (114, 144). The overall efficiency of tidal power plants is 80–90%.

Other tidal power plants are planned. The economically feasible potential for tidal energy in a number of countries (e.g. France, Great Britain, the USA, Canada and the USSR) is rising (174–176). It can be assumed, nevertheless, that tidal energy will remain a matter of local significance, and that it will never assume a significant position as a source of energy on the world-wide scale.

4.5 *Secondary energy from geothermal sources*

To date, the utilization of geothermal energy has been limited to use of dry steam or water-steam mixtures for generation of electricity, space and water heating. (Table 4-4 shows the fraction of electricity generated from geothermal sources in several countries (138)).

Dry steam sources are especially favorable for generation of electricity, but they are rare. Some examples are Larderello, Italy (500 MW), The Geysers, USA (500 MW), El Salvador in Central America (90 MW), and Onikobe, Japan (25 MW) (177, 178). The source at Larderello has been used for generating electricity since 1904, and the steam from The Geysers, in Northern California, since 1960. Due to the low pressure and temperatures (about 7 bar and 480 K), the conversion efficiencies with geothermal sources are generally only 6 to 18%. In spite of this, the power plants in California are cheaper than conventional or nuclear power plants, both in construction and in operation.

Most geothermal sources produce mixtures of steam and water. Before the steam can be used to generate electricity, it must be separated from the water. There are wet steam sources in Wairakei, New Zealand (193 MW), Cerro Prieto, Mexico (75 MW), Onuma (10 MW), Otake (11 MW) and Matsukawa (22 MW) in Japan, and small power plants at Pauzhetsk, USSR (5 MW) and Iceland (3 MW) (198). Electricity can also be generated from hot water, but in this case the heat must be transferred by heat exchanger to a secondary liquid with a low boiling point (e.g. freon or isobutane). The advantage of this technique is that the turbine is less subject to corrosion, and undesirable substances are not released into the environment.

The geothermal sources now in use are economically competitive. The installation and operation costs of power plants at different geothermal sources vary considerably. The installation cost at The Geysers (USA) was $ 110/kW, based on a 110 MW unit, while the cost at Cerro Prieto (Mexico) was $ 80/kW, based on a 75 MW unit (in 1973 dollars) (179). The installation costs for generating electricity from hot water sources are higher, due to technical and environmental problems.

Because of high transport costs, the hot water can only be used in the immediate neighborhood of the hot spring, but on a local scale, it can make a very significant contribution to the energy supply. In Reykjavik, for example, about 90% of the houses are heated by natural hot water.

4.6 *Secondary energy carriers*

4.61 Electricity

4.611 Generation

4.611.1 Generation of electricity in thermal power plants

Electricity can be generated from any source of primary energy. This extremely convenient form of secondary energy, which creates essentially no environmental problems at the point of consumption, will presumably never be replaced by another equivalent form of secondary energy. In particular, it is irreplaceable for lighting and the electrochemical industry.

The capacity for electric generation has risen over the past decades in all parts of the world (see Table 4-6). The trend will continue, at different rates in different areas, as the industrialization of many countries proceeds.

In 1978, for example, about 72% of the electricity generated in the world was based on the conversion of primary energy into heat in conventional thermal power plants. The efficiency of power plants run on fossil fuels is limited by the laws of thermodynamics to about 40%, i.e. only this fraction of the energy in the primary fuel is converted to electricity. The remainder, which is the larger part, generally goes as waste heat into the coolant, which is often a river, or into the atmosphere (see 2.23). In 1978, nuclear energy contributed 8% of the total electricity. (LWR power plants have an efficiency of around 33%). Geothermal energy contributed only 0.13% and hydroelectric plants 21% (the overall efficiency of running-water power plants is 80–90%). (As discussed in 4.312, solar-generated heat can also be used to generate electricity.)

By far the largest fraction of the world's electricity is thus generated in fossil-fueled thermal power plants. The fuel can be hard or soft coal, oil, gas or even garbage; the steam turbines are essentially the same, and only the burners differ with respect to the details of fuel feed and ash and smoke removal, and in the temperature and pressure at which the steam is generated. In every case, the heat generated by combustion is used to generate steam, which drives a turbine. The rotational energy is converted to electricity in a generator in which a rotating magnet interacts with stationary magnets, according to the electrodynamic principle (see 2.23). Except for the method of generating heat, a nuclear power plant operates in the same way as a normal steam plant. The kinetic energy of the nuclear fragments arising from fission is absorbed as heat by the coolant, i.e. the reactor (and heat exchanger) replaces the boiler of a normal steam power plant. The steam temperatures and pressures attained with the LWRs in current use are not as high as in fossil-fuel plants, so that the LWR has a total efficiency of only about 33%.

Table 4-6: Electricity generated in GWh (Total generation, including the power consumed by the generator)

Year	1948	1955	1960	1965	1970	1975	1979	kWh/capita[1] 1979
World	800 800	1 540 100	2 300 900	3 379 900	4 908 400	6 461 600	7 993 596	1 860[2]
United States	336 808	629 010	844 188	1 157 583	1 639 771	2 003 002	2 451 700	11 036
Canada	47 262	82 816	114 378	144 274	204 723	273 392	348 500	15 130
Germany, F.R.	34 093	76 542	116 418	172 340	242 612	301 802	372 765	6 145
France	28 851	49 627	72 118	101 442	140 708	178 514	241 124	4 547
United Kingdom	48 036	94 076	136 970	196 495	249 193	272 082	299 960	5 340
Italy	22 694	38 124	56 240	82 968	117 423	147 333	180 522	3 175
Japan	35 579	65 193	115 472	188 377	350 590	475 794	599 600	5 210
USSR	66 341	170 225	292 300	506 700	740 900	1 038 600	1 239 998	4 751
India	5 725	10 877	20 123	36 755	61 212	85 926	109 100	170
China, P.R.	n.a.	n.a.	n.a.	n.a.	102 000[3]	150 000	260 500	273

[1] Author's calculations.
[2] In 1976, the values for the individual regions (in kWh/capita) were: Africa, 338; North America, 10 070; South America, 866; Asia, 417; Europe, 4020; Australia, 4479. (The values for the USSR are not included in the values for Asia or Europe.)
[3] This figure was for 1972.

Sources: United Nations, Statistical Yearbook 1959–1977, New York 1960–1978.
Statistik der Kohlenwirtschaft, Essen und Köln, September 1980.

Liquid and gaseous fuels can also be converted to electricity by combustion in gas turbines or combustion engines. In a gas turbine, a compressor sucks in air from the environment, compresses it, and feeds it into the combustion chamber where it combines with the oil or gas. The hot, highly compressed combustion gases expand through the turbine and thus drive it. The rotational energy of the turbine is then converted to electricity in a generator. The efficiency of an open gas turbine is about 28%; the efficiency can be improved somewhat with a closed cycle. Because gas turbines can be brought quickly from a standstill to full power, and the installation costs are relatively low, they are especially suitable for covering the demand at peak power loads. In general, gas turbine power plants are built with capacities up to about 200 MW. It is also possible to couple gas turbines with thermal steam generating equipment. The exhaust gas from the turbine, which has a temperature of 700 to 800 K, can be used in a heat exchanger to generate steam.

An electric generator can also be driven by a combustion engine (Otto or Diesel). The diesel engine is especially suitable for this purpose, due to its relatively high efficiency (40%). However, the specific installation and maintenance costs are relatively high, so that this type of generator is generally used only for emergency installations.

Since the world coal reserves are large, generation of electricity from coal will continue to be very significant in the future. (In 1977, for example, coal was used to generate 65% of the electricity in Great Britain, 57% in the Federal Republic of Germany, and about 46% in the USA.) The technology for conventional coal power plants has reached a very high level. The developments in the last few years have increased the unit power outputs and the operation time per year, and have decreased the specific fuel consumption. It has also been possible to reduce considerably the emission of noxious wastes. The dust emission of an anthracite power plant at the beginning of the sixties was about 10 g/kWh of electricity, but in modern conventional coal power plants, the level is only about 0.5 g/kWh. This has been achieved by the use of filters and electroprecipitators. The emission of gaseous substances like sulfur dioxide and nitrogen oxides has also been reduced, the former by removing sulfur from the smoke. At the beginning of the sixties, the emission of sulfur dioxide was typically about 11 g/kWh, but modern conventional plants emit about 3 g/kWh.

Conventional coal power plants (of the so-called first generation) have grate or coal dust firing. In both cases, the temperature in the combustion chamber is well above 1300 K. These high temperatures lead to the production of relatively large amounts of gaseous pollutants. The sulfur in the coal is oxidized to sulfur dioxide, while the nitrogen in the air is converted to nitrogen oxides. For this reason, the more recent development work has been concentrated on finding methods to burn coal without producing so much pollution, as well as on utilizing the primary energy more efficiently and on reducing the installation costs of the plants. New techniques have been developed which prevent most of the sulfur and nitrogen oxide emissions (so-called second-generation power plants). Two lines of development seem particularly promising: the use of the fluidized bed combustion principle, and the development of a coal power plant with a Lurgi pressure coal gasifier.

Fluidized bed firing is based on the fact that finely divided coal material (anthracite or lignite coal with a particle size up to about 10 mm diameter) can be lifted by air blown in from below, and finally becomes fluidized – i.e. the air suspension of coal particles resembles a fluid in many respects. In burners of this type, the pipes carrying water or steam to be heated can be laid through the vortex of fluidized coal. This close contact allows more efficient heat transfer than is possible in the conventional heat-exchange areas in the chimney of a boiler, which in turn means that more heat can be extracted from less fuel. Furthermore, the heat-exchange pipes can be placed fairly close together, which leads to a considerable saving of space, and a concomitant saving in the investment required per kW capacity. Because the combustion temperatures are relatively low, 1100 to 1250 K, only small amounts of nitrogen oxides are formed. In addition, the sulfur can be removed from the fuel in the combustion chamber by adding limestone ($CaCO_3$) to the fuel. The sulfur in the coal combines with oxygen and the calcium to form gypsum ($CaSO_4$), which can be used to make plaster. (A 230 MW coal power plant produces about 14000 tons of salable gypsum per year.) Another advantage of the fluidized bed firing is that coal with a high ballast content and low heat content and waste from coal processing can be used. Because this combustion technique produces small amounts of pollutants, this type of plant can be built close to densely populated areas. In other words, the plant can be used to generate both electricity and usable heat (see 4.62). The efficiency of fluidized-bed power plants for electricity alone is 42%, and the total efficiency with heat coupling is 75%.

The fluidized-bed concept is being further developed, primarily in the USA (Rivesville, West Virginia), Great Britain (Renfrew) and the Federal Republic of Germany (Völklingen, Saarbergwerke). The present plants work with atmospheric air pressure, and have capacities up to about 200 MW. The use of higher air pressure has the further advantage that the gas/steam turbine process is then feasible for a coal-burning plant. It also increases the efficiency and, due to the reaction kinetics of oxide formation, reduces the emission of pollutants. The first plant with a pressurized fluidized bed firing is currently being built at Grimethorpe, Great Britain, with support from the International Energy Agency. It will have a capacity of 80 MW thermal output.

In addition to the principle of fluidized-bed firing, the Lurgi pressure coal gasifier applied to coal power plants is promising. This type of installation has been working successfully for eight years at Lünen in the Federal Republic of Germany (165 MW). In this type of power plant, coal is gasified, then fed into an integrated gas/steam turbine system. The conversion to electricity is 42%, and the process has low pollutant emission. Gasification of the anthracite coal makes it possible to remove the sulfur before the fuel is burned in the gas turbine, and thus avoids the elaborate removal of sulfur oxides from the smoke.

4.611.2 Generation of electricity from mechanical energy

The conversion of mechanical to electrical energy has already been discussed in 4.321 to 4.324 and in 4.4. In hydroelectric generating, the kinetic energy of the water in a river or the ocean is converted to electricity; in addition, the potential energy of water in lakes or artificial storage ponds can be utilized. In any of these techniques, the water is used directly to turn a turbine and generator. Water turbines consist essentially of a conduit, which gives the water directional velocity, and a bladed wheel which is turned by the flowing water and is connected to the generator (180). Hydroelectric plants using running water, stored water, pumped stored water and tidal flows have been built on a commercial scale.

Running water power plants utilize the kinetic energy of river water, which is often impounded behind a dam. In storage power plants, the water of a natural lake is led downhill through a pipe to the turbine. Pumped storage plants consist of two (artificial) ponds, connected by a pipeline (see Table 4-5). When electricity is being generated, the water flows from the upper to the lower pond, and thus drives the turbine and generator. When the load on the network is lower, the generator can be run as a motor to turn either the turbine (in the reverse direction) or a pump to return the water to the upper pond. This stored energy can then be used when the load on the network peaks. The total efficiency of pumped storage plants is about 75%. They are a practical method of storing electric energy on a large scale.

Tidal energy plants utilize the differences in water level between low and high tides at suitable locations on coastlines (see 3.38 and 4.4).

At the present time, no other methods of generating electricity from mechanical energy, e.g. wind or wave energy, are being used on a large scale.

4.611.3 Direct generation of electricity

4.611.31 Fuel cells

As we have seen in 4.611.1, by far the majority of electricity generated in the world is produced in thermal power plants from the chemical energy of fossil fuels or the physical energy of atomic nuclei. The energy is first converted to heat, then to mechanical energy, and finally to electricity. Because of the intermediate steps, the process involves relatively high losses (low efficiency), and this has led to reseach and development work on methods of direct generation of electricity. In general, this heading includes any method which bypasses at least one of the intermediate steps. The main processes are photoelectric conversion, fuel cells, magnetohydrodynamic, thermoelectric and thermionic conversion, and radionuclide batteries. The direct conversion of sunlight to electricity via the photovoltaic effect was discussed in 4.313. In fuel cells, chemical energy is converted directly to electricity. In both processes, the conversion avoids the detour through heat and mechanical energy. In magnetohydrodynamic, thermoelectric and thermoionic conversion, heat is directly converted to electricity, without the intermediate step of mechanical energy.

Like other electrochemical systems, a fuel cell consists in principle of two electrodes separated by an electrolyte. In the oxidation reaction, the fuel delivered to the anode[1)] (e.g. hydrogen) gives up its electrons to the electrode, which acts as a catalyst (181). The oxidizing agent introduced at the cathode takes up electrons from the cathode, and the ionized reaction partners then combine to form electrically neutral compounds (water, if the fuel is hydrogen). Most of the chemical reaction energy of a fuel is converted into electrical energy in the process; the practical efficiencies of fuel cells are about 60%, and even higher efficiencies might be attained. In addition, this method of generation is much cleaner than conventional methods.

In conventional methods of generating electricity, the chemical energy is first converted to heat in a combustion process, and this heat is then converted to mechanical energy in a heat machine. Finally, the mechanical energy is converted to electricity. Because a fuel cell avoids the conversion of chemical energy to heat, it is also often called "cold combustion".

The most advanced fuel cell is based on the reaction between hydrogen and oxygen. In it, gaseous hydrogen is continuously supplied to the anode as fuel, and gaseous oxygen to the cathode as oxidizing agent. (The reaction chambers are separate.) The hydrogen and oxygen molecules coming into contact with the electrodes are disassociated into atoms (H_{ads}, O_{ads}) by adsorption. The following reactions occur in an H_2/O_2 fuel cell, which has acid electrolytes:

Anode reaction: $2\,H_2 \rightarrow 4\,H_{ads} \rightarrow 4\,H^+ + 4\,e^-$
Cathode reaction: $O_2 \rightarrow 2\,O_{ads} \rightarrow 2\,O^{--} - 4\,e^-$

Overall reaction: $$2\,H_2 + O_2 \rightarrow 2\,H_2O \qquad (18)$$

The fuel (hydrogen) is oxidized at the anode, giving up its electrons, while the oxidizing agent (oxygen) is reduced at the cathode, by absorbing electrons. The product is water. (If the cell has alkaline electrolytes, the partial reactions at the anode and cathode are different.) The increasingly positive charge of the cathode and the increasingly negative charge at the anode produces a potential difference between the electrodes, the maximum value of which depends on the difference between the energy of ionization of the two reaction partners. The charge is equalized by the flow of electrons through a metallic conductor between the half-cells; it is this circuit which includes the consumer resistance element. If fuel and oxidant are continuously

[1)] It is customary in electrochemistry to call the positive electrode the cathode, and the negative electrode the anode. In a fuel cell, the electrode at which the fuel is introduced is thus the anode (– pole) and the electrode at which the oxidizing agent is introduced is the cathode (+ pole).

supplied, the system can deliver a continuous current to the consumer. The processes in fuel cells are actually complicated, and are influenced primarily by catalysts, the electrodes, the electrolytes and the operating temperatures.

The most suitable catalysts for the hydrogen electrode are nickel, platinum, palladium and alloys of these metals. Platinum and palladium can be used with either alkaline or acid electrolytes, but their high price is a disadvantage. Nickel can be used only with alkaline electrolytes. When hydrogen is prepared from hydrocarbons, such as methane, or by gasification of coal, it contains contaminants like CO and CO_2. If these are not removed by elaborate purification procedures, they form carbonates in fuel cells with alkaline electrolytes. With acid electrolytes, it is not necessary to remove the CO and CO_2 from the hydrogen, and for this reason, there is considerable interest in the use of acid electrolytes. However, the metals suitable for use with acid electrolytes are too expensive for large-scale application, so that systems based on other catalysts have been developed. Tungsten carbide, and cobalt, nickel or iron phosphides can be used as anode catalysts, and activated charcoal, with a large specific surface area (about 1300 m^2/g) can be used at the cathode.

Because the electrochemical reactions take place occur where the fuel or oxidizing agent catalyst and electrolyte are all in contact, i.e. at the three-phase interface between a gas, a liquid and a solid, electrodes with a large three-phase interface (porous electrodes) are desirable. Some of the suitable forms are gas-diffusion, double-layer and multiple-layer electrodes.

H_2/O_2 fuel cells have been used in space flights (Apollo and Gemini projects). Because the voltage from a single cell is rather low, about 1 V, a number of cells have to be connected in series to form a battery. Fuel-cell installations with outputs in the kilowatt range have been used for years on the earth's surface, for example to provide current for meteorological stations (decentralized energy supply). There have also been experiments in which fuel cell batteries have been tested for use in vehicles (see 4.673). Their use in stationary installations is also promising. For example, the United Technology Corporation in the USA has developed an FCG-1 fuel cell power plant with a 26 MW output to provide current at peak consumption periods. Because it does not pollute the environment, this type of plant can be built close to the consumer (182, 183).

Because there are many possible applications for fuel cells, a number of different types have been developed and tested. They can be classified according to their working temperature, the oxidant used, the electrolytes or the fuel.

Depending on the working temperature, fuel cells are classified as low-temperature (up to about 500 K), medium-temperature (500–800 K) or high-temperature (to more than 1300 K). Air, oxygen or hydrogen peroxide can be used as oxidant, and there have been some experiments with chlorine. The choice of electrolyte depends primarily on the working temperature. The low temperature cells, which have been furthest developed, use liquid acid or alkaline electrolytes. As mentioned above, acid electrolytes such as dilute sulfuric or phosphoric acid are generally preferred, because with alkaline electrolytes (e.g. dilute potassium hydroxide), any CO or CO_2 impurities in the fuel lead to carbonate formation.

The fuels which have been most widely tested are the gases hydrogen, hydrocarbons (ethane, propane or butane), natural gas and ammonia; liquids such as hydrazine and methanol, and solids (coal). Because hydrogen (see 4.66), hydrocarbons (see 4.642) and methanol (see 4.643.1) are secondary energy carriers which may play an important role in the future energy economy, fuel cells which burn these fuels are of particular interest (184). This is the reason that methane and methanol cells, along with hydrogen peroxide cells, have been the most thoroughly tested types. Coal-burning fuel cells might make possible a non-polluting utilization of the world's large coal reserves, and that with an efficiency of 50% conversion to electricity. The development of coal-burning fuel cells is being pursued in the USA (Argonne National Laboratory, Illinois); here the coal is converted to gas before it is used in the fuel cell.

4.611.32 Magnetohydrodynamic generators

Another interesting technique for direct production of electricity is based on the principle of magnetohydrodynamic generation. With an MHD generator coupled to a conventional steam power plant, efficiencies of 60% should be attainable. It would be possible with this technique to generate electricity from coal or natural gas more effectively, and also with less environmental damage, than is possible with conventional thermal power plants (185).

In an MHD generator, hot ionized gas (plasma with a temperature of about 3000 K) or liquid metal is pumped at high velocity through a high induction magnetic field (1 to 10 Tesla) which is perpendicular to the direction of flow. This produces a voltage perpendicular to the magnetic field and the flow direction; it can be picked up with suitable electrodes. The power density of a MHD generator depends mainly on the electric conductance and flow velocity of the working medium, and the magnetic inductance in the MHD channel.

In order to increase the electric conductivity of the working medium – if it is a gas – efforts are being made to increase the working temperature or to include easily ionized substances e.g. alkali metals. Another possibility is to use a liquid metal as the working medium. However, there are serious difficulties in accelerating the liquid metal to the velocity required, so that the problems with liquid metals are greater than those with a gaseous medium. (It is not especially difficult to achieve the necessary velocity with a gaseous medium in the MHD channel.)

Because MHD generators would be considered primarily for the MW range, the cross section of the channel must be quite large, and the generation of the necessary magnetic fields is difficult. The generation of a field of several Teslas is possible with superconducting magnets, and has in fact been achieved, but in this case there is the additional difficulty that the temperatures in the MHD channel can be as high as

3000 K, while the superconducting magnet coils must be kept about 4 K. The result is that the expenditure for thermal insulation is very high.
The materials for the electrodes and the walls of the channel must be able to fulfill exacting requirements. The electrodes should be good conductors of electricity, but poor heat conductors, and furthermore, they must withstand high temperatures but not be oxidized. Zirconium oxide (ZrO_2) doped with yttrium oxide (Y_2O_3) has proved to be suitable for temperatures up to about 2500 K. Because the channel lining is exposed to similar thermal and chemical conditions, the same materials can be used for them as for the electrodes, namely zirconium oxide (116).

At present, the MHD concept exists in three forms: open systems with gaseous working media and closed systems with gaseous or liquid working media.

In the open MHD generator, the working medium (combustion products) are not returned to the combustion chamber. After leaving the combustion chamber, the hot gas flows through the MHD channel, then an air-preheater, a heat exchanger, and a scrubbing tower, and finally escapes into the atmosphere. The combustion takes place in a chamber, in which ionization nuclei are added to the fuel in order to increase the electrical conductivity of the combustion products. (The ionization nuclei are provided by elements with relatively low ionization energies, such as sodium or potassium, or compounds containing them, such as K_2CO_3.) High gas temperatures ($\approx$ 3000 K), which are needed to make the gas a good conductor, are achieved by preheating the air to about 2000 K by heat exchange with the gas leaving the MHD channel. In the heat exchanger, through which the gas passes after it has preheated the air, as much of the remaining heat is removed as possible and fed into a conventional steam system. In principle, it would be possible to use a gas turbine to convert the energy of the hot gas to electricity, but there would be serious corrosion problems to solve. After the combustion products have been scrubbed and the ioniziation nuclei removed, they are released after further cooling into the atmosphere.

Projects based on the MHD principle have been subsidized for years in the Soviet Union and the USA. In the USA, the Mark V MHD generator of the Avco Everett Co. achieved an output of 32 MW of electricity. The USSR has a large MHD generator, the U-25 at the Institute for High Temperature Physics in Moscow (186, 187). The U-25 uses natural gas for fuel. Atmospheric air is enriched to 40% oxygen, compressed to a pressure of about 3 bar, and then heated to about 1500 K in the preheater. It is fed into the combustion chamber together with K_2CO_3 and the natural gas. The conductive combustion products, a plasma at a temperature of about 2800 K, leaves the combustion chamber through a jet and enters the MHD generator channel with a velocity of about 1000 m/s. The dimensions of the channel are 0.38 m $\times$ 0.77 m at the entrance and 0.38 m $\times$ 1.88 m at the exit, and the working section is about 5 m long. The generator channel is surrounded by a normally conducting magnet system with a field strength of about 2 Tesla. The hot gas leaving the MHD generator channel passes through a heat exchanger and generates steam at about 810 K and 100 bar. After scrubbing and removal of the ionization nuclei

(K_2CO_3), the combustion products are further cooled and released into the atmosphere. The K_2CO_3 is returned to the combustion chamber (116). At the beginning of 1977 this installation was run at an output of 12 MW electric power for 250 h. For a time the output reached 20 MW. The installation of superconducting magnets is expected to produce a large increase in power. The USSR is now planning to construct a 500 MW plant of this type.

The other two types of MHD generators, closed MHD systems with either gaseous or liquid (metal) working medium, are much farther behind in their development. In the type with a gaseous medium (e.g. helium) the gas is to be kept in a closed circulation system. The advantage of using liquid metals as working media is that their electric conductivity is several powers of ten higher than that of gases. However, there are serious problems with the use of liquid metals, particularly that of accelerating the metal to the required high flow velocity.

4.611.33 Thermoelectric energy conversion

Direct generation of electricity can only play a significant part in the energy economy if the technique is suitable for large power outputs, under economically justifiable conditions. Thermoelectric and thermionic conversion, and radionuclide batteries are not suitable for large-scale use, and will thus find limited application.

Thermoelectric conversion is based on an effect discovered by T. Seebeck. If two conductors of different materials are in contact at two different points, and if the two contacts are kept at different temperatures T_1 and T_2 ($T_1 > T_2$), a thermal voltage U arises. $U = \alpha_s (T_1 - T_2)$, where α_s is the Seebeck coefficient. (The thermal voltage for a bismuth-antimony circuit is 10^{-4} V/K.) This type of circuit is called a thermoelement. A larger voltage can be attained by connecting a number of thermoelements together in a thermocolumn. This type of thermoelement has no practical applications because of the low voltage, the low power and the poor efficiency. However, with suitable semiconductor materials and optimally designed thermogenerators, efficiencies of up to 11% can be obtained.

Thermoelectric materials should have a high Seebeck coefficient, high electrical conductivity, and low thermal conductivity. Different materials can be used for the generator elements, depending on the temperature of the "hot solder joint". For temperatures up to about 600 K, bismuth-antimony-tellurium alloys are suitable for the p-conductor, and bismuth-tellurium-selenium alloys for the n-conductor; for temperatures from 600 K to 800 K, lead telluride, which can be made as either a p- or an n-conductor, is good; and in the range from about 800 K to 1300 K germanium-silicon alloys, which can also be made as either p- or n-conductors, can be used (116).

Since a single thermoelement yields only a low voltage, a thermogenerator for practical purposes has to consist of many elements electrically connected in series. Any heat source can be used to drive a thermogenerator, so long as it supplies the necessary temperatures. However, due to their relatively low efficiencies of about 11%, thermogenerators are not suitable for large installations, but only for certain applications such as supplying electricity to an unmanned extraterrestrial measurement station. Because this application requires a long and maintenance-free operation, radionuclides are a good heat source (see 4.611.35); for higher outputs up to a few kW, small nuclear reactors might be used (116).

4.611.34 Thermionic energy conversion

Heat can be converted directly into electricity by the "radiant emission" effect, in which electrons are emitted from the surface of hot electrodes.

A thermionic converter (thermionic diode or element) consists of two electrodes. The emitter or cathode is heated, which leads to emission of electrons from its surface. The electrons travel to the cooler electrode (collector or anode), which produces a voltage U between the two (116). If the circuit is closed by a resistance, the electrons can return to the emitter via the external circuit. To obtain a useful level of electron emission (current density), temperatures between 1300 and 2800 K are needed, depending on the type of converter. (In the thermionic elements, the transport of heat to the emitter is a problem, due to its small dimensions. One possibility is to use a heat pipe.)

Thermionic diodes (e.g. a Cs diode) differ from normal diodes in having an extremely small gap between the electrodes (about 0.01 mm). This prevents the buildup of a space charge and insures that nearly all the electrons reach the collector. The usable voltage is about 0.5 V, so the number of elements required to achieve any given voltage is much smaller than for thermoelectric elements. In addition, thermionic elements have somewhat higher efficiencies (up to about 20%) than thermoelectric elements. As with thermoelectric elements, any source of heat can be used which gives the required temperatures. Thermionic generators using radionuclides, nuclear reactors (20 kW electric output) or solar energy are built (116).

4.611.35 Radionuclide batteries

Radionuclide batteries convert the radiation energy released by the decay of a radionuclide (radioactive isotope) into electric energy. There are various methods of conversion. The kinetic energy of the α or β particles can be converted to heat when they collide with the walls of a metal capsule, and the heat can be converted into electricity with thermoelements (thermoelectric radionuclide batteries, see 4.611.33) or with thermionic elements (thermionic radionuclide batteries, see 4.611.34). One of the advantages of radionuclide batteries is that they require no maintenance.

Most radionuclide batteries work with thermoelements. In order to provide as nearly constant a voltage as possible, the half-life of the radionuclide chosen should be several times longer than the desired operational lifetime of the battery. (The longer the half-life of the radionuclide, however, in general, the lower the power density.) The half-lives of the nuclides used range up to 100 years, depending on the application. Some examples are ^{90}Sr, a β emitter with a half-life of 28 years, and a specific power output of the oxide, SrO, of 0.23 W/g; and ^{238}Pu, an α emitter with a half-life of 86 years and a specific power output of the oxide, PuO_2, of 0.44 W/g. A radionuclide battery based on ^{90}Sr, as is used to supply weather stations, for example, is heavy, due to the necessary shielding. A battery with an electric output of about 50 MW would typically weigh about 1000 kg and have a lifetime of 5 to 10 years. Small radionuclide batteries based on ^{238}Pu are used, among other things, for heart pacemakers implanted in the body (116).

4.612 Transport

There are two disadvantages of electricity in an energy supply system: transport of electricity is, per unit of energy, an expensive form of energy transportation; and the storage of electric energy is difficult (see 4.613).

In industrial countries, up to 30% of the primary energy consumed is used to generate electricity. For economic reasons, and on account of pollution problems, most electricity is generated in large units at a distance from population centers. Therefore the methods for transport of electricity are very important, and their further development is being encouraged in many countries. The efficiency of transport is increased by high voltages and currents, but for both there are upper limits imposed by economical and technological considerations.

Power lines can be either above or below the ground; the aboveground lines will continue to be preferred for long-distance transport, because they are relatively inexpensive to build, and because damage to them can be more quickly found and repaired. One disadvantage is that they require a large amount of space. With time, rotary current systems have replaced alternating current for transport, and with these systems, the practical limits of current strength for transmission have been reached. (For example, the cross section of the lines cannot be increased indefinitely, due to the weight which must be suspended, the amount of material required, and the cost.) Therefore, the voltages used for long-distance transmission have been raised higher and higher. The long-distance network in the Federal Republic of Germany is now largely operated at 380 kV. In the USA, more than 8000 miles (22%) of the network carries 500 kV, and in 1970 a 1000 mile line was built to carry 765 kV. Experiments have already been done with transmission voltages of 1100 kV, 1500 kV and 2250 kV (189, 190). In the past few years, high-voltage constant-current transmission has also been increasing in significance. However, this type of transmission is not economically feasible for distances less than 1000 km, because the current must be converted from alternating to direct and back again, which requires elaborate equipment (183).

Because of the space required for aboveground transmission, the power lines in population centers are more and more frequently being laid underground. However, the transmission cost in these cables is 10 to 20 times greater than in aboveground lines. In large cities, like New York, Berlin or Tokyo, underground cables capable of carrying 1000 MVA are needed (1, 183). The same considerations apply to efficiency of transmission in underground and aboveground power lines, but the voltages which can be applied to underground cables are also limited by the insultation materials available. Two solutions to this problem are the oil/paper cable and water- or gas-cooled cables. SF_6 (a gas) has been used in a special purpose cable which is 700 m long, and carries 1000 MW at 420 kV (183).

Intense efforts are being made to develop a superconducting cable for transmission of electricity. At present, all superconducting materials have to be cooled to the temperature of liquid helium (4 K), but the installation of a 4 K transmission cable would be very expensive. It would appear that superconducting cables would only be practical at much higher transmission voltages than are in general use at present (2–10 GW), and then only if superconducting materials can be developed which can be used at higher temperatures, e.g. that of liquid hydrogen.

In principle, hydrogen could also be used for the transport and storage of electrical energy. Electrolytically generated hydrogen (see 4.661) could be reconverted to electricity, for instance in a fuel-cell power plant (see 4.611.31) near the site of consumption. However, this is not presently economically feasible.

Liquid energy carriers (e.g. methanol) (see 4.643.1) and gases of the sort resulting from gasification of coal (see 4.642) are also in principle suitable for the transport and storage of energy. They can be shipped by pipeline, which is more economical than transmission in the form of electricity. In addition, energy can be shipped over much greater distances in this form (latent heat gas, *synthetic gas/methane*) (see 4.65). With this form of energy transport, in contrast to district heating, it is also possible to generate electricity in the neighborhood of the consumer centers, as well as to provide space heat (see 4.611.1 and 4.611.31).

4.613 Storage

Storage of electrical energy is important for two main reasons. One is that the cost of power plants has risen so high, that it is ever more important to obtain the maximum output from them. The other is that the demand varies diurnally, monthly and seasonally (see 4.645). It is therefore necessary to be able to provide for peak loads from stored energy; in addition, overloads must be prevented from causing the entire network to collapse and to leave all the consumers without power. Electrical energy can in principle be stored hydraulically, pneumatically, mechanically, electrically, chemically or thermally. The following treats some of the useful methods of storing electrical energy. It is a difficult form of energy to store, and in general additional installations are needed.

Pumped storage installations have proved useful for covering peak loads. In them, electricity is used to pump water uphill into a storage basin; when needed, it flows back downhill through a turbine and generates electricity. The overall efficiency of pumped storage plants is presently about 75%, and they can be used to store electrical energy economically and on a large scale. In the USA, the total capacity of pumped storage installations is about 8100 MW (2% of the total installed electric capacity) (see Table 4-5) (183).

Large pressure containers are needed for the pneumatic storage of electrical energy, which is used in connection with gas turbine power plants. Underground caverns in salt mines, for example, can be filled with air under high pressure (40–60 N/m^2) during slack times. At times of peak demand, the compressed air is fed through the combustion chamber into the turbine, which can raise the total capacity of the power plant to about double the normal capacity. Such plants are another economical, large-scale method for indirect storage of electrical energy. The first air-storage gas-turbine power plant was put into operation in December, 1978, in Huntorf, near Bremen, Federal Rep. of Germany. This plant has a capacity of 290 MW electric power. About $2.7 \cdot 10^6$ m^3 pressurized air is stored daily in two caverns in salt deposits. At times of peak current demand, the pressurized air is used for the combustion process in the gas turbine (see 4.611.1) (192).

One way to store electric energy as mechanical energy is to store the steam which has been generated in the power plant in pressure vessels. (A 67 MWh steam reservoir with 3300 m^3 capacity has been in operation in Berlin (Charlottenburg) since 1929. With modern cast pressure containers, 8000 m^3 of steam at 60 N/m^2 can be stored in a vessel 71 m high and 12 m in diameter (183).

Storage of electricity in batteries is only suitable for small and medium capacities, because batteries are still too expensive. In many countries, there are programs for the development of battery systems for load leveling and for mobile consumption (see 4.673).

Fuel cells based on the "cold" combustion of hydrogen can also provide a means of storing electrical energy. Hydrogen could be generated by hydrolysis of water during times when the load is low, stored, and used in the fuel cells to supply the peak demands (see 4.611.31).

Magnetic storage could be accomplished in superconducting coils. The rotary current from the network is rectified and fed into the superconducting coil. Because the efficiency is 85–90%, this form of storage would be especially interesting. However, for the present, it is still too expensive (183).

Equalization of the load can also be achieved by storage by the consumer, as in night-storage heaters for dwellings. However, considering the amount of primary energy required, this is not an economical form of space heating (see 4.325).

The basic function of energy storage systems is to absorb a secondary energy carrier (or primary energy carrier) and to release it at need. They can be used for storing a reserve, leveling the fluctuations in the energy demand, and optimizing the utilization of the primary energy used (see 4.62).

As discussed in 4.611, electrical energy can be generated from any primary energy source. This means that any liquid or gaseous energy carrier is in principle a suitable medium for energy storage. (The same holds for solid energy carriers, but because they cannot be pumped from one place to another, their further use is less advantageous.) In recent years, there has been a trend toward the use of underground storage for gaseous and liquid energy carriers. The reservoirs can be empty gas and oil fields, aquifers, caverns created in salt deposits by dissolving the salt in water, or in caverns in solid, non-porous rock, such as natural caves, tunnels and abandoned mine shafts.

In general, empty oil or gas wells are especially suitable respectively, for storage of liquid and gaseous energy carriers. It is an advantage that they can usually be used without great expense. Aquifers are most suitable for the storage of gases, if the geological conditions are met (porous, permeable rock formations sealed off by non-porous layers). This type of reservoirs, which can store on the order of 10^9 m^3 of gas, have been in use for several years in the USA, Canada, France, the Federal Republic of Germany and the USSR. Caverns created artificially in salt deposits are very tight and are suitable for storage of either gas or oil. Oil reservoirs are used primarily for storing reserve supplies, while gas reservoirs are suitable, due to the speed with which they can be tapped, for covering the demands for electricity at times of peak load. As mentioned earlier, pressurized air can also be stored in salt caves, and experiments are being done on the storage of liquid natural gas (LNG). Gas storage in solid rock formations has so far been unsatisfactory, because the stone is not sufficiently tight. For this reason, only oil reservoirs have been established (e.g. cave reservoirs in Norway and Sweden) (116).

4.614 Economic aspects

The demand for electricity varies widely with the time of day, month and year. Fig. 4-23 shows typical daily load curves, for a workday, a Saturday and a Sunday in January and in July, in the Federal Republic of Germany. These curves illustrate the differences on different days of the week and in summer and winter. Because large-scale and economical storage of electrical energy is only possible in indirect form, and in general requires additional installations (e.g. pumped storage or air reservoir plants), the generating capacity required to meet the demand cannot be immediately inferred from the load curves. Rather, the capacity must be distributed among the following load ranges; base load[1)] (constant operation), medium and peak load

[1)] The base load is that amount of current which is used 24 h a day, plus 10%. The peak load is the range between the minimum demand between 8 a.m. and 8 p.m. and the peak daily load. The medium load is the range between the base and peak loads.

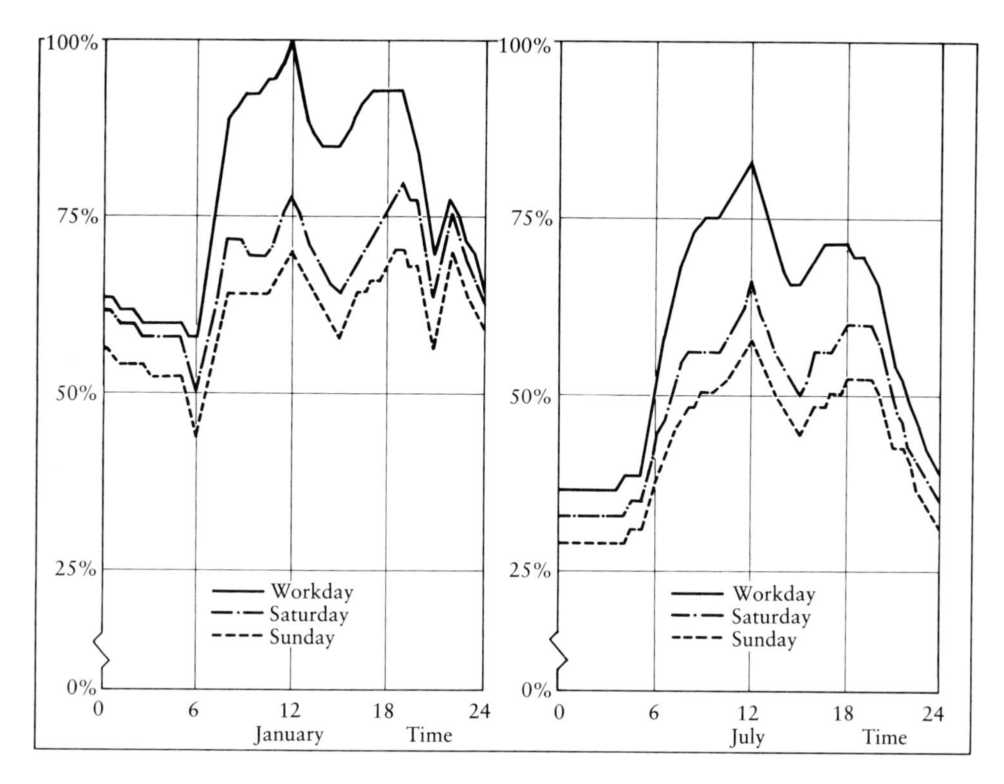

Fig. 4-23: Typical daily load curves in January and July (in percentages of the winter peak load)

Source: E. Pestel et al., Das Deutschland-Modell, Stuttgart: Deutsche Verlags-Anstalt 1978.

ranges (194). Because power plants, as businesses, should operate as cost-efficiently as possible, different types of plant are used to supply current in different load ranges. This means that the capacities needed for the base, medium and peak loads have to be determined separately, and the power company must plan in such a way that even at the annual peak load, the demand can be met and the network prevented from collapsing.

The most suitable types of power plants for the base load are hydroelectric, nuclear and coal plants, which function best in continuous operation. (Nuclear power plants, for example, are normally in operation 7000 to 8000 hours per year.) Coal, oil and gas plants are used for the medium load – some of these can also operate on more than one fuel. (Anthracite plants, for example, when used for the medium load, are operated 3000 to 4000 hours per year.) As far as the technology is concerned, medium-load plants can cover part of the base load, and vice versa. The peak loads are covered by storage and gas turbine power plants, which may be in operation only a few hours per day.

Given the amounts of current required in the individual load ranges (work) and the usual operation times (time), one can calculate the capacity needed (power) to meet the demand in each load range, since power = work/time (see 2.21). When plans are made for the construction of new power plants, the capacity in relation to the different load ranges must be considered.

Considerations of economy dictate that the cost of current be kept as low as possible, which can be achieved in part by using the different types of power plant in the load range in which they operate most economically. The price of electric current is based essentially on the costs of generating, transmission and distribution. For any type of power plant, the generating costs can be divided into installation (cost of building the plant, including such costs as interest on capital) and operation costs. The latter can be further subdivided into variable (fuel and, for nuclear plants, fuel cycle and decommissioning) and fixed operational costs. The latter include personnel, maintenance and repairs.

These cost components vary with time, and in different ways, so that it is difficult to determine the absolute costs of generating electricity. It is more to the point to analyse the cost structure of power plants to see how changes in the costs of individual components affect the total cost of generation.

The following is a comparison of the typical cost structures for generation in an anthracite coal and a nuclear (LWR) power plant in the Federal Republic of Germany. The anthracite coal plant is assumed to be a double unit, with a net capacity of 2 × 677 MW, a net efficiency of 36.5%, installation costs of 1075 DM/kW, and fuel cost (including delivery to the plant) of 182 DM/tce. The LWR nuclear power plant is assumed to have a net capacity of 1228 MW, a net efficiency of 32.0%, installation costs of 1634 DM/kW, and fuel costs which can be subdivided as follows: uranium, US-$ 27/lb U_3O_8; separation of fissionable isotope, US-$ 110/kg; preparation of fuel elements, 340 DM/kg; recycling, 900 DM/kg (195, 196)[1].

Based on these data, the cost structure for an anthracite plant in operation 6500 hours/year, starting in spring 1977, is made up as follows: installation, 24%, operation, 76% (variable operational costs, 68%, fixed operational costs, 8%.) The cost structure for the nuclear power plant, also operating 6500 hours per year and starting in spring 1977, consists of installation, 48% and operation, 52% (variable costs 35%, fixed costs 17%.)

In other words, the installation costs for an anthracite plant are low compared to the variable operation costs, due to the high fuel costs (this is true to some extent for

[1] These data are typical for power plants in any part of the world, except for the prices of fuel (especially anthracite) which vary from one region to the other. At the 10th World Energy Conference, Istanbul, the installation costs for a 1000-MW anthracite power plant with an SO_2 removal system were estimated to be US-$ 675/kW, and US-$ 858/kW for a 1000 MW PWR nuclear plant (start of design and construction July 1, 1978; start of commercial operation, July 1, 1984). The investment required per kilowatt declines sharply with increasing plant size.

any fossil fuel plant), while the installation costs for the nuclear power plant are high compared to the variable operational costs.

Based on these data, the cost of generating electricity (assuming the plant is in operation 6500 h per year) is 4.1 cts/kWh in an anthracite coal plant and 2.4 cts/kWh in a nuclear plant. (If the plant is in operation for 4000 h/year, the cost is 5.55 cts/kWh in a coal plant and 3.55 cts/kWh in a nuclear plant). The Atomic Industrial Forum (AIF) in the USA found, from a survey of 43 power companies, that the generating costs were as follows: nuclear, 1.5 cts/kWh; coal, 2.3 cts/kWh; and oil, 4.0 cts/kWh. The fuel costs accounted for 22% of the cost of nuclear power, 55% of the cost of anthracite coal power, and 59% of the cost of petroleum-generated electricity. (However, because the fuel contributes so heavily to the cost of generating power from coal, the cost is lower than that of nuclear power in some plants located near coal mines from which the coal can be extracted cheaply (see 3.312). This applies to some areas in the USA.) It must be recognized that the costs for recycling nuclear fuel and for decommissioning these plants are still not exactly known, because no large-scale commercial recycling plant has yet been built, and there is still no permanent burial site for highly radioactive waste from these plants.

4.62 District heating

Because there is a large demand for low-temperature heat, hot water is a very important energy carrier (see 2.331). In densely populated areas, centralized heat production is a practical method of supplying low-temperature heat. With a reasonable amount of insultation, depending on the temperature of the hot water, it is practical to extend district heating over a range of up to 30 km. The major advantages of this form of heating are the relatively low specific expenditure of primary energy (compare Fig. 4-21), the low pollutant emission in comparison to the output of many individual furnaces, the possibility of using a number of different energy carriers efficiently (e.g. anthracite), and, by using the waste heat from generation of electricity, it is possible to achieve efficiencies of about 75%. This means an optimal utilization of the primary energy source and a relatively low level of heat pollution (197).

The first district heating system went into operation in Lockport (New York) in 1882. In 1979, there were about 600 km of district-heat pipes in the Federal Republic of Germany alone – Hamburg, with 342 km, has the largest network in the country. Since the combustion of coal can be made much less polluting by burning it in a fluidized-bed burner, it would be advantageous to build such burners in the neighborhood of urban centers and to use them for district heating. In other words, district heating could make it feasible to replace light heating oil with coal or another primary energy source for space and water heating. In less densely populated areas, it is not practical to use district heating to provide low-temperature heat, but in these areas, solar heating would be more practical than in urban areas.

The source of heat for a district heating network can be either a pure heating plant, or a combined electric power/heating plant, in which the primary product is electricity, and the heat is simply a byproduct. This combination makes particularly efficient use of primary energy. When a thermal power plant is operated only to produce electricity, the temperature of the cooling water is so low that it is not practical to use it for space heating. However, if the plant is planned so that the heat is removed at a temperature between 80°C and 130°C, and distributed in a district heating system, the efficiency of electricity production is somewhat reduced, but the total efficiency of the plant may be as high as 75%.

Industrial installations for the production of both electricity and process heat have been operated for decades. In this case, the primary product is process steam, and the electricity is a byproduct. Correspondingly, this mode of operation is called heat/power coupling, although the distinction is usually no longer made, and all such systems are referred to as power/heat coupling.

The following considerations apply to the choice of the temperature at which the water enters the district-heating network: The heat loss for a given amount of insulation is lower at lower temperatures, and the amount of electricity which can be generated in a power/heat installation is greater. However, at lower temperatures, more water is required to transport a given amount of heat to the consumer, which leads to higher costs for pipelines and pumping stations. The temperature at which the water leaves the heating plant is typically between 80°C and 130°C (116). Since the amount of heat required by the consumer varies with the season, the amount transported must be regulated, for example by changing the water temperature.

The principle of power/heat coupling is independent of the primary source of heat; it can be applied either to fossil-fuel or nuclear power plants (198). The technology for such systems is being developed and tested in the Federal Republic of Germany in the Ruhr district-heating supply line (197), and in other countries, including the USA, Japan, France, Great Britain, The Netherlands, Belgium, Poland and Czechoslovakia, there are programs to increase the installation of district heating. In a few highly urbanized countries, it would be possible, given the necessary investment, to supply up to 30% of the demand for low-temperature heat (space and water heating) by district heating.

To minimize heat losses, it is sensible to build the heating plant as close as possible to the population center, but this is problematic in the case of nuclear heating plants – heating reactors with up to 100 MW thermal output or nuclear power plants – which should be built as far as practical from population centers for safety reasons. The thorium high-temperature reactor (THTR) would be especially well suited for power/heat coupling (see 4.213.3), and it is in fact planned to use high-temperature reactors both for generation of electricity and to provide heat for distric heating.

4.63 Petroleum refinement products

Crude petroleum is not directly usable, but must first be converted in refineries to various usable products, including light distillates (gasoline), medium distillates (diesel and heating oil), heavy distillates (heavy heating oil), and other products (e.g. asphalt). The relative amounts produced in a given refinery are technologically determined and, if it has no cracking facility (hydroskimming refinery), they can only be changed within certain limits.

Refined petroleum products have a special place in all energy economies, due to their versatility and the low cost of transporting (pipeline or tankers) and storing liquid energy carriers. They have a dominant position in the transportation sector and as industrial raw materials. The liquid hydrocarbons have the advantage, with reference to their versatility, that they can be transported to the consumer in mobile containers (tank trucks), in contrast to gaseous hydrocarbons, which must be transported in a pipe, or electrical energy, which also requires a distribution network.

Heavy heating oil is the petroleum product which can be most easily replaced by another energy carrier, such as coal, gas or nuclear fuel in a nuclear power plant. It is relatively difficult to find substitutes for the light and medium distallates, and for this reason, efforts are made to obtain as large a fraction of the light and medium distillates as possible from crude petroleum.

Most of the 130 West European and 50 Japanese oil refineries are of the hydroskimming type, because there is a relatively large demand for medium distillates and heavy heating oil in these areas. The latter accounts for 34% of the petroleum consumption in Western Europe and 40% in Japan, while in both regions, gasoline represents only 20%. In the USA, by contrast, 40% of the petroleum is consumed in the form of gasoline. The apprixomately 260 refineries in the USA have cracking towers with which they increase the fraction of gasoline produced. One of the reasons for this development in the USA was the very low price of natural gas, which led to its use for space, water and process heating (the latter in industry) and for electric generating. It has thus taken over some of the uses to which light and heavy heating oil are put in other countries.

A hydroskimming refinery produces about 16% gasoline, 39% medium distillates, 39% heavy heating oil and 6% other products. These relative amounts vary somewhat, depending on the type of crude. If more gasoline is needed than is present in crude oil, it must be "cracked", which means that some of the larger hydrocarbon molecules are converted into smaller ones. Because of the large demand for gasoline in the USA, many of the refineries in that country include cracking towers. (The cracking capacity of the USA is about 7 times that of Western Europe.) The most important conversion processes are thermal, catalytic and hydro cracking.

In thermal cracking, the large hydrocarbon molecules are split simply by the action of heat. At temperatures above 630 K, they begin to vibrate so violently, that the bonds between the carbon atoms break. This process takes place in the pipes of a cracking furnace at temperatures which may be as high as 1150 K. The temperature and the passage time through the furnace depend on the composition of the crude oil and the desired products. A refinery with a thermal cracking furnace produces about 19% gasoline, 45% medium distillates, 30% heavy heating oil and 6% other products.

In catalytic cracking, the large hydrocarbon molecules are split at lower temperatures in the presence of catalysts. The process yields both more light products and gasoline with a higher octane rating. A refinery with catalytic cracking produces about 25% gasoline, 49% medium distillates, 17% heavy heating oil and 9% other products.

In the hydrocrack process the heavy starting materials are almost completely converted to light hydrocarbons. The conversion takes place catalytically, in the presence of a hydrogen atmosphere. Depending on the operating conditions (pressure, temperature) and the catalyst, one obtains primarily gasoline and medium distillates. However, the higher efficiency of conversion is bought at the price of higher investment and operating costs. A hydrocracking refinery produces about 39% gasoline, 39% medium distillates, 10% heavy heating oil and 12% other products (199).

There have recently been suggestions that the energy required by refineries might be supplied by process steam and electricity from nuclear power plants (200).

Since heavy heating oil is the petroleum product which is most easily replaced by other energy carriers, the demand for it can be expected to decline. On the other hand, the world demand for gasoline and diesel fuel, and for naphtha for the chemical industry can be expected to increase, especially since it is difficult to find substitutes for them. Furthermore, in the foreseeable future the refineries will be receiving more and more heavy crude oil with relatively small amounts of light products (e.g. oil from oil shales and oil sands), because the major part of the petroleum reserves are heavy crudes.

At present, the majority of the petroleum refineries in the world, with a capacity of about $3.5 \cdot 10^9$ t, have no cracking capacity. Therefore, in the coming years, it will be necessary to equip many of them with cracking facilities, and to build more efficient refineries which produce a larger fraction of light hydrocarbons than before. The worldwide investment which will be necessary up to the turn of the century has been estimated as US-$ $130 \cdot 10^9$ (199).

4.64 Coal refinement products

4.641 The significance of coal refinement

In light of the world's coal reserves, efforts are being made to compensate for the disadvantages which have led, in many countries, to the decline in the use of coal. By more efficient conversion processes, it is hoped to increase the versatility of coal and to simplify its use.

Coal was created from biomass which was generated by photosynthesis about $265 \cdot 10^6$ years ago, during the carboniferous period. There are also coal deposits dating from the tertiary period, about $60 \cdot 10^6$ years ago. In the course of millions of years, the biomass was covered by other materials, and the process of carbonification began. The biomass was first converted to peat, then to brown and hard coal, as oxygen and hydrogen were gradually lost.

Coal can be roughly classified as brown or hard, depending on the stage of carbonification. Brown coal is less carbonified, and has a heat value of 0.2–0.9 tce/ton. Some of the subdivisions of brown coal are soft brown coal, lignite, and subbituminous coal. Brown coal is typically 58–72% carbon by weight, and 6–9% hydrogen. Hard coal, with a heat value of 0.9–1.1 tce/ton, includes gas coal; bituminous coal, which is 82–87% carbon and 5% hydrogen (by weight); semibituminous coal, which about 95% carbon and 3% hydrogen (by weight); and anthracite, which is more than 95% carbon and less than 3% hydrogen (by weight). Coal also contains varying amounts of sulfur and nitrogen, minerals and water.

The atomic ratio of hydrogen to carbon in coal is roughly 0.8. In petroleum this ratio is almost 2, and in natural gas, about 4. Conversion of coal to oil or gas can be achieved by increasing the hydrogen content and removing the sulfur, nitrogen and water.

Coal has a high potential for refinement. It can be converted either to electricity or to gaseous or liquid fuels. It can also be used as a raw material in the chemical industry, or converted to metallurigical coke and activated carbon. (The methods for generating electricity from coal are treated in 4.611.1.)

Gaseous and liquid fuels have a number of advantages over coal, including suitability for cheap transport in pipelines or ocean-going tankers. The world trade in crude oil has played a large part in the world economy (compare Fig. 3-7), and transport of liquid natural gas in tankers is becoming increasingly important (see 3.333). Gaseous hydrocarbons, however, can only be transported to the consumer in a pipeline network, i.e. their utilization is dependent on the presence of the appropriate infrastructure. In contrast, the fact that liquid hydrocarbons can be transported in mobile containers (e.g. tank trucks) is a great advantage. In addition, gaseous and liquid hydrocarbons, especially the latter, can be stored economically in large quantities, and neither type of fuel, in contrast to coal, produces ash when burned. An important aspect of coal refinement is that sulfur-containing coal can be used, because the sulfur can be removed in the processing.

In light of the above, it can be seen that liquid and gaseous hydrocarbons from coal have the potential to replace natural gas and petroleum on the world energy market. Fig. 4-24 is a summary of the processes for coal refinement and the possible products, along with their potential uses in various sectors (201). The most interesting of these may be synthetic gas (a mixture of CO and H_2), synthetic natural gas (SNG), liquid hydrocarbons (e.g. gasoline), and methanol.

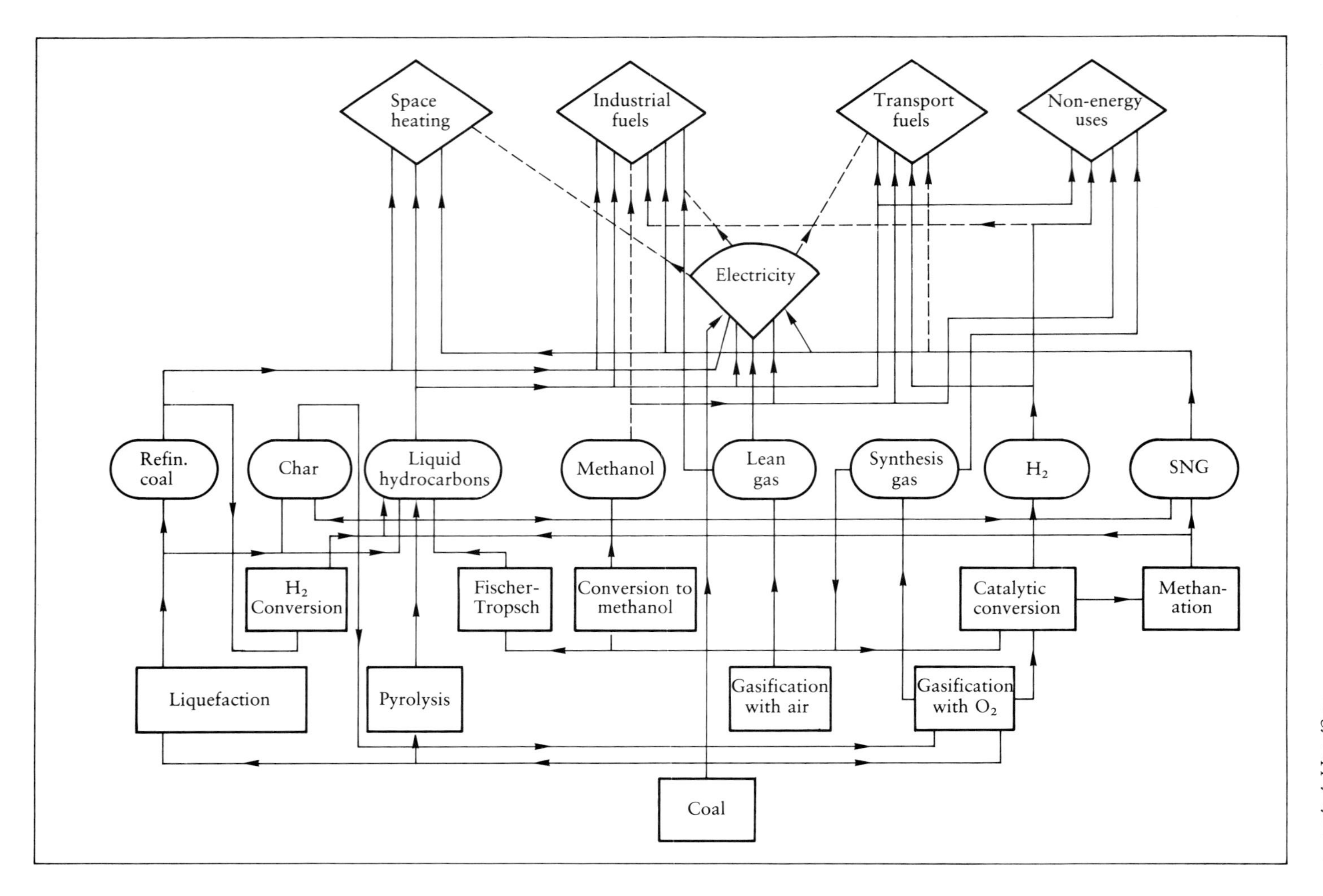
Space heating
Industrial fuels
Transport fuels
Non-energy uses
Electricity
Refin. coal
Char
Liquid hydrocarbons
Methanol
Lean gas
Synthesis gas
H_2
SNG
H_2 Conversion
Fischer-Tropsch
Conversion to methanol
Catalytic conversion
Methanation
Liquefaction
Pyrolysis
Gasification with air
Gasification with O_2
Coal

4.642 Gasification of coal

Until quite recently, coal gas was used in the industrial countries to supply energy. It was obtained both by degassing of coal during the conversion to coke and by adding a gasifying agent to coal, but was displaced by cheap natural gas. Because the energy efficiency of generating gas from coal is high, and gaseous fuels have advantages over coal, it is understandable that large-scale economical gasification of coal is very interesting for countries with large coal reserves.

There are about 35 methods of converting coal to gas (202–204). In principle, gases are generated form brown or hard coal by addition of a gasifying agent such as steam or hydrogen. The composition of the mixture of gases produced – mainly carbon monoxide CO, hydrogen H_2 and methane CH_4 – can be varied considerably to meet the needs of the consumer. Depending on whether the carbon is allowed to react with steam or hydrogen, the process is referred to as steam or hydrogenating gasification. In general, however, the starting materials are carbon, in the form of coal, and liquid water.

The following equations represent some of the basic reactions which occur in coal gasification. The heat of vaporization of water, $\Delta H = +44.0$ kJ/mol, is included in the values of ΔH.

Synthesis of methane: $2C + 2\,H_2O_{liq} = CH_4 + CO_2;\ \Delta H = +79.2$ kJ/mol (19)

Synthetic gas: $C + H_2O_{liq} = CO + H_2;\ \Delta H = +162.9$ kJ/mol (20)

Synthesis of hydrogen: $C + 2\,H_2O_{liq} = CO_2 + 2H_2;\ \Delta H = +165.5$ kJ/mol (21)

These equations show that the conversion of carbon into gaseous products is always endothermic when coal and water are used as starting materials. The composition of the gas produced depends on the reaction conditions.

The processes which actually take place are much more complicated than is indicated by the net equations (19)–(21). The thermodynamic system containing carbon, steam, methane, carbon monoxide, carbon dioxide and hydrogen can undergo a number of mutually independent reactions. The laws of thermodynamics and kinetics require that a number of reaction steps lie between the starting and end products shown above.

The methods for gasification of coal can be classified in a number of ways. One possibility is to arrange them according to the product, e.g. lean gas, synthetic gas, SNG. Another is to classify according to the method of heat generation: in au-

---

Fig. 4-24: Summary of coal refinement methods and possible products

Source: A. Ziegler, R. Holighaus, Technical possibilities and economic prospects for coal refining, ENDEAVOUR, Vol. 3, No 4, 1979.

tothermal processes, the heat required for the endothermic reactions is provided by partial burning of the coal, while in allothermal processes, the heat is generated outside the gasifier and transferred to the reaction chamber by a suitable heat carrier.

Not all the methods are equally suitable for brown and hard coal, and differences in the properties of the coal (e.g. the particle size, reactivity and ash composition) lead to different requirements for the process conditions. Brown coal, for example, is more reactive and can therefore be gasified at lower temperatures than hard coal.

In the following, a few of the promising methods for coal gasification will be discussed: the Lurgi process, the Winkler fluidized-bed process, the Koppers-Totzek process, the Bigas process from Bituminous Coal Research, Inc., the Synthane process of the U.S. Bureau of Mines, and the Hygas process of the Institute of Gas Technology (Chicago).

In the Lurgi process, either synthetic gas (a mixture of CO and H_2) or SNG (gas with a high methane content) can be made. Ground and dried coal, either brown or hard, is distributed over the grate of a fixed-bed gasifier. The gasification temperature is between 900 and 1000 K, the pressure is between 25 and 100 bar, and the process duration is 1 hour. The heat is generated by burning some of the coal with oxygen. A rotating grate in the floor of the reactor holds the coal and removes the ashes. The gasifying mixture (steam and oxygen) is added to the lower part of the gasifier. Depending on the type of coal used, the raw gas leaves the reactor at a temperature between 650 and 900 K, and is purified by removal of sulfur, naphtha, unsaturated hydrocarbons and carbon dioxide. 730 kg brown coal, 700 kg steam and 90 Nm^3 oxygen are consumed in the production of 1000 Nm^3 synthetic gas (1 Nm^3: 0°C; 1.01325 bar). The purified gas typically contains the following (in vol %): 54.0% H_2, 25.2% CO, 17.6% CH_4, 2% (CO_2 + H_2S), 1.2 % N_2 and other gases. Methane can be produced from the CO/H_2 mixture according to the equation

$$CO + 3\,H_2 = CH_4 + H_2O;\ \Delta H = -205\ \text{kJ/mol} \tag{22}$$

After removal of the CO_2 and drying, the product is SNG. There are many plants based on the Lurgi pressurized gas process in operation around the world (201–203). One plant at Sasolburg (South Africa) produces about 22500 Nm^3/h of synthetic gas for the production of liquid hydrocarbons by the Fischer-Tropsch process (see 4.643). In the USA, there are plants with outputs up to about 50000 Nm^3/h, and in the Federal Republic of Germany, the "Ruhr 100" plant in Dorsten (Ruhr Valley), which was intended to improve the Lurgi process, went into operation at the end of 1979. This plant is intended to gasify about 28000 t coal/year, and to produce about $15 \cdot 10^6$ m^3 SNG/year, to be fed into the pipeline network of the Ruhrgas AG.

Synthetic gas can be synthesized by the Winkler fluidized-bed process, in which ground and dried coal (brown or hard) is mixed with oxygen and steam in a fluidized bed, under normal pressure, and gasified at 1100 to 1300 K, depending on the type of coal. At the higher temperature, all tar and heavy hydrocarbons are gasified. The steam and oxygen are added in the lower part of the fluidized bed, and the gas is withdrawn from the upper part of the generator. The ash is drawn off from the bottom. Winkler gas generators have high capacities: with brown coal, outputs of 60000 Nm^3/h are reached. At present there are about 16 industrial plants based on the Winkler process. To generate 1000 Nm^3 CO/H_2 synthetic gas, for example, they consume 660 kg brown coal, 360 kg steam and 320 Nm^3 oxygen.

Synthetic gas can also be produced with the Koppers-Totzek or air-stream process. Coal is ground, dried and pulverized; the coal dust and the gasifying agent (oxygen and steam) are blown into the generator. Gasification occurs at a pressure of about 30 bar and a temperature of about 1800 to 1900 K. The advantage of this method is that all types of coal and oil coke can be gasified, even those with high ash, sulfur and water contents; the disadvantage is that it uses more oxygen than the Lurgi pressurized gas or the Winkler fluidized-bed process. Depending on the type of coal, the Koppers-Totzek process uses 550 kg coal, 230 kg steam and 430 Nm^3 oxygen per 1000 Nm^3 CO/H_2 synthetic gas produced. A typical composition of the purified gas is (in vol %) 25.5% H_2; 65.2% CO, 1.1% (CO_2 + H_2S), 8.2% N_2 and other gases. Since 1950, about 15 Koppers-Totzek plants have been built around the world. A pilot plant with a daily consumption of 6 t coal/day was built in the Shell laboratory in Amsterdam, and has been in operation since 1976. A large Koppers-Totzek plant built by Shell International Petroleum Maatschappij (Den Haag) in Hamburg went into operation at the end of 1979. Fig. 4-25 is a flow scheme for the Shell Koppers process (205). Typical data for this process are a daily consumption of 150 t coal, and a raw gas production of about 2000 Nm^3 per ton of hard coal. The consumption of oxygen and steam depend on the quality of the coal. About 0.9–1 t oxygen is used per ton of water- and ash-free coal. With oxygen as the gasification medium and removal of the sulfur and CO_2 from the gas, a high-quality synthetic gas is produced. The heating value is about 11300 kJ/Nm^3, and the overall efficiency of the process is 77%. (This is a pilot plant from which data are being collected for the planning of a demonstration plant, which will process 1000 t coal per day.) The high quality synthetic gas from this process is suitable for industrial heat production and as a starting material for the production of various liquid and gaseous products. Some examples of these are hydrogen or hydrogen-rich gas (made by operating the plant under the appropriate conditions), reduction gas for the direct reduction of iron ore, methane or SNG (methane synthesis according to equation 22), methanol, and synthetic liquid hydrocarbons, e.g. gasoline or light to medium distillates by the Fischer-Tropsch synthesis (see 4.643.1).

In the Federal Republic of Germany there are now 8 pilot plants for gasification of coal in operation. 6 of these produce synthetic gas, 1 makes SNG, and 1 generates electricity from the coal gas.

In contrast to the processes discussed so far, the methods developed in the USA (Bigas, Synthane and Hygas) are intended for the immediate production of SNG.

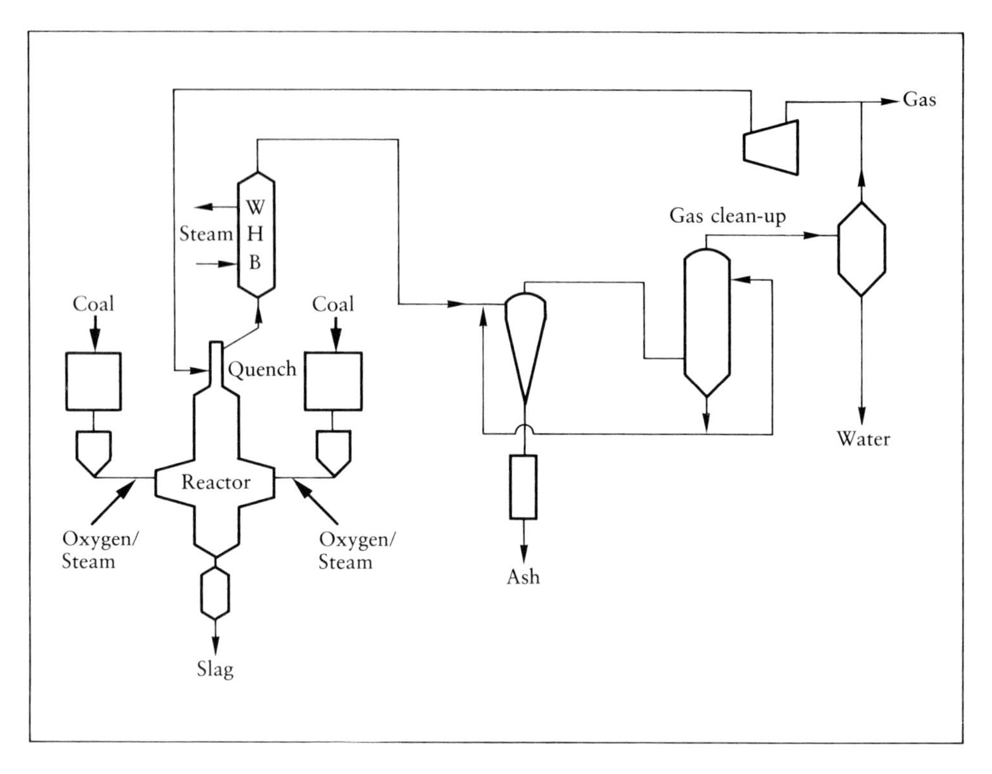

Fig. 4-25: Flow diagram of the Shell-Koppers Process for coal gasification

Source: H. K. Völkel, M. J. van der Burgt, Shell Oil Company, Hamburg 1981.

About 15 SNG plants are now in operation in the USA (207). In the Bigas process developed by Bituminous Coal Research, Inc., coal is ground, dried and pulverized. Coal and steam are fed into the upper part of the reactor through four concentrically arranged feed pipes. Here the coal is heated to about 1200 K by hot gas from the lower part of the reactor, and a raw gas is produced. Unreacted coal is separated from the raw gas and fed into the lower part of the gasifier; at the same time, oxygen and steam are added. In this part of the reactor, gasification occurs at roughly 1700 K. Next the CO_2 and H_2S are removed, and then the gas is methanized and dried. A pilot plant with a daily coal processing capacity of 120 t per day is now in operation and produces about 70000 Nm^3 gas per day.

In the Synthane process developed by the U.S. Bureau of Mines, ground and dried coal is fed into a low-temperature carbonizer (working temperature about 700 K), where part of the coal is gasified. (About 12% of the total steam and oxygen is fed into the carbonizer.) The remainder of the coal is fed into a gasifier, where it is gasified at 1300 K and a pressure between 35 and 70 bar. The gas is then purified, methanized and dried. A pilot plant with a coal processing capacity of 70 t per day produces around 40000 Nm^3 per day.

In the Institute of Gas Technology's Hygas process, the coal is ground and dried, then rubbed to a paste with oil. This paste is fed into a 3-stage gasifier. In the first stage, the light coal components are distilled off; in the next two, a hydrogenating gasification takes place. The hydrogen required for this is generated from the coal remaining after the third stage of gasification. The gas is purified, methanized and dried. The advantage of this technique is that it can be used for any type of coal. A pilot plant in Chicago processes about 75 tons per day and produces about 50000 Nm^3 per day of a gas which is of a comparable quality to natural gas. There are plans to build plants with capacities of 200000 Nm^3 per day (206).

In addition to these relatively mature techniques, there are several which are just at the beginning of development. Several examples are the Hydrane process from the U.S. Bureau of Mines, the Molten-salt process of the M.W. Kellogg Corporation, and the ATGAS process of the Applied Technology Corporation.

In the conventional gasification methods discussed above, the process heat is obtained by burning part of the coal (autothermal processes). There are plans to use the heat from a high-temperature nuclear reactor to gasify coal (see 4.213.3 and 4.65). This would yield more gas from the coal, which would conserve the coal reserves (207).

For the long term, there are efforts to develop methods of using the world's enormous coal reserves without conventional mining. The method of underground gasification seems especially promising (208, 209), in part because it would make economical the use of seams which cannot be economically mined. This would multiply several times the available coal reserves (see 3.31). There are extensive projects aimed toward this end in the USA, the Soviet Union and Europe. The Belgian "Institut National des Industries Extractives" (INIEX) is planning a large field experiment of this nature. In Manna, Wyoming (USA), an underground gasification operation is producing 56000 Nm^3 of gas daily, with an energy content between 3.7 and 5.6 MJ/Nm^3 (210, 211).

4.643 Liquefaction of coal

4.643.1 Fischer-Tropsch synthesis

There are basically two methods of producing liquid fuels from coal: the gasification of coal to carbon monoxide and hydrogen (synthetic gas) followed by the synthesis of liquid products from the gas (Fischer-Tropsch synthesis), and the direct hydrogenation of the coal according to F. Bergius.

Fig. 4-26 shows the steps in the Fischer-Tropsch synthesis of liquid fuel from coal (201). Depending on the reaction conditions (pressure, temperature) and the catalysts used, this process can yield a wide variety of hydrocarbons, from methane

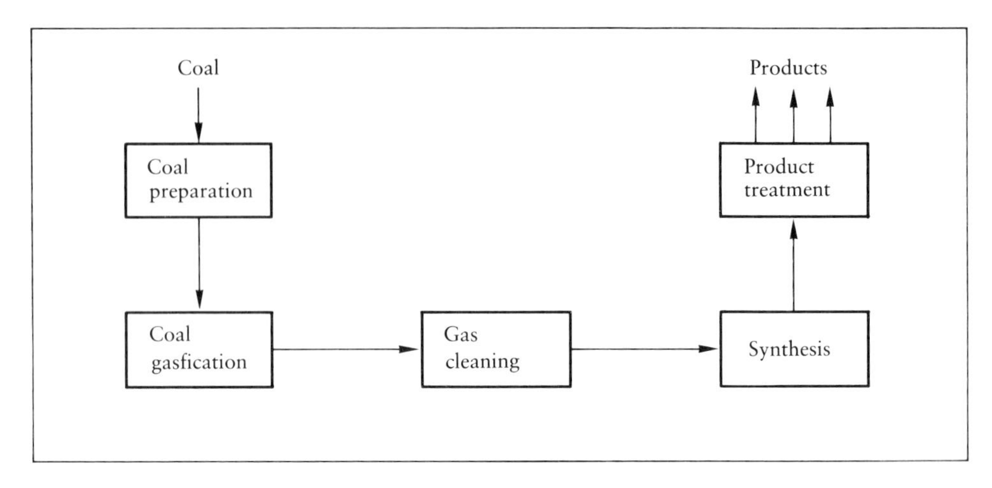

Fig. 4-26: Coal liquefaction by Fischer-Tropsch synthesis (block diagram)

Source: A. Ziegler, R. Holighaus, Technical possibilities and economic prospects for coal refining, ENDEAVOUR, Vol. 3, No 4, 1979.

to high-molecular-weight substances like paraffins and olefins, and also C, H, O compounds like alcohols, aldehydes and ketones. In other words, it is possible to make liquid fuels for vehicles, heating oils in all boiling point ranges, and important starting materials for the chemical industry from coal (202).

F. Fischer and H. Tropsch first demonstrated the synthesis of liquid hydrocarbons from carbon monoxide and hydrogen at 200°C and normal pressure, using metallic catalysts (iron, cobalt). Synthesis plants were built in Germany between 1935 and 1939, some of which used normal pressure, and some working at medium pressure. They had a total capacity of about 700000 tons per year of liquid hydrocarbons. Even after 1945, there was considerable interest in the synthesis. Starting in 1949, for example, the U.S. Bureau of Mines carried out systematic investigation at a large-scale test plant with a daily production of about 7 tons of liquid products. In addition, the Hydrocol fluidized-bed technique was developed in Brownsville, Texas (USA) by the Hydrocarbon Research Corp. and the U.S. oil companies, and in 1950 a large-scale plant with a yearly production of about 400000 tons liquid hydrocarbons was built. However, the advent of cheap petroelum and natural gas in the mid-fifties resulted in a decline in the interest in the Fischer-Tropsch synthesis in all parts of the world (except South Africa). Since the price of oil began to rise rapidly, there has again been interest in the Fischer-Tropsch process. One advantage of the method is that it is relatively easy to remove the sulfur contained in the coal after it has been converted to gas, and as a result, the products of the synthesis are relatively sulfur-free. Another advantage is that even ash-enriched coal can be used as a starting material.

The basic reaction of the Fischer-Tropsch synthesis is the formation of aliphatic chains from carbon monoxide and hydrogen (catalytic hydrogenation of carbon monoxide):

$$nCO_2 + 2nH_2 = (-CH_2-)_n + nH_2O \qquad (23)$$

At 250°C, $\Delta H = -158$ kJ/mol (CH_2); i.e. the reaction is exothermic. The metals iron, cobalt, nickel and ruthenium can be used as catalysts. The best yields are obtained if the amounts of CO and H_2 in the synthetic gas are adjusted to be the same as the amounts consumed in the reaction. The ratio of CO to H_2 consumed depends on the composition of the catalyst and the reaction conditions (pressure, temperature) and mean length of time the gas remains on the catalyst. Depending on the catalyst, reaction conditions and composition of the gas phase, the water formed in the reaction may be able to react with unreacted CO to form CO_2 and H_2 according to the equation $H_2O + CO = CO_2 + H_2$.

In most cases, the reaction products are paraffins and olefins, but the conditions can be chosen so that oxygen-containing organic compounds, in particular alcohols (e.g. methanol), are the main products. The formation of oxygen-containing compounds is favored by low temperature, high pressure and a high ratio of CO to H_2. The synthesis of methanol from CO and H_2 takes place (with catalysis, e.g. by zinc oxide/chromium oxide) at a pressure of about 300 bar and a temperature of about 350°C according to the equation

$$CO + 2\,H_2 = CH_3OH; \Delta H = -\,91 \text{ kJ/mol} \qquad (24)$$

The reaction is exothermal and causes a reduction in volume of the reactants. As mentioned above, methanol is an important raw material for the chemical industry and, because it contains no sulfur, it is a non-polluting liquid fuel.

The fixed-bed reactor (Arge process), developed by Lurgi-Ruhrchemie (Federal Republic of Germany) and the Kellogg-SASOL coal dust reactor (Synthol process), developed by M.W. Kellogg (USA) and improved by SASOL (South African Coal, Oil and Gas Corp.) have been especially satisfactory for the large-scale application of the Fischer-Tropsch synthesis (202).

SASOL is the only company which operates Fischer-Tropsch plants at present. SASOL I employs both the Arge and the Synthol processes to produce synthetic hydrocarbons. Since 1960, it has had 5 fixed-bed reactors in operation, which are now producing about 100000 tons liquid hydrocarbons per year, and three coal dust reactors, which produce about 250000 tons liquid hydrocarbons per year. (The synthetic gas needed for the process is produced from coal by the Lurgi process.) The Synthol process, for example, produces 75% olefins, 10% paraffins, 7% aromatics and 8% oxygen-containing compounds; about 65% of the end product is gasoline which corresponds to the usual specifications for Otto cycle engines. At present, a large plant, SASOL II, is being built to operate the Synthol process. When completed, SASOL II will produce, among other things, about $1.5 \cdot 10^6$ t vehicle fuel per year.

4.643.2 Coal hydrogenation

Coal hydrogenation is the catalytic addition of hydrogen to coal at high temperature and pressure. Coal was liquified in 1913 by F. Bergius by adding hydrogen to coal at high temperature and pressure, in the presence of a catalyst. He obtained oil-like compounds. The development of coal hydrogenation to a large-scale method was carried out under M. Pier of the Badische Anilin- und Soda-Fabrik (BASF) in Ludwigshafen/Rhein. Of the coal consumed by coal hydrogenation, only about one third is hydrogenated to the liquid products, while the other two thirds are used to produce the hydrogen and the necessary energy. In order to obtain 4 tons of oil, one must consume about 12.5 tons of coal. (The production of 1 ton gasoline requires about 2000 Nm^3 of hydrogen, the exact amount depending on the type of coal and the method of production, and the amount of the heavy oil components which are converted to gasoline.)

In 1943, there were 12 hydrogenation plants in operation, converting brown and hard coal and tar. They produced about $4 \cdot 10^6$ tons of fuel per year, and in 1943, they supplied about one third of the total German oil demand. (50% of the gasoline and diesel oil and 90% of the aviation gasoline was supplied by these plants.) There were also coal hydrogenation plants in England (hard coal hydrogenation plant of the Imperial Chemical Industries, Ltd.), France, Japan and the USA. (The Standard Oil Co. of New Jersey, for example, built two hydrogenation plants in 1930. All of these plants were closed. In the fifties and sixties, the trend was toward the hydrogenation and cracking of heavy oils, and especially in the USA in the 1960's, many large plants for producing vehicle fuel from heavy oils (hydrocracking plants) were put into operation (see 4.63).

The hydrogenation of coal takes place in two steps, the first in liquid phase. Coal is converted at 450°C and a pressure of 700 bar, in the presence of finely divided catalysts, into an intermediate product (raw gasoline, medium and heavy oil). In the presence of catalysts like iron, tin and molybdenum compounds, hydrogen becomes bound to carbon, and some of the heteroatoms in the coal (sulfur, oxygen, nitrogen) are converted to hydrides (H_2S, H_2O, NH_3) which are relatively easily removed. The first step of coal liquefaction can also be an extraction of the coal with solvents which provide reactive hydrogen. The extraction consists of treating the coal with a solvent, under pressure, which causes the coal to depolymerize and combine with hydrogen atoms from the solvent. The second step takes place in the gas phase. The intermediate products, which have been separated from the ash and used catalysts, are refined over fixed catalysts at a temperature of 400°C and 300 bar, to produce gasoline (for example). Although the products obtained in liquid phase are fairly crude, it is possible to choose the reaction conditions (pressure, temperature, catalyst) so that within limits, the desired end products represent a fraction of the crude mixture.

Fig. 4-27 shows the important steps of coal hydrogenation (201). In the past few years, the further development of coal liquefaction by hydrogenation has been going on in a number of countries, especially the USA and the Federal Republic of Germany (212, 213). In the latter country there are a number of small experimental installations in operation, and two pilot plants are being built. (These are based on a modified IG-process). The plants are run by the Saarbergwerke AG (using 6 tons/day hard coal, they produce raw gasoline and medium distillates) and the Ruhrkohle AG/VEBA-Öl (using 200 tons/day of hard coal, they also produce gasoline and medium distillates.)

The following are further developments based on the experience with coal liquefaction plants: the SRC II process (Solvent Refined Coal II) of Gulf Oil Corp., the EDS process (Exxon Donor Solvent) of the Exxon Corp., the Synthoil process of the U.S. Bureau of Mines, the H-Coal process of Hydrocarbon Research Inc., the CSF process of the Consolidation Coal Co., the COED process (Char Oil Energy Development) of the Food Machinery Corp. (212).

4.644 Production of coke

If coal is heated in the absence of air, it produces gas (coke-oven gas), liquids (e.g. heavy and light coal-tar oils) and solids (brown and hard coal coke). The amounts, types and compositions of the decomposition products are determined both by the raw material and the conditions under which the *dry distillation* process takes place. At temperatures up to about 900 K, one speakes of *low-temperature carbonization,* while at temperatures above 1200 K, the process is called *coking*.

Only the coking of hard coal is of economic significance. Coke is used as the reducing agent in the smelting of iron from various iron ores. Table 4-7 shows the figures for world production and coking of hard coal (214). Although the world production of coke has increased in the past few years, the amount produced in the major industrial countries has decreased. This is mainly due to the fact that the iron and steel industry, which is the major consumer of coke (up to 80%) has been able to reduce the coke required per unit of iron. In 1960, for example, the specific coke requirement was about 830 kg per ton of pig iron, while in 1974, it was only 500 kg. This has been made possible by improvements in blast furnace technology, and also by the partial substitution of heating oil for coke. It can be assumed, however, that the iron industry will continue to need metallurgical coke for some time to come.

4.645 Economic aspects

At present, the gasification and liquefaction of hard coal is being done on a large scale only in South Africa, where cheap hard coal is available. In the USA, Japan and

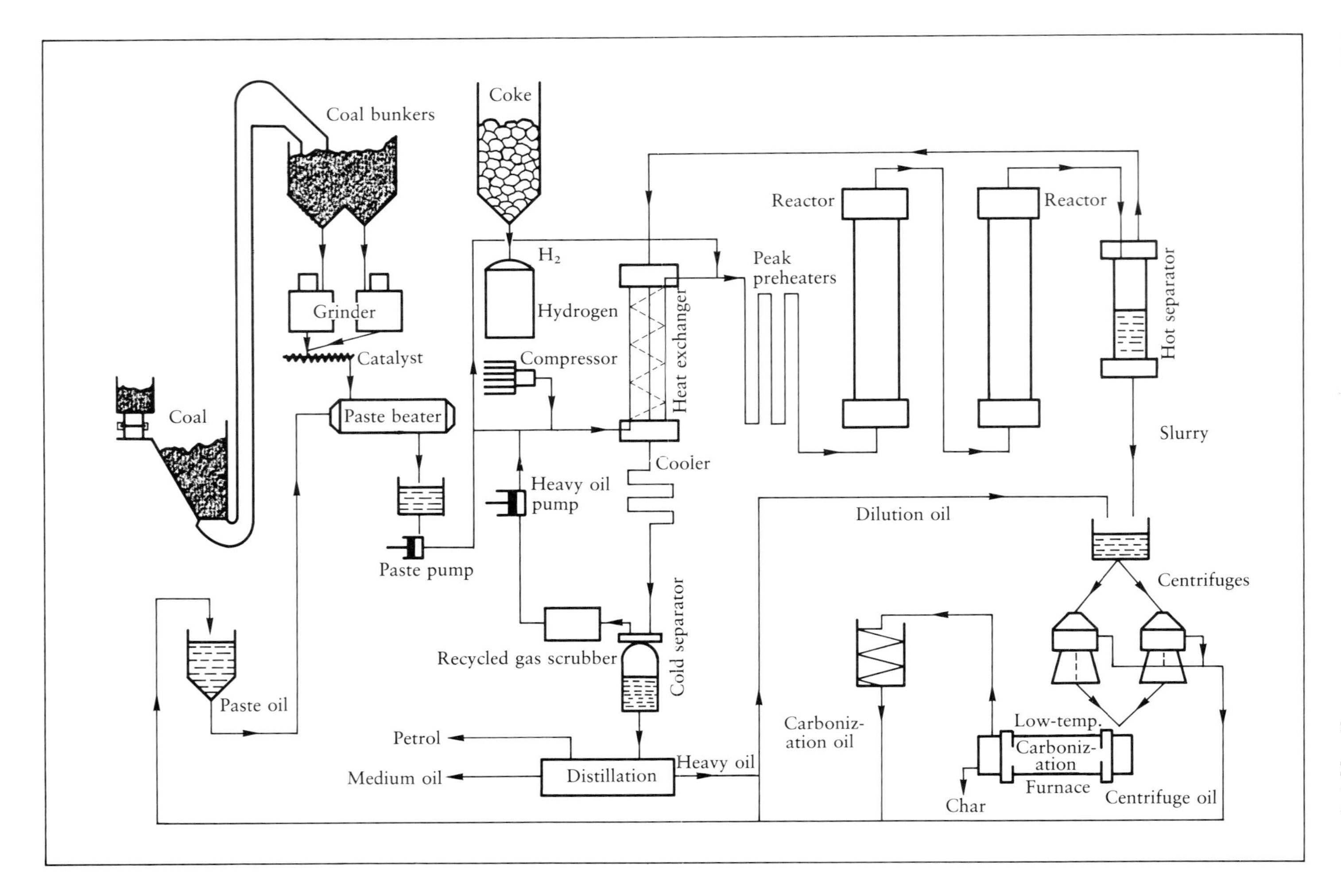

Coal bunkers
Coke
Coal
Grinder
Catalyst
Paste beater
Paste pump
Paste oil
H_2
Hydrogen
Compressor
Heavy oil pump
Recycled gas scrubber
Heat exchanger
Cooler
Cold separator
Petrol
Medium oil
Distillation
Heavy oil
Peak preheaters
Reactor
Reactor
Hot separator
Slurry
Dilution oil
Centrifuges
Carboniz-ation oil
Low-temp. Carboniz-ation Furnace
Char
Centrifuge oil

Table 4-7: World production of hard coal and hard coal coke (in 10^6 t)

Country	Year	Hard coal mined 1966	1973	1977	1979	Hard coal coke produced 1966	1973	1977	1979
USSR		n.a.	524[1]	556	554	n.a.	83[1]	85	87
USA		493	530	606	655	65	58	49	48
China, P.R.		n.a.	450[1]	550	663	n.a.	28[1]	28	31
Germany, F.R.		132	104	91	93	40	34	27	27
United Kingdom		177	130	121	121	19	18	14	13
France		50	26	21	19	13	12	11	12
EC		389	270	240	239	89	82	68	67
Europe[2]		565	478	477	490	122	123	112	114
World		2093	2238	2538	2792	315	367	357	360

[1] These figures were for 1974.
[2] European totals, excluding the USSR.

Source: Statistik der Kohlenwirtschaft, Essen und Köln, September 1978 and September 1980.

some European countries (the Federal Republic of Germany, Great Britain), development work is being done on the realization of large-scale commercial coal gasification and liquefaction.

The energy efficiency of autothermal gasification of coal (conventional gasification) is about 70%, while that of allothermal gasification (gas from coal by nuclear gasification) is 87%; the efficiency of coal liquefaction (direct hydrogenation of the coal) is on the order of 60%, with the exact value dependent on the type of coal and the technique used (203). Coal gasification is technologically most advanced, and is closest to the threshold of economic feasibility.

Large installations for coal gasification and liquefaction will be complex and expensive. The Rand Corporation, USA, has estimated that the investment for a plant which would produce either $5 \cdot 10^6$ t *coal oil* (syncrude) or $6 \cdot 10^9$ m^3 *coal gas* (SNG) per year would be about \$ $3 \cdot 10^9$ (US, 1979 prices). Given the present state of the art, either type of plant would consume about $15 \cdot 10^6$ t coal per year.

The economic feasibility of producing synthetic fuels from coal, especially hydrocarbons, depends on a number of factors, including the availability and price of coal, the technological development of the processes for coal refinement, and the availability and price of petroleum. It follows that the threshold of economic feasibility will differ widely from country to country (215, 216).

Fig. 4-27: Diagram of coal liquefaction by hydrogenation

Source: A. Ziegler, R. Holighaus, Technical possibilities and economic prospects for coal refining, ENDEAVOUR, Vol. 3, No 4, 1979.

The US Department of Energy estimates that the cost of producing synthetic fuels from coal will vary, depending on the product and the location, from $ 27 to $ 45 per barrel of oil or, in the case of gas, per barrel equivalent (1979 prices). The cost of producing "coal gas" is at the lower end of the spectrum. As discussed in section 3.341, the US has extremely large deposits of oil shale. It is less expensive to make liquid fuels from shale than from coal: the Department of Energy estimates the cost at $ 25 to $ 35 per bbl (1979 prices). In other words, the relative costs are such that oil from oil shale will probably be the first fuel which will become competitive with conventional oil. The production of methanol from coal (see 4.643.1) and of ethanol from biomass could in principle contribute to the supply of liquid fuels. Either alcohol, mixed with gasoline (Gasohol, a mixture of 90% gasoline and 10% alcohol) can extend the supply of oil-based fuel and improves the anti-knock quality of the fuel. In terms of energy content, the production of alcohol from coal or biomass is still too expensive, compared to petroleum fuels. The economic threshold for "synthetic" fuels from coal will also depend on the extent to which the USA is able to utilize its enormous potential of nonconventional natural gas sources, e.g. geopressured gas (see 3.331), by developing technologies with which to make its extraction economically attractive.

The Federal Republic of Germany, in contrast to the USA, has essentially only coal, brown and hard, as an energy source. Furthermore, its hard coal is extremely expensive, for geological reasons, which led to a decrease in hard-coal production in the sixties and seventies, and which means that imported coal is cheaper in the FR of Germany than domestic coal. According to the Federal Ministry of Economics, Bonn, the conversion of domestic hard coal to "coal oil" would be possible at a price of $ 50 to $ 55/bbl, while for imported coal, the price would be $ 35 to $ 40/bbl (1979 prices). It follows that coal liquefaction will not become economically feasible in the intermediate future for reasons of cost alone. However, gasification of coal, especially of brown coal, is cheaper than liquefaction. It is expected that gasification of coal, that is the production of SNG and especially of synthetic gas (as a raw material for chemical syntheses) will become competitive in the Federal Republic of Germany in the course of the eighties. (At present, the Federal Republic annually converts about $30 \cdot 10^6$ tce brown coal to electricity.) It is significant that *coal gas* (SNG or synthetic gas) can be used as a substitute for petroleum products in many areas.

The situation in Great Britain is that domestic hard coal is less expensive to mine than in the Federal Republic of Germany. In addition, Great Britain has considerable reserves of oil and gas in the North Sea, so it can be expected that here the high-quality products of coal refining will be late in entering the energy market, and especially the raw material market.

The current volume of world trade in coal is about $240 \cdot 10^6$ tce/year; it is expected to increase to about $800 \cdot 10^6$ tce/year by 2020 (see 3.312). (This is about 30% of the present volume of world trade in oil.) The imports of coal into countries which have few energy raw materials, e.g. France, Italy, and Japan, will probably increase, as coal gas should become increasingly competitive with petroleum in certain areas. As mentioned, the conversion of coal into *coal oil,* at a cost of $ 35 to $ 40/bbl from imported coal (1979 prices), is not yet competitive with conventional oil.

The first conventional coal liquefaction plants will probably be built near large coal deposits which can be cheaply mined, because the transport of liquid fuels, even over large distances by sea, is economically practical. In other words, one can expect that as the reserves of petroleum and natural gas are exhausted, the demand for liquid and gaseous fuels will be filled by oil and gas from "artificial" sources.

When and in what areas coal refinement products will become competitive with petroleum products will depend heavily on the availability and price trends of petroleum. In the last few years, the time horizon for the reaching of economic feasibility (in competition with oil products) by coal refinement products has been steadily receding (217). One important reason is that the price of coal, like that of other energy carriers, has been rising. The substitution of coal for heavy heating oil in industry and power plants, and the conversion of the heavy heating oil into gasoline and light heating oil may continue for a long time to be more economical than the liquefaction of coal. If the construction of high-capacity conversion plants for petroleum is accelerated, it will be possible to produce a larger fraction of lighter petroleum products than at present (see 4.63). Petroleum products thus might supply the demand for vehicle fuel and chemical raw materials for a long time to come (218).

4.65 Long distance energy

The basic idea behind a system of long distance energy is to transport heat as chemical energy, in other words, "cold", over large distances to the consumer. The heat in such a latent-heat gas (e.g. synthetic gas/methane) has the advantage over the heat in a distance-heating system that it is not lost in transport, and the consumer can obtain higher temperatures (up to 600°C) than are feasible with district heating. This greatly widens the area of applicability of the system. For one thing, it can be used to supply a part of the industrial demand for process heat, and for another, it can be used to generate electricity close to the consumer, i.e. is relatively non-polluting.

The system of long distance energy makes it possible to transport heat from a high-temperature nuclear reactor over large distances, and thus to separate the reactor site from the consumer[1]. In contrast to coal gasification, a long distance energy system uses fossil energy carriers only as transport media, in a closed circulation. After the energy has been removed from it, the medium is returned to the reactor for recharging. The advantage over electricity is the possibility of a higher efficiency in

[1] The heat might be supplied from other sources, such as solar collectors, as well.

the conversion of the primary to the secondary energy for heating purposes (heat can be directly stored or withdrawn) and in the ease of storing the transport medium.

Among the many chemical systems which might be used, the latent heat gas (synthetic gas/methane) favoured by R. Schulten, Jülich, Federal Rep. of Germany, is especially suitable (32, 219). The basic equation for this reaction is

$$CO + 3H_2 = CH_4 + H_2O;\ \Delta H = -205\ \text{kJ/mol} \tag{25}$$

This cycle has been tested with conventional technology in both directions, synthesis and decomposition of methane. Fig. 4-28 shows the flow diagram of the synthetic gas/methane cycle which has been given the working title *ADAM/EVE system*. The energy coupling step takes place at the site of the nuclear reactor, and the energy decoupling near the consumer. The heat required for energy coupling through methane decomposition is supplied at 825° C by a high-temperature reactor. The helium coolant from the reactor gives up its sensible heat in the pipe still (EVE), where the methane is catalytically split in the presence of steam. The product gas ($3H_2$ + CO), which is made at a pressure of 40 bar, still contains carbon dioxide and unreacted methane. Outside the reactor, it is cooled, dried, and condensed to 64 bar.

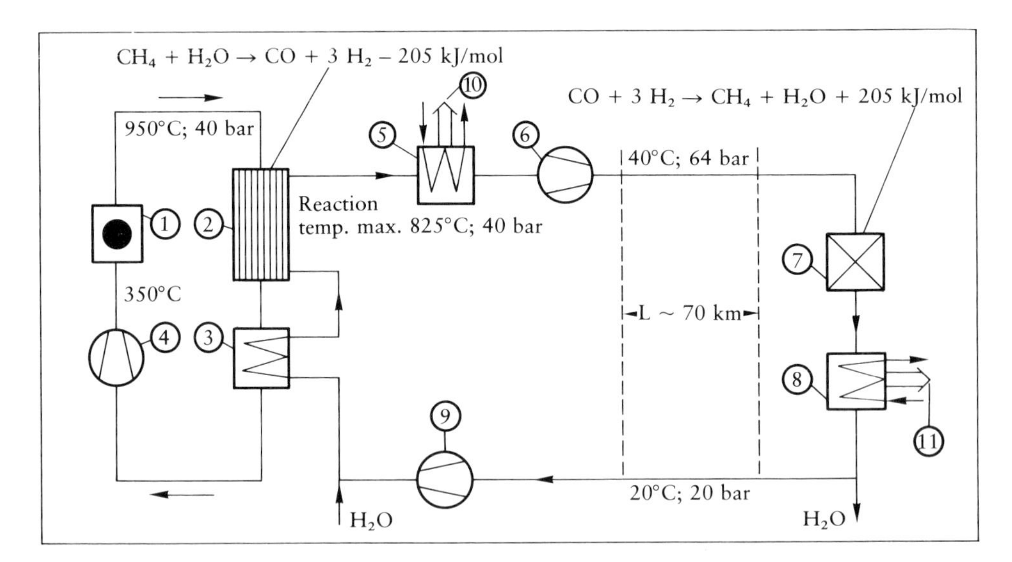

Fig. 4-28: Long distance energy system
1 = Nuclear reactor; 2 = Pipe still (EVE) (Reaction temp. max. 825° C, 40 bar); 3 = Pre-heater; 4 = Bellows; 5 = Utilization of waste heat; 6 = H_2, CO-compressor; 7 = Methanization (ADAM); 8 = Heat exchanger; 9 = CH_4-compressor; 10 = Vicinity heat + electrizity + waste heat; 11 = Heat (Steam 600° C).

Source: K. J. Euler, A. Schramm (Eds.), Energy Supply of the Future, Munich: Karl Thiemig 1977.

From the high-temperature reactor site, it is transported about 70 km in underground pipes to a population center, and is there used to generate heat at a central plant. The energy decoupling takes place in a methanizing reactor (ADAM), where the pre-warmed synthetic gas is led over a catalyst and reacts to form methane and steam. The released latent heat (chemically bound heat) can then be delivered to the consumer. The temperature at which the methanization reaction occurs is high enough to allow heat/power coupling, so that ADAM produces both heat and electricity. The water is separated from the methane in the reacted gas, and the methane is pumped back to the high-temperature reactor site in a pipeline laid parallel to the one in which the synthetic gas arrived. There it is again split in EVE to close the cycle (220).

It should also be mentioned that in a so-called "open system", part of the synthetic gas ($CO + H_2$) can be used for a raw material by the chemical industry (32, 206).

4.66 Hydrogen as energy carrier

4.661 Production

Hydrogen is a versatile secondary energy carrier. It is available from sea water in practically unlimited amounts, it is suitable for transport and storage of energy, and it can be converted either to electricity and heat or into a fuel. (Electricity can be generated either thermally, e.g. by gas turbine and generator (see 4.611.1), or in an H_2/O_2 fuel cell (see 4.611.31), or for example in a MHD generator (see 4.611.32). Hydrogen is an extremely clean fuel, and it is also an important raw material, both for chemical synthesis and for reduction of metals. There is much talk in this connection of a possible future "hydrogen economy" (171, 221–223).

The hydrogen which is now produced is used, aside from experimental purposes, almost entirely as a raw material. In 1938, the world production was about $70 \cdot 10^6$ Nm^3, in 1973 it was $250 \cdot 10^{12}$ Nm^3 ($\triangleq 22.5 \cdot 10^6$ t) and in 1977, it was about $330 \cdot 10^{12}$ Nm^3 ($\triangleq 30 \cdot 10^6$ t). About 80% was produced from petroleum, about 15% from coal, and the rest by electrolysis (116).

Large-scale and economical production of hydrogen for use as a secondary energy carrier is an unsolved problem. Production of hydrogen as an energy carrier from fossil raw materials is not economically feasible, due to the unused carbon. For a large-scale hydrogen economy, water would be the suitable raw material. The reaction is given in section 4.326.1 (eq. 17). Hydrogen and oxygen are generated from water upon application of energy. The amount of energy which must be used (ΔH) to split liquid water H_2O_{liq} into gaseous H_2 and O_2 at 25°C ($\triangleq$ 298.2 K) and 1 bar pressure (standard conditions) is $\Delta H = 286$ kJ/mol. The energy can be supplied in the form of electricity, heat or radiation. (This energy is released when the H_2 and O_2 recombine to form water.) There are basically three methods of production: conventional electrolysis, high-temperature steam-phase electrolysis, and cyclical thermochemical processes. (The electricity or heat required can in principle be supplied by solar energy (see 4.31)). Photolytic decomposition of water would be an extremely favorable method, which was discussed in 4.326.3.

Because conventional electrolysis techniques require electricity, a rather expensive secondary energy carrier, the H_2 so produced is also expensive. The methods have been known since the 19th century, and have long been used to produce pure hydrogen (more than 99.5% pure). When a constant voltage is applied across two electrodes which are submerged in an electrolyte, H_2 is formed at the cathode, and O_2 at the anode. About 5 kWh is requried to produce 1 Nm^3 H_2. An electrolysis plant generally consists of three basic units, the electric system (voltage generator, rectifier, etc.), the electrolysis unit with the individual cells and their components (electrodes, diaphragms, etc.) and additional process units (pumps, heat exchangers, pressure regulator, compressors, etc.). In each plant, these units have to be optimized and adjusted to each other. In addition, the materials used for the electrolysis units must meet high standards of corrosion resistance. Conventional electrolysis plants use 25 to 30% potassium hydroxide as electrolytes at temperatures of 60 to 90°C. Some typical values for an electrolysis plant (bipolar arrangement) are: current density, 0.2 to 0.4 A/cm^2; operating voltage, 1.87–2.1 V/cell; H_2 output per individual cell, 1000–4000 m^3H_2/h; H_2 output of the plant, up to 200000 m^3H_2/h; energy consumption, 4.3–4.6 kWh/m^3H_2; pressure, 0.1–1 bar.

In high-temperature steam-phase electrolysis, steam is decomposed on a solid, temperature-resistant electrolyte. The significance of this method is that the current requirement is reduced by the input of more heat energy. This improves the overall efficiency. A promising method for the future is the production of hydrogen by cyclic thermochemical processes. Unlike electrolysis methods, thermochemical processes use only process heat to produce hydrogen, and thus avoid the detour through electricity. The process heat can be provided by a high temperature reactor (see 4.213.3), for example, or by the sun. (It has already been mentioned that hydrogen, because it is more easily stored and transported, "fits" solar energy better than electricity does.) Cyclic thermochemical processes are needed, because water does not dissociate in appreciable quantities below 3000 K, and its dissociation at technologically feasible temperatures cannot be achieved in a single step. A cyclic thermochemical process consists of a series of chemical reactions which result in the splitting of water.

$$\begin{array}{rcl} H_2O + X = XO & + & H_2 \\ XO = X & + & \frac{1}{2}\,O_2 \\ \hline H_2O = H_2 & + & \frac{1}{2}\,O_2 \end{array} \tag{26}$$

In addition to water, hydrogen and oxygen, this series includes as a reactant the so-called reaction mediator X, which, however, is neither consumed nor produced, but is "recycled". The amount of water needed for hydrogen production can easily be supplied; the annual heat consumption by a typical apartment could be supplied by decomposition of only 6 m^3 of water, which is only about 2.4% of the annual household consumption.

4.662 Transport

Hydrogen, if it is to be used as a secondary energy carrier on a large scale, will in general be produced at a distance from the consumer and will therefore have to be transported. Pipeline transport has a number of advantages over transport by rail or transport in the form of electricity. The most important are that the roadbed, locomotion and vessel are a single unit, the supply is independent of traffic and weather, the rate of transport can be quickly changed, lower transportation costs per unit of energy, and low environmental impact.

It has been shown experimentally that hydrogen has flow properties, both as a gas and a liquid, similar to those of natural gas. Only the pumping power and the size of the condensor need to be increased, due to the lower volume-specific heating value of hydrogen (224). Table 4-8 compares the important physical properties of H_2 and CH_4 gas (225).

There are few problems of constructing a pipeline which are peculiar to hydrogen. In addition, experience has been accumulated with hydrogen supply networks. In the Ruhr Valley, for instance, there is a hydrogen supply line with a total length of 204 km used to supply the chemical industry with this raw material. (At a transport pressure of 15 bar, about $3 \cdot 10^8$ Nm^3 per year is passed through pipes with nominal

Table 4-8: Physical properties of H_2 and CH_4 gases

Physical properties	H_2	CH_4
Boiling point at 1 bar	20.4 K	112 K
Heat of vaporization	0.45 MJ/kg	0.51 MJ/kg
Gas density (0°C, 1 bar)	0.08987 g/l	0.717 g/l
Diffusion coefficient in air	0.63 cm^2/s	0.20 cm^2/s
Ignition range in air	4–76 vol %	5–15 vol %
Ignition range in oxygen	4–95 vol %	5–61 %
Explosion range in air	18–59 vol %	6–14 vol %
Ignition temperature	850 K	807 K
Ignition energy	0.02 mJ	0.3 mJ
Flame temperature	2400 K	2200 K
Flame velocity	2.75 m/s	0.37 m/s
Heat content	143 000 kJ/kg	55 700 kJ/kg

Source: C. Keller, Wasserstoff: Energieträger mit Zukunft, Bild der Wissenschaft, Vol. 13, 10 (1976).

widths of 10 to 30 cm (224). In addition, there has been much experience with supply lines for coking and city gas, which contain up to 80% hydrogen. There are estimates which show that even with present day technology, transport of hydrogen over distances of more than 400 km is cheaper than the transmission of an equivalent amount of electricity. Based on the cost of underground transmission of electricity, hydrogen transport is cheaper over distances of about 30 km or more. (The cost of security measures which might be required by law have been ignored here) (1).

It can be assumed that the investment required to convert the present gas pipelines in the Federal Republic of Germany, which are made of welded steel pipe, to a pure hydrogen transport would be relatively low. The net could be kept in essentially its present form (224), and the habits of the consumers would not have to change much, if hydrogen were introduced.

Transport of hydrogen to the consumer in a mobile container is practical only within certain limits. Even at high compression, hydrogen has a relatively low energy density. It must therefore be liquified. (Liquid hydrogen can be kept liquid at 20 K and 1 bar, or at temperatures up to 32 K at 11 bar pressure.) However, liquefaction of hydrogen requires a relatively large expenditure of energy. It can therefore be assumed that mobile transport of hydrogen will only be suitable for areas which are not connected by pipeline to the producer.

If hydrogen is generated at sea, it could, like natural gas, be transported in special tankers in liquified form (LH_2). The only practical experience with pipeline transport of liquified gases has been gained in rocketry. There are small supply pipelines for LH_2 at Cape Canaveral Space Center and at Los Alamos in the USA. They are 500 m long, have a diameter up to 40 cm, and an operating pressure up to 150 bar. The maximal transport capacity reached was 3.8 m^3/s (about 10^7 Nm^3/h) (224).

4.663 Storage

In addition to the specific transport costs for a secondary energy carrier, the possibility of storing it has become increasingly significant in the last few years. One reason is that the demand fluctuates, but the most efficient use of the plants which convert primary to secondary energy requires that they operate at full capacity at all times. It has been mentioned that the storage of electricity is difficult, and requires additional elaborate installations. In this respect, hydrogen, like other gaseous, liquid or solid secondary energy carriers, has a great advantage over electricity. Although the efficiency of a storage system for electricity based on hydrogen would only be about 56%, compared to 75% for pumped storage systems, the hydrogen system would have the advantage of availability, in any geographical situation. In other words, hydrogen offers the possibility of strategic reserves.

In principle, hydrogen can be stored in either stationary or mobile containers. Because of the very low volume-specific energy content of hydrogen (even liquid hydrogen has only about 23% of the energy density of heating oil), the main problem with storing it is one of volume (compare Table 4-9). This, however, is only a serious difficulty with mobile storage systems.

For stationary storage of gaseous hydrogen, it is possible to use high-pressure containers, either above or below ground, exhausted oil or gas fields, or pipelines. In addition, hydrogen can be stored as a liquid in cryotanks, above or below ground, or as metal hydrides. Some of these stationary storage methods could also be used, within limits, for mobile storage (pressure tanks, cryotanks and metal hydrides) (see 4.672).

High pressure gas containers with capacities up to 330000 Nm^3 are already in existence for the storage of city gas, which may contain up to 80% H_2. Exhausted oil and gas fields would be of interest for the building up of strategic reserves. However, since hydrogen diffuses much more rapidly than natural gas, such underground reservoirs would have to be tighter to hold H_2. The difficulties do not seem to be insurmountable, however, since large amounts of helium are stored in empty gas fields in the USA. (In the Federal Republic of Germany, there are at present only two natural gas reservoirs in use, with capacities of $100 \cdot 10^6$ Nm^3 and $300 \cdot 10^6$ Nm^3 (224)). Within limits, pipelines can be filled with more than the planned load, and thus offer a certain storage capacity for the leveling out of peak demands (short-term storage). The total storage capacity of a 500 km pipeline (1 m diameter, 70 bar operating pressure) is about $2.7 \cdot 10^7$ Nm^3 gas. In the case of hydrogen, this corresponds to an energy of about 10^8 kWh; with natural gas, it is $3 \cdot 10^8$ kWh (224, 225).

Liquid hydrogen (LH_2) can be stored in double-walled tanks with an evacuated Perlite insulation layer. This method is used for the reserve storage at Cape Canaveral ($2.4 \cdot 10^5$ kg LH_2 capacity) and Los Alamos ($1.3 \cdot 10^5$ kg LH_2 capacity). In Europe, there are three smaller reservoirs at CERN (Geneva), each with a capacity of 50 m^3.

Table 4-9: Specific energy contents of various fuels

Fuel	Energy/mass [MJ/kg]	Energy/volume [MJ/Nm^3]
Gasoline	42.5	31.250
Diesel oil	41.4	35.820
Methanol	22.8	18.200
Ethanol	29.8	23.500
Methane	55.7	0.040
H_2-gas	143.0	0.013
LH_2	143.0	7.560

Source: Auf dem Wege zu neuen Energiesystemen, Teil III, Wasserstoff und andere nichtfossile Energieträger, Federal Ministry of Research and Technology (Ed.), Bonn 1975.

4.664 Safety problems

Safety is probably the most disputed problem standing in the way of the use of hydrogen as a secondary energy carrier. (The explosion of the hydrogen-filled zeppelin *Hindenburg* on May 6, 1937 in Lakehurst, USA, has not been forgotten) (227). Hydrogen is more dangerous than methane. The main problems are that it can be ignited and explodes under a wide range of conditions, and its ignition energy, 0.02 mJ in a mixture with air, is only 1/10 that of a mixture of natural gas and air (compare Table 4-8). However, the greater ease of ignition of hydrogen must be set against the fact that its diffusion rate is about three times that of methane, which means that in case of a leak, the hydrogen is diluted below the ignition concentration much more rapidly than methane. With modern gas distribution networks consisting of welded steel pipes, there is essentially no greater danger in distributing hydrogen than city gas, which contains up to 80% H_2.

The safety of hydrogen storage depends greatly on the nature of the reservoir. For example, underground reservoirs would in general be safer than above-ground storage, because the entry of oxygen from the air can be practically prevented in the former. The danger of sabotage to above-ground tanks must be taken very seriously.

Safety problems will be especially acute with the storage of hydrogen in mobile consumers, such as airplanes or ships. Here, however, the danger can be considerably reduced by appropriate training of the crews. Because it is likely that liquid hydrogen will be used, the experience gained with LNG-driven vessels can serve as a starting point (227).

4.665 Environmental aspects

Hydrogen used as a secondary energy carrier has relatively little impact on the environment. Its production is free of pollutant emission, i.e. there are practically no polluting products of electrolysis or thermochemical processes, aside from waste heat. The transport and storage of hydrogen can also be free of environmental impact.

As a fuel to provide space and process heat in population centers, hydrogen has considerable advantages over other secondary energy carriers, because the emission of pollutants from a hydrogen flame is very low. Carbon monoxide, carbon dioxide, sulfur and metal compounds, unburned hydrocarbons and soot particles cannot be made by a hydrogen flame. In addition to water, nitrogen oxides (NO_x) and small amounts of ammonia can be formed, depending on the air supply, and thus the combustion temperature (224). Because of the wide combustion limits of hydrogen, however, the burners can be adjusted to minimize the emission of pollutants (the ideal mixture is about 35% air).

A final point: if one assumes that hydrogen will be produced from water, it would be possible to save non-renewable resources like fossil raw materials. The energy required to produce the hydrogen could be supplied by nuclear energy (nuclear fission or nuclear fusion) or solar energy.

4.67 Alternative drive systems for mobile consumers

4.671 Methanol, Ethanol

Both the heavy dependence of transportation on petroleum products (see 2.332) and the desire to reduce the pollution caused by automobile exhaust, especially in population centers with high traffic density, make the development of alternative fuels a necessity. (By alternative fuels, we do not mean synthetically produced conventional fuels, such as gasoline from coal.) Mobile consumers (e.g. automobiles) have to carry the necessary energy with them, and thus differ from stationary consumers, such as households, which can be supplied by wires or pipes.

At present, road, air and inland shipping is nearly 100% dependent on petroleum products, in all countries of the world. Although it is thought that about 35% of this fuel can be saved by the year 2000, assuming an equal distance traveled, the demand for fuel can be expected to rise, due to increasing motorization, expecially in the developing regions (compare also Table 2-7b and 2-8b) (228). For example, the streets of the world (excluding the Eastern block) are now traveled by about $260 \cdot 10^6$ cars and pick-up trucks, which consume about $700 \cdot 10^9$ liters ($\triangleq 530 \cdot 10^6$ t) gasoline per year. The production of this fuel accounts for about 25% of the annual consumption of crude oil. The number of trucks in the world (excluding the Eastern block) is about $70 \cdot 10^6$. About 47% of the motor vehicles in the world (excluding the Eastern block) are in North America, 36% in Western Europe, 11% in Australia and the Far East, 4% in South America, 1% in the Near East and 1% in Africa (229). They contribute to the total emission of pollutants, in some areas, very considerably. For example, motor vehicles in the Greater Cologne area contribute 49.2% of the CO, 10.4% of the nitrogen oxides, 5.7% of the unburned hydrocarbons C_nH_m, 2.9% of the soot particles, and 0.5% of the sulfur dioxide. In addition, an internal combustion engine burning leaded gasoline emits lead compounds (224).

The fraction of vehicles with diesel engines has been low, especially in the USA. It has been shown that the diesel engine compares favorably with a gasoline engine of the same power in a number of respects. For example, a 37 kW diesel engine uses 11.47 liters fuel/100 km, while a gasoline engine uses 12.4 liters/100 km, and it should be remembered that diesel fuel can be more cheaply produced, and with less energy expenditure, than high quality gasoline. As far as pollution goes, the CO emission of a diesel engine is lower than that of a gasoline engine, but its emission of NO_x and soot particles is higher. The desire to reduce environmental pollution by automobile exhaust has resulted, in many countries, in stricter exhaust laws. In the USA, for example, for the model year 1979, the legal upper limits for exhaust emission from either diesel or gasoline engines are 15 g CO/mile, 1.5 g C_nH_m/mile, and 2.0 g NO_x/mile. For the model year 1981, the limits are 3.4 g CO/mile, 0.41 g C_nH_m/mile, and 1.0 g NO_x/mile. For diesel engines, the particle emission limits for 1981 are 0.6 g/mile, and 0.2 g/mile for 1985. The results of tests by the EPA (Environmental Protection Agency) show that no diesel engine is presently able to meet the 1985 standards for both particle emission and NO_x emission, but it is assumed that the problem will be solved in the time available. It can be assumed that the fraction of vehicles with diesel engines will increase in the coming years. The National Highway Traffic Safety Administration (NHTSA) has predicted that in the USA, the fraction of passenger cars with diesel engines will rise to 25% of the new vehicles licensed in 1985.

The suitability of a fuel for use in a motor vehicle depends on a number of factors. To be considered a possible substitute for gasoline or diesel oil, a fuel must be available in sufficient quantity, and able to be produced for a relatively long time. It must produce as few pollutants as possible, and it must fulfill criteria which are of less importance for stationary consumers: it must have a high mass and volume-specific energy content (compare Table 4-9). The higher these values are, the lighter and smaller the tank can be for a given driving range. Finally, the process of filling the tank has to be uncomplicated.

In principle, there are four groups of fuels which at present appear potentially suitable for internal combustion engines: inorganic hydrogen compounds, hydrocarbon compounds, organic compounds containing oxygen, and hydrogen. The best known inorganic compound is ammonia (NH_3). It can be synthesized from air and water, but the problems of storing the toxic substance in a mobile tank and the combustion products (nitrogen oxides) are grave disadvantages, so that its use in motor vehicles is at present out of the question (230). The hydrocarbons under consideration are fuels which are gases under normal conditions, such as propane, which is liquefied (LPG, liquefied petroleum gas), or liquefied natural gas (LNG). The liquefaction is needed to obtain a higher energy density, but it requires an additional expenditure of energy, which increases the cost. For this reason, and because the supplies are limited, it can be assumed that liquefied hydrocarbons will not become a major alternative fuel, except in a few countries like The Netherlands, whose only energy raw material is gaseous hydrocarbons. Even now, liquefied fuels account for 5% of the Dutch vehicle fuel consumption.

The oxygen-containing compounds methanol and ethanol are promising alternative vehicle fuels. These alcohols have already been subjected to many long-term, large-scale tests with slightly modified conventional vehicles. It is an important consideration that the large-scale production of methanol from coal or biomass is already technically feasible, and ethanol can be produced from biomass. In other words, these fuels would be available independently of petroleum.

Methanol is produced in a two-step process from coal. The first step is production of synthetic gas (carbon monoxide and hydrogen), and the second is the catalytic synthesis of methanol from the gas (see 4.643.1). The total cost of producing methanol depends heavily on the type and price of the coal used as starting material, and on the cost of the required process heat. At present, it costs about twice as much to produce methanol from coal as to produce gasoline from crude oil (the comparison is based on equal amounts of energy.)

The heat content per unit volume of methanol is only half that of gasoline (compare Table 4-9). However, since the total efficiency of methanol-powered engines is greater, the capacity of the fuel tank of a methanol-driven vehicle would need to be only 70% larger than that of a gasoline vehicle with the same driving range. Since the danger of fire or explosion is no greater for methanol than for gasoline, the present safety features for gasoline tanks could be adopted. The operation of filling the fuel tank would also be no problem. However, since some plastics swell or become brittle on exposure to methanol, some auto parts would have to be replaced with more resistant materials.

In principle, there are no technical problems with running reciprocating combustion engines on methanol, although there must be some adjustments in the carburetor, ignition and timing. Although the operating conditions are similar in a gasoline-burning and a methanol-burning engine, there are problems with cold starts with the latter, especially when the fuel is pure methanol (M-100). This is because there are no components of methanol with low boiling points, and starting at temperatures below 0°C is not possible without some additional measures. One method is to use a starter like gasoline or ether, and another is to preheat the fuel line electrically, as is done with diesel engines. Methanol-driven Otto engines compare favorably to gasoline driven engines with respect to power, fuel consumption and exhaust gases. The nitrogen oxide emission is only a third that of a gasoline engine, and the exhaust from methanol contains much less carbon monoxide and unburned hydrocarbons. Finally, due to the fact that there is less carbon in methanol, the exhaust contains almost no soot particles. Likewise, methanol contains no sulfur, in contrast to gasoline and especially to diesel fuel. Because methanol does not tend to pre-ignite, it does not need lead additives, even for high compression, and its exhaust is thus also free of lead.

In 1979, the Volkswagen company, with the support of the Federal Ministry of Research and Technology, Bonn, started a three-year large-scale test program with more than 1000 automobiles. The test includes methanol, hydrogen (see 4.672) and electric (4.673) cars. The major emphasis of the program is the testing of about 800 passenger cars with methanol fuel. 600 cars are being driven on a methanol/gasoline mixture called M 15, which is 15% methanol and 85% super gasoline; 100 cars are using pure methanol, M 100, and it is planned that 100 cars will, in a later phase, be driven with a mixture of methanol and diesel. The infrastructure needed for the large-scale test includes about 30 gas stations in Berlin (West), Hamburg, Hannover, Cologne, Dortmund, Frankfurt, Stuttgart, Munich and Nürnberg. It has been found that the M 15 fuel is as easy to use as normal gasoline; the engines are the same as in normal cars, except for a few parts in the fuel system. The cost of converting an engine to M 15 is about $ 100. Converting to M 100 is more elaborate.

As a fuel, ethanol has essentially the same properties as methanol. It can be produced from any plant material containing sugar (e.g. sugar beets or cane), starch (e.g. potatoes or grain) or cellulose (e.g. wood). The starch or cellulose must be converted to glucose, enzymatically or otherwise, then the sugar solution can be fermented to alcohol. The alcohol is distilled, concentrated and dried.

In Brazil, ethanol is already being used on a large scale as a substitute for conventional gasoline. The ethanol is produced from sugar cane; in 1979 about $2.5 \cdot 10^6$ m^3 ethanol was produced, which supplied about 15% of the demand for vehicle fuel. It is planned that the annual production should increase to about $11 \cdot 10^6$ m^3 by 1985. According to Brazilian estimates, 70 liters of ethanol can be produced from a ton of sugar cane. In order to reach this production goal, the present area of sugar plantations must be doubled, and about 120 distillation plants built. At present, the ethanol is mixed with conventional gasoline at the refineries and depots to give about 20% alcohol. This mixture is suitable for use in normal Otto engines. Beginning in the early 1980's, about 200000 new Brazilian-built vehicles per year will have engines which burn pure ethanol. There are also plans in the USA to replace some of the conventional fuel with ethanol. According to the Department of Energy, starting in the middle of the 80's, about $7 \cdot 10^9$ liters will be produced, which will supply about 3.5% of the demand for vehicle fuel.

The production of ethanol (or hydrocarbons) from biomass is not yet economically competitive. The energy utilization effect of ethanol produced from agricultural materials depends on the costs of growing these materials and of the conversion process. In particular, the achievement of a positive energy balance depends on the extent to which it is possible to avoid the use of valuable fuels (like natural gas or heating oil) to run the distillation process (231). (For this reason, there are studies on the separation of alcohol from water through membranes.) The cost of ethanol in the USA, under the present agricultural conditions, is three to four times that of synthetic gasoline from coal, which in turn is much more expensive than conventional gasoline from petroleum. Even in Brazil, the ethanol produced from sugar cane is considerably more expensive than conventional gasoline. Nonetheless, domestic ethanol production is of great economic and political significance for Brazil, because 50% of that country's import budget goes for crude oil.

Although the properties of methanol and ethanol as fuels are favorable, conversion of vehicles to their use can only proceed step by step, as the production and distribution systems are built up.

4.672 Hydrogen

As discussed above, the problem of large-scale production of hydrogen for use as a secondary energy carrier has not yet been solved (see 4.661). It can be assumed that the cost of hydrogen, although it depends to some extent on the type of raw material and the process energy, will be at least three times that of gasoline. However, given the ever-increasing prices of fossil fuels, it can also be assumed that in the long run, the balance will shift in favor of hydrogen. If hydrogen were used as a vehicle fuel, it would be possible to conserve the fossil fuels and use them only as raw materials.

Although the operation of filling a fuel tank with methanol and storage of the alcohol in a mobile tank are, as mentioned, relatively simple, these operations would be difficult in the case of hydrogen. The main problem is storing an adequate supply of energy in the fuel tank of a vehicle. There are basically three possible ways to store hydrogen: as pressurized gas, as a liquid (cryogenic reservoir) and as a metal hydride. The volume and masses of the reservoirs – for a given energy content – vary widely. If the volume of a gasoline tank is used for comparison, then to store the same amount of energy, a methanol tank would have to be almost twice as large, a liquid hydrogen tank, five times as large, a titanium-iron hydride reservoir, three times, and a pressurized hydrogen tank (230) seven times as large (230). Comparing masses of the storage tanks, the methanol tank is twice as heavy, the liquid-hydrogen tank, four times, the titanium-iron hydride tank, fifteen times, and the pressurized hydrogen tank (400 bar), 32 times as heavy (230).

From the above, it can be seen that hydrogen tanks need to have a very large volume and are heavy. The high-pressure tanks are completely out of the question, due to their volume, their large mass, and above all, due to the danger of explosion in the case of accident. Liquid hydrogen is not feasible because it is not energetically economical to liquefy it, but also because of its expense and for safety reasons. (The cost of liquefaction would have to be added to the cost of production.)

For hydrogen storage in a vehicle, it would thus appear at present that metal-hydride storage is most promising. In this type of storage, the hydrogen forms a compound with a metal or alloy at a particular pressure and temperature. Suitable metals for this purpose are titanium, iron, aluminum, magnesium, or alloys of these (e.g. titanium-iron). The metal is melted, and poured into an ingot, which is then granulated or pulverized. This material is then poured into a container, where it is hydrogenated under pressure. The uptake or release of hydrogen, like other chemical reactions, is associated with energy turnover. This means that by regulating the heat added to a titanium-iron hydride reservoir, the amount of hydrogen needed for a desired power output can be released (230).

The dependence of the desorption rate on the temperature is of critical importance to the safety of the reservoir. If it is damaged by an accident, the hydrogen is not released all at once, as would be the case in a pressurized gas reservoir, but is gradually desorbed. The system can be constructed so that in case the container is ruptured, the engine coolant, which transfers heat to the hydride, will run out. In this way, the amount of hydrogen released will be so small that there is no danger of explosion. (In an experiment in the USA, a metal hydride reservoir was subjected to gunfire, but it was not possible to make it burn or explode.)

Refueling a hydride reservoir could not be done in the manner to which consumers are accustomed, and the entire present distribution system would have to be revised. The present half-manual process of filling a tank would presumably have to be replaced by a completely automated process.

Compared to other fuels, hydrogen has little environmental impact. Carbon monoxide, carbon dioxide and hydrocarbons can only form if lubricating oil gets into the combustion chamber. Because the combustion temperature is lower than that of gasoline, there should be less nitrogen oxide formation. Finally, there can be no formation of soot particles or lead or sulfur compounds.

Hydrogen also has the natural advantages of a gaseous fuel. (Before it can be burned in a conventional internal combustion engine, a liquid fuel must first be prepared in a carburetor or injection system which has the function of creating as homogeneous an air-fuel-vapor mixture as possible.) Hydrogen could be burned with air, with oxygen, or in a fuel cell (electric drive) (see 4.611.31 and 4.673). If it is burned in air, the relatively high diffusion coefficient of hydrogen favors rapid formation of a homogeneous fuel-air mixture. Also, hydrogen will ignite under a wider range of fuel-air ratios than other liquid and gaseous fuels, which makes it possible to control the power output over the entire range of operation by regulating the amount of fuel injected. This means that the hydrogen engine can operate with higher efficiency (about 20%) than a gasoline engine. It is also possible to make an engine which burns a gasoline-hydrogen mixture, which would be more efficient and have lower pollutant emission than a pure gasoline engine. As mentioned above, combustion of hydrogen in the air generates nitrogen oxides. Engines which burn hydrogen in pure oxygen would avoid this problem, but they would have to carry a supply of oxygen in addition to their fuel, or else they would have to extract it from the air as they went.

Experiments with vehicles with hydrogen engines and hydride reservoirs have been promising. On the whole, the problems are not so much in the engine, as in the economical production of hydrogen. It can be assumed that hydrogen-fueled vehicles will first be used where the reduction of pollution from automobile exhaust is expecially important, and where electric vehicles are being considered for this reason (inner city areas). However, hydrogen, like alcohol, can also be used as an addition to gasoline, so that conversion of the vehicle fleet from gasoline to hydrogen, a fuel which is not dependent on fossil supplies, could be made gradually.

A hydrogen-fueled vehicle has a decided advantage over an electric vehicle with respect to the volume and mass (weight) of the energy reservoir, assuming that the same amount of energy is carried in each. The lead-acid batteries would occupy 16 times the space and weigh 6 times as much as the hydrogen reservoir. In addition, it is likely that the hydrogen reservoir could be more quickly charged than a lead-acid battery – for an average medium-sized car the process would probably requrie a few minutes – and would have a longer lifetime.

One more application of hydrogen will be discussed: the use of liquid hydrogen as an airplane fuel. It has been used for years as a rocket fuel; the success of the Apollo programs is proof that the technology of hydrogen fuel has been mastered. The problems of storage and fueling are more easily solved for airplanes than for land vehicles, partly because the former need to be fueled at many fewer locations. The specific heat content of hydrogen is 3 times that of kerosene (143 000 kJ/kg for LH_2 compared to 42 800 kJ/kg for kerosene). The use of liquid hydrogen would increase the range of airplanes, or would increase the payload for a given range. A disadvantage, however, is that because the energy density per unit volume of LH_2 is only 1/4 that of kerosene, more space would be required for the fuel tanks.

Although there are no fundamental difficulties in designing an engine to run on hydrogen (or methanol or ethanol), hydrogen vehicles are not likely to be introduced in the immediate future, because the problems of storing fuel in the vehicle have not been satisfactorily solved, and hydrogen cannot yet be produced competitively on a large scale. Finally, the introduction of hydrogen as a vehicle fuel would require the development of a new infrastructure for storage and distribution.

4.673 Electric drive

The special advantages of electric vehicles are that they are quiet and emit no exhaust gases. Although electric vehicles have been in use for decades for special purposes, relatively little progress has been made on the development of electric drive systems. Considering output, weight and original cost, lead acid batteries have been the best solution to the drive problem. A typical electric vehicle with lead acid batteries has storage capacity for about 20 kWh and weighs about 850 kg. Its range is between 60 and 100 km, its top velocity is about 80 km/h, and it can accelerate from 0 to 50 km/h in 12 seconds.

One important criterion for judging batteries is the energy density, or the amount of electric energy which can be stored per unit of battery mass. The theoretical energy density is based only on the mass of the electrochemical reactants, while the practical energy density includes the other battery components such as the electrolyte, the housing, etc. It follows that the practical energy density is less than the theoretical. The future of electric vehicles depends more on the development of batteries with higher energy density than on developments in motors or electronics. Research and development work on battery systems have not yet yielded the desired results, and in light of the limited capacity of current battery systems, it is not likely that electric vehicles will be used outside city center areas. In order to be practical not only for delivery trucks, but also for passenger cars, battery systems with about 100 kWh/kg capacity and delivery of 10 kW/kg are needed (183).

A number of battery systems are being developed. The most likely candidates for the positive pole of the battery are lithium, sodium, magnesium, aluminum, iron and hydrogen; those for the negative pole are fluorine, chlorine, oxygen (or air) and sulfur. The electrolytes may be melted salt mixtures, such as lithium chloride and potassium chloride, or solid electrolytes like a ceramic which conducts sodium ions ($Na_2O-8\ Al_2O_3$), and which is as good a conductor of electric current as liquid electrolytes (232, 233). Table 4-10 shows some new high-energy systems for electric vehicles (183). The high-temperature cells based on sodium/sulfur and lithium/iron sulfide promise to provide systems of very high capacity. Their working temperatures are 300°C and 400°C, respectively. In the Na/S cell, liquid sodium is the negative electrode, the electrolyte is a sodium-conducting ceramic, and liquid sulfur is the positive electrode. The operating temperature is maintained by combining a large number of cells in a battery, which is surrounded by an extremely effective heat insulation. The heat lost to the surroundings is replaced by the heat losses arising from charging and discharging. For economic reasons, the lifespan of a battery must be as long as possible, which is equivalent to a maximum of charging and discharging cycles. Iron-nickel batteries have a relatively long lifespan.

It can be assumed that a fully developed Na/S battery system will be far better than a lead-acid battery for an electric vehicle. In the volume which is needed for a lead-acid system, it will probably be possible to put a Na/S system which weighs half as much, can store twice as much electricity, and can deliver more power. In other words, Na/S batteries would increase the range, the highest velocity and the acceleration attainable by an electric vehicle. It is also likely that they will be less expensive to produce than lead-acid batteries (234).

Recently, electric vehicles with two independent drive systems have been tested. They have a gasoline or diesel engine for use in long-distance highway driving, and an electric drive for use in city traffic (235).

Table 4-10: New high-energy batteries for electric vehicles and peak current storage (compared to Pb/PbO_2 and Fe/NiOOH)

System	Rest potential [V]	Working temp. [°C]	Energy density [Wh/kg] Theor.	Pract.	Charge/ discharge cycles
Pb/PbO_2	2.0	– 20 to + 60	161	40	~ 1200
Fe/NiOOH	1.2	– 20 to + 45	260	50	~ 2000
Na/S	1.8	300	660	120	~ 1000
$Li(Al)/FeS_2$	1.8	400	650	120	150 to ~ 1000
$Na/SbCl_3$	3.0	200	780	?	?
Ca/CuF_2	3.4	450	1290	?	?
$H_2/0_2$	1.2	80–120	3670	140–200	–

Source: F. de Hoffmann, Energy storage and methods of energy transport, Die Naturwissenschaften, Vol. 64, No 4 (1977).

Electric vehicles can in principle be driven by fuel cells as well as batteries. However, it is difficult with either hydrogen or carbon-containing fuels to attain an adequate power density, because the reactions at the electrodes are not rapid enough. At present there are no economically feasible fuel cells for cars[1].

---

[1] In principle, nuclear drives would also be possible for mobile consumption. However, the power plants (light water reactors with conventional steam turbines) are too large for use in air or land vehicles, so their use has been limited to ships. At present there are about 300 atomic ships, almost all of which are warships.

5. Environmental impact and safety problems

5.1 Problems associated with release of energy

It is becoming ever clearer that the problems of energy supply – aside from the problems of distribution and costs of individual energy carriers – are not the available energy resources, but the environmental and safety problems associated with the release of energy. The magnitude of the expected world demand for primary energy is such that, considering the options now available, effects which have been negligible when small amounts of energy were consumed will become global environmental problems if the projected future demand is satisfied (1, 2).

Every energy carrier has an impact on the environment which becomes significant if it is used in sufficient quantity. This is especially true of fossil fuels and nuclear fission, which are likely to provide all but a few percent of the growing energy consumption until the turn of the century (see 2.333). Even geothermal and hydroelectric power cause environmental problems, although they are local, and direct utilization of the solar energy falling on the earth's surface, if it can be achieved on a large scale, would be only approximately "environmentally neutral".

Environmental impact factors can be divided into those which are common to all sources of energy (with the possible exception of solar energy) and those which are specific to fossil and nuclear energy carriers. All energy sources generate waste heat, which can damage the environment. The additional impact of pollutants, especially carbon dioxide and aerosols (finely divided solid or liquid substances in the air), is specific to the fossil fuels. The factors specific to nuclear fuels are all those problems of safety and disposal associated with radioactivity, including the problem of preventing proliferation of nuclear weapons.

The nature and extent of their environmental impact as well as possible safety and disposal problems will necessarily become increasingly important criteria in the choice of energy carriers. These matters will be discussed, to the extent it is possible within the limits of this book, in this chapter.

5.2 Direct anthropogenic heat release

All energy sources have in common the fact that they stress the environment with waste heat. The ratio η of the useful work done by a machine (e.g. a thermal power plant) to the energy it consumes is given by

$$\eta = 1 - \frac{T_2}{T_1} \tag{1}$$

where T_1 is the temperature at which the heat is supplied to the machine, and T_2 is the temperature at which heat leaves it (both are absolute temperatures). The non-utilizable heat is transferred to the environment via the condensor. (Thermodynamic losses, such as heat loss through the walls, are being ignored.) Modern fossil fuel power plants have efficiencies around 40%. Light-water-reactor power plants have about 33% efficiency, and the thorium high-temperature reactor is reported to have 40% efficiency. This means that about 2/3 of the primary energy used is lost as waste heat, and only 1/3 is converted to electricity.

Because of its heat capacity, water is a good coolant. There are three basic types of cooling systems for power plants: fresh water cooling (flow-through cooling), evaporation cooling and dry cooling towers (direct air cooling) (3). The simplest of these is fresh water cooling, and for this reason, power plants are preferentially located on large rivers. The cooling capacity of these rivers in some heavily populated industrial areas has reached the "stress limit", however. (The amount of oxygen which will dissolve in water decreases as the temperature rises.) In order to avoid the serious effects of overheating the river water, heatload plans have been drawn up for the large rivers. In addition, evaporative cooling has been introduced. In these towers, large amounts of water are evaporated, i.e. the waste heat is released to the atmosphere in the form of heat of evaporation. However, since the water is taken from the rivers, there are limits to this technique as well. It is then necessary to turn to dry cooling, in which the coolant flows through an air-cooled grille, but does not come into direct contact with the air. However, in the end all techniques add heat to the earth/air system (3).

One should remember that waste heat from power plants is only a part of the thermal stress on the environment. According to the second law of thermodynamics, every form of energy conversion leads to dissipation of heat. (Even the mechanical potential energy which is temporarily stored in the construction of a building is eventually converted to heat.) The second law states that the entropy of a closed system can increase or remain constant, but it cannot decrease. This means that not only waste heat, but the entire amount of primary energy consumed is eventually added to the environment as heat (see 2.22).

Let us compare the relation between the direct anthropogenic heat added to the earth/air system and the equilibrium temperature. The earth/air system receives

$1.78 \cdot 10^{14}$ kW in solar radiation, and re-radiates the same amount, so that an e-quilibrium temperature (about 290 K) is established. The absorbed radiation S_s is equal to the thermal radiation S_p, or $S_s = S_p = S$. According to Boltzmann's Law,

$$S = \sigma T^4 \tag{2}$$

where S is in Wm^{-2}, T in K, and $\sigma = 5.7 \cdot 10^{-8}\ Wm^{-2}K^{-4}$. (For the moment, the influence of the atmosphere can be ignored.) When the power converted to heat at the earth's surface is increased by dS due to combustion of fuels, the equilibrium will be displaced to a higher temperature, T + dT. By differentiation of equation (2) we obtain

$$\frac{dT}{T} = \frac{dS}{4S} \tag{3}$$

This means that a 1% increase in S increases the temperature by $dT = 2.5 \cdot 10^{-3} \cdot T = 0.75$ K. If the effect of the atmosphere is taken into consideration, the increase in temperature is about doubled, because the atmosphere inhibits the radiation of heat into space (4). As an approximation, then, an artificial heat production equal to 1% of the absorbed solar radiation will raise the average temperature of the earth by 1 degree.

At present, the heat added to the earth/atmosphere system by human activity is slight in comparison to the natural solar radiation (compare Table 5-1). The world consumption of primary energy in 1980 corresponded to an energy release of about 0.007% of the mean solar energy absorbed (235 W/m²), averaged over the total earth surface. In the USA and the Federal Republic of Germany, the value is considerably higher, especially in heavily populated areas. In the Ruhr Valley, for example,

Table 5-1: Direct anthropogenic heat release compared to solar radiation

Region	Primary energy consumption per year in W/m² 1980	2000	2176	Mean solar energy input in W/m²	Col. 1 / Col. 4 %	Col. 2 / Col. 4 %	Col. 3 / Col. 4 %
Column	1	2	3	4	5	6	7
Earth's surface	0.016	0.047	0.25	230	0.007	0.02	0.1
Land surface	0.054	0.16	0.87	230	0.023	0.07	0.4
USA	0.27	0.36	–	220	0.12	0.16	–
Germany, F.R.	1.6	1.9	–	150	1.07	1.27	–

Source: Author's calculations based on the prognoses given in 2.333.

the value is about 10% (5). If one assumes a world consumption of primary energy of about 600 q/a in the year 2000 (see 2.333), the heat released will be about 0.02% of the mean solar energy input over the entire earth, and about 0.07% of the input to the land areas. If one assumes a primary energy consumption of about $3.6 \cdot 10^3$q/a – this is the level predicted for the year 2176 (compare Table 2-13) – the "artificial" heat input would be about 0.1% of the mean solar input, over the total surface, and 0.4% of the input to the land surface. The result would be a global increase in temperature of about 0.1 degree (6).

It can therefore be assumed that, outside heavily populated areas, the direct anthropogenic heat release and thus the global temperature increase due to the increase in world primary energy consumption will be small compared to the indirect anthropogenic heat stress caused by an increased CO_2 concentration in the air (see 5.42).

5.3 Environmental impact specifically due to fossil energy sources

5.31 Pollutant emissions

Heat is released from fossil fuels by combustion (oxidation). This causes the direct heat stress to the environment which has just been discussed (section 5.2), and in addition, the combustion processes release a large number of different pollutants into the atmosphere. Some examples are gaseous combustion products like sulfur dioxide (SO_2), nitrogen oxides (NO_x), carbon monoxide (CO), hydrocarbons (C_mH_n), carbon dioxide (CO_2), and dust or aerosols. Different amounts of these pollutants are released by combustion, as shown in Table 5-2, which shows the mean values for pollutant emission associated with generation of electricity from different fossil fuels (7). The non-poisonous CO_2 is not a significant pollutant (see 5.32, however).

Table 5-2: Pollutant emissions (means) for generation of electricity

	kg Pollutant/tce fuel				
Fuel	SO_2	NO_x	C_mH_n	CO	Dust
Heating oil	23	7	0.2	0.1	1.0
Gas	–	5	–	–	–
Hard Coal	26	7	0.1	0.5	3.5
Brown coal	23	8.5	0.1	0.1	4.5

Source: Zur Friedlichen Nutzung der Kernenergie, Federal Ministry of Research and Technology (Ed.), Bonn 1978.

Comparable pollutant emission is caused by combustion by the final energy consumers (industry, motor vehicles, households and other small consumers), and these are released into the air as smoke. The emission of carbon monoxide is particularly high in the motor vehicle sector (8); in the Federal Republic of Germany, for example, motor vehicles are responsible for nearly 50% of the CO emission (see 4.671).

Natural ("pure") air consists of 78.07 vol.% nitrogen (N_2) and 20.95 vol.% oxygen (O_2). The remaining 0.98% (on the average) is made up of trace gases and other compounds, such as H_2O, CO_2, CO, O_3, noble gases (helium, neon, argon, krypton, xenon), CH_4, SO_2 and aerosols. In other words, the substances released by combustion are already present in nature.

The role of human sources cannot be estimated until the cycles in the natural, "pure" atmosphere are known. In fact, relatively little is known about these cycles, and there is no place on earth where there are no traces of substances of human origin.

The SO_2 added directly to the atmosphere is nearly all from human sources, and already accounts for a significant fraction of the global sulfur balance. Sulfur is present in the atmosphere as the gases hydrogen sulfide (H_2S) and sulfur dioxide (SO_2), as droplets of sulfurous and sulfuric acids (H_2SO_3 and H_2SO_4), and as sulfate aerosols. The human sources of SO_2 are not evenly distributed over the earth: 93% is emitted in the Northern Hemisphere, and 7% in the Southern. At present, the anthropogenic sulfur emission (as SO_2) is 60 to $80 \cdot 10^6$ t/year. Fig. 5-1 shows the SO_2 concentration measured over the Atlantic (horizontal profile) between 60°N and 10°S latitude. (The SO_2 concentration is generally higher over the continents than the oceans, and reaches its highest values above industrial centers or large cities.) In spite of increasing rates of SO_2 production from combustion, the majority of the atmospheric sulfur comes from natural sources such as volcanoes (SO_2), microbes (H_2S) and the oceans (sulfate) (9). (Sea salt particles contain 7.7 weight % sulfate.)

The nitrogen oxides found in the atmosphere are N_2O, NO and NO_2. NH_3 has an important role in biological cycles. Like sulfur, nitrogen compounds are present both as gases and in aerosols, as ammonia and nitrate, and they come from both natural and human sources. The natural sources produce about $1.2 \cdot 10^9$ t NH_3/year; the amount from human sources is relatively small. The human emission of NO and NO_2 is about $50 \cdot 10^6$ t/year, compared to a natural production of about $450 \cdot 10^6$ t/year (9). The N_2O content of the air is being increasingly affected by the use of fertilizers.

Incomplete combustion of fuels produces hydrocarbons (C_mH_n) at about the same concentration as CO. Almost 75% of the anthropogenic hydrocarbons come from the combustion of vehicle fuels. It is estimated that the world emission of hydrocarbons due to human activities is now about $90 \cdot 10^6$ t/year. For instance, the methane content of the air is about 1.4 ppm at the ground, and 0.25 ppm 50 km out. CH_4 is also produced by microorganisms in the course of anaerobic decomposition of or-

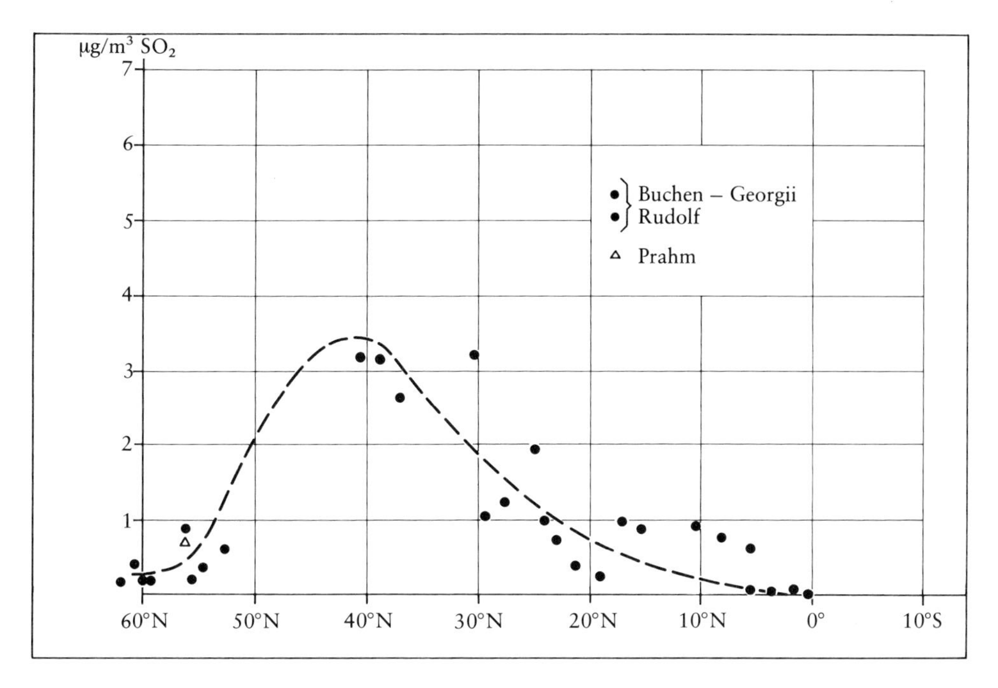

Fig. 5-1: Concentration of SO_2 over the Atlantic

Source: H.-W. Georgii, Large-Scale Distribution of Gaseous and Particulate Sulfur Compounds and its Impacts on Climate, in: W. Bach, J. Pankrath, W. W. Kellogg (Eds.), Man's Impact on Climate, Amsterdam, Oxford, New York: Elsevier Publishing Company 1979.

ganic materials, but like CO, CH_4 is present in higher concentrations in the Northern Hemisphere than in the Southern, due to human emissions.

CO also is produced by natural and human sources. For example, oxidation of methane by OH radicals leads to CO formation. (The OH radicals are formed in the troposphere and stratosphere[1]) by photochemical processes.) The human emission of CO is estimated to be $300 \cdot 10^6$ t/year, worldwide. The CO content of the Northern Hemisphere is about twice that of the Southern Hemisphere, which can only be due to human emission.

The sources of aerosols are found mainly over the continents. The natural emission of aerosol particles is about $2.3 \cdot 10^9$ t/year for the whole earth. The sources are dust, sea salts, volcanic eruptions and forest fires. The estimated global human emission of aerosol particles is $3 \cdot 10^6$ t/year. The sources of these are, for example, soot particles, particles condensed from gaseous pollutants, sulfate from SO_2 or nitrate

[1]) The lower part of the atmosphere, up to 11 km altitude, is called the troposphere. Above it is the stratosphere, which extends to about 45 km. The mesosphere extends from 45 to 80 km altitude, and is surrounded by the ionosphere.

from NO_x (9). About 99% of the aerosol particles are found at altitudes less than 5 km. In the air close to the surface, about 10^4 particles/cm^3 are found above the continents, and about 300 to 600 particles/cm^3 over the oceans.

The human health problems which can be caused by excessive pollutant concentrations are now widely known. For example, SO_2 can be adsorbed to aerosols, which then enter the fine lung channels and cause bronchial diseases. The effects can be measured with concentrations as low as 100 μg SO_2 per m^3. According to the National Academy of Sciences in the USA, a 620 MW coal power plant located in New York, which emits about 43 000 tons of sulfur annually, is responsible for about 42 premature deaths per year (10). Nitrogen monoxide (NO) can cause brain damage. Carbon monoxide (CO) forms a compound with hemoglobin (carboxyhemoglobin) which blocks the function of hemoglobin as the oxygen carrier in the blood. It has been shown in animal experiments that some hydrocarbons, e.g. benzpyrene ($C_{20}H_{12}$) are carcinogenic. As mentioned, aerosols often acts as carriers of poisonous substance.

Chlorofluorocarbons (used in pressurized spray cans – CF_2Cl_2 and $CFCl_3$, for example), nitrogen oxides (including those released directly into the stratosphere by aircraft), and CH_4 affect the ozone balance in the stratosphere. According to a report of the National Academy of Sciences, the world production of chlorofluorocarbons in 1977 was 680000 t). Ozone (O_3), which is present at roughly 1 ppm in the atmosphere, and shields the surface from short-wave UV light, is photochemically decomposed, and the process is catalysed by chlorofluorocarbons, hydrocarbons and nitrogen oxides which diffuse into the stratosphere. There are no known decomposition reactions for the chlorofluorocarbons, i.e. they have a relatively long lifetime (about 30 years) and can build up in the stratosphere. This leads to a disruption of the natural chemical equilibrium in the ozone layer of the stratosphere (decomposition of O_3). However, there are also mechanisms which form O_3, and it has not been possible to determine an unequivocal trend in the O_3 concentration.

Finally, there is the release of natural radioactive materials by the combustion of fossil fuels. A modern hard-coal power plant with 300 MW electric output releases about 500 t fine ash per year. This ash contains a number of radioactive minerals which were present in the coal, including about 30 mCi ^{210}Pb and 4 mCi ^{226}Rn. When ingested with the food, these substances are deposited in the bones, which are then subject to irradiation. The level can reach about 19 mrem per year. (A 600 MW nuclear power plant, for example, emits an everage of 10 mCi ^{131}I. This isotope, which has a half-life of 8 days, is also ingested with food and can lead to irradiation of the thyroid at a dose of 0.4 mrem per year (11)). This radiation load, however, is considerably below the fluctuation range of natural background radiation (see 5.841) (12).

The pollutants mentioned above can be reduced or eliminated, at reasonable cost, by technological improvements. For example, the emission of sulfur and nitrogen oxides and dust can be largely prevented by the appropriate technology in modern coal power plants. However, there are no practical methods to avoid or reduce the emission of carbon dioxide (CO_2). One reason for this is the amounts produced. The combustion of hard coal yields 3.4 t CO_2/tce, and natural gas yields 1.9 t CO_2/tce. (The atomic weights of carbon and oxygen are such that 12 g carbon is equivalent to 44 g CO_2.) The world annual anthropogenic emission of CO_2 is about $20 \cdot 10^9$ t per year (14). If one were to attempt to precipitate the CO_2 with calcium, for instance, for each mole of carbon (12 g C) one would obtain one mole calcium carbonate (100 g $CaCO_3$). Using the figures for a coal power plant, the amount of $CaCO_3$ produced would be about ten times the amount of coal burned, and it would have to be disposed of. C. Marchetti has suggested that industrially produced CO_2 could be frozen out or liquefied, and then dumped into the ocean depths (15). Because the circulation is slow, the CO_2 might be expected to remain in place for about 500 years, and this would reduce the CO_2 problem in the atmosphere. However, since the molecular weight of CO_2 is 44, the problem of transporting it as a solid (liquid CO_2 could be transported by pipeline) would not be much less difficult than the $CaCO_3$ problem, and it would probably be economically less favorable. In other words, the cost of disposing of CO_2 would probably be prohibitive with today's technology.

So far, there has been no significant decrease in the atmospheric oxygen due to the combustion of fossil fuels. As mentioned in section 4.326.1, oxygen makes up 20.95 vol.% of the atmosphere, after removal of the water. This amounts to $1.2 \cdot 10^{15}$ t O_2. Plants produce about $2 \cdot 10^{11}$ t O_2 per year by photosynthesis. The combustion of all fossil fuels consumes about $15 \cdot 10^9$ t O_2 annually, which is only about 7% of the amount produced by photosynthesis, or about 0.001% of the oxygen present in the atmosphere (16).

The global assimilation of CO_2 is about $27.5 \cdot 10^{10}$ t per year, or about 5% of the total present in the atmosphere, but experience has shown that the CO_2 released by combustion of very large amounts of fossil fuel is not entirely assimilated (see 5.32). The enrichment of atmospheric CO_2 is a global problem, in contrast to other emissions like SO_2, CO and dust, which affect primarily the locality in which they are released (17, 20).

5.32 The carbon dioxide problem

The CO_2 problem is receiving more and more attention, due to the relatively large increase in the CO_2 content of the atmosphere caused by human activities (21).

The natural (pre-industrial) CO_2 concentration is not exactly known; in 1860, the atmospheric concentration was about 295 ± 5 ppm, or a total of $2300 \cdot 10^9$ t CO_2, corresponding to $627 \cdot 10^9$ t C. The CO_2 concentration did not rise detectably until the end of the 19th century, when human activities began to make a noticeable difference. According to estimates by the UN, about $500 \cdot 10^9$ t CO_2 have been released by human activities since 1860, which corresponds to about $135 \cdot 10^9$ t carbon. Most of the production is from combustion of fossil fuels; a much smaller part comes from cement factories, lime burners, etc. The present human emission is about $20 \cdot 10^9$ t CO_2 per year (≙ $5.4 \cdot 10^9$ t C). Fig. 5-2 shows the annual human CO_2 production in tons carbon equivalent and in volume percent of the atmosphere (22).

Since 1958, there have been systematic measurements of the atmospheric CO_2 concentration on Hawaii and at the South Pole (23, 24). (The Mauna-Loa Observatory, Hawaii, run by the US National Oceanic and Atmospheric Administration, NOAA, is especially suitable for measurement of global environmental parameters, because the nearest continent is about 3000 km away.) Fig. 5-3 shows the measurements of CO_2 concentration at Mauna Loa since 1958 (24). The points are the monthly average concentrations; the variation in the course of the year is caused by the changes in assimilation by vegetation.

The present average CO_2 concentration is 333 ppm, which corresponds to a total of about $2600 \cdot 10^9$ t CO_2, or $703 \cdot 10^9$ t carbon (14). This means that the CO_2 concentration has increased by about 13% since 1860. The rate of increase is now about 1.3 ppm/year. This increase, however, accounts for only about half the CO_2 emitted since 1860. In other words, the present atmospheric CO_2 concentration is much

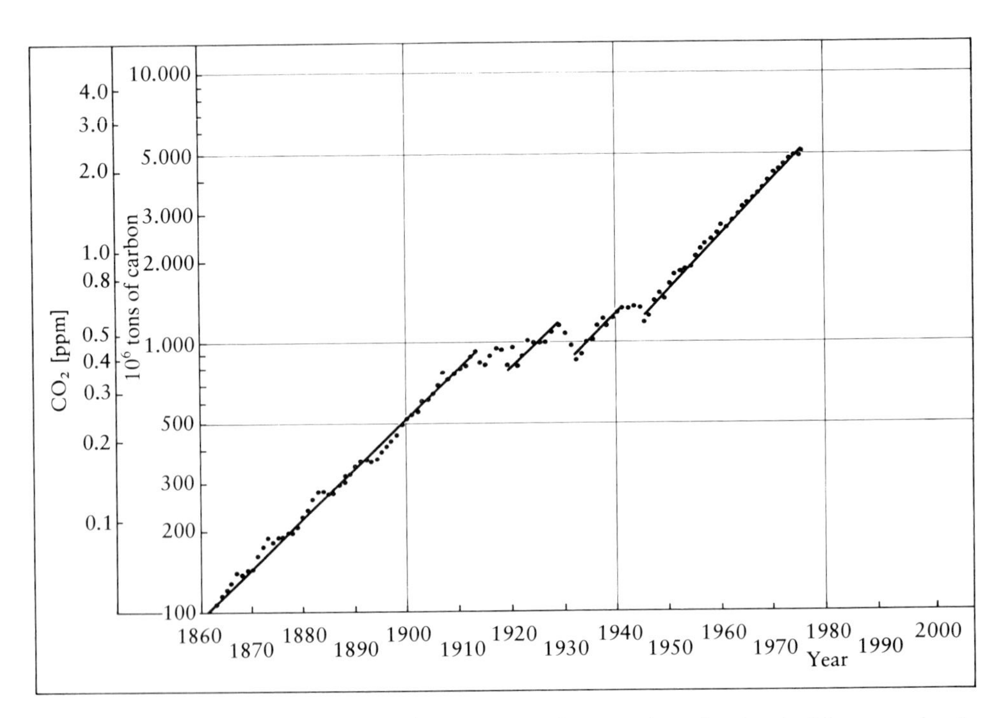

Fig. 5-2: Annual anthropogenic CO_2 production (in t carbon or volume fractional of the atmosphere)

Source: D. F. Baes, H. E. Groeller, J. S. Olson, R. M. Rotty, The global carbon dioxide problem, Oak Ridge National Laboratory, ORNL-5194, 1976.

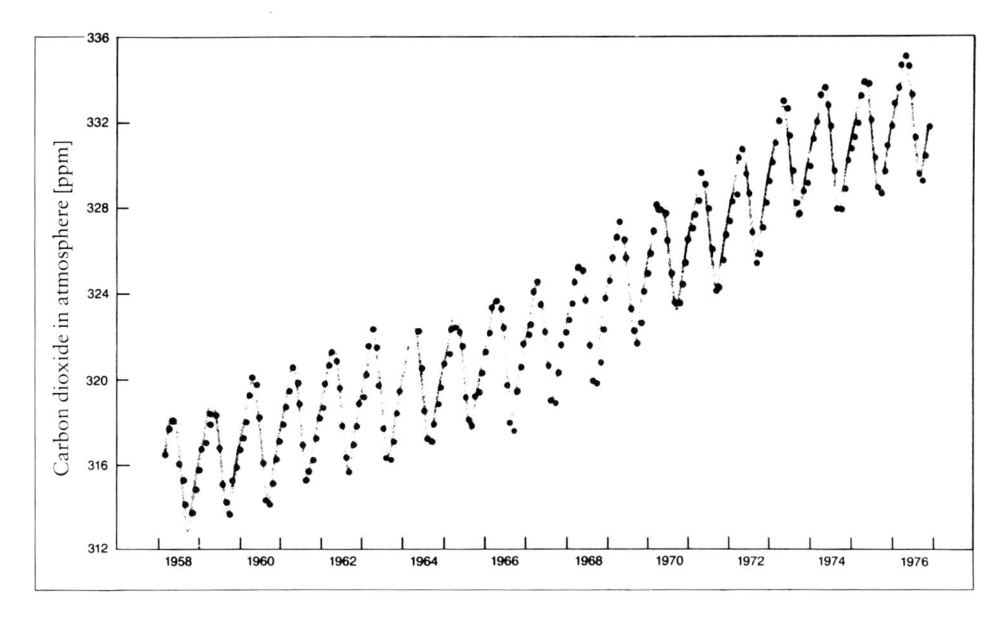

Fig. 5-3: Carbon dioxide concentration in the atmosphere

Source: G. M. Woodwell, The Carbon Dioxide Question, Scientific American, Vol. 238, No 1, 1978.

lower than would be expected from the known human emission. It is assumed that the remainder of the CO_2 has been absorbed by the oceans and the biosphere (25, 26).

The effect of CO_2 is due to the fact that it is transparent to visible light, but not to the long-wave infra-red (with a maximum at 10 μm) radiated from the earth's surface. This infra-red radiation carriers off the majority of the solar radiation absorbed by the surface, and it can be expected that increasing amounts of atmospheric CO_2 will lead to an increase in the temperature of the troposphere (the greenhouse effect). (The heating effect decreases with increasing altitude, and becomes a cooling effect in the stratosphere.)

The natural carbon cycle, which includes assimilation, respiration, anaerobic decomposition of organic substance, and the amounts of CO_2 in the air and ocean water, has been in equilibrium for thousands of years. This natural cycle is extremely complex, and contains a number of reservoirs which can exchange carbon according to certain rules. Our knowledge of this system, especially of the size of the reservoirs, the nature of the compounds exchanged, and the rates of exchange, is very incomplete (25, 26).

In the following, a few essential aspects of the carbon cycle will be discussed (27, 28). The subsystem relevant to the climate, i.e. the atmosphere-ocean-biomass-detritus, will be treated. Outside this subsystem, there are huge amounts of carbon present in the carbonate of sedimentary rocks, and in more or less concentrated form as organic carbon in various deposits. The natural exchange of carbon between the climatologically important subsystem and the sedimentary formations can be ignored for the present. Fig. 5-4 shows the carbon cycle of the earth (except for rocks and ocean sediments) (27). The entire carbon content of this system is $40\,930 \cdot 10^9$ t C (lower limit). (Other estimates may vary slightly from this value (14)).

The atmospheric CO_2 exchanges with the carbonate cycle in the oceans and the biological cycle. CO_2 is present in the ocean not only as the dissolved gas, but also as bicarbonate (HCO_3^-) and carbonate (CO_3^{2-}). The three forms are in equilibrium, according to the following equation:

$$CO_2 + H_2O + CO_3^{2-} = 2\,HCO_3^- \qquad (4)$$

The oceans contain $38\,680 \cdot 10^9$ t C, but of this only about $280 \cdot 10^9$ t (0.7%) is in the form of dissolved CO_2. The rest is in HCO_3^- and CO_3^{2-} ions. The concentration of these anions is determined by the concentrations of H^+ and Ca^{2+} in the sea water. Thus the CO_2 content of the atmosphere is influenced by the chemical behavior of the ocean/atmosphere georeservoir (9).

The annual biosynthesis of dry biomass is the basis for consideration of the exchange of atmospheric and biospheric CO_2. The land plants of the world photosynthesize about $109 \cdot 10^9$ t biomass (see 4.326.2). Some of the carbon in the biomass is

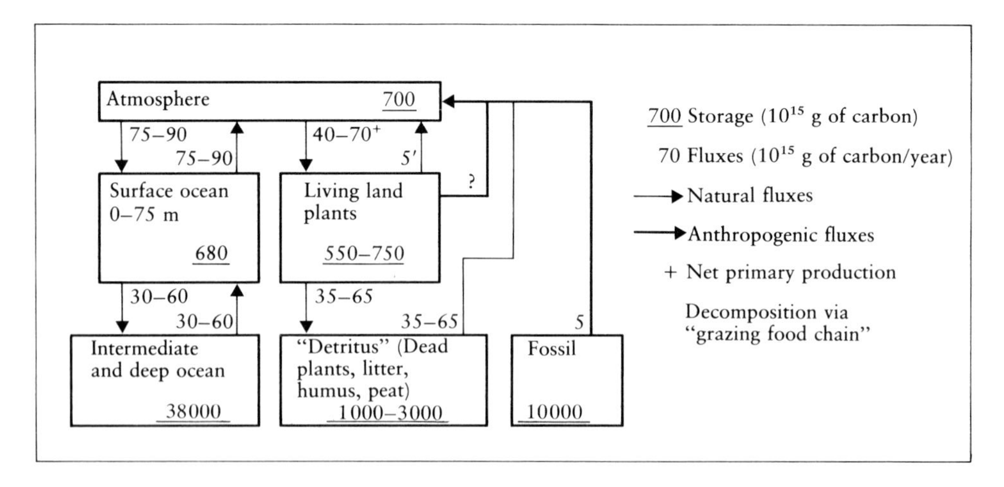

Fig. 5-4: The carbon cycle of the earth in 1977 (except rocks and ocean sediments)

Source: U. Hampicke, Man's impact on the Earth's Vegetation Cover and its Effects on Carbon Cycle and Climate, in: W. Bach, J. Pankrath, W. W. Kellogg (Eds.), Man's Impact on Climate, Amsterdam, Oxford, New York: Elsevier Publishing Company 1979.

returned directly to the atmosphere by respiration; the rest passes through the detritus reservoir.

The carbon in the deep sea reservoir stays there an average of 500 to 2000 years; the average turnover time for carbon in the other reservoirs (atmosphere, biosphere, surface ocean water) averages between 10 and 35 years. (The dynamics of the CO_2 system, including rates of exchange between reservoirs, can be studied using radioactive carbon (30, 31).)

It follows from the above that the increase in the CO_2 concentration caused by human activities is determined essentially by the capacities and uptake rates of the buffer reservoirs in the ocean and biosphere (32–34). Not all the details of the carbon cycle are known, but it has been definitely established by measurement that the atmosphere is currently being enriched in CO_2 at the rate of 1.3 ppm per year (4, 19, 35).

Further increases in the atmospheric CO_2 concentration will depend on various factors, and can thus be predicted only within certain limits. One important parame-

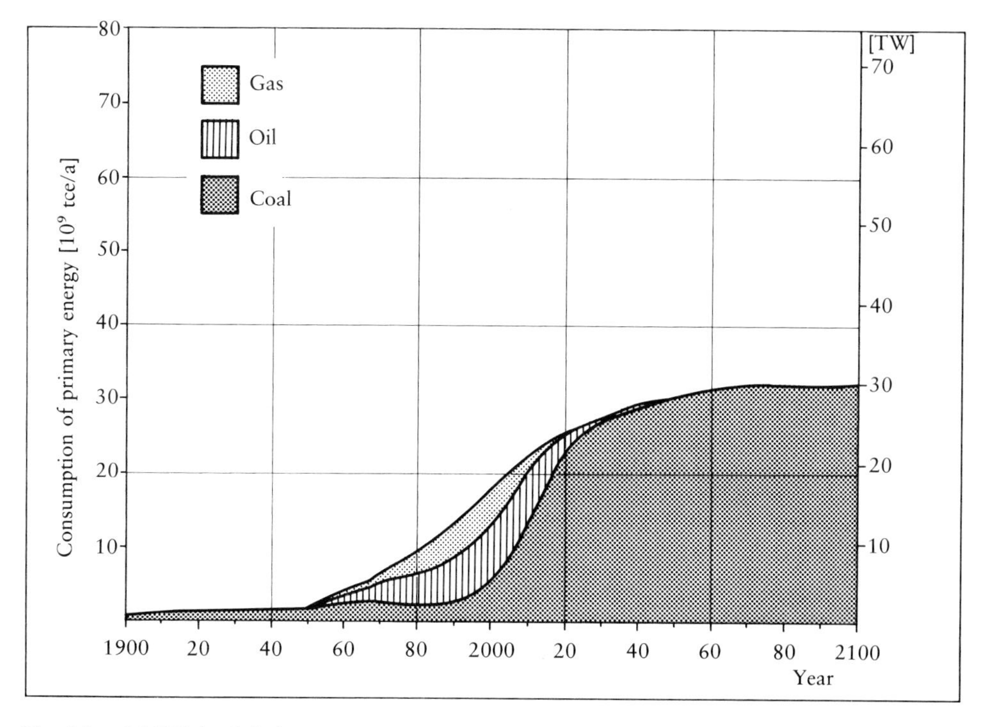

Fig. 5-5 a: 30 TW fossil fuel energy strategy

Source: F. Niehaus, Carbon Dioxide as a Constraint for Global Energy Scenarios, in: W. Bach, J. Pankrath, W. W. Kellogg (Eds.), Man's Impact on Climate, Amsterdam, Oxford, New York: Elsevier Publishing Company 1979.

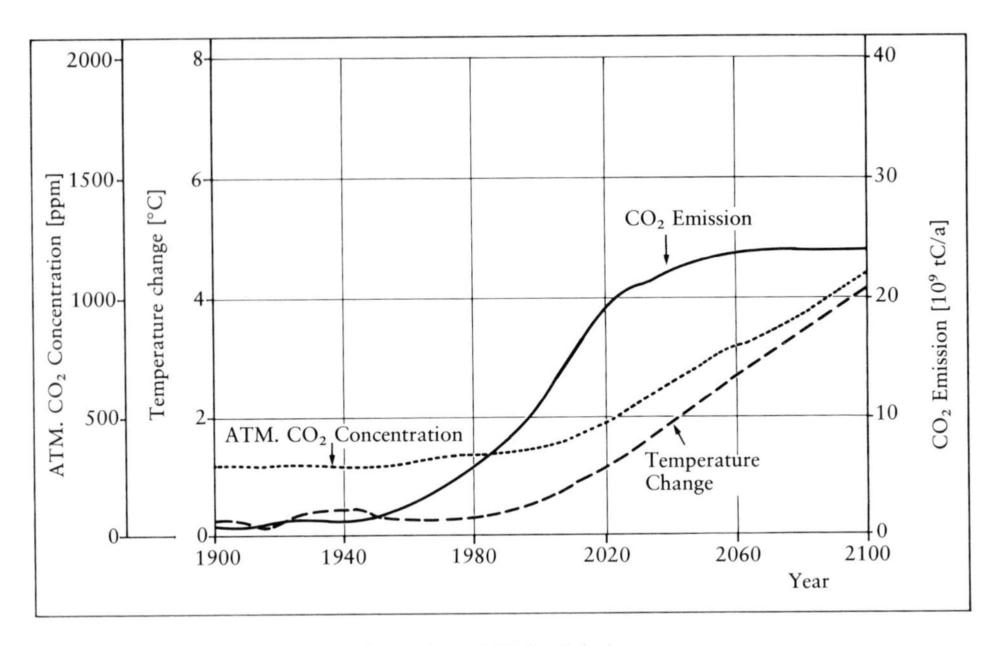

Fig. 5-5 b: Simulation of the CO_2 effects of a 30 TW fossil fuel energy strategy

Source: F. Niehaus, Carbon Dioxide as a Constraint for Global Energy Scenarios, in: W. Bach, J. Pankrath, W. W. Kellogg (Eds.), Man's Impact on Climate, Amsterdam, Oxford, New York: Elsevier Publishing Company 1979.

ter is the human CO_2 production, which in turn depends critically on the amount of fossil fuels used to meet future energy demand. Based on what is now known, it can be assumed that a steady annual increase of 3–4% in the consumption of fossil fuels will result in a doubling of the atmospheric CO_2 concentration by about the middle of the coming century (36–39).

Fig. 5-5 b shows a simulation of the atmospheric CO_2 concentration, assuming the 30 TW energy strategy ($\triangleq 32.4 \cdot 10^9$ tce/a) shown in Fig. 5-5 a, using only fossil fuels. In this strategy, the cumulative consumption of coal by the year 2100 is $2800 \cdot 10^9$ tce; of petroleum, $280 \cdot 10^9$, and of natural gas, $170 \cdot 10^9$ tce. The human CO_2 emission would be increased from the present $5.4 \cdot 10^9$ t carbon/year to $24 \cdot 10^9$ t carbon/year, and the atmospheric CO_2 concentration would rise from the present 333 ppm to more than 1000 ppm in 2100.

If the climatic effects (see 5.42) of the increased CO_2 concentration should be considered intolerable, it might be decreased in the following ways: The total consumption of primary energy can be correspondingly limited, or the CO_2 concentration can be regulated through the use of the appropriate mixture of fuels which do or do not yield CO_2. As discussed in section 2.3, it is realistic to assume that the world demand for primary energy will continue to rise for many years to come. Fig. 5-6 a shows a

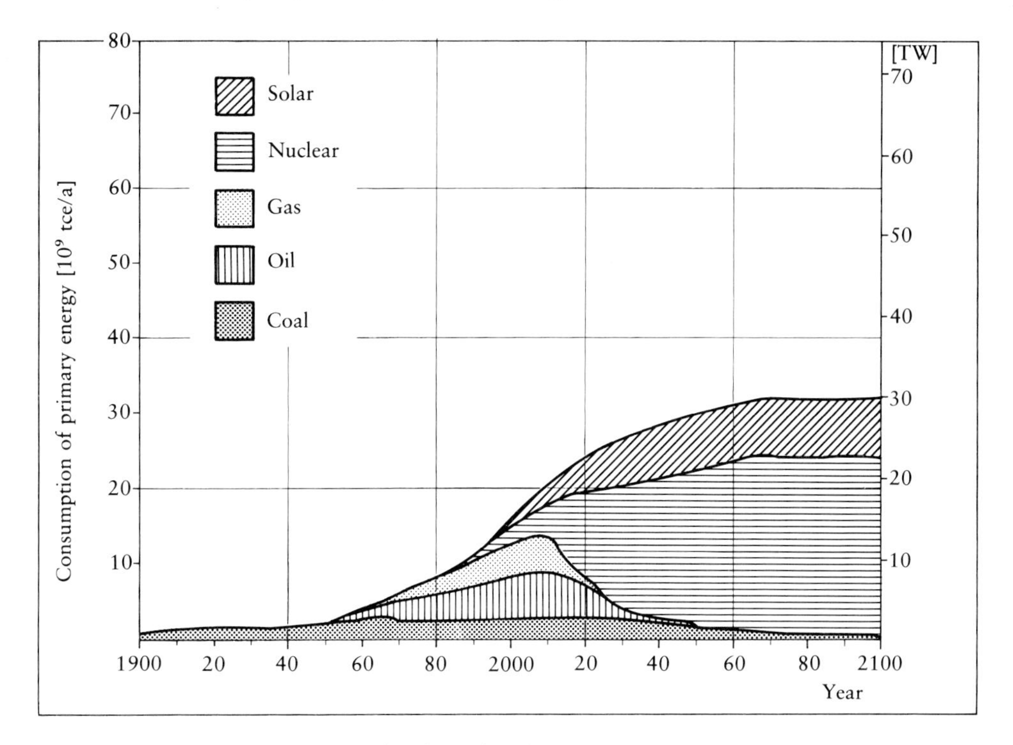

Fig. 5-6a: 30 TW energy strategy with solar and nuclear energy

Source: F. Niehaus, Carbon Dioxide as a Constraint for Global Energy Scenarios, in: W. Bach, J. Pankrath, W. W. Kellogg (Eds.), Man's Impact on Climate, Amsterdam, Oxford, New York: Elsevier Publishing Company 1979.

30 TW energy strategy in which the use of fossil fuels is reduced, and Fig. 5-6b shows the simulated atmospheric CO_2 concentration for this strategy (40). In this strategy, the present level of coal consumption is maintained until the year 2030, the cumulative consumption of petroleum, $240 \cdot 10^9$ tce, is less than in Fig. 5-5a, and the cumulative consumption of natural gas is $150 \cdot 10^9$ tce. The remainder of the energy demand will be met by nuclear and solar energy. With this strategy, the human CO_2 emission would rise from the present $5.4 \cdot 10^9$ t carbon/year to about $8 \cdot 10^9$ t carbon/year, and the atmospheric CO_2 concentration would rise to about 400 ppm.

It should be kept in mind, with reference to these simulations, that not all the details of the global carbon cycle are known, so that they may be quantitatively inaccurate. However, they are useful for recognizing trends (40). In any case, since the CO_2 stays in the atmosphere for an average of 10–13 years, a lower CO_2 concentration (or even the pre-industrial CO_2 level) can only be attainded after a time lag of several decades. In other words, even in the hypothetical case that after a certain date, no more CO_2 were produced, there would be a slowly declining CO_2 excess for decades,

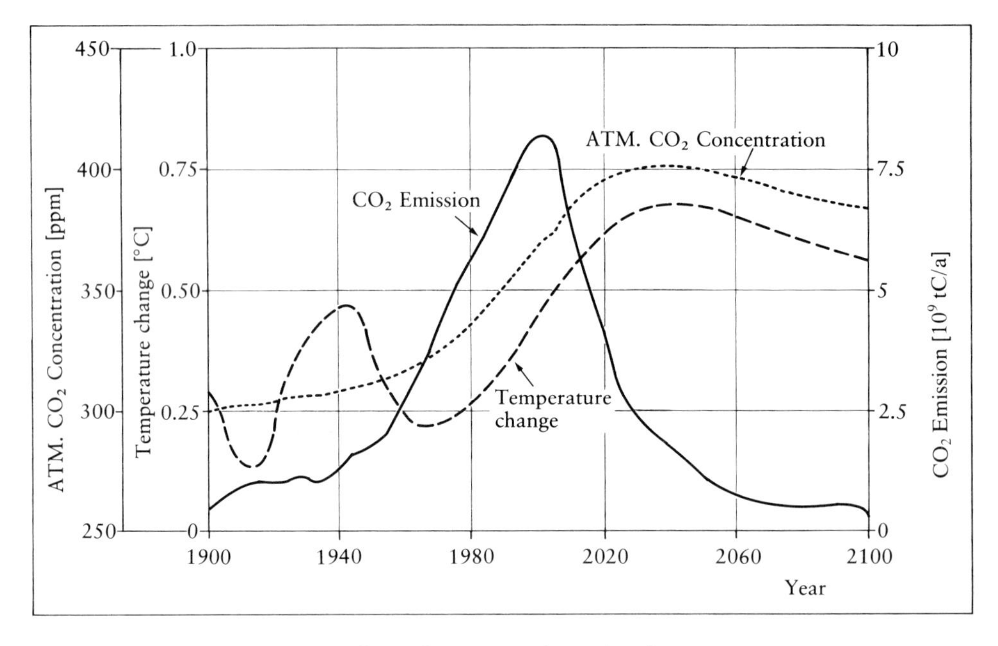

Fig. 5-6b: Simulation of the CO_2 effects of a 30 TW solar and nuclear energy strategy

Source: F. Niehaus, Carbon Dioxide as a Constraint for Global Energy Scenarios, in: W. Bach, J. Pankrath, W. W. Kellogg (Eds.), Man's Impact on Climate, Amsterdam, Oxford, New York: Elsevier Publishing Company 1979.

and probably the new equilibrium would not be reached for several hundred years. This would be considerably higher than the pre-industrial level. The possible climatic effect of an increased atmospheric CO_2 concentration is a world-wide warming of the troposphere (see 5.42).

5.4 *Climatic changes*

5.41 Climatic changes in the past

By *climate*, one means the statistical behavior of the atmosphere over long, but limited periods of time. The climate is caused by interactions, mostly non-linear feedback interactions, of various factors within a geophysical climatic system. The latter consists of the following five subsystems: the atmosphere, the hydrosphere (oceans, lakes, rivers and ground water), the cryosphere (continental and ocean ice, the ice on lakes and rivers, glaciers and snow cover), the lithosphere (the land masses including the mountains, rocks and soil) and the biosphere (all plants, animals and humans).

There are external influences in addition to the internal interactions of the five subsystems: changes in the earth's orbit (excentricity and precession) and the resulting changes in the intensity of the solar radiation, and large volcanic eruptions. The changes in the transparency of the atmosphere caused by large eruptions can affect the climate for several years (41). Fig. 5-7 is a schematic representation of the climatic system. (The outline arrows are examples of internal processes, and the solid arrows are examples of external processes (42).) The processes which can cause climatic changes can last from days to about 10^9 years, and can be either internal or external in nature.

In order to have some idea of the significance of temperature changes of a few degrees, one should study climatic excursions of the past. Fig. 5-8 shows the mean temperatures in middle Europe over the last 60 million years (43). The temperature was well above the present average for the entire time, but it has decreased sharply in the last million years (Quaternary Period) and now oscillates a few degrees with a periodicity of about 100000 years (cold and warm periods). Because of the extensive glaciation occuring in the cold periods, they are frequently called *ice ages*.

The last great ice age, the Würm ice age, began about 75000 years ago. (This ice age is called the Weichsel in Germany, the Wisconsin in North America, and the Devensian in England.) Before the Würm ice age there was an interglacial period, similar to the present, which was called the Eem period; after the Würm ice age came the present neo-warm period. The Würm ice age can be subdivided into relatively colder periods, the stadial periods, and somewhat warmer, interstadial periods. However, even during the interstadial periods, it was much colder than at present in the time from about 65000 to 12000 years ago. The stadials occured 59000 (High Glacial A), 41000 (High Glacial AB), 28000 to 29000 (High Glacial B), and 17000 to 18000 (High Glacial C) years ago. High Glacial C was probably the apex of the last ice age; during this time, it is thought that the average temperature were about 5 K lower than they are at present. All of Scandinavia, Iceland, Ireland, Scotland, Wales, the northern part of Germany (Schleswig-Holstein), the northern part of Poland and the Alps and their foothills (to Munich) were covered by thick layers of ice. North America was also covered by a sheet of ice which extended south of the present border between the USA and Canada. Because so much water was frozen, the sea level was about 150 m lower than it is now, and much of the North Sea, for example, was dry land. England and France were joined by land (44–46).

The withdrawal of the great inland ice sheets of northern Europe and America lasted about 10000 years. The glaciers retreated from Scandinavia about 8500 years ago, and from North America only 6500 years ago. In the period between 6000 and 4500 years ago, the temperature in the medium and high latitudes was 1 to 2 K higher than it is now (climatic optimum, or altithermum). During this time, even the arctic pack ice had disappeared from northern Greenland, and the present dry belt

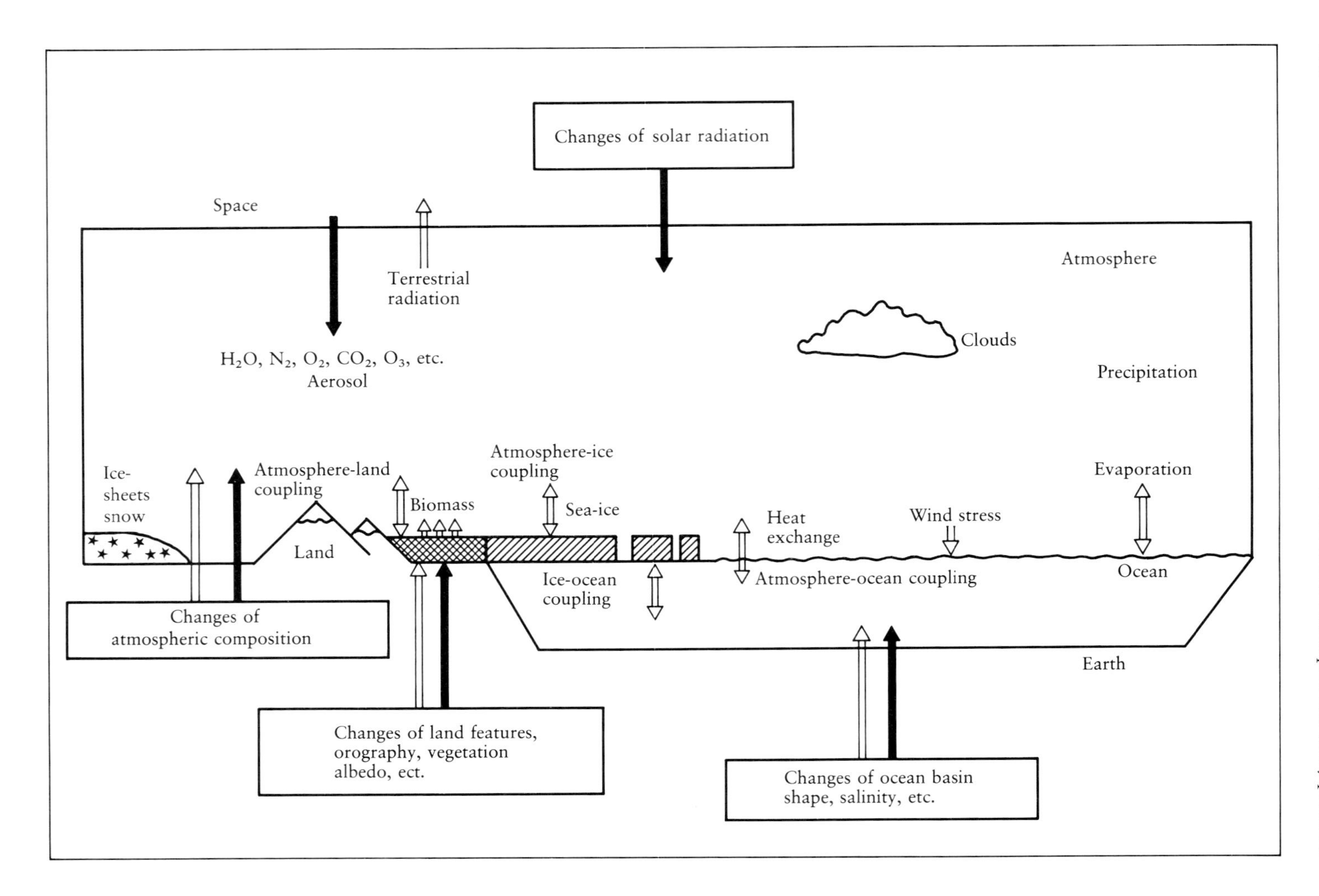
Changes of solar radiation
Space
Terrestrial radiation
Atmosphere
Clouds
H_2O, N_2, O_2, CO_2, O_3, etc.
Aerosol
Precipitation
Atmosphere-ice coupling
Evaporation
Ice-sheets snow
Atmosphere-land coupling
Biomass
Sea-ice
Heat exchange
Wind stress
Land
Ice-ocean coupling
Atmosphere-ocean coupling
Ocean
Changes of atmospheric composition
Earth
Changes of land features, orography, vegetation albedo, ect.
Changes of ocean basin shape, salinity, etc.

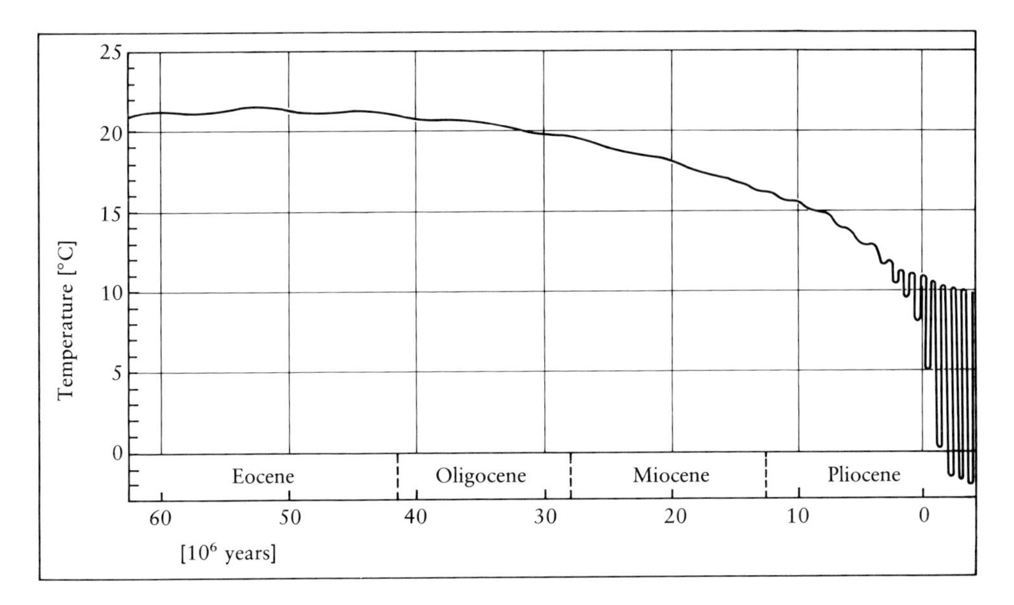

Fig. 5-8: Temperatures in middle Europe in the last 60 million years

Source: O. Haxel, Beitrag der Physik zur Klimageschichte, Die Naturwissenschaften, Vol. 63, No 1 (1976).

from North Africa to India was very wet. Lake Chad, for instance, was about 50 m deep and had an area of 350000 km^2. At the height of the ancient culture in Egypt, the climate there was much more favorable than it is now.

Since the climatic optimum, there have been several colder and warmer climatic periods, and even in the last centuries, slight temperature variations have been observed. For example, the *Roman climatic optimum* occured between 300 B.C. and 400 A.C.; during this time, several Alpine passes were probably passable throughout the winter. The period was relatively rich in precipitation, and the climate only began to get drier after 400 A.C.

There was another climatic optimum in the Middle Ages, roughly from 900 to 1050 A.C. During this time, the average yearly temperatures in Europe were 1 to 1.5 K higher than at present. Important information can be inferred, for example, from the history of the Viking settlement of Iceland and Greenland and their advances to Labrador and Newfoundland. About 900 A.C., the Vikings found Iceland about 60% forested, and were able to raise sheep and grain there.

Fig. 5-7: Components of the coupled atmosphere-ocean-land-ice-biomass climate system

Source: A. H. Murphy, A. Gilchrist, W. Häfele, G. Krömer, J. Williams, The Impact of Waste Heat Release on simulated Global Climate, IIASA, Laxenburg/Vienna, RM-76-79, December 1976.

Between 1550 and 1850, there was a cold period in Europe which has been called the *Little Ice Age*. At that time, the mean annual temperatures were about 1 K lower than they are at present. There is evidence in all the high mountains of the Earth for glacier advances during that time (47, 48).

The data on climate in the past is obtained from a number of physical methods in addition to conventional paleoclimatology. These physical measurements, which give relatively precise indications of the climatic conditions, are methods for determining the age of specimens and the temperatures prevailing during their lives.

The ^{14}C method developed by W. F. Libby is used for dating samples from the past 20000 years, or in some cases, somewhat older samples. ^{14}C is formed from the nitrogen of the air by cosmic rays. It is radioactive and has a half-life of 5730 years. The radioactive carbon is assimilated by plants as $^{14}CO_2$, and is incorporated by animals through their food. After an organism dies, it no longer incorporates atmospheric CO_2, so the amount of ^{14}C in its body begins to decrease with a half-life of 5730 years. The age of any organic substance can be determined from the amount of radioactive carbon it still contains. The natural radionuclides of protactinium (half-life 33000 years) and ionium (half-life 75000 years) can be used in a similar way to date samples up to 100000 years old. However, there are no comparable methods for dating older ice-age sediments or fossils (57).

The ratios of stable isotopes of hydrogen, carbon and oxygen in sediments or fossils indicate the temperatures at which they were laid down or grew. The isotope ratio of an element is the same in compounds and the pure element, and it can be determined with a mass spectroscope. With this instrument, it is possible to show that samples of different origin differ in their isotope ratios. For example, evaporation of a body of water leads to a change in the ratios of ordinary hydrogen (H) and deuterium (D) because a molecule containing a deuterium atom is slightly less readily evaporated than one containing only the light isotope. In other words, in a closed vessel containing liquid water and steam at equilibrium, the concentration of deuterium in the liquid phase is slightly higher than in the vapor phase: $[D/H]_{liquid} > [D/H]_{vapor}$. The difference amounts to a few parts per thousand. The heavy isotope of oxygen, ^{18}O, has the same relationship to the common isotope ^{16}O.

Similar isotope effects are observed when CO_2 is dissolved in water. The ^{18}O isotope is enriched in the bicarbonate (HCO_3) which is formed from CO_2. If calcium carbonate is precipitated in the presence of excess calcium, the concentration of the ^{18}O is higher in the calcium carbonate than in the CO_2 of the air. The decisive point, however, is that the degree of enrichment of ^{18}O depends on the temperature. By analysing the carbonate from deep-sea borings, C. Emiliani demonstrated from the $^{18}O/^{16}O$ ratios the above-mentioned 100000-year rhythm of warm and cold epochs. Also, mollusk shells which grow at lower temperatures have higher ^{18}O contents than those which grow at higher temperatures. In either case, higher ^{18}O contents indicate colder periods, and lower ^{18}O contents, warmer ones.

No definitive statement can be made about the causes of climatic changes, because causal relationships in the climatic system are extraordinarily complex, and are not all known (compare Fig. 5-7). The geophysical climate system not only includes all the subsystems and their internal interactions, and the external influences, but it is still further complicated by some of the properties of the components.

For example, parameters like the water vapor have different turnover times in different components. The water vapor in the troposphere turns over in a few days, but it spends several months to years in the stratosphere. Water in the shallow layers of the ocean stays there a few months, but spends several centuries in the depths of the ocean. It remains a few years to decades in the glaciers, but thousands of years in the great continental ice sheets. In addition, the components of the climatic system have different heat capacities. One result is that much more heat must be absorbed by an ocean than by a continent to produce a given increase in temperature, i.e. water has a larger heat capacity than land. This also means that water must give up more heat than land to cool by a certain amount. (This is why maritime climates are more moderate than continental climates.) The climatic system also contains buffers which absorb some variation, and reflect the changes only after a relatively long time or when the variation is especially great. The deep ocean, for example, acts as a CO_2 buffer.

Some of the complexity of the climatic system is due to feed-back phenomena. Feed-back can be either positive (self-amplification) or negative (self-damping). The following is an example of positive feed-back: Cooling of the troposphere over relatively long periods will increase the proportion of the precipitation which falls as snow. One result of this, especially in the polar land areas, is that the area covered by snow and ice expands, which leads to further cooling of the earth, because snow and ice reflect more of the solar radiation than do water and soil which are not covered by snow and ice. In other words, when the earth is covered with snow and ice, it is less warmed by the sun, and this leads to further expansion of the ice areas. In addition to this type of self-amplification, there are damping mechanisms which tend to stabilize the conditions. For example, if the temperature of the earth rises, e.g. through more intense solar radiation, the concentration of water vapor in the atmosphere increases, and this leads to the formation of more clouds. The thicker cloud blanket reduces the amount of sunlight reaching the surface, and thus cools it. In this way the original increase in temperature is damped out.

The natural causes of climatic variation must be the internal and external factors mentioned above. The majority of short-term changes in climate and climatic anomalies are caused by internal, non-linear interactions which lead to a redistribution of heat and precipitation within the system. For instance, arctic sea ice plays an important role in large-scale climatic processes (48), in a typical case of interaction within the system. Cooling causes the formation of more ice, which spreads and leads to more cooling. (At present, the arctic sea ice has a surface of about $11.8 \cdot 10^6\ km^2$ in spring and about $8.2 \cdot 10^6\ km^2$ in late summer.)

In addition to this sort of internal interactions in the climatic system, external factors might be responsible for variations in climate. Some examples would be (hypothetical) changes in the amount of solar radiation, variation in the earth's orbital parameters, and volcanic eruptions. Variations in the amount of solar radiation could be caused by changes in solar activity. The variations in the orbital parameters are the variation in the times of year at which the earth passes through the perihelion and aphelion, the points of its orbit nearest and farthest from the sun, respectively, the inclination of the earth's axis with respect to its orbit, and the eccentricity of the orbit. The variation in the dates of passage through the perihelion and aphelion has a period of about 21000 years; the inclination of the axis varies in 40000 year rhythm, and the mean period for the variation in the eccentricity of the orbit is 96000 years. The superposition of these three variations results in a periodic change in the effective solar radiation. The effects of large volcanic eruptions on the climate have been documented. When large amounts of volcanic ash are blown into the stratosphere, as in the eruption of Agung on Bali in 1963, and of Fuego in Guatemala in 1974, they absorb some of the sunlight and also scatter a fraction of it back into space. This can cause a cooling of the entire earth by 1 to 1.5 K, an effect, however, which only lasts 1 to 3 years. Such an effect was observed after the eruptions of Tambora on Sumbawa in 1815; the summers of 1816 and 1817 were unusually cool in Europe, North America and Japan. The frequency of large volcanic eruptions varies widely. In some periods, like the end of the 17th century or 1815 to 1835, there were many eruptions, but in other long periods, like 1912 to 1948, there have been few or none (50).

The causes of climatic variation are not yet fully understood (51). For instance, there are two groups of hypotheses concerning the origin of ice ages, one which assumes external factors are responsible, the other, internal factors. The first group includes those which suggest that changes in the earth's orbital parameters, resulting in changes in the amount of solar radiation falling on the earth, are the cause of the ice ages. One of the arguments against this hypothesis as the sole cause of ice ages is that the changes in temperature, given in Fig. 5-8, do not support it, since it is not likely that the earth's orbital parameters only began to vary in the last million years, but were constant before that.

The (more likely) hypothesis assume that the ice ages result from internal factors. H. Flohn argues that the Arctic Ocean was pushed into its present position by continental drift about the middle of the Pliocene. Continental ice began to build up on the land around the ocean, and it grew thicker and thicker until, under pressure and the heat of freezing, it began to flow outward. Because ice has a higher albedo than water, this led to a general cooling of the earth. The lower temperature caused the snow limits in the mountains to go lower, and a lower equilibrium temperature was established (ice age).

There is growing concern that human activities may also affect the climate. The activities which have especially to do with energy consumption are the production of

CO_2 and trace gases, air pollution, especially the emission of aerosol particles, direct heat stress due to release of energy, and changing the albedo of the earth's surface.

5.42 Possible effects of carbon dioxide on climate

Our present knowledge suggests that of the human influences on climate, the increase in atmosphere CO_2 concentration is of particular importance (52, 53). However, in order to predict what effect the increasing atmospheric CO_2 concentration will have on climate, one would have to know which energy supply strategy will actually be followed, and to understand the global carbon cycle as well as the causes of climatic changes.

A number of models of the long-term increase in the CO_2 content of the atmosphere due to human CO_2 emission have been developed (54–58). There are also several climatic models describing the temperature increase caused by the rise in CO_2 (59–62). The climatic result of an increase in the CO_2 content is a warming of the troposphere (greenhouse effect) which decreases with altitude and becomes a cooling of the stratosphere. The increase in the temperature of the troposphere is also related to the increase in water vapor and in cloudiness. The models developed by S. Manabe and R. T. Wetherald, and by T. Augustsson and V. Ramanathan, which take into account the water vapor effect, agree that a doubling of the CO_2 content of the atmosphere would cause an increase of 2 to 3 K in the average temperature of the troposphere; for the polar areas, the models predict that positive feed-back mechanisms would increase the temperature by 7 to 10 K (60).

Fig. 5-5 b and 5-6 b show the effects of increased CO_2 concentrations on the temperature for different energy-supply strategies, assuming an energy demand of 30 TW. With a 30 TW energy strategy based exclusively on the use of fossil fuels (Fig. 5-5 a), the CO_2 emission would rise from the present $5.4 \cdot 10^9$ t C/year to $24 \cdot 10^9$ t C/year and the atmospheric CO_2 concentration would rise from the present 333 ppm to more than 1000 ppm by the year 2100. The average global temperature would probably rise by 4 K as a result (compare Fig. 5-5 b). (It should be emphasized here that if energy were obtained from biomass, the amounts of CO_2 and other gases (such as NO_x see 5.43) released would be less than with fossil fuels, and the CO_2 would simply circulate through the carbon cycle.) If one assumes a 30 TW energy strategy which used only reduced amounts of fossil fuels, and more nuclear and solar energy (compare Fig. 5-6 a), the CO_2 emission would rise from the present $5.4 \cdot 10^9$ t C/year to about $8 \cdot 10^9$ t C/year, and the atmospheric CO_2 concentration would rise correspondingly from the present 333 ppm to about 400 ppm. In this case, the result would probably be an increase of about 0.5 K in the global average temperature (compare Fig. 5-6 b).

As mentioned above, the results of these simulations must be treated with some caution, since not all the details either of the global carbon cycle or of the climatic factors and interactions are known (63–66). However, such simulations are very important as aids in recognizing trends, although one must remember that the simulations shown in Fig. 5-5b and 5-6b did not take into account other human effects besides the CO_2 effect (see 5.43).

5.43 Other human effects which might influence climate

There are a number of other human effects besides CO_2 which might influence climate, most of which have to do with energy. These are changes in the composition of the atmosphere due to trace gases and aerosol particles, direct heat load (due to energy release) and the changes in the albedo of the earth's surface.

The composition of the atmosphere is affected mainly by dinitrogen oxide, N_2O, which is released by the use of artificial fertilizers; chlorofluorocarbons (e.g. CCl_2F_2 and CCl_3F), which are used in pressurized spray cans and refrigeration; methane, CH_4; ammonia, NH_3; and ozone, O_3. These gases absorb in the near infrared, up to 12 μm, and amplify the greenhouse effect of CO_2.

In this respect, N_2O in particular is becoming increasingly significant, because it is an end product of nitrogen-containing fertilizers. In the last few years, the use of these fertilizers has increased at a rate of about 10% per year; in 1979 alone, $54 \cdot 10^6$ t nitrogen were used as fertilizer (67, 68). About 100 years ago, the wheat yields in temperate climates were about 120 t/hectare; today they are about 500 t/hectare and the peak yields are 1000 t/hectare. The increases are due not only to plant breeding, better preparation of the ground and pesticides, but are in large measure due to the use of fertilizer. Because the world population is growing, it must be assumed that the world-wide use of fertilizer will continue to increase.

It has been estimated that the gases produced by industrial processes, especially N_2O, CCl_2F_2, CCl_3F, CH_4 and NH_3, increase the CO_2 greenhouse effect by about 50% (69, 70). Also, as discussed in 5.31, chlorofluorocarbons, nitrogen oxides and methane disturb the ozone balance in the stratosphere (71), and ozone shields the surface from short-wave ultraviolet radiation from the sun.

Emission of krypton $^{85}_{36}Kr$ which is released by fuel recycling from nuclear reactors, may effect the distribution of precipitation. The radioactive noble gas is created by nuclear fission, and remains enclosed in the fuel elements. It is released when the spent fuel elements are dissolved in the process of recycling. The half-life of $^{85}_{36}Kr$ is 10.76 years. Since the beginning of commercial application of nuclear fission, the $^{85}_{36}Kr$ content of the atmosphere has increased steadily; from 1960 to 1975, it was quadrupled. The main sources are in the Northern Hemisphere; in 1975, the concentration in the Northern Hemisphere was 17 picocurie/m^3 (72). Because the solubility of $^{85}_{36}Kr$ in water is negligibly low, essentially all of the released $^{85}_{36}Kr$ remains in the atmosphere. The only mechanism which removes it is radioactive decay. $^{85}_{36}Kr$ is a beta-emitter, which means that it emits electrons, and these increase the natural ionization rate of the air due to natural radionuclides and cosmic radiation. One result is a decrease in the electric resistance between the earth and the ionosphere. Our present knowledge is not sufficient to say what effect changes in the atmospheric electric field might have on the climate. A number of effects are under debate, especially the effects of a rapid increase in the $^{85}_{36}Kr$ concentration. The ionization caused by $^{85}_{36}Kr$ is still less than 1% of the natural rate, but it is possible that a large increase could affect the world-wide distribution of precipitation. In several countries, efforts are being made to contain the $^{85}_{36}Kr$ from fuel recycling, for radiological reasons. It may be that these efforts are entirely justified for non-radiological reasons as well.

Not only trace gases, but, as mentioned in section 5.31, aerosol particles as well can change the composition of the atmosphere (73). Satellite photographs have shown that the atmosphere is clouded for about 100 km around industrial centers, large cities or forest fires. In dry areas, dust is picked up by the wind, and this is partly a result of the centuries of disruption of natural vegetation by human activities. The effects of aerosol particles on the climatic system are very complex: larger particles (> 10 μm) absorb mainly the longer wave radiation from the earth, and thus warm the troposphere for a number of days, so that these aerosols are far less effective than the volcanic ash mentioned in section 5.41, which is ejected into the stratosphere and can remain there for several years. Smaller particles (< 2 μm) absorb sunlight, but at the same time scatter some of it back into space. The ratio of absorption to back scattering depends in part on the composition of the aerosol particles. In general, the warming effect of aerosols outweights the cooling, and this is observed especially in the areas around large cities where the aerosol concentrations are high. It should be mentioned that as a result of anti-pollution laws, the aerosol concentrations in many large cities have declined (73).

As mentioned in section 5.2, the heat added to the earth/atmosphere system as a direct result of human activities is still small (0.016 W/m^2) compared to the mean solar energy budget (230 W/m^2) (compare Table 5-1). It cannot be concluded from this, however, that the direct anthropogenic heat release is climatologically insignificant. In many cases, the energy turnover is concentrated in large industrial and population centers, and can reach local levels of 10–50 W/m^2. This causes the cities to become "heat islands" with additional thermal circulation. It has been found that there is a statistically significant increase of heavy local rainfal over the cities (74).

One of the most important parameters in the heat budget of the earth is the albedo (reflectivity) of the earth's surface (73). For this reason, changes in the surface can influence the climate. The reflection depends on the color of the vegetation: the albedo for forest is about 0.12, for steppe and grassland, 0.20 to 0.25, for light desert sand, 0.35, for fresh snow, 0.80, and for the ocean, 0.05 to 0.25, depending on the height of the sun and the wave conditions. The mean albedo of the continents, taking into account the seasonal variations in the snow cover, is 0.260 ± 0.02; the mean for the oceans is estimated to be 0.100 ± 0.005. For the whole earth, the albedo is thus 0.147 ± 0.010 (75). From radiation models, it has been calculated that an increase in the mean albedo of 1% will result in a decrease in the equilibrium temperature of 1.3 K (60). It is estimated that the albedo of the earth has changed in the last 6000 years from 0.138 to 0.147; and the corresponding decrease in temperature has been 1 K (75). In other words, the destruction of vegetation in dry areas, and the conversion of forests to cultivated fields since the climatic optimum 6000 years ago were responsible for an increase in albedo (see 5.41). In general, changes in or destruction of the original vegetation increases the albedo, and this is one intervention by humans which can lead to a cooling of the earth.

The increasing rate of desert formation in several regions is another aspect of this problem. In the Maghreb (Tunesia, Algeria and Morocco), Lybia and the Sudan, the desert is advancing an average of 1 to 2 km annually, which for the Sahara alone means an annual increase of about 20000 km^2. This is not the result of a change in climate, but of centuries of human intervention in the natural ecosystem. It can happen when the number of grazing animals exceeds the support capacity of the vegetation, which is then destroyed. The recent events in the Sahel zone have been an example: between 1949 and 1968, the population of grazing animals was increased six-fold. In addition, the human population consumes the last trees and bushes for firewood. Model calculations show that the increase in albedo caused by destruction of the vegetation causes a further decrease in precipitation, and intensifies the process of desert formation. To be sure, the great deserts of the earth were not created by humans, but the increase in desert formation is a result of human intervention.

In summary, all of the human activities which might affect the climate, with the exception of the changes in the ozone content of the stratosphere and the albedo, tend to warm the earth. According to W. Bach, the sum of all the human effects could lead to a global temperature increase of 0.8 to 1.2 K in the year 2000, and of 2 to 4 K by the year 2050. Furthermore, the increases in the polar areas could be a few degrees higher (76). All that we know today indicates that temperature increases of a few degrees are a danger to be taken very seriously, since they can cause global climatic changes (see 5.41).

From all this, it is clear that the release of energy cannot be increased without limit. The possibility of increasing the use of coal to meet the world's energy demand, about the beginning of the next millenium, will be limited, because the possible climatic effects of increased atmospheric CO_2 could be catastrophic. It can be taken for granted that only a part of the world's coal reserves can be used to meet a growing energy demand. The CO_2 problem also shows, however, that effects which can be neglected when an energy carrier is used on a small scale become decisive factors when it is used on a large scale. It is also reasonable to ask, with respect to ecologically disruptive factors, whether it should be tolerated that a smaller part of humanity places many times more stress on the environment than the majority (see 2.32).

H. Häfele has compared the climatic-risk problem to the risks of nuclear technology: we cannot afford to learn by trial and error. If the climate changes, it can be assumed that the results will be disastrous and irreversible, at least within a time period relevant to our culture. Therefore it must be emphasized that a long-term energy concept must take into account the problem of possible changes in the climate.

5.5 Environmental impact of solar energy

Solar radiation is the energy source whose utilization will have the least impact on the environment. This applies, for example, to the use of solar energy for decentralized energy supply using collectors, photocells, etc. (see 4.311 and 4.313). However, even with this source of energy, it can be shown that side effects which are negligible for small-scale use can become problems with large scale use. For example, direct methods of utilization of solar energy can change the albedo of large areas, and thus affect the climate (see 5.43) (77).

It can also be assumed that, due to the requirement for large surface areas, the large-scale collection of solar energy will be done in areas with low population densities (e.g. the Sahara), and the secondary energy carrier (e.g. hydrogen or electricity) will be transported long distances into other regions. The energy will thus be released in an area distant from the area in which it was collected, and can cause environmental stress in this way.

Ecologically damaging factors can also be associated with indirect methods of utilizing solar energy. A few will be mentioned here. A hydroelectric power plant (see 4.321) frequently changes the natural drainage of water. For example, the Aswan Dam represents an intervention in the ecology of the Nile Valley. The Nile has always had a double function in Egyptian agriculture, as it provides both water and fertilizer. Since prehistoric times, the Nile has flooded in August and September, spreading a part of its rich, alkaline mud (up to 10^8 t/year) over the flooded fields. The water which now passes through the turbines of the dam is poor in nutrients, because the mud settles at the south end of the lake, where the current drops off. Furthermore, the relationship between surface and ground water has been disturbed in many parts of this region.

Utilization of wave energy (wave power plants, see 4.322) could reduce the orbital motion of the water, and thus disturb the exchange of oxygen and transport of plankton between the upper and lower layers of water. It could thus disrupt the local biological equilibrium (78).

If the vertical temperature gradient in tropical seas were used to supply energy (see 4.323), it can be supposed that this would lead to an increase in the rate of release of dissolved CO_2 in the deeper water (see 5.32); this could affect the climate (76).

Utilization of wind energy could lead to changes in air circulation; in addition, if it were used on a large scale, the amount of land required could become a problem (see 4.324).

Even the use of stored solar heat by means of heat pumps is not ecologically neutral. If the heat stored in the ground water or earth were used, there would be a danger, for large scale use, that the ground water could be contaminated, and that the heat balance over the entire area would be affected (see 4.325).

Aside from the question of feasibility, large-scale utilization of photochemical conversion would also affect the environment (see 4.326). For example, if large areas were planted with sugar cane or some other suitable plant (biomass) in monoculture, there would be ecological changes. In section 5.32 it was mentioned that the details of the carbon cycle are not completely known, but that when energy is extracted from biomass, the CO_2 circulates through the subsystem atmosphere-ocean-biomass-detritus, rather than being released from long-buried deposits, as is the case when fossil fuels are burned (compare Fig. 5-4).

5.6 Environmental impact of tidal energy

Because geographical factors determine the economic feasibility of tidal energy, there are relatively few sites suitable for tidal power plants (see 3.38 and 4.4). However, these plants do have local effects on the environment (78), for instance, by changing the ocean currents.

5.7 Environmental impact of geothermal energy

At present, technological utilization of geothermal energy is only possible when geothermal deposits can be tapped. The environmental effects of utilization of geothermal energy are usually considerable (79, 80).

When electricity is generated from geothermal heat, the low thermodynamic efficiency of only 8% to 16% leads to a much larger proportion of waste heat than is generated by other power plants. This waste heat can be rejected in essentially the same way as from other power plants (see 5.2).

Hot springs are often salty. The hot water in the neighborhood of the Salton Sea, California, may contain up to 20% salt. (Sea water contains about 3.3% salt.) The daily flow of spent water from a 1000 MW geothermal power plant, if it were operated at Cerro Prieto, Mexico (salt content 2%) would contain about 12000 t salt (81). Therefore, it will frequently be necessary to return the spent water to the wells. This can also help to prevent sinking of the ground, which can occur when large amounts of water are removed from underground reservoirs. Removal of large quantities of water can also cause shifts in the rock layers, which are the cause of "seismic effects", i.e. local earthquakes.

Geothermal vapors contain non-condensable gases, which cause serious difficulties in operating the plants. The gas composition of "The Geysers", in the USA, is 63.4% CO_2; 15.3% CH_4; 14.7% H_2; 3.5% Ar; 1.7% H_2S; 1.3% NH_3; 0.1% H_2BO_3. The gas at Wairakei, New Zealand, consists of 92.1% CO_2; 4.2% H_2S; 1.8% H_2; 0.9% CH_4; 0.6% NH_3; 0.3% Ar; 0.1% H_2BO_2 (78).

A geothermal power plant also requires a large area of land. At The Geysers, a well yields about 7 MW, so for a 1000 MW plant (about 150 wells), a surface of about 30 km^2 is needed (78).

5.8 *Environmental and safety problems specific to nuclear fission*

5.81 Introduction

The role of nuclear energy has become an important issue in many countries, although nuclear power plants are in operation in 22 countries, and some of them have been operated for years. In 14 other countries, nuclear power plants are on order or under construction (as of May 1, 1979, compare Table 4-2). An important reason for the occasionally violent controversy is that, partly because of Hiroshima and Nagasaki, many people are deeply afraid of nuclear energy.

Because the fraction of electricity generated from nuclear energy is becoming significant in some countries, it is understandable that some aspects of the peaceful utilization of nuclear energy are only now becoming apparent to many people. As with every other source of energy, there are problems resulting from the use of nuclear energy which only become important when it is used on sufficiently large scale. (In 1978, for example, nuclear energy supplied 10% of the electricity generated in the Common Market countries, and about 12% of the electricity in the USA, compare Table 4-4.)

If one observers the development of the discussion of the peaceful uses of nuclear energy, one can see that attention is no longer focused only on the nuclear power plants themselves, but also on problems arising from the entire nuclear fuel cycle. It has apparently become clear to the general public that the nuclear reactor is only one element of the fuel cycle, and that long-term planning for nuclear energy must take into account the entire cycle. (The term *nuclear fuel cycle* denotes all the processes and stages involved in supplying fuel to the reactor and disposing of wastes and spent fuel.)

The supplying of fuel to nuclear reactors, i.e. the mining of natural uranium, enrichment of the ^{235}U content to about 3% in an enrichment plant and preparing fuel elements from uranium (and reprocessed plutonium) is highly developed. By contrast, some of the problems of disposal, i.e. development of storage facilities for spent fuel elements prior to their reprocessing to retrieve plutonium, the reprocessing itself, preparation of the wastes for permanent storage, and their final disposal, have not been completely satisfactorily solved, although it is said, by the competent authorities (geologists and hydrologists) that the problems are "solvable".

It appears that more and more people consider the final disposal of radioactive wastes to involve a new dimension. If the spent fuel elements are reprocessed, the highly radioactive waste for disposal must be stored in such a way that it cannot contaminate the biosphere for hundreds of years; if the fuel elements are not reprocessed the storage site must guarantee safety for many thousand years (the half-life of plutonium 239 is 24000 years). In other words, the present generation is enjoying the energy released, but it is passing on to later generations the atomic waste in its permanent storage depots. Such a depot will probably always require a certain amount of watching. In this connection, people sometimes speak not only of "environmental damage", but of "damage to posterity", and the question is raised whether the "responsible party principle", i.e. inclusion of the cost of disposal in the price of nuclear fuel, can be applied here.

A special problem of political security arises from the fact that uranium and plutonium can be used not only as fuels, but as explosives for weapons. It is in principle possible to build nuclear explosives from highly enriched uranium. The method is elaborate, however, and requires more time than is needed with plutonium, which can be extracted from spent fuel. Therefore, questions of peaceful use of nuclear energy in reactors are closely related to the non-proliferation of nuclear weapons.

As H. Häfele, Laxenburg/Vienna, has pointed out, the risks associated with nuclear technology differ from other technical risks in that they cannot be found out by trial and error. In this respect, this technology is of a "new quality". According to C. F. von Weizsäcker, "for recognized technically caused dangers, there are in general also technically feasible means to provide relative security. The difficulty is to recognize the dangers in time" (82). The incalculable "human factor" can never be completely excluded as a possible source of danger. It must be emphasized that with respect to technology, there is no such thing as absolute security, but only a greater or lesser degree of reliability, or probability of accident. There will never be an absolutely safe nuclear power plant. A so-called reactor safety study can therefore make no prediction whether the risk associated with nuclear energy will be accepted.

It would sometimes appear that the faith of many people in what is generally termed "technological progress", a faith that in the past has often been very strong, has been shaken. If this is the case, humanity is undergoing a deeply significant historical change (83, 84).

A comprehensive discussion of all the environmental and safety problems specific to nuclear fission would exceed the limits of this book, so a few selected aspects will be presented in the following sections.

5.82 On the nuclear fuel cycle

5.821 Provision of nuclear fuel

The provision of nuclear fuel is the part of the fuel cycle which is furthest developed. Fig. 5-9a shows the fuel cycle for a light water reactor (LWR) (see 4.212.2). The fuel cycle for 2nd generation reactors, high temperature gas reactors (HTGR, see 4.213.3) and liquid-metal-cooled fast breeder reactors (LMFBR, see 4.214) differ at several points from the LWR fuel cycle, and require additional development work. Fig. 5-9b shows the fuel cycle for the HTGR, and Fig. 5-9c, the cycle for the LMFBR (85).

The provision of nuclear fuel can be divided into the following sectors: obtaining natural uranium (prospecting, exploration, mining), conversion to UF_6, enrichment in ^{235}U, and the preparation of fuel elements.

The geographical distribution of the uranium and thorium reserves, the centers of production and consumption, and the predicted development of the demand were discussed at length in section 3.35. Whether the predicted demand can be met will probably depend heavily on the price people are willing to pay for extracting uranium from lower grade ores or from sea water. The forced development of breeders in many countries (compare Table 4-3) could be taken as an indication that the economically recoverable uranium reserves are thought to be very limited, and that some countries consider this uranium only conditionally available.

The conversion of natural uranium oxide (U_3O_8) to uranium hexafluoride (UF_6) is a relatively straightforward chemical process. This is not a critical point in the fuel cycle, because the chemical industries of many countries have mastered fluorine technology. The West has large conversion capacities in Great Britain, France, Canada and the USA. It can be assumed that the large-scale technical conversion could be begun in the Federal Republic of Germany at any time.

Light water reactors require enriched uranium (about 3% ^{235}U) as fuel. At present, this enrichment is done in Great Britain, France, the USA and the USSR. The demand for enriched uranium will continue to rise, and it is expected that starting in 1982, more than half the European demand will be met by the European enrichment facilities Eurodiff (countercurrent process) and Urenco (centrifugation process) (see 3.352) (86–90). For example, the original charge of a Biblis type LWR (el. power ca. 1200 MW) requires about 100 t of uranium enriched to about 2.5% ^{235}U. For refueling, about 30 t per year of 3% ^{235}U is needed (91).

The preparation of uranium oxide from fuel elements for the present LWRs can be considered a mature technology. The first step is the conversion of the UF_6 from the enrichment plant to a sinterable UO_2 powder; then UO_2 pills are produced. To protect the fuel from the coolant, and to prevent the escape of fission products into the coolant circulation, the pills are put into metallic cladding tubes (fuel rods) and sealed in by gas-tight welding. A bundle of filled fuel rods forms a fuel element. Zir-

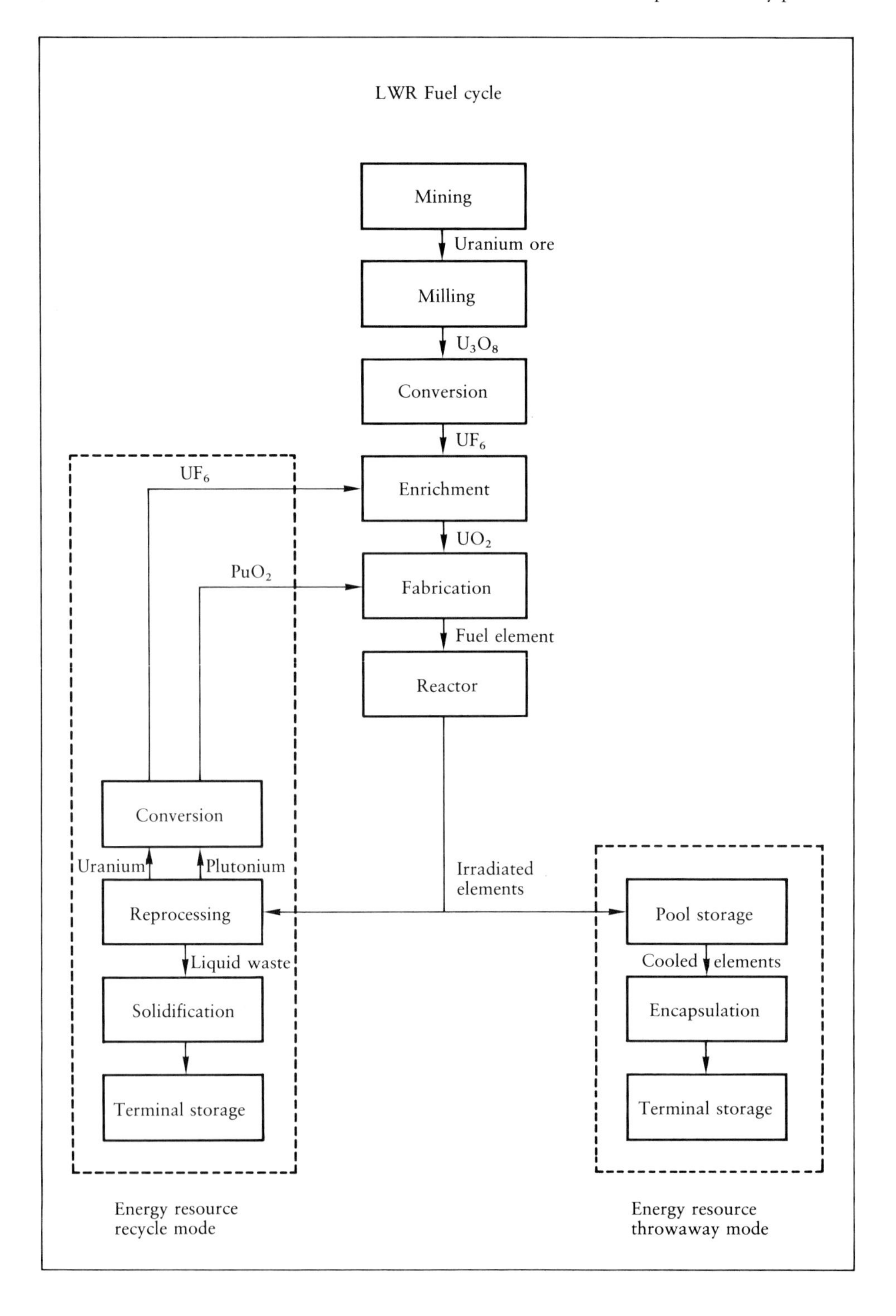
LWR Fuel cycle
Mining
Uranium ore
Milling
U_3O_8
Conversion
UF_6
Enrichment
UO_2
Fabrication
Fuel element
Reactor
UF_6
PuO_2
Conversion
Uranium
Plutonium
Reprocessing
Liquid waste
Solidification
Terminal storage
Irradiated elements
Pool storage
Cooled elements
Encapsulation
Terminal storage
Energy resource recycle mode
Energy resource throwaway mode

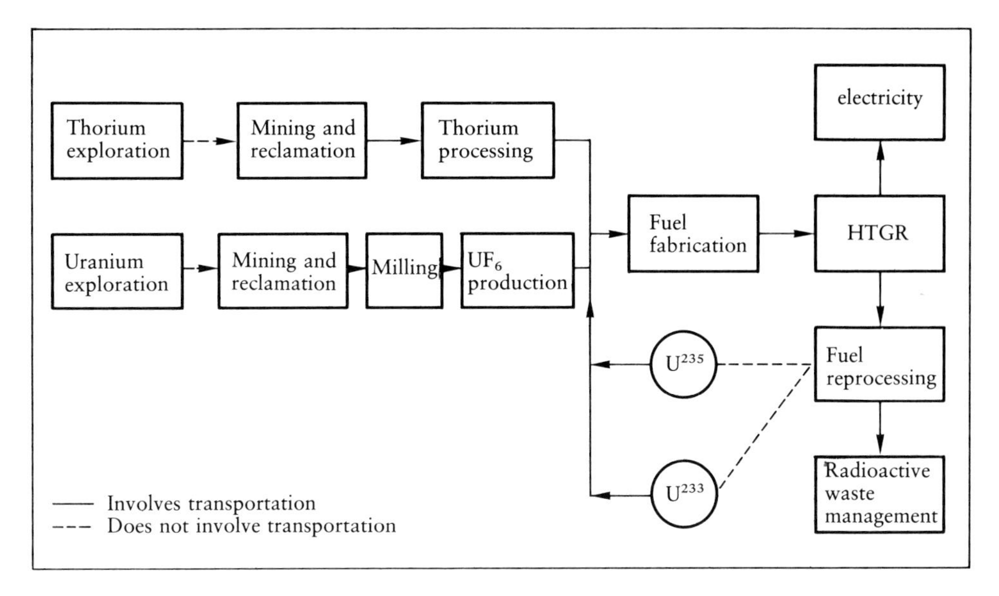

Fig. 5-9b: High temperature gas reactor fuel cycle

Source: Project Interdependence: US and World Energy outlook through 1990, A Report printed by the Congressional Research Service, U.S. Government Printing Office, Washington D.C., November 1977.

conium alloys have proven suitable for cladding materials for the fuel rods, as they have a low absorption of thermal neutrons, and display good corrosion resistance as well as adequate strength.

Plutonium can also be used as a fuel in an LWR (see 5.822.22). In terms of quantity, ^{239}Pu and, to a lesser extent, ^{241}Pu are significant (see 4.214). By the beginning of 1977, about 15 t of plutonium had been worked up into SFBR and LWR fuel rods in the western world, and it had been demonstrated that the safety requirements necessitated by the radiotoxicity of plutonium can be met (92). The main reasons for using plutonium as a fuel in LWRs are that recycling reduces the amount of plutonium which comes into final disposal depots, i.e. the long-term storage problem can thus be considerably lessened, and the use of plutonium improves the efficiency of uranium utilization. (The fuel element concepts of the various reactors have been treated under the corresponding reactor type: 4.212, 4.213, 4.214 and 4.215.)

Fig. 5-9a: Light water reactor fuel cycle

Source: Project Interdependence: US and World Energy Outlook through 1990, A Report printed by the Congressional Research Service, U.S. Government Printing Office, Washington D.C., November 1977.

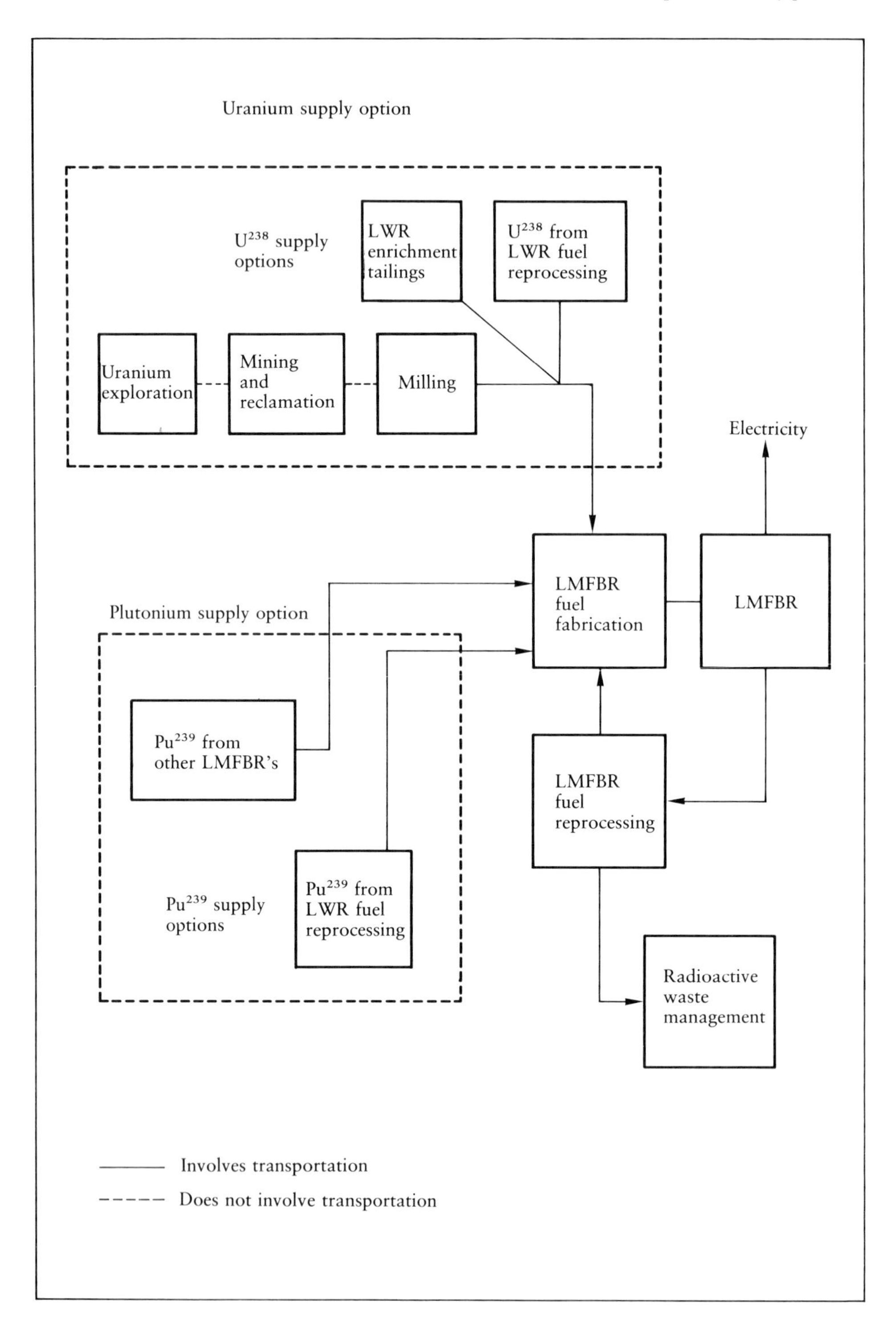

Uranium supply option
U^{238} supply options
LWR enrichment tailings
U^{238} from LWR fuel reprocessing
Uranium exploration
Mining and reclamation
Milling
Electricity
LMFBR fuel fabrication
LMFBR
Plutonium supply option
Pu^{239} from other LMFBR's
LMFBR fuel reprocessing
Pu^{239} supply options
Pu^{239} from LWR fuel reprocessing
Radioactive waste management
Involves transportation
Does not involve transportation

5.822 Disposal

5.822.1 Disposal concepts

While a fuel element is in the reactor, the fissionable isotopes (e.g. ^{235}U) in the fuel are consumed, and neutron-absorbing fission products are formed. The limited mechanical stability of the cladding also limits the time the element can stay in the reactor. In other words, for both physical and security reasons, the fuel elements have to be changed about every 3 years. It has proved convenient to exchange 1/3 of the elements each year. After this time in the reactor, the composition of the fuel elements has been considerably changed. For a PWR type LWR, the starting composition of the fuel is 3.3% ^{235}U and 96.7% ^{238}U, burn-up 33 MWd/kg. 1000 g of spent fuel of this type contains 945.0 g ^{238}U, 4.2 g ^{236}U, 8.6 g ^{235}U, 5.3 g ^{239}Pu, 2.4 g ^{240}Pu, 1.2 g ^{241}Pu, 0.4 g ^{242}Pu and about 0.4 g other actinide elements. It also contains 32.5 g fission products. In other words, the amount of ^{235}U in spent fuel is comparable to the amount in natural uranium (0.86% ^{235}U in spent fuel and 0.72% ^{235}U in natural uranium). The spent fuel elements, which are removed after about 3 years in the reactor, are thus not yet radioactive waste. In addition to the highly radioactive fission and activation products, the actual waste, they still contain considerable amounts of usable fission and breeding materials. If these are returned of the reactor, one speaks of a closed fuel cycle; if not, one sometimes speaks of an open or throwaway fuel cycle.

The disposal of nuclear power plants begins with the treatment of the spent fuel elements, which are highly radioactive and which therefore must be constantly cooled because of the heat produced by the radioactive decay. In general, the spent fuel rods are first put into intermediate storage in a water tank, generally inside the reactor building. The walls and bottom of the tank are made of stressed concrete, and the fuel elements are placed into racks under the water. The normal racks are built so that the arrangement of fuel remains sub-critical; there are also compact racks in which the fuel elements are closer together, but are separated by a neutron absorber. The use of compact racks can increase the storage capacity in a power plant severalfold, and is now practiced in several countries.

After about 12 months, the radioactivity has decreased to about 2% of its original value. In this time, the short-lived isotopes have decayed. (For example, the gas ^{131}I has a half-life of about 8 days.) Consequently, the heat of decay is also greatly diminished.

Fig. 5-9 c: Liquid-metal fast breeder reactor fuel cycle

Source: Project Interdependence: US and World Energy Outlook through 1990, A Report printed by the Congressional Research Service, U.S. Government Printing Office, Washington D.C., November 1977.

Because the specific activity and heat production of the fuel elements decrease by orders of magnitude in the first few years after their removal from the reactor, but only slowly thereafter, it is practical to put them in intermediate storage for a few years. In general, they are stored 5 to 7 years before they are recycled. (Intermediate storage can be either at the site of the power plant or elsewhere.) This step considerably simplifies the subsequent disposal steps.

There are basically three things which can be done with spent fuel elements: they can be kept in intermediate storage, with the intention of recycling them later (long-term intermediate storage); they can be conditioned for final disposal (conditioning spent fuel elements); or they can be recycled to retrieve the energy raw materials still contained in them, with a concomittant conversion of the radioactive waste to a form suitable for final disposal.

Long-term intermediate storage means that the fuel elements which are to be recycled at a later date are packed in containers and stored, either above or below the ground. There are schemes for both dry and wet intermediate storage, and it has been demonstrated that it is possible to build dry intermediate storage facilities which are inherently safe. It is possible, in other words, to store spent fuel elements for relatively long periods, certainly for several decades, in air-cooled depots. The cooling is thus accomplished by convectional exchange with the surrounding air, and is independent of the functioning of technology or human reliability. It is also considered possible to build an inherently safe wet intermediate storage depot, but it must still be shown how a leaking tank can be avoided with absolute certainty, and whether the heat will still be carried off even if the cooling fails (93). Wet or dry depots can be built at the site of the nuclear power plant, the recycling plant, or elsewhere. The present depots are either above the ground or close to the surface, but it is in principle possible to build them in a deeper geological formation, e.g. in a mine, where the heat can be dispersed through the geological medium. However, as in any form of intermediate storage, the fuel elements stored in a geological formation would have to be easily accessible.

The positive aspect of long-term intermediate storage is that, although the valuable fissionable and breeding materials are not immediately returned to a reactor, neither are they lost. In this way, too, no possible option for a final disposal scheme is precluded, so the plan allows time for the research and development which still need to be done in this area. However, long-term intermediate storage in no way solves the problem, but only postpones the further treatment of the spent fuel (either its final disposal or its recycling). In the meantime, however, large quantities of spent fuel elements would be collected in the intermediate depots, which could be a long-term security problem. Thus it would appear, from our present standpoint, that in the long run, only the other two plans (conditioning for direct disposal or recycling) are practicable (compare Fig. 5-9a).

5.822.2 Treatment of spent fuel elements

5.822.21 Conditioning

As mentioned in 5.822.1, spent fuel elements are generally put into intermediate storage for 5 to 7 years before further treatment, regardless of the nature of that treatment, because it considerably simplifies the subsequent steps. At the end of the intermediate storage, the spent fuel is transported to a conditioning plant for further treatment.

There are strict international safety regulations for the transport of irradiated fuel elements and radioactive waste, established by guidelines from the IAEO. The transport containers must also meet strict standards of stability, temperature resistance and impermeability. The number of times the radioactive material must be transported can be reduced by spatial concentration and partial integration of the individual steps of disposal: conditioning and final disposal of the radioactivity, or reprocessing synthesis of new fuel elements (refabrication), conditioning of the radioactive wastes, and final disposal at a single site.

In contrast to the situation for recycling (see 5.822.22), there is no standard method for conditioning spent fuel elements for final disposal, and furthermore, no experience with industrial-scale conditioning has been accumulated (94).

In principle there are a number of methods for conditioning spent fuel elements. The simplest consists of packing whole fuel elements into suitable containers, without any further treatment, such as removal of the gaseous fission products or mechanical breaking into smaller pieces. Another method is to pack individual fuel rods, which allows a higher packing density, and thus reduces the number of containers which must be transported to the final disposal site. Lead can be used as filler between the fuel elements or fuel rods, because it is an effective shielding material for gamma rays, and has a relatively high heat conductivity. It also serves as an additional barrier to penetration by ground water. Another complex method of conditioning involves the conversion of the fuel elements into stable glass or ceramic products. In this case, the fuel elements are broken up and the fission gases are removed. In other words, it is conceivable that methods for conditioning spent fuel elements will have a certain similarity to some of the steps in reprocessing and waste treatment (see 5.822.22).

To date, no spent fuel elements have ever been conditioned to a product suitable for final disposal, and furthermore, there are still no binding and internationally accepted criteria for suitability. One reason for this is that there are different types of fuel elements from different kinds of reactors (e.g. light or heavy water reactors), and besides, for a given type of fuel element, the amount of heat produced depends on the degree of burn-up and the position in the core, so that there are large differences from one element to the next. Finally, the necessity for using multiple barriers (immobility barriers, container barriers, geological barriers) and the consequences for the safety of the depot depend heavily on the geological formation and on site-specific properties of the depot (99). There are now demonstration projects for the conditioning of spent fuel elements in Canada, the USA and Sweden (100–102).

Experiments on the conditioning of spent fuel elements have not yet revealed any scientific or technical reasons why the direct final disposal of these elements should be impossible. However, it will no doubt be necessary to continue the work on the development of optimal methods and to demonstrate the long-term stability of the containers under the conditions which occur in the final depots (94).

The potential radiological danger from conditioned fuel elements is higher, due to their higher uranium and plutonium contents, than from the highly active waste which would be generated if they were recycled (see 5.822.22). (Naturally occuring plutonium is practically nonexistent in the earth's crust, although traces of ^{242}Pu have been detected in bastnaesite from California. It was present when the earth was formed, and has not yet decayed in the intervening $5 \cdot 10^9$ years.) Direct disposal of spent fuel elements would entail the burial of large amounts of plutonium (103), which presents a genuinely long-term problem. The half-life of ^{239}Pu is 24400 years, and the technical/geological barriers would have to be planned accordingly (see 5.822.3). In addition, direct final disposal of the spent fuel would be tantamount to throwing away valuable fissionable and breeding materials (compare Fig. 5-9a). (This throwaway cycle is also sometimes called an "open" fuel cycle.) However, fast breeders can only be employed if the fuel elements from light water reactors are recycled to recover the ^{239}Pu and much lower quantity of ^{241}Pu which is bred from the ^{238}U (see 4.214). Likewise, thorium high-temperature reactors (uranium-thorium cycle) cannot be optimally utilized without recycling the fissionable ^{233}U bred from the ^{232}Th (see 4.213.3) (95).

5.822.22 Reprocessing

In the course of recycling, it is possible to process the radioactive waste, separately from the valuable materials, and to prepare it for disposal. In other words, a reprocessing plant serves to separate uranium, plutonium and radioactive waste. An advantage of this is that, as mentioned in 5.822.21, the long-term problem of final disposal which would arise in the permanent storage of large amounts of plutonium, is reduced by about three orders of magnitude. In addition, the reprocessing of fissionable and breeding materials makes possible an optimal utilization of valuable energy raw materials (uranium and thorium). However, it is a widely-held opinion that recycling of fuel would simplify the misuse of plutonium for construction of nuclear weapons, even though the International Nuclear Fuel Cycle Evalution Study came to the conclusion that the problem of non-proliferation of nuclear weapons is primarily a political problem for which there is no technological or scientific solution (see 5.83).

A light water reactor using fuel which is initially 3.3% ^{235}U will produce about 25 to 30 t spent fuel annually for each 1000 MW electric power. The mean thermal burn-up is 30000 to 40000 MWd per t uranium (some of the plutonium is already "burned" during the burn-up), and the radioactivity in the reactor after several months of operation is about 10^{10} curies (91).

As mentioned in 5.822.1, the fuel elements are removed from the core after about 3 years and put into intermediate storage. From there, they are brought to the recycling plant. It is possible, by concentrating and integrating the individual steps of disposal to reduce the number of times radioactive material must be transported. This is especially important for plutonium, which, by integration of the individual steps, would only need to be transported in the form of new or spent fuel elements. Also, location of the recycling plant at the site of the final disposal depot would save the transport of radioactive wastes removed from the spent fuel.

Of a number of reprocessing methods which have developed, the PUREX (Plutonium and Uranium Recovery by Extraction) process, which was developed in the USA, Great Britain and France for military nuclear programs, has been most successful. (It is reported that about 1300 t plutonium has been synthesized and worked by the world's military establishments) (94). In several countries, spent fuel is being recycled by the PUREX process, e.g. the plant at Cap de la Hague, France, has a capacity of 800 t uranium per year, and at Karlsruhe, Federal Republic of Germany, a demonstration plant with a capacity of about 40 t uranium per year has been recycling different kinds of experimental and power reactor fuel elements since 1971.

The PUREX method is based on solvent counter-current extraction (also called liquid-liquid extraction). The technique depends on the fact that the nitrates of uranium and plutonium, which are formed when the fuel is dissolved in nitric acid, are easily extracted into organic solvents, while more than 99% of the salts of the fission products remains in the aqueous phase. The solvent used is tributyl phosphate (TBP).

There are three main parts of the PUREX process: the head end, the extraction and the tail end. In the first step (the "head end"), the fuel elements are mechanically or chemically broken up. After mechanical breaking (which has become most common), the sections of fuel rod – which are about 5 cm long – fall directly from the cutting machine into hot nitric acid. As the mixture is cooked, uranium, plutonium and the solid fission products are dissolved, while the zircaloy or stainless steel cladding is unaffected. In the process of cutting and dissolving the fuel rods, the gaseous fission products are released. The head end process thus produces solids (cladding and structural material), liquids (the nitric acid solution and a condensation of tritiated water) and gases (nitrogen oxides, krypton, xenon, tritium and iodine). Only a part of the tritium is released as gaseous HT or T_2; the larger part is generated during the solvation product as HTO. The most important of the gases are ^{85}Kr, with a half-life of 10.76 years (see 5.43) and the long-lived ^{129}I, wiht a half-life of

$1.72 \cdot 10^7$ years. The recycling plants presently in operation generally release these radioisotopes into the atmosphere, but with large-scale plants, this will no longer be possible, because the amounts would exceed the limits of radiological safety regulations (98). However, adequate methods of retaining these gases have been developed. It is planned, for instance, that the iodine will be precipitated from the waste-gas flow by special filters; krypton can be trapped in a cold trap and then stored in gas bottles until it is mostly decayed. The entire process of cutting and dissolving the fuel rods must be done by remote control in "hot cells" which have concrete shielding walls up to 2 m thick.

In the first step of the extraction process, the uranium and plutonium are transferred together to the organic phase, while somewhat more than 99% of the fission products remain behind in the aqueous phase. This *high active waste*, HAW, is then sent on for further treatment. The next step is the separation of the uranium and plutonium and the removal of the remaining fission products. Extraction can be done in the pulsed colomn, the mixer-settler, and the centrifugal extractor (94).

The final phase, or tail end, is the production of the uranium and plutonium end products, concentrated uranyl nitrate solution and either concentrated plutonium nitrate solution or solid plutonium oxide. The fuel solutions from the extraction cycles are concentrated in evaporators. If plutonium oxide is required, the metal is precipitated from the nitrate solution by adding oxalic acid. The solid plutonium oxalate is converted to the oxide by heating; the process is known as "calcining".

The reprocessing technology has been developed to the point that more than 98% of the uranium, and more than 99% of the plutonium can be retrieved; thus less than 1% of the plutonium goes into the radioactive waste or is left behind in the plant machinery. This residual plutonium is also measured and accounted for; plutonium is not lost (flow control of fissionable materials) (91).

The Zwisint process is one way to produce mixed-oxide fuels from uranium and plutonium (91). First the two metals are mixed in a specific ratio. The mixture of powders is then converted to pellets in a cold press; finally they are sintered at about 2000 K. The specifications for the cladding (the material and the welding process) are the same for fuel rods with $(U, Pu)O_2$ as for UO_2 fuel.

The purpose of waste treatment is to bind the radioactive waste produced in the reprocessing and production of mixed-oxide fuel elements to solid, stable matrix materials. This is to protect the radionuclides for a sufficient time against external (leaching, mechanical stress) and internal (radiation, thermal stress) influences. In addition, because the volume of the radioactive waste is much increased by the recycling process (e.g. due to the volume of added chemicals), waste treatment is intended to reduce the volume. In other words, the treatment of radioactive waste is intended to concentrate it and to convert it to a form suitable for final disposal (see 5.822.3).

At the end of 1979, the following amounts of fuel elements had been recycled: about 800000 t from the military, about 28000 t from gas-graphite reactors (magnox), and about 900 t from light water reactors (94). The plutonium content of military fuels is about 0.3% rather than about 1% as in LWR fuel, but the difference is not significant for purposes of recycling, so that experience gained in military plants can be applied to the recycling of LWR fuels.

The PUREX process is also suitable for recycling oxide fuels from breeder reactors (SFBRs). The spent fuel elements differ from those of LWRs in the following ways:
1. SFBR fuel contains much more plutonium (see 4.214); this affects the solution process in the head end stage and makes it necessary to take extra precautions against the creation of a critical configuration during processing.
2. The burn-up and the decay heat of spent SFBR fuel elements are greater, which necessitates more cooling during transport and handling. In order to limit the irradiation of the organic solvent, tributylphosphate (TBP), during the process, it is possible to dilute the highly active elements from the inner fission zone of the reactor with less active elements from the radial breeding zone during work up. By the choice of the proper ratio of highly active to less active elements, the radioactivity of the mixture can be kept down to values which have been mastered in the recycling of LWR fuel elements.
3. The outer walls of the SFBR fuel elements and the cladding of the fuel rods are made of stainless steel, which means that they are much harder than the zircaloy cladding of the LWR fuel rods. This requires modification of the mechanical cutting process. However, previous experience with the recycling with oxide breeder fuels (e.g. in Dounray, Scotland, with an annual capacity of 10 t), indicates that SFBR fuel can be recycled by the PUREX method, with slight modifications (104).

Thorium high-temperature reactors (uranium-thorium cycle) also require recycling for optimal utilization, because the fissionable ^{233}U bred from the ^{232}Th is recovered during the reprocessing of spent fuel. Due to the structure of THTR fuel elements and the presence of thorium as breeding material, the THOREX process is used to recycle them. This process differs in a few steps from the PUREX process. In the uranium-thorium cycle, the heavy-metal content of spent fuel is mostly ^{232}Th (93.1–96.1%). The fissionable material is ^{233}U and ^{235}U (the ^{233}U was bred), which makes up 2.6% to 3.3% of the heavy metals. The plutonium isotopes and actinides make up less than 1% of the heavy metal. In the head end phase, the graphite is removed by combustion. The subsequent steps are dissolving the mixture of fissionable and breeding materials and fission products, solvent extraction by the THOREX process, and final purification of the solutions of fissionable and breeder materials. The products are solutions of very pure uranium, thorium and plutonium nitrate, and of radioactive wastes. The radioactive waste solutions are converted to products suitable for final disposal. The fissionable and breeding materials can be refabricated into fuel elements. At present a pilot plant is being built in the Federal Republic of Germany, called JUPITER (Jülicher Pilotanlage für Thorium Element Reprocessing). It will have a capacity of 2 kg heavy metal oxide per day, taken from the spherical fuel elements of the AVR (see 4.213.3) (104).

5.822.3 Final disposal of radioactive materials

There are three classes of radioactive waste, distinguishable by their specific activity:
- Low active waste, LAW, with less than 10^{-1} Ci/m^3;
- Medium active waste, MAW, with 10^{-1} to 10^3 Ci/m^3;
- High active waste, HAW, with more than 10^3 Ci/m^3.

The waste is also classified by its physical state – as solid, liquid or gaseous waste (105).

Radioactive waste is generated at every stage of the nuclear fuel cycle in which radioactive materials are worked or used. A reprocessing plant generates essentially every kind of waste which is generated in other parts of the fuel cycle, although the amounts, composition and specific activities differ.

Radioactive material will also have to be disposed of after a nuclear facility is shut down. It is estimated that the operational lifetime of a nuclear power plant will be 3 to 4 decades, depending on the economy and further technical developments. (Research and prototype reactors will be shut down after a shorter period.) The activity inventory of a nuclear power plant at the time it is shut down consists mainly of fuel elements and operational supplies (e.g. filters), media (e.g. cooling water) and corrosion products, as well as buildings and components. After the final shut down of a reactor, the fuel elements, and operational supplies and media are removed. (After forty years of operation and one year of cooling off, a 1100 MW power plant has about 10^7 Ci (91). Depending on the technical and economic conditions, the shut-down plant can be either securely sealed off or demolished. Modern installations are built so that individual components, except for the pressure vessel, can be removed intact from the reactor building, i.e. demolition within the installation can largely be avoided. There has been little experience with the closing and disposal of nuclear power plants or other nuclear installations, but no insoluble difficulties are expected.

The purpose of waste conditioning is to bind the radioactivity in a solid and stable matrix. Depending on the type of waste, the suitable matrix material can be glass, cement, bitumen or PVC (106-108).

Compared to the disposal of highly active wastes (from reprocessing, or the unprocessed fuel elements), the disposal of low and medium activity wastes is relatively easy and is already a routine practice in many countries, e.g. the Federal Republic of Germany, France, Great Britain, Italy, Canada, Sweden, the USA and the USSR. Low-activity wastes can be solidified in bitumen and brought in barrels to the permanent depot. After a cooling period of 2 to 3 years, the same can be done with medium-activity wastes. In the Federal Republik of Germany, for instance, between 1967 and 1976 60500 barrels (200–400 liter capacity) of low-activity waste were buried at 750 m depth in the Asse salt mine (in Braunschweig). Between 1972 and 1976, about 1300 barrels of medium activity waste were added to this, at 511 m depth (94). Certain kinds of low-activity waste can also be sunk in deep oceans, with international supervision (London Convention of 1972). The NEA of the OECD has conducted such internationally supervised sinkings, almost yearly since 1967, at deep places (400–5000 m) in the North Atlantic (109).

In contrast to the disposal of low and medium activity wastes, the permanent disposal of highly active waste is a difficult problem, and no country has yet undertaken it. The highly active materials not only have more than 10^3 Ci/m^3, but they are also thermally hot, due to the heat of radioactive decay. Another problem is that some of the isotopes have relatively long half-lives.

The conversion of highly active wastes to glass is internationally considered the most suitable method of stabilizing them. The process consists essentially of the following steps: denitrification, i.e. removal of the free nitric acid in the HAW; drying and calcining of the wastes, i.e. conversion of the fission-product nitrates to the oxides; melting of the calcinate (heating to about 1500 K); pouring the glass melt into stainless steel containers; and welding the containers.

Vitrification (borosilicate or phosphate glass) of the highly active reprocessing waste produced from a 1000 MW (el. power) reactor yields about 5 m^3 vitrified HAW per year (102). (The highly active waste contains more than 99% of the total radioactivity.) There is also on the order of 200 m^3 solidified medium-activity waste and 150 m^3 low-activity waste (94). The development of methods for solidification of wastes continues.

A geological formation in which highly active wastes can be buried must meet a number of requirements. It can have no contact with ground water, and must be tectonically stable. It must have adequate heat conductivity, hardness and plasticity, and sufficient depth, and the overlying layers must be impenetrable to radionuclides (94). Since ground water is practically the only route for contamination of the environment by radionuclides from the permanent disposal site, it should be placed in a geological formation which has no contact with ground water, and in an area where there is little tectonic activity, and the probability of earthquakes is low. (If possible, there should be no valuable raw materials in the neighborhood of the permanent depot.) The geological formation should be capable of absorbing and dispersing the heat generated by the highly radioactive waste, without any serious loss of mechanical strength. The depth of the permanent depot should be chosen so that certain events, such as a meteorite impact or an ice age, would not release radioactivity. Finally, the overlying layers should be such that if, at some later time, ground water should enter the depot, the transport of the ground water to the surface would be sufficiently delayed that only inconsequential amounts of radioactive elements would enter the biosphere (94). There is no known geological formation which meets all these requirements in optimal fashion simultaneously. Salt, granite and clay deposits are considered favorable formations (110, 111).

As mentioned above, the permanent disposal of highly active waste is not being practiced in any country, i.e. vitrified highly active wastes are being put into intermediate storage. By the end of 1979, the Atelier Vitrification Marcoule plant, in France, had vitrified 140 m^3 of HAW, and the resulting 60 t of glass, in 180 containers, is in an air-cooled intermediate depot. In France, salt and granite deposits are being considered for final disposal. In the USA, it was suggested at the end of the fifties that radioactive wastes be permanently buried in geological salt deposits. A demonstration experiment with spent fuel elements was begun in 1969 in a salt mine in Lyons, Kansas (Project Salt Vault). In the following years major emphasis was placed on studies of the properties of salt mines as permanent depots, and recently, the program was extended to other geological formations, such as granite, basalt, tuff and clay. In the CLIMAX project, 14 encapsulated fuel elements were deposited in a granite formation. (The experiments in the CLIMAX mine are limited to 5 years.) A similar demonstration is planned for the basalt formation in Hanford, Washington. In the Federal Republic of Germany, the concept for permanent disposal of radioactive wastes in salt mines was solidified in the early sixties. Experiments to determine the effects of heat on rock salt are being carried out in the Asse salt mine in Braunschweig. In Great Britain, granite, clay and rock salt are considered suitable geological formations for permanent disposal of radioactive wastes. Canada plans to dispose of spent fuel elements or reprocessing wastes in granite. The major emphasis of all present work is on geological experiments. The USSR has an broad research and development project in which a large number of possibilities for disposal of radioactive wastes in geological formations is being examined (94).

There is a large difference in the volume of waste produced by conditioning of spent fuel elements for direct disposal and reprocessing. The conditioned fuel elements have about 9 times the volume of the highly active reprocessing wastes; however, reprocessing produces a much larger amount of medium and low activity waste (94).

The comparison of the amounts of waste produced by these two methods is not complete without consideration of the amounts of waste produced in the remainder of the fuel cycle. According to the INFCE report, the amount of low and medium activity waste from the reactors is much greater than from reprocessing. Furthermore, abstention from reprocessing and recycling would lead to much larger amounts of waste from the additional mining and processing of uranium which would be required to supply the reactors; and the radiological effects of these activities make up a significant fraction of the total radiological load (102).

As discussed in 5.822.21, the separation of plutonium and uranium during reprocessing and their recycling as fuel means that a far smaller amount of these elements comes into the permanent disposal site than if the elements are not reprocessed. However, since it is precisely these two elements and their fission products

---

Fig. 5-10: Relative radiotoxicity of the highly active reprocessing waste from 1 t uranium/plutonium nuclear fuel, compared to that of the cladding and uranium ore tailings

Source: Federal Ministry of Research and Technology (Ed.), Zur friedlichen Nutzung der Kernenergie, Bonn 1978.

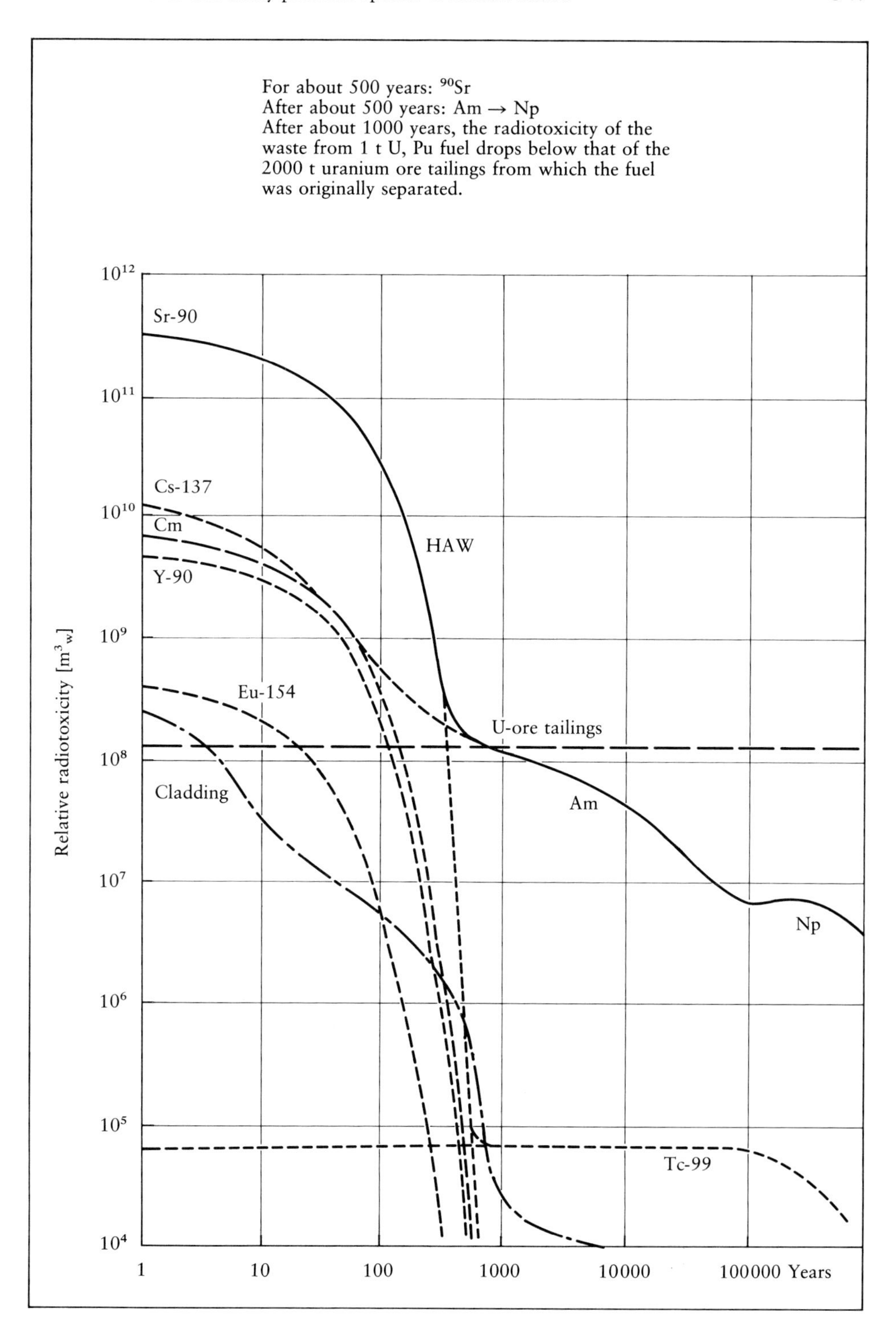
For about 500 years: ^{90}Sr
After about 500 years: Am → Np
After about 1000 years, the radiotoxicity of the waste from 1 t U, Pu fuel drops below that of the 2000 t uranium ore tailings from which the fuel was originally separated.
Sr-90
Cs-137
Cm
Y-90
HAW
Eu-154
U-ore tailings
Cladding
Am
Np
Tc-99
Relative radiotoxicity [m^3_w]
10^{12}
10^{11}
10^{10}
10^9
10^8
10^7
10^6
10^5
10^4
1
10
100
1000
10000
100000 Years

which are responsible for the long-lived radiotoxicity of spent fuel elements, the non-reprocessed fuel elements contain much more long-lived activity than the wastes from their reprocessing.

Fig. 5-10 shows the relative radiotoxicity[1)] of the highly active reprocessing wastes from 1 t uranium/plutonium fuel (91). For the first 500 years, the fission products, especially ^{90}Sr and ^{137}Cs dominate. The relative radiotoxicity declines to about 10% of the original level in the first 100 years, and to less than 0.5% after about 500 years. Thereafter, the long-lived alpha-emitting transuranium elements dominate, especially americium, and later its alpha-decay product neptunium. As a comparison, the figure also shows the relative radiotoxicity of the 2000 t uranium ore tailings which result from the extraction of 1 t of nuclear fuel – assuming recycling of uranium and plutonium. Here the predominant elements are the daughter nuclides of the uranium decay scheme, especially ^{226}Ra and ^{230}Th. The composition of the ore tailings reflects the equilibrium concentrations of the uranium decay series, and its activity and radioactivity, like those of natural uranium, are practically unchanging (112).

Comparing the relative radiotoxicity of highly active reprocessing wastes and conditioned fuel elements, the former has decayed after about 1000 years to the level of the uranium ore tailings, while the latter does not reach this level for approximately 10^6 years. In other words, if the uranium and plutonium are not removed from spent fuel elements, the potential danger to the environment from the permanent waste depots extends over a far greater time span.

A comparison of the relative radiotoxicities resulting from the two concepts for permanent disposal is not a sufficient basis, however, for judging the relative long-term risks. The amount and types of radionuclides, their relative radiotoxicity and their half-lives undoubtedly play a role. The physical-chemical state of the fission products and the radionuclides, and the effectiveness of barriers between the wastes and the biosphere will also be important in the event of their being leached out, which is a possible danger (94). It has been shown that the time required for radionuclides to be transported from a disrupted permanent depot into the biosphere is very long, on the order of 10^6 years. Because of the extremely low diffusion rates of the radionuclides through the earth, however, the radiological hazard from an accidental release of radioactive material from a sealed depot would not come immediately from the original radionuclides, but from their decay products. (In this amount of time, not only the fission products but also important actinides like plutonium and americium would be essentially decayed.) The radiological load on the generations living at that time would be almost completely due to the ^{226}Ra, a daughter isotope of ^{238}U. It follows, too, that unprocessed fuel elements would lead

[1)] The volume of water (m_w^3) or air (m_a^3) which would be needed to dilute a radioactive material to the maximum permissible concentration, MPC, in water or air, respectively, is often taken as a measure of the relative radiotoxicity of the material. The relative radiotoxicity changes as the material decays, but does not disappear until it has decayed completely.

to higher radiation doses for the biosphere at that time than would reprocessing wastes, which contain very little uranium. However, regardless of the form in which fuel elements are handled, the radiation doses in the distant future ($> 5 \cdot 10^5$ years) which might be released by damage to the permanent waste depots would lie considerably below the natural background radiation. The higher load mentioned above due to ^{226}Ra would thus be slight.

In addition to the two concepts for disposal discussed so far, conditioning the elements for direct disposal or reprocessing them, a number of unusual possibilities for disposal of highly active wastes have been proposed. These include the transmutation of radioactive wastes, for instance, neutron activation of long-lived fission products to convert them into short-lived or stable isotopes; or fission of the heavy elements (e.g. actinides) into short-lived isotopes. Chemical separation of the isotopes from the highly active waste would be a prerequisite to their transmutation; in addition, the process (e.g. in reactors) would have to be essentially complete, or it would again generate long-lived, highly active waste (see 4.222). Another thought is to shoot highly active waste into space (113). However, for the foreseeable future, this will probably be too expensive and, given the possibility of accidents, too risky. Other possibilities which have been discussed are sinking wastes in polar ice and burying it in wells drilled in the sea floor.

The available data suggests that the problem of permanent disposal of highly active material can be solved. However, it will require intensive development work, especially in those countries which already have comprehensive nuclear energy programs (94).

5.83 The problem of non-proliferation of nuclear weapons

Since the oil crisis of October, 1973, the opinion is becoming more widespread in many countries that the problem of energy supply could be solved by nuclear energy. This is associated, however, with the fact that the problem of non-proliferation of nuclear weapons is still an important issue in international politics (114–118). The Treaty on the Non-Proliferation of Nuclear Weapons (TNP or NPT) is of central importance to the international efforts to prevent the further proliferation of nuclear weapons (see 3.343.1).

The NPT (see 7.2) was signed simultaneously in Washington, London and Moscow on July 1, 1968, by the executives of the three signatory powers, the United States of America, the United Kingdom of Great Britain, and the Union of Soviet Socialist Republics. The treaty went into effect on March 5, 1970. The federal government of the Federal Republic of Germany signed the treaty on November 18, 1969; the treaty was ratified on March 8, 1974, and went into effect for the Federal Rep. of Germany on May 2, 1975, after the copies had been deposited in Washington and London. (No copy was deposited in Moscow, because the Soviet Union refused to accept the inclusion of Berlin (West) in the treaty.) With its copies of the ratification document, the government of the Federal Rep. of Germany made a statement including the point that "no provision of this treaty is to be interpreted in such a way as to hinder the further development of European unification, in particular the creation of an European Union with the corresponding competencies" (the so-called European option). (It should be mentioned that the Federal Republic of Germany had already signed the Brussels Treaty in October 1954, renouncing the production of nuclear, biological or chemical weapons.)

The goal of the NPT is to prohibit the transfer of nuclear weapons from nuclear-weapon countries to non-nuclear-weapon countries, and to prevent the latter from producing or procuring nuclear weapons (120). Article I of the NPT states that the *nuclear powers* which are signatory to the treaty will not transfer any nuclear weapons to the *non-nuclear powers* (see 7.2). Article II prohibits the production or acceptance of nuclear weapons by the non-nuclear powers. On the basis of this treaty (Articles I and II), the world is divided into signatory and non-signatory powers, and both groups include nuclear and non-nuclear powers.

The fact that India exploded a nuclear bomb on May 18, 1974 led to questioning of the goals of the NPT. Although three of the six nuclear powers, France, the Peoples' Republic of China and India, have not signed the treaty and do not intend to, the following has been achieved: According to the IAEA, by May 1, 1980, 113 countries had signed the NPT (120). These are listed in Table 5-3. (The IAEA does not consider India a nuclear power.)

In Article III of the NPT, the non-nuclear powers agree to submit all nuclear installations in their territories to the safeguards of the IAEA. (Nuclear powers do not have to submit to these.) These safeguards shall not, according to Article III,3 and IV, inhibit the economic or technological development of the signatory parties, nor the international cooperation in the area of peaceful nuclear activities. The goal of the safeguards is to prevent theft or misuse of fissionable material for non-peaceful purposes. Basically the IAEA control measures are intended to oversee the flow of fissionable material (entry and exit) through a limited number of strategic points in a nuclear installation (121).

The so-called "sensitive installations", especially the enrichment and reprocessing installations, have come into the center of attention in connection with the problem of non-proliferation of nuclear weapons (122–127). With an enrichment plant, a country which has natural uranium has the capacity to produce nuclear fuel (uranium enriched to about 3% ^{235}U content) (see 5.821). The production of highly enriched "nuclear-weapons grade" ^{235}U is extraordinarily laborious, however, and can only be accomplished by a highly developed industry (128). A reprocessing plant, as discussed in 5.822.22, serves to separate uranium, plutonium and radioactive waste. It has been estimated that it would be possible, given a functional reprocessing plant, to make a nuclear weapon within a few weeks, if the accompanying preparations had been made (129). T. B. Taylor has warned of the danger that ter-

Table 5-3: The international nuclear system

States Party of NPT with NPT Safeguards Agreements in force, and NPT nuclear weapon states		
Afghanistan	Indonesia	New Zealand
Australia	Iran	Nicaragua
Austria	Iraq	Norway
Belgium	Ireland	Paraguay
Bulgaria	Italy	Peru
Canada	Jamaica	Philippines
Costa Rica	Japan	Poland
Cyprus	Jordan	Portugal
Czechoslovakia	Korea, Republic of	Rumania
Denmark	Lebanon	Samoa
Dominican Republic	Lesotho	Senegal
Ecuador	Libyan Arab Jamahiriya	Singapore
El Sàlvador	Liechtenstein	Sudan
Ethiopia	Luxembourg	Surinam
Fiji	Madagascar	Swaziland
Finland	Malaysia	Sweden
Gambia	Maldives	Switzerland
German Dem. Rep.	Mauritius	Thailand
Germany, Fed. Rep. of	Mexico	USSR*
Ghana	Mongolia	UK*
Greece	Morocco	USA*
Holy See	Nepal	Uruguay
Honduras	Netherlands	Yugoslavia
Hungary	(including Neth. Antilles)	Zaire
Iceland		
States Party to NPT for which NPT Safeguards Agreements are not yet in force		
Bahamas	Guinea-Bissau	Sierra Leone
Bangladesh	Grenada	Somalia
Barbados	Guatemala	Sri Lanka
Benin	Haiti	St. Lucia
Bolivia	Ivory Coast	Syrian Arab Republic
Botswana	Kenya	Togo
Burundi	Lao People's Dem. Rep.	Tonga
Central African Republic	Liberia	Tunisia
Chad	Mali	Turkey ■
Congo	Malta	Tuvalu
Democratic Kampuchea	Nigeria	United Rep. of Cameroon
Democratic Yemen	Panama	Upper Volta
Egypt	Rwanda	Venezuela ■
Gabon	San Marino	Taiwan ■
Non-NPT States in which IAEA Safeguards Agreements are in force on all nuclear activities		
Argentina	Chile	Cuba
Brazil	Colombia	Dem. People's Rep. of Korea

Table 5-3 (continued): The international nuclear system

Non-NPT States having no significant nuclear activities (except for France* and P.R. of China*)		
Albania	Guyana	Sao Tome and Principe
Algeria	Kuwait	Saudi Arabia
Angola	Malawi	Seychelles
Bahrain	Mauritania	Trinidad and Tobago
Bhutan	Monaco	Uganda
Burma	Mozambique	United Arab Emirates
Cape Verde	Nauru	United Rep. of Tanzania
Comoros	Niger	Viet Nam
Djibouti	Oman	Yemen Arab Republic
Equatorial Guinea	Papua New Guinea	Zambia
Guinea	Qatar	
States in which certain nuclear activities are not under IAEA Safeguards		
India	Pakistan	Spain
Israel	Rep. of South Africa	

■ All nuclear activities currently under IAEA safeguards (non-NPT).
* Nuclear Weapon States.
Author's notes: In the IAEA-Bulletin India is not a nuclear power.

Sources: International Atomic Energy Agency, Bulletin, Vol. 22, No 3/4, 1980.
International Atomic Energy Agency, Bulletin, Vol. 23, No 2, 1981.

rorists might make homemade plutonium bombs. Others are of the opinion, however, that the mechanical difficulties of this, aside from the difficulty of obtaining the several kg of plutonium, would be too great (130–132).

It is clear that the NPT (Articles III,3 and IV) does not inhibit the development of nuclear energy for peaceful purposes. However, it is a widely held opinion that the proliferation of "sensitive installations" tends to blur the line between "military" and "civilian" nuclear technology (133–136). In addition, more than 20 countries have nuclear power plants in operation, and plants are under construction or on order in still more countries (compare Table 4-2).

Because of the danger of proliferation of nuclear weapons, several fuel cycles have been discussed which are of interest from this point of view. In other words, technical solutions have been sought which would prevent or at least lessen the danger of proliferation (137–139). The uranium-thorium cycle is significant in this context (see 4.213.3). If one used a medium uranium enrichment (about 20% ^{235}U) in a thorium high-temperature reactor, only about 10% as much plutonium would be produced as in a LWR, due to the lower ^{238}U content of the fuel (assuming the same power outputs) (140–144).

F. von Hippel of Princeton University (USA) has suggested a novel fuel cycle concept: the fissionable ^{233}U bred from thorium should be mixed (denatured) with the practically non-fissionable ^{238}U, and this mixture would be used to make new fuel elements. Since this denatured fuel cannot be separated into its components by chemical methods, the method would have certain advantages with respect to proliferation resistance (139, 140, 145). However, in highly enriched form, not only ^{239}Pu, but also ^{233}U and ^{235}U can be used to make nuclear weapons. Fuel consisting of ^{238}U enriched with 12% to 20% ^{233}U or ^{235}U can be relatively easily enriched to 90% by modern methods, e.g. with ultracentrifuges, and thus used to produce nuclear weapons (146, 147).

Because of the danger of further proliferation of nuclear weapons, the Nuclear Suppliers Groups (NSG) met in June, 1975, at the suggestion of the United States, and agreed on binding regulations for the export of nuclear installations (148). These regulations are intended to insure the non-proliferation of nuclear weapons. The NSG originally included the USA, Canada, Japan, the Soviet Union, Great Britain, France, and the Federal Rep. of Germany. In 1976, Sweden, Belgium, Italy, Switzerland, The Netherlands, the German Dem. Rep., Poland and Czechoslovakia were included.

In addition, between October 1977 and February 1980, the International Fuel Cycle Evaluation (INFCE) conference studied established and alternative fuel cycles, not only with respect to the specific hazards of misuse of the technologies and materials, but also with respect to the energy-political significance of these technologies (149, 150). All interested countries and those international bodies concerned with such matters were invited; 66 countries and five international organizations made use of the opportunity. Both industrial and developing countries took part, nuclear and non-nuclear powers, countries with nuclear energy programs at various stages of development, countries with free-market and centrally planned economies, countries which are parties to the Euratom Treaty, the NPT, the Tlateloco Treaty, and countries which have signed none of these, and countries from every geographical region of the earth.

The result of the conference was an extensive technical and analytical study which made no political recommendations (150). In addition to a better understanding for the different approaches of the individual countries to nuclear politics, the INFCE reported essentially the following:

1. The INFCE made it clear that there is no nuclear fuel cycle which would be absolutely resistant to misuse for the production of nuclear weapons. After analysis of all the facts, it was not possible to reach a decision which would also be valid in the future, as to whether one fuel cycle is more dangerous than another with respect to the proliferation of nuclear weapons. (This also applies to the denatured uranium-thorium cycle.)

2. INFCE therefore emphasized that the problem of non-proliferation of nuclear weapons is primarily a political, and not a technical problem. This is particularly

true when governments decide to misuse installations or materials intended for peaceful purposes. In other words, no combination of technical measures can prevent misuse of fissionable material, or serve as a substitute for political actions which would remove the motivation for construction of nuclear weapons. Although the installations for the peaceful use of nuclear energy can be misused to produce "weapons grade" nuclear materials, the construction of facilities specifically for the production of these materials is probably cheaper.

3. For these reasons, INFCE emphasized measures which could be taken at various stages of the fuel cycle to reduce the risk of proliferation of nuclear weapons. These include the further development of technical and institutional measures and of the international safeguards of the IAEA. In practice, all three types of measure should be combined in such a way as to oppose the siphoning off of sensitive materials from nuclear installations.

The misappropriation of nuclear material from reprocessing plants can also be hindered, in principle, by institutional measures and the IAEA safeguards. Technical controls are generally effective, but their primary function is to prevent the theft of material from which "weapons grade" material can be made, rather than to prevent government misappropriation. (It is expected that with further improvements in technology, efficient control, even of large installations will be possible at reasonable cost.) Great stress is also laid on the development of suitable institutional measures and IAEA safeguards, for instance, for the storage of excess plutonium (120).

The storage and transport of spent fuel elements should be carried out according to national and multinational guidelines, and in agreement with the IAEA. For all types of waste from the reprocessing and refabrication, the suitable safeguards can probably be applied and completed before the final disposal. However, this is not true for spent fuel elements, which are directly disposed of. To be sure, the high activity would make misappropriation extremely difficult at first, but the effect would be considerably reduced after several decades, because the activity and the thermal output of spent fuel elements decay rapidly in the first years after they are removed from the reactor. The safeguards for spent fuel elements are relatively simple during transport, deposit in the permanent depot and after the depot is sealed. However, such a depot would have to be guarded for an indeterminant length of time (94, 102).

4. INFCE came to the conclusion that although the non-proliferation of nuclear weapons is an important consideration, it is not the only consideration. Other factors, such as the security of the supply, protection of the environment and economy are also important criteria for the development and use of nuclear energy systems. Different weightings of the individual factors, corresponding to the national situation, will lead to different solutions.

In addition to these basic points of agreement, there were differences of opinion in individual areas. For example, the future uranium demand was estimated at different levels, and countries with large uranium reserves assign a different priority to the employment of uranium-saving reactor systems, such as the fast breeders, than countries without significant uranium reserves, including Japan and several European countries. In any case, the large-scale use of commercial fast breeders is not expected before the turn of the century – or a bit earlier in some industrial countries.

In sum, INFCE came to the conclusion that nuclear energy can be made available to the whole world at the same time that the risk of nuclear weapons proliferation is reduced.

The Second NPT Review Conference, which was held in 1980 in Geneva, related to the problem of non-proliferation. About 80 member countries participated in this Second Conference, which received considerable attention because the participants were not able to agree on a common concluding document, unlike the participants in the First Conference in 1975. One reason was the criticism of some of the developing countries of the London Nuclear Suppliers Group and of the behavior of individual supplier countries, which placed new requirements – sometimes unilaterally, and often ex post facto – on the recipient countries. They not only made access to nuclear technologies more difficult, but in some cases denied it. In short, a number of countries felt that Article IV of the NPT, which guarantees the transfer of technology, was being slighted in practice. In addition, several states criticized the apparent lack of willingness to disarm shown by the superpowers, and the lack of successful disarmament (disarmament is called for by Article IV of the NPT).

In spite of some criticism of the NPT, it remains true, even after the Second Review Conference, that the overwhelming majority of countries in the world community consider the NPT a key factor in the non-proliferation of nuclear weapons.

At this point, it should be mentioned that one advantage of a fusion reactor, should it be realized, is that it would not produce any materials which could be used for the production of weapons (see 4.22 and 5.9).

5.84 The safety of nuclear installations

5.841 Normal operation

Years of generally satisfactory experience with more than 250 nuclear power plants in 22 countries (see Table 4-1 and 4-2) have led to the opinion generally held at present, that nuclear power plants in normal operation are less damaging to the environment than those run on fossil fuels (see 5.3) (130, 151–156). In various countries, nuclear power plants have been in operation for about 30 years, and the total operating time, according to the IAEA, adds up to about 2000 reactor operating years.

Even in normal operation, nuclear power plants and other installations release small amounts of radioactive substances into the air and water. These so-called operational releases cause a certain increase in background radiation, which must be kept below certain values (157, 158).

As far as is presently known, both natural and human sources of radiation contribute to the total risk associated with exposure to ionizing radiation. The extent of the natural radiation dose is an important basis for evaluating the effect of radiation exposure due to modern civilization, because human beings evolved under the influence of background radiation, and are adapted to it. The natural exposure to radiation is generally said to be that amount of radiation from natural sources to which a person is exposed in the place he chooses to live and in the course of his normal activities. The artificial exposure to radiation is the radiation from natural or artificial sources to which humans are exposed as a result of technology (12).

Exposure to natural radiation can be divided into the external and the internal dose. The external dose is caused by cosmic and terrestrial radiation. The former consists of a galactic component and a solar component, and is highly altitude dependent. For example, the annual whole-body equivalent dose in the Federal Rep. of Germany for a continuous outdoor existence varies between 31 mrem[1)] at sea level and 160 mrem at 2964 m (mountain peaks). Terrestrial radiation comes from radioisotopes present in the earth's crust, and it depends on the local geology and minerology. For example, the annual whole-body equivalent dose in the Federal Rep. of Germany due to terrestrial radiation varies locally between a few mrem and about 310 mrem. (In other parts of the world, this value is sometimes exceeded, e.g. in Kerala, India, where it can be 500 mrem or more.) The total external natural radiation dose in the Federal Rep. of Germany, due to cosmic and terrestrial radiation, varies between 31 and 360 mrem annual whole-body equivalent (12). The internal natural radiation load comes from the ingestion and inhalation of naturally radioactive materials, especially potassium ^{40}K, carbon ^{14}C, radium ^{226}Ra and ^{222}Ra, and the fission products of the last two. Potassium and carbon are relatively evenly distributed throughout the human body, so that incorporation of radioactive isotopes of these two elements produces a whole-body irradiation of about 20 mrem/year from ^{40}K and 1.5 mrem/year from ^{14}C. ^{226}Ra and its decay products, however, are concentrated in bones and bone marrow, where they cause an exposure of about 40 mrem/year. The inhalation of gaseous isotopes, especially of radon and its decay products, also leads to an uneven exposure, although here it is primarily the air passages which are affected. The bronchi receive the highest dose, about 80 to

---

[1)] Two important parameters in dosimetry are the energy dose D (radiation absorbed dose) and the equivalent dose H; H takes into account the differences in biological effectiveness of different types of radiation. The unit of the energy dose is the rad (1 rad = 0.01 J/kg), and the unit of the equivalent dose is the rem (radiation equivalent man). Ionizing particles with high ionizing density along their paths (e.g. alpha particles) have a larger biological effect at the same energy dose than particles with lower ionization densities (e.g. gamma and X-rays). This difference is adjusted by introducing a quality factor Q, which is 20 – according to the ICRP – for alpha particles and 1 for gamma and X-rays and beta particles. The product of the energy dose D and the quality factor Q is the equivalent dose H for the particular type of radiation in the tissue under consideration. $H\,[\text{rem}] = Q \cdot D\,[\text{rad}]$.

200 mrem/year. This high dose is a result of the large amount of air passing through the bronchi and the fact that some of the decay products of radon emit alpha particles. The total natural radiation dose, external and internal, is obtained by addition of the individual doses, and comes to about 110 mrem/year at sea level (12, 155).

The radiation exposure from technological sources should also be subdivided into the external and internal doses. The largest contribution to the external dose comes from the use of X-rays for diagnostic purposes and the use of materials which have an above-average content of naturally radioactive isotopes for the construction of buildings. The genetically significant equivalent dose to the gonads is 50 mrem/year in the Federal Rep. of Germany, for example (12). (This is an average, per capita value for the entire population, which also includes people who have not been X-rayed. Since only a fraction of the population is X-rayed in the course of a year, the actual dose to the affected individual is much higher.) The use of building materials which are in some cases enriched in radioactive isotopes adds a gonad-equivalent dose of 45 mrem/year from the time spent in these buildings. Compared to these two contributions, all other individual contributions are negligibly small. These include, for example, the use of radioactive materials and ionizing radiation in research and technology, flight at high altitudes, the distribution of phosphate fertilizers and the associated specific activity of the ground, the burning of fossil fuels and the resulting release of radioactive matter in the airborne ash (see 5.31), and the operation of nuclear power plants. Each of these contributes less than 2 mrem/year (155). In particular, the operation of 20 nuclear power plants in the Federal Rep. of Germany (each with 1000 MW el. power), and of the necessary fuel processing plants, would add only about 0.23 mrem/year to the average gonad equivalent dose to the external technological dose (12). The internal radiation load from technological sources comes from the combustion of fossil fuels and the resulting emission of radioactive materials with the airborne ash, and from the operation of nuclear plants. These materials are ingested or inhaled. Nuclear power plants and their support facilities release tritium ^{3}H (half-life 12 years) iodine ^{129}I (half-life $1.72 \cdot 10^{7}$ years) and iodine ^{131}I (half-life about 8 days), which are responsible for the internal radiation dose. The operation of 20 nuclear power plants (each with 1000 MW el. power) and their support facilities within the Federal Rep. of Germany would add less than 0.01 mrem/year to the equivalent dose to the gonads, and about 0.03 mrem/year to the thyroid dose (12, 159).

As mentioned at the beginning, the natural radiation dose serves as an important reference for evaluating the technological radiation dose. The effects of ionizing radiation in either living or non-living matter depends on the interactions between the radiation and the matter, and these are controlled by physical laws, not by the source, be it natural or artifical (160, 161). The interactions are essentially excitation and ionization of atoms or molecules. The biological effects of radiation depend on the energy and the type of radiation, and also on the time during which the dose is delivered. In general, the damage from a particular dose is lower when the period of time in which it is received is longer. The effect also depends on whether the dose is applied to the whole body or only to a part of it. (Different organs of the human body have different sensitivities to radiation.)

In summary, the natural radiation exposure varies considerably from one place to another, depending on the geographical latitude, the altitude and the geology of the region. Under normal operating conditions, the additional radiation dose due to nuclear power plants, including their support facilities, is small compared to the natural background radiation, and is less than local variations in the natural background. (The additional radiation is approximately comparable to that received by a person who moves from sea level to a location at 800 m altitude (162).)

The extent to which radioactive materials are released when spent fuel elements are conditioned or reprocessed was discussed in section 5.822.2. In either case, the emissions of gaseous and easily vaporized radionuclides from the corresponding installations are below the allowable limits. The emissions from conditioning are generally lower than from reprocessing. (It has been shown experimentally that most of the gaseous or volatile fission products remain bound to the fuel during intermediate storage, conditioning and final disposal.) The (calculated) whole-body exposure in the neighborhood of a reprocessing plant with a capacity of 700 t/year is, at the least favorable points for waste and air, less than 10 mrem/year (94). The radiation protection laws allow 30 mrem (157, 158). The results of measurements of the dispersion and uptake of radioactive iodine in the environment of the reprocessing plant at Karlsruhe, Federal Rep. of Germany, show that the actual organ dose is only 0.5% of the calculated dose (163).

It was mentioned in 5.822.3 that no country has begun permanent storage of highly radioactive wastes. Therefore the emissions from such a dump (either from reprocessing wastes or from spent fuel) under normal conditions can only be estimated from extrapolation of the results from experimental programs or on the basis of models (94). This is also true for emissions during the process of disposal. In other words, no experimentally verified figures can be given for the radiation dose to the population resulting from the operation of a permanent disposal facility. However, there would be differences between the radiation resulting from disposal of conditioned fuel elements and reprocessing wastes. As mentioned in 5.822.3, the former scheme would result in larger volumes of highly active wastes, while the latter would produce more medium and low-activity wastes. Also, reprocessing releases gaseous and volatile fission product (e.g. ^{85}Kr), which can be kept from entering the atmosphere by appropriate techniques. The major radioactive gas is ^{85}Kr, with a half-life of 10.76 years (the activity of the others, e.g. ^{3}H, ^{14}C, ^{129}I, is several orders of magnitude lower than of ^{85}Kr). The krypton can be stored in gas bottles separately from the other highly active reprocessing wastes until it decays. Although studies have shown that most of the gaseous or volatile fission products are firmly bound to the fuel, under the conditions of final disposal of conditioned fuel elements, it is likely that there would be some additional radiation, especially during the normal replacement phase (94).

5.842 Malfunctions and accidents

The quality of the safety technology of nuclear installations is dictated by the inherent potential hazard, i.e. the possible release of the radioactive inventory. The theoretical potential hazard from a 1000 MW light water reactor after several months of operation, for example, is about 10^{10} Ci.

As discussed in 4.212.2, the most common type of 1st generation light water reactor is the pressurized water reactor (PWR). It has a 20-year development history; the first PWR for commercial electric generation (136 MW) was built in 1957 in Shippingport, USA. The general plan of commercial PWRs is now essentially standardized, although there are differences in the way individual suppliers handle the details of some components. In other words, the most experience, internationally, has been had with PWRs. For this reason, the following, after a general discussion of reactor safety technology, will refer mainly to PWRs. Afterwards, a few special safety aspects of the boiling water reactor (BWR) will be discussed and finally, as far as it lies within the limits of this discussion and the present state of knowledge, some information on the safety problems associated with advanced reactors like the sodium-cooled fast breeder reactor (see 4.214) and high temperature gas reactors (see 4.213.3) and with other points in the fuel cycle (reprocessing, final disposal) will be presented.

The purpose of reactor safety technology, for any type of reactor, is to prevent the release of radioactive substances into the environment in the event of reactor malfunction or accident. About 95% of the radioactive materials are in the fuel (including fission products); the rest are in the reactor coolant circulation, the storage basin for spent fuel elements, the (loaded) transport containers for fuel and in support facilities like the waste gas and waste water systems. In the intact facility, these fission products are enclosed in several layers of containment structures, the fission-product barriers. Fig. 5-11 shows the arrangement of these barriers in a PWR (64). The first barrier is the crystal lattice of the fuel itself, which holds back about 95% of the fission products. It is surrounded successively by the fuel rod cladding, which is sealed with a gas-tight weld; the reactor pressure vessel, together with the completely closed reactor coolant circulation; and the gas-tight and pressure-resistant containment vessel, which surrounds the reactor coolant system. The outer stressed concrete shell has only a limited containment function; it makes it possible to pump up coolant leaks from the containment vessel and protects the installation from external forces.

Fission products cannot escape into the environment as long as the fuel elements are intact, which is why overheating must be avoided. Therefore the basic strategy for reactor safety is to prevent overheating of the fuel elements, i.e. a disequilibrium

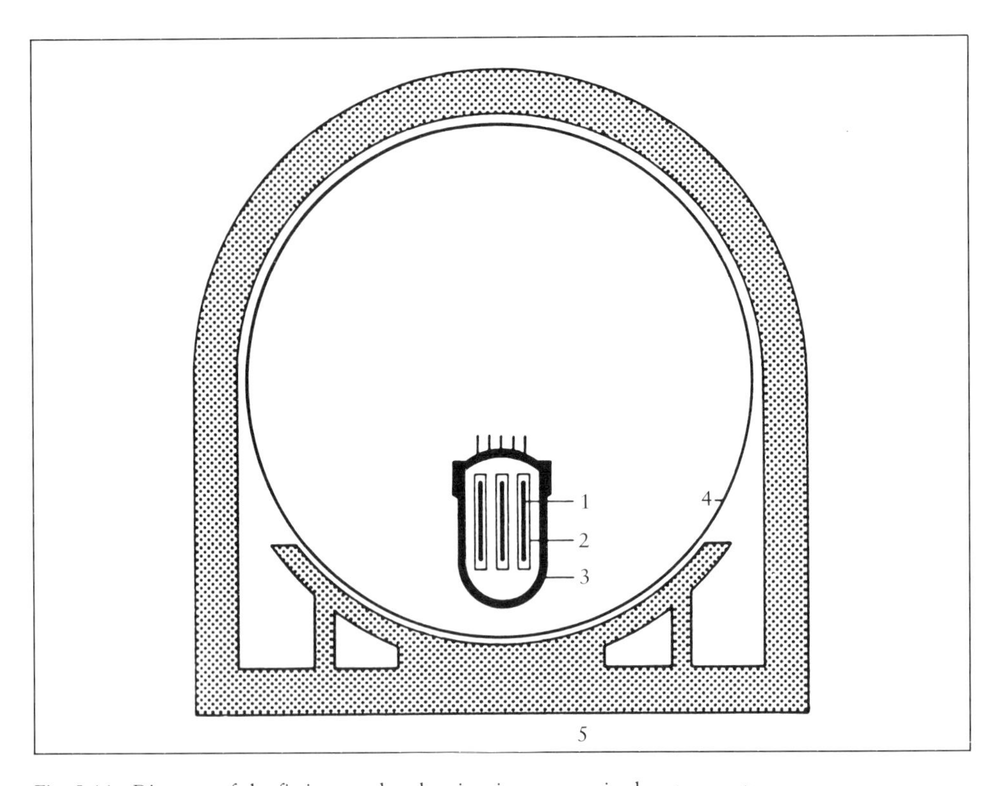

Fig. 5-11: Diagram of the fission product barriers in a pressurized water reactor
1 = Crystal lattice of the fuel; 2 = Fuel rod cladding; 3 = Reactor coolant circulation; 4 = Containment vessel; 5 = Stressed concrete shell.

Source: Federal Ministry of Research and Technology (Ed.), German Reactor Safety Study, Bonn 1979.

between the rates at which heat is produced and removed. Such a disequilibrium could occur either while the reactor was operating or after a shutdown. While the reactor is critical, one distinguishes between transients, in which the heat production is above the norm or the rate of heat removal is below the norm; and loss of coolant, in which the cooling medium escapes through a leak (154). (Transients can also be considered those malfunctionings which are not caused by leaks or breaks in the reactor coolant circulation.) In contrast to the situation with the reactor critical, after it is shut down, only the decay heat must be removed.

For purposes of safety technology, it is customary to classify non-normal functioning of nuclear power plants as upset conditions, malfunctions (see below) and accidents (154).

Upset conditions are deviations from normal operation which are absorbed by the system and do not prevent further normal operation. They thus have no effect on the operation or safety of the plant and the amounts of radioactivity released into the air and water do not exceed the values permitted for normal operation (see 5.841) (157,158).

Malfunctions are a series of events which make it necessary to shut down operation, for safety reasons, but which do not lead to the release of more radiation than is allowed by the radiation regulations for one-time events (157, 158). (In American usage, malfunction events are subdivided into emergency and fault conditions.)

Accidents are events whose consequences are not contained within acceptable limits by the safety systems. (The distinction between "malfunction", for an event against which safety measures exist, and "accident", for which this is not the case, is made only in the German language. In English, the word "accident" is used for both (154)).

The purpose of reactor safety systems is to prevent accidents (malfunctions) or, if this is not possible, to limit the consequences. To do this, a multi-level safety plan has been developed (defense in depth). There are basically three levels of security measures which are employed to optimize the reliability of individual systems within nuclear power plants: 1) Quality control; 2) Prevention of accidents; 3) Limiting the effects of accidents. Quality control reduces the probability of upset conditions, measures for accident prevention should keep upset conditions from becoming emergency or fault conditions, and if an accident occurs, its effects should be limited (154).

Quality control is primarily effected during the manufacture of components and the construction of the installation. Important components like the reactor pressure vessel, coolant pipes and the containment vessel are subjected to multiple independent safety checks; in addition, the control system for shutting off the power output are very thoroughly tested. All quality control measures are intended to prevent upset conditions from occuring in the first place.

Nuclear power plants have multiple and redundant control and safety systems to prevent upset conditions from becoming more serious. The most important is the reactor protection system, which monitors all the important parameters of the installation, such as reactor power, the pressure in the reactor coolant circulation, and the speed at which the main coolant pumps are turning, and if one of these exceeds certain values, it automatically activates safety measures, e.g. a reactor shut down.

The safeguards for limiting the effects of accidents are elaborate. They are activated by the reactor protection system and their function is largely automatic. They are intended to contain the radioactivity and to limit the damage caused by a malfunction. They are arranged in such a way that they can effectively cope with a large number of possible malfunctions (164–166).

The safety systems generally consist of a number of subsystems, which in turn are made up of a number of components, e.g. valves, pumps, pipes, measuring instruments, etc.

Because the failure of an individual component cannot be completely excluded, the regulations for reactor construction require that the subsystems, or at least the total system, must be built in such a way that the failure of one or even several components will not interfere with the function of the safety system. Both independent and common-mode failures have to be considered here.

Independent failures of individual components can also be regarded as internal failures, and they are caused by defects in the component itself, due for example to aging or defective manufacture. Because independent failures are not, by definition, correlated with other failures, they occur as statistically independent individual events. The most important insurance against them is to provide several parallel components for the same function. Then an independent failure has no effect, because another component immediately takes over its function. This is the principle of redundance (see 4.214). Redundance means that for every safety function, more components or systems are provided than are absolutely necessary. In addition, redundant systems are generally specially separated and specially protected by the building, against fire or flooding, for example.

Common-mode failures can be divided into two groups, those which are causally related, and those which are systematic. If the occurrence of A causes B to occur, then A and B are causally related. However, if several components of the same type fail simultaneously or within a short time due to faulty construction or calibration, the failure is systematic. Redundance does not offer adequate protection against common-mode failures, because they can occur simultaneously in the redundant systems. Protection against common-mode failures is offered by the presence of several systems which serve the same end, but are independently constructed, operate on different principles, and are triggered by different physical signals. This is the principle of diversity (see 4.214) (154).

The *fail-safe* principle is another important principle for preventing damage from independent or common-mode failures. It dictates that safety measures be planned so that in the event of a breakdown, they automatically revert to the safe condition. For example, the control rods may be held above the reactor core by electromagnets, so that if their current is interrupted, they fall into the core by gravity and shut down the reaction.

We shall now discuss the most important safety systems of a pressurized water reactor (PWR). Fig. 5-12 shows the relevant systems in a PWR nuclear power plant (only two of the four primary coolant circulations are shown) (165). The reactor

---

Fig. 5-12: Systems relevant to safety in a nuclear power plant with a pressurized water reactor
1 = Reactor pressure vessel; 2 = Steam generator; 3 = Pressure reservoir (emergency cooling); 4 = Loading and unloading basins (not safety systems); 5 = Containment vessel; 6 = Stressed concrete shell; 7 = Combination safety valve-shut-off armature; 8 = Shut-off armature; 9 = Safety valve; 10 = Pressure-release valve; 11 = Back-pressure valve; 12 = Emergency water supply system; 13 = Emergency system; 14 = Emergency and shut-down cooling system; 15 = Control system (not a safety system); 16 = Reactor protection system; 17 = Normal and emergency power supply systems; 18 = Volume-regulation system (only partly a safety system); 19 = Bellows; 20 = Filter; 21 = Chimney; 22 = To turbine; 23 = From the main water feed pumps. The safety valves on the primary circulation are on the pressure vessel, which is not drawn here.

Source: D. Smidt, Reaktor-Sicherheit und menschliche Unzulänglichkeit, Die Naturwissenschaften, Vol. 66, 593–600, 1979.

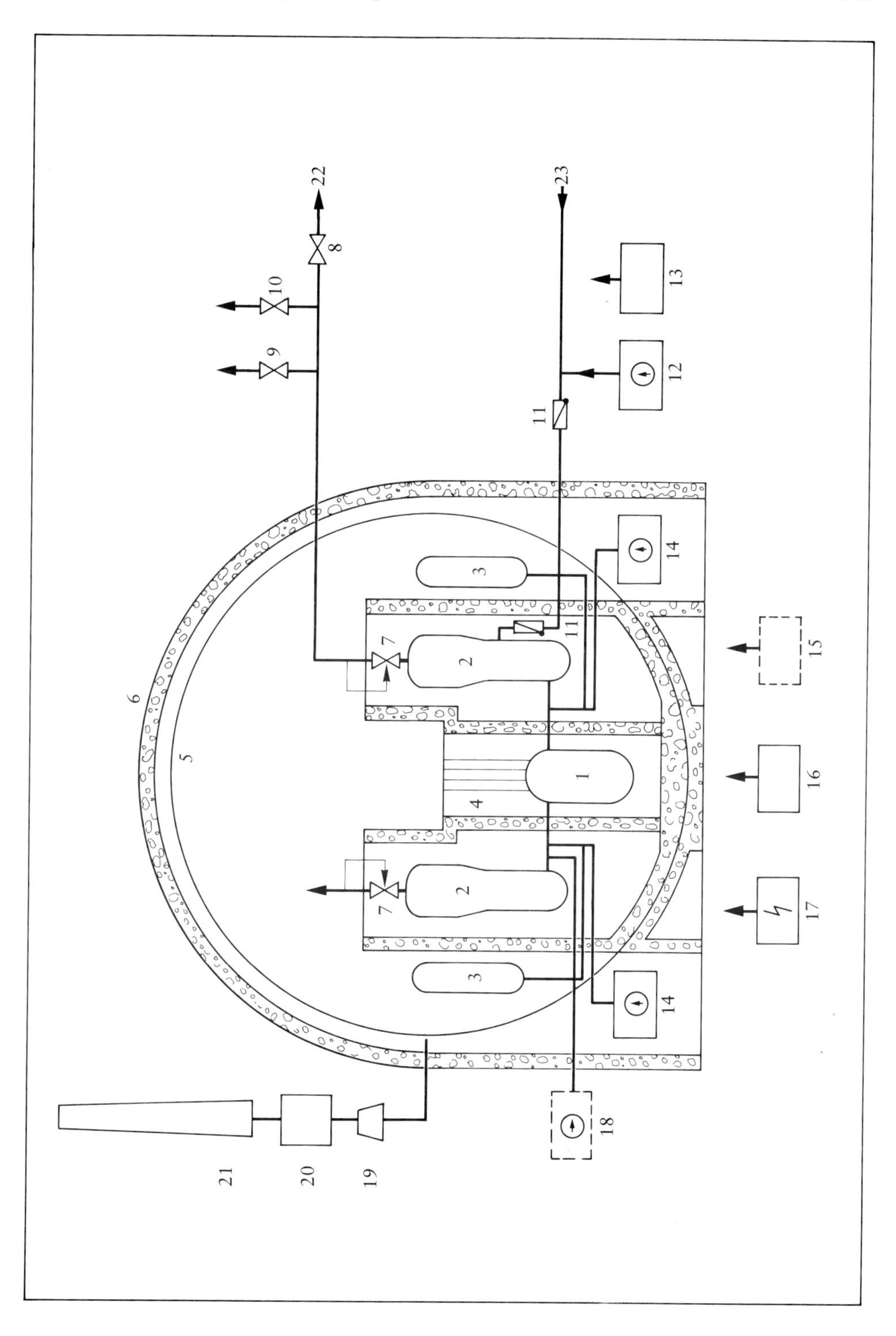
22
23
10
9
8
13
12
11
6
5
3
14
7
2
11
2
15
1
4
16
7
2
17
3
14
18
21
20
19

pressure vessel surrounds the core, which produces the heat under operating conditions and decay heat even after shut-down; it also contains by far the largest part of the radioactive inventory of the power plant. The primary system (pressure containment) consists of the reactor pressure vessel, the primary side of the steam generator, the primary coolant circulation, the primary coolant pumps and the pressure containment system. Its function is to guarantee the presence of cooling water in the core and to prevent the escape of active substances into the environment. The control system serves to maintain an equilibrium between the rates at which heat is generated and removed from the reactor under normal operation and mild upset conditions. The emergency system is designed to bring the plant into a safe configuration in event of an external emergency. To do this, it must provide for continuous heat removal from the shut-down reactor. The emergency water supply system provides the secondary side of the steam generator with water, whenever the main water supply system fails, and can be used to remove decay heat and to bring down the temperature of the primary coolant. In principle, it consists of a water reservoir, which is sufficient to maintain cooling for 10 to 15 hours, and the pumps required to pump the water into the steam generator. (In German power plants, each of the four steam generators has its own emergency water system, and the output of two of the four is sufficient to remove decay heat.) The emergency and shut-down cooling system has both operational and safety functions. When the plant is shut-down, after the temperature has dropped sufficiently, the emergency and shut-down coolant system takes over the removal of decay heat and the further cooling of the primary reactor coolant (shut-down cooling). In loss-of-coolant accidents, this system has the task of refilling the reactor pressure vessel and insuring adequate cooling of the reactor core (emergency cooling) (164). Like the emergency water supply system, the emergency and shut-down cooling system consists of four individual systems, which feed into the primary coolant circulation, and the output of two indivudual systems is sufficient to remove the decay heat. The reactor protection system monitors all the relevant data and, when certain thresholds are reached, it automatically activates the appropriate safety systems. In particular, it can activate the emergency shut-down of the reactor. The emergency power supply system supplies current, if the normal power fails, to the components which have safety functions, for example the components for shutting down the reactor or removing decay heat. The reactor containment vessel encloses most of the radioactive parts of the plant, which minimizes the release of radioactivity either during normal operation or in an accident. The surrounding stressed concrete shell protects the reactor containment vessel against external influences, and also makes possible a controlled removal of leaked activity through the waste-gas system (bellows, filter, chimney).

Human error was a crucial element in some of the accidents which have occurred in nuclear power plants. For example, the cable fire at Browns Ferry, USA, on March 22, 1975, was caused by a lighted candle, and the serious damage to the reactor core at Three Mile Island (Harrisburg, Pennsylvania, USA) on March 28, 1979 was caused by an incorrect response to a deviation from normal operation (167–169). There are a few points which must be observed when the operation of a nuclear power plant is to be made as resistant as possible to human error. Actions such as rapid shut-down, activation of the emergency water supply system or emergency and shut-down cooling system, etc., which must be taken within a few seconds to prevent damage to the reactor, must be automatic. Humans should be used essentially only in situations where the time factor is of secondary importance. Experience has shown that there are two basic reasons for this. First, people tend to base their actions, especially when they are under stress and have little time, primarily on the information which fits their understanding of the situation. There have been cases in which instrumental data (e.g. indication of excessive temperature) have been ignored because the personnel wanted to see another situation, especially when data from different instruments appeared to be mutually contradictory. Second, the activation of certain safety systems requires a certain sequence of switching events. Since humans, especially when under time pressure and stress, tend to make mistakes, these systems must be automatically activated (165).

On the other hand, human flexibility is an advantage in some situations. Therefore, the human operators must be clearly informed of the condition of the plant at a glance, so the important parameters (neutron flux, temperatures, etc.) must be presented in suitable form (165). And, of course, the plant personnel must be especially well trained.

Quantitative accident analysis for a nuclear power plant is concerned with the study of possible accidents, their effects, and the probability of their occurrence. The analysis yields a so-called accident spectrum. Two important methods are error tree and sequence of events analysis.

In error tree analysis, one starts with an assumed event, such as loss of power for a security system, and examines the causes which might lead to this event. The causes are followed back until one comes to individual components whose failure rate is either known or can be determined experimentally (154, 166). Where this is not possible, the inadequacy of the available data is compensated by "safety margins". For most of the components and materials, however, data is available from conventional power plants. In the sequence of events analysis, one proceeds in the reverse direction, and examines the results of failure of a system component (accident-causing event), and the probability of dangerous results.

The product of the probability of a damaging event W_i (in events per unit of time) and the effects of a damaging event S_i (in damage per event) is the risk R_i (in damage per unit time) of the accident under consideration:

$$R_i = W_i \cdot S_i \tag{5}$$

For example, the number of accidental deaths/year is equal to the number of accidents/year times the number of deaths/accident. The same risk can be the product of a high probability of occurrence and a low amount of damage; or of a low probability of occurrence and a large amount of damage. The total risk is the sum of the partial risks for each possible cause of accident, or $R = \Sigma R_i$.

One then establishes, in the last analysis by definition, which accidents will be considered limiting in planning for the plant. Limiting accidents must be controlled by safety systems in such fashion that the corresponding limits of radiation escape are not exceeded. There are thus automatic safety devices to prevent these limiting accidents[1)].

It can be seen from the above, that accidents which can be controlled by the safety systems do not cause damage outside the plant. Risk analysis therefore concentrates mostly on events in which a failure of the safety systems is postulated. For LWR plants, the most serious of these are accidents which could lead to melting of the core (164), so that the determination of risk must include a study of the probability of occurrence or such events, and under what circumstances accidents could lead to melting of the core in spite of the safety systems.

In the following we shall give a qualitative description of the events which would have to be analysed in the event of core melting in a PWR. (The events in a BWR are not significantly different, as will be discussed below.) If the core cooling fails, for example through a large leak and a simultaneous failure of the emergency cooling system, the decay heat causes the temperature of the reactor core to rise, and the water in the pressure vessel to boil (compare Fig. 5-11 and 5-12). The steam escapes through the leak in the containment vessel. It is assumed that the fuel rod cladding is damaged and gaseous fission products escape into the reactor pressure vessel and from there into the containment vessel. If the fuel is heated to its melting temperature, larger amounts of fission products which were trapped in the crystal lattice are released and escape via the leak in the reactor coolant circulation into the containment vessel. It is also to be assumed that the decay heat released when the core melts is sufficient to cause the bottom of the reactor pressure vessel to melt, and the hot mass to melt into the concrete structure below it.[2)]

The containment vessel and the surrounding stressed concrete shell are the last barriers to fission products in the event of core melting (compare Fig. 5-11). Therefore, the impermeability of the containment vessel determines whether and to what degree fission products can escape from the plant. Permeability or failure of the con-

---

[1)] A large *loss of coolant accident*, LOCA, used to be considered the limiting accident for a LWR. The common term "largest assumable accident" is no longer used, because it is misleading. A number of other accidents and a few external influences are now included among the events which must be controlled by the safety systems (154).

[2)] There are a number of different sequences of events which could conceivably lead to core melting, but various core-melting accidents would be phenomenologically similar. Only the timing of the various events can be quite different, because it depends on the amount of decay heat available. The amount of decay heat decreases with time, so that if the emergency cooling failed shortly after the accident, there would be a large amount of decay heat, while later failure would leave a smaller amount of decay heat (164).

tainment vessel could be due either 1) to defects in the vessel which make it permeable, or 2) to the presence of stresses which exceed the strength of the vessel. The containment vessel is a gas-tight welded spherical stell shell (the diameter of the containment vessel at the Biblis B reactor, Federal Rep. of Germany, is 56 m). There are a number of points at which pipes and electric cables penetrate the steel shell; these are required for supplying systems inside the containment vessel. Each conduit is supplied with armatures – e.g. two in succession – which can close it off. If, however, both armatures fail to close, the conduit cannot be sealed off and the containment vessel can leak through that conduit. A number of other physical and chemical processes can also lead to an increase in the pressure and temperature in the containment vessel. Melted core material would evaporate or boil the water remaining in the reactor pressure vessel, or, under some circumstances, might cause a steam explosion. Hydrogen is generated by an exothermic chemical reaction between metal and steam, and also by radiolysis of water. The hydrogen contributes to the pressure in the containment vessel, especially if it burns; or an explosive mixture of hydrogen and oxygen might form. If the melt penetrates the concrete, water and hydrogen are released. The extent of the damage depends critically on the nature of the failure of the containment vessel (compare Fig. 5-11 and 5-12): leakage, failure under pressure or destruction due to an (extremely unlikely) steam explosion in the reactor pressure vessel. Model studies of a core melt accident have shown, for instance, that a melt through of the reactor pressure vessel is probable about 2 hours after the initial event, and a high-pressure failure of the contaiment vessel, about 21 hours after the initial event (164).

The concentration of the fission products in the containment vessel is continously reduced, up to the time of release from it, by a number of active and passive processes, e.g. gravitational settling, and by radioactive decay. In some cases, the reduction is large, so that the longer the fission products remain in the containment vessel, the better it prevents their deleterious effects. In a core-melt accident, the following four phases can be distinguished on the basis of physical, chemical and thermodynamic conditions: 1) The gap release when the fuel-rod cladding is breached; this allows mostly gaseous and easily volatilized fission products to escape. 2) The melt-down release when the fuel becomes hot enough to melt. 3) The vaporization release due to the interaction of the melt and the concrete foundation. 4) The release due to a steam explosion (164). The physical and chemical properties of the individual fission products dictate that different fractions of the radioactive inventory are released in different stages. The elements can be divided into the following 7 groups on the basis of their release behavior: noble gases (Kr, Xe), halogens (Br, I), alkali metals (Rb, Cs), tellurides (Te), calcium earth metals (Sr, Ba), noble metals (Ru), and non-volatile metal oxides (La). The amount of release in a core-melt accident depends on the way in which the containment vessel fails. The most fission products are released in a steam explosion in the reactor pressure vessel, if the containment vessel is subsequently ruptured. There are two reasons for this. First, it is assumed that core-melt accidents which lead to a steam explosion intrinsically release more fission products than those which do not, and second, most of the products would be released immediately after the core melted, and because they would spend little time in the containment vessel, the settling-out effects would be slight.

The containment vessel is surrounded by a shell of reinforced concrete which forms the outer wall of the reactor building (compare Fig. 5-11 and 5-12). This wall protects the reactor against external influences and, in the case of accident, shields the environment against the direct radiation from the containment vessel. The space between the containment vessel and the concrete wall is kept below atmospheric pressure by suction pumps. Activity released by small leaks can be removed in controlled fashion through filters and the chimney. In the extreme case that the containment vessel is ruptured, however, the mixture of gas and steam and its load of radioactive elements will escape directly into the environment. The models of the consequences of an accident are based on the amount and type of radioactive materials which, in the most unfavorable case, would be released into the atmosphere in the course of a postulated reactor accident (e.g. a core-melt accident followed by a steam explosion). It is useful to subdivide such a model into four sections: 1) a model for the atmospheric dispersion and deposition of radioactivity, 2) a dose model, 3) a model for protective countermeasures, 4) a model for the health problems which would develop as a result of radioactivity (164).

The activity-carrying "cloud" would be carried away from the site by the wind, and its intrinsic heat would cause it to rise in the air. The initially compact cloud would also expand perpendicular to the wind direction due to turbulence. Thus as the cloud moves away from the site, it covers an ever larger area. This dilution of the cloud, along with the decay of the radioactivity with time and fallout of particles and water-soluble substances with precipitation, will reduce the concentration of activity in the cloud, but the fallout of course contaminates the area over which the cloud passes.

Humans who are present under the radioactive cloud are exposed to direct radiation and to an internal dose from inhaled and ingested substances. The air and ground concentrations of radioactivity can be determined from the model for atmospheric dispersion and fallout as functions of time and place, and from this, the radiation exposure of the affected population can be estimated. Radiation which is released into the atmosphere reaches people via *exposure pathways*, which are essentially as follows: external radiation from the cloud passing overhead, external radiation from the activity which has reached the ground as fallout, and internal radiation from substances inhaled from the air or ingested with food. The energy absorbed by the body from external radiation is primarily determined by the distribution of the activity in the air and on the ground. It is nearly evenly distributed over the whole body. The amount of energy absorbed internally in individual organs depends not only on the amount and type of material ingested, but also on the metabolism of this organ. The bone marrow, bone surfaces, lungs, thyroid and breasts are especially susceptible; for the other organs, the damage tends to be proportional to the whole-body exposure. The genetically significant dose is based on the exposure of the gonads.

The degree of exposure and the number of people affected, and thus the possible damage, depend not only on the activity concentrations to be expected, but also on the effectiveness of protective countermeasures. It is therefore useful to estimate first the potential doses which could be received by persons staying out of doors continuously, and the expected doses received if the protective countermeasures were taken. For example, the population could be directed to go into cellars or to take iodine tablets; the population from an area A (e.g. a full circle around the reactor site with a radius of 2.4 km and a wedge in the wind direction with $r = 8$ km and an angle of 30°), where high doses are to be expected could be evacuated; the population from an area B, which would extend about 30 km in the direction travelled by the radioactive cloud, could be relocated after evacuation of area A was complete; the area might be decontaminated; or a temporary injunction could be issued against the consumption of local agricultural products.

To determine the risk of radiation damage as a function of the dose, one must know the relationship between the dose in a particular tissue and the probability of occurence of radiation damage in that tissue (170, 171). The absorption of radiation energy in living cells sets in motion a series of physical, chemical and biological reactions which can be deleterious to the health of the individual or, if the gonads are irradiated, to his offspring. However, since cells and organisms have effective mechanisms for the repair of the primary biological effects of radiation, the absorption of radioactivity does not necessarily lead to later radiation damage; it is only a possibility. Radiation exposure may lead either to early somatic damage (e.g. death from acute radiation sickness), late somatic damage (e.g. death from leukemia or other cancer), or genetic damage in later generations as a result of specific mutations in gametes after irradiation of the gonads of the parents (e.g. skeletal anomalies, mental retardation, genetic diseases). Mutations occur spontaneously as natural events at a natural rate. According to the ICRP, a dose of 75 rem would lead to an average doubling of the natural mutation rate. The results of a reactor accident can be estimated from the expected doses received by the population, assuming that protective counter measures were taken (170 – 172). To determine the early or late somatic damage, the individual probability of occurrence for each type must be calculated. In other words, one would determine the probability of death for individuals at each location, on the basis of the expected dose. (For example, a dose of 500 rem to 600 rem will kill 50 out of 100 people exposed to it.) The collective damage to the affected population is then determined; this is the number of fatalities to be expected (173). Estimates of the probability of radiation damage are generally based on the hypothesis that the dose-effect relationship is linear. If x is the number of people exposed, and the total dose received by each is y, it is assumed that the product $x \cdot y$ is indicative of the number of additional deaths to be expected in the 40 years after the accident (e.g. from cancer). It is assumed that one additional death results from each 10000 man · rem. In other words, if 1000 people receive 10 rems each, or if 100 people receive 100 rems each, on the average, one will die (173).

There is relatively little experimental data on radiation damage. Some information on the radiation-induced risk of cancer in humans has been gathered from studies of cancer frequency among the survivors of the Hiroshima and Nagasaki bombs, who where exposed for a short time to a relatively high whole-body irradiation by neutrons and gamma rays. There are also data on the risk of radiation cancer in patients who have received radiation treatment for medical reasons, especially those who have received diagnostic or therapeutic X-ray treatment. There are also observations of cancer frequency in persons occupationally exposed to radiation. For example, lung cancer is more frequent in miners working in mines with high concentrations. However, populations living in areas with high natural background radiation have not yet been shown to have higher than normal cancer rates. The estimates of genetic radiation damage are supported by animal experiments.

The first comprehensive reactor safety study in the USA was prepared under the direction of N. C. Rasmussen (172). The goal of this report was to determine the risk to the population of the USA of having 100 LWRs in operation. The nuclear power plant risks were also compared with other risks due to technology and natural causes. Fig. 5-13 a shows the risk associated with the presence of 100 nuclear power plants, compared with other risks of a technological civilization, including airplane crashes, fire, explosions, bursting dams and chlorine release from the chemical industry. (The "expected frequency" – in general only the word "frequency" is used for the sake of brevity – of accidents resulting in a given number of deaths is plotted.) Fig. 5-13 b compares the risk from 100 nuclear power plants with natural risks, such as earthquakes, tornadoes and hurricanes. It shows that the expected frequency of non-nuclear disasters is 500 to 10000 times greater than that of nuclear events. Of the natural events, only the extremely rare impact of a meteorite has a comparably low risk (compare Fig. 5-13 b). The figures show that the studies in WASH 1400 indicate an expected frequency of 10^{-9}/year for a single plant, or 10^{-7}/year for 100 power plants. The probability of a core-melt accident is given in the Rasmussen Report as 1:20000 per year and reactor, which means that for 100 reactors, the annual probability is 1:200. However, the containment vessel would prevent release of fission products in the vast majority of cases, so that the damage to the population would not be acute. For the worst accident, with a probability of $1:10^9$ per year and reactor, the effects were estimated to be 3300 immediate deaths, 4500 immediate illnesses, and \$ $14 \cdot 10^9$ property damage. (The number of acute illnesses to be expected from a reactor accident would be about 10 times as high as the number of deaths given in Fig. 5-13 a and b.)

The U.S. Nuclear Regulatory Commission authorized another study by the Risk Assessment Review Group, under the direction of H. W. Lewis, to analyse some of the criticisms which had been raised against the Rasmussen Report (173). Although

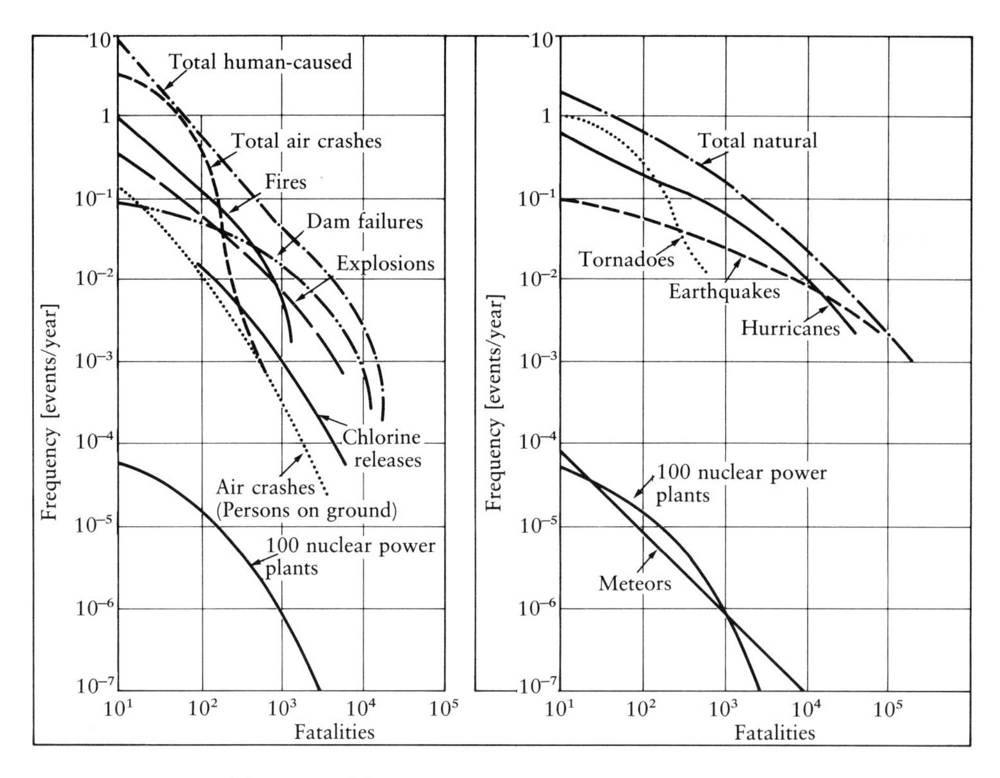

Fig. 5-13 a: Estimated frequency of fatalities due to human-caused events

Fig. 5-13 b: Estimated frequency of fatalities due to natural events

Source: N. C. Rasmussen, Reactor Safety Study: An Assessment of Accident Risks in U.S. Commercial Nuclear Power Plants, U.S. Nuclear Regulatory Commission (Ed.), WASH-1400, NUREG-75/014, Oct. 1975.

this group found that the Rasmussen Report did contain unjustifiable statements, the Lewis Report expressly emphasized the significance of the Rasmussen Report with respect to the determination of nuclear risk (174).

A number of parameters affect the damage to health resulting from a reactor accident, e.g. the population density, weather conditions, and protective countermeasures. In addition, the reactors on which the Rasmussen Report was based are somewhat different in basic concept from the reactors which are in operation in Western Europe. It follows that the results of the Rasmussen Report cannot necessarily be applied to other countries. For this reasons, the Federal Ministry of Research and Technology commissioned a German Reactor Safety Study under the direction of A. Birkhofer (164). This study determined the risk associated with the operation of 25 nuclear power plants (the reference plant is Biblis B, with a PWR and a el. power of 1300 MW) at 19 sites in the Federal Rep. of Germany. This study, using basically the same methods as the Rasmussen group, yielded essentially comparable results, if the parameters like the size of the power plants and the number of sites are normalized (155, 164, 172). One point of agreement was that the containment vessel of a nuclear power plant has a high probability of greatly reducing the consequences of a core-melt accident. It gives the probability of a core-melt as 1:10000 per year and reactor. In 93% of all core-melt accidents, the containment vessel would prevent acute damage due to the release of fission products; in 99% of all core-melt accidents, no premature deaths would be expected. This is due in part to the fact that even in case of a serious accident, there is time to initiate emergency protection measures. In the event of the worst accident, a steam explosion which destroys the containment vessel – the probability of occurence for 25 reactors is $1:2 \cdot 10^9$ per year – the effects are estimated to be 14500 immediate deaths and 104000 delayed illnesses (additional cancer and leukemia cases over 30 years) (164).

It should be mentioned here that the *nuclear explosion* of a nuclear reactor, the kind of explosion generated by an atom bomb, is physically impossible, for several reasons. The common LWR differs in many ways from a nuclear explosive, one of them being that the fuel is enriched only to 3% ^{235}U.

It should also be mentioned that when the nature and extent of the preventative measures against damage in the case of accident are established, a risk calculation is always involved, both with nuclear power plants and other technological installations. There will always be thinkable, physically possible events, which cannot be prevented or neutralized by the safety system. These events are extremely unlikely, to be sure, but there is always a residual risk (175, 176).

The safety features of the boiling water reactor (BWR) are in principle the same as those of the pressurized water reactor (PWR). The Rasmussen study determined essentially the same risk for the BWR as for the PWR (172). Although the emergency cooling systems of the BWR on which the study was based are somewhat more reliable than those of the PWR studied, because they are more highly redundant, the containment vessel of the BWR is smaller, and is thus less capable of restraining the materials released by a core-melt accident than the PWR containment vessel. As mentioned earlier, the sooner the containment vessel gives way, the higher the damage to the environment, because the active and passive activity-reducing mechanisms have less time in which to work. Therefore, the risk for the two types of reactor is about the same (154).

At this point, we shall mention a few aspects of the accident which occurred on March 28, 1979 about 4 a.m. at the nuclear power plant at Three Mile Island 2 (about 15 km southwest of Harrisburg, Pennsylvania). The details are contained in references 154, 165, 167–169 and 173. The plant had a PWR with a rated capacity of about 880 MW el. power. The accident began with a normal excursion in the secondary coolant circulation, which led to the correct shut-down of the turbine. Since no more heat was being removed, the temperature and pressure in the primary circulation rose, as was fully normal. The signal for rapid shut-down was given by the increased pressure in the primary circulation, and after the shut-down had occurred, the normal operating pressure was exceeded in the pressure vessel. Therefore, an automatic pressure-release valve on the pressure vessel opened and released coolant. Up to this point, everything had functioned,according to plan. Now a technical failure occurred: the pressure-release valve did not close when the pressure in the primary system had fallen below the value at which it should have closed, which happened after about 13 seconds. The loss of pressure through the leak caused the emergency cooling system to turn on, as it should, and to pump water at high pressure into the primary circulation. Now the water level in the pressure vessel rose, so after 5 minutes the personnel turned off one of the high-pressure pumps, and after 10 minutes, the second pump was turned off, i.e. the emergency cooling system was turned off. This reaction – which appeared correct to the personnel – was wrong, however, because coolant was flowing out of the pressure vessel and into the containment vessel through the open valve. The personnel apparently did not realize that the primary circulation was constantly losing coolant, because the water level in the pressure vessel simulated a full primary circulation. (The steam escaping from the primary circulation prevented the water in the pressure vessel from coming in.) At this time, however, the personnel apparently ignored other signals, like the high temperature readings on the in-core thermoelements, which clearly indicated a loss of coolant. Because the emergency cooling was off, and because a second valve – which could have been closed by pressing a button – in front of the defective one was not closed, there was a relatively large loss of coolant. (Two hours and 20 minutes after the beginning of the incident, the second valve was finally closed and stopped the loss of coolant; during this time between 150 and 200 t of reactor coolant flowed into the containment vessel.) The result was that the core was overheated to the point that the fuel-element cladding was damaged and fission products were released. Some of this radioactivity, mainly xenon and krypton, escaped into the containment vessel, from there into the reactor building, and from there, via the building ventilation – and controlled – it was released into the air. The temperature in the upper part of the core was high enough to support the generation of hydrogesn by the zirconium-water reaction between the zirconium cladding and the coolant. Some of the hydrogen collected in the upper part of the pressure vessel, and some of it escaped through the leak in the pressure vessel into the containment vessel. The fear that there could be radiolytically generated oxygen mixed with the hydrogen was unfounded, as any free oxygen would have reacted with the core metals. In summary, the effects were caused by mechanical failure, greatly compounded by human error.

There are as yet no comparably comprehensive safety studies for the advanced reactors now under development, such as the sodium-cooled fast breeder (SFBR) and the high temperature gas reactor (HTGR). The American Nuclear Society and the European Nuclear Society co-sponsored the International Meeting on Fast Breeder Reactor Safety and Related Physics in Chicago in 1976 (177). Accident analyses for the Clinch River Reactor, USA (178), the SNR-300, Federal Rep. of Germany (179) and the Super-Phénix, France (180, 181), have been published.

The general comments on reactor safety also apply to advanced reactors. The following aspects of the SFBR, which were discussed in 4.214, are advantages with respect to safety: the low coolant pressure and the large temperature difference (400 K) between the operating temperature and the boiling point of sodium prevent the core from going dry in the event of a leak, so that the emergency cooling system does not need to replace coolant, but can be limited in its function to removing excess heat from a coolant under conditions similar to operating conditions (154). This is also mentioned in the Rasmussen Report, according to which a large part of the risk associated with LWRs comes from loss of coolant accidents in which the safety systems subsequently fail. However, as also mentioned in 4.214, a reactivity excursion, i.e. an overpower transient, is possible. It could lead both to core melting and to a release of mechanical energy. This causes a risk which is intrinsically potentially greater than for an LWR, but this risk can be compensated by measures to prevent the excursion or to limit the accident. To prevent core damage, SFBRs are equipped with diversified and redundant rapid shut-down systems. If one postulates the failure of these systems, a core-disruptive accident could result. In the event of such an accident, the containment vessel would have to withstand the release of mechanical energy, continue to remove large amounts of decay heat, in the absence of the coolant system, encapsulate the fission products, and prevent the outbreak of a sodium fire. With the SFBR, there are several layers of containment. The SNR-300, for example, has a primary or inner containment of concrete, which is filled with inert nitrogen, and which surrounds the entire primary circulation. There is an external core-catcher on the floor below the reactor tank; this has two sodium/potassium circulations and is intended to catch a melted core and remove decay heat from it over a long time period. The secondary or outer containment is filled with air and encloses the inner containment. The outer containment is surrounded by a gas-tight steel container, and the air space between the concrete and the steel is kept below atmospheric pressure. The reactor building outside the steel shields the plant from external influences (182, 183).

According to D. Smidt, the main difference in safety technology between an LWR and an SFBR is that in an LWR, the destruction of the reactor core is an accident against which no particular safeguards are planned, while in an SFBR, there are safeguards even against a core melting. The result of core melting, in an LWR, is the release of a large amount of activity into the environment; the effect of the containment vessel is to delay the release for a longer or shorter time, and thus to decrease the damage by allowing the activity to settle out or decay. (The Rasmussen Report shows that this is possible, to a large extent.) In the SFBR, a melted core might cause a large power excursion, so that the containment vessel and primary circulation are designed to withstand the stresses which would occur; the melted core can, if need be, be caught in the cooled core catcher (154).

As mentioned in 4.213.3, the high temperature gas reactor (HTGR) has properties which make it inherently safer in the event of accident than other types of reactor.

The use of ceramic materials (e.g. graphite) with high melting and sublimation temperatures in the core means that the normal operating temperatures can be exceeded by a large amount. (The thermal stability of graphite guarantees an intact core structure almost up to 3700 K, its sublimation temperature.) The combination of the large heat capacity of the core and its low power density has the effect that temperature transients develop very slowly, so that in the event of an accident, there is a relatively long time in which suitable countermeasures can be taken. The gas-tight pyrolytic carbon layer coating the particles ensures that the primary coolant has a low amount of radioactivity in it; in addition, the helium used as coolant is chemically inert and does not easily absorb neutrons. All the primary components of the installation are enclosed in a stressed concrete container, which cannot burst under pressure, so that damage to the reactor building can be excluded (184, 185). It is of particular significance that the construction of the HTGR core and the fuel are such that the temperature coefficient of reactivity is strongly negative; so much so, that an accidently caused increase in temperature would cause the reactor to become subcritical (see 4.211). HTGRs can be built inherently safe. Even if the cooling fails after shut-down, the reactor system would suffer no damage for several hours, so there is a possibility for repairing the system. For up 5 hours after the start of an accident, the temperature of the core is below the threshold at which the fuel elements break down, so that there can be no increased release of fission products. Heating experiments have shown that the fuel-containing coated particles can withstand temperatures up to about 2700 K; above this temperature there is a delayed release of fission elements – and the delay is great enough, so that short-lived fission products have largely decayed by the time they are released (186).

A risk study for an HTGR with block-shaped fuel elements has been done in the USA (187). This study was analysed at the Jülich Nuclear Research Station, also with respect to transfer to the site conditions in the Federal Rep. of Germany (188). It was found that the "hypothetical accidents" due to overheating of the core or as the result of water breaking into it make up the greatest part of the risk. (It is an advantage, in case water should get into the core, that the reaction between graphite and water is endothermal.) The risk studies for HTGR show that the possible effects on the environment of accidents in these reactors are distinctly less serious than the possibilities for core-melt accidents in LWRs indicated by the Rasmussen study (155, 189).

Studies on possible accidents in conditioning or reprocessing plants have shown that these would be controllable (93, 94). That is, in the event of accident, the radiation exposure of the public could be kept below the legally allowed values. However, higher radiation levels could prevail for a short time; in this case, it would have to be decided whether the control range around the plant would have to be evacuated temporarily or permanently (93). It is possible, for example, to make the storage facility of a reprocessing plant, in which the majority of the radioactive inventory is to be found, inherently safe, so that the cooling is not dependent on the functioning of technical apparatus or human reliability. Comprehensive studies have showed, however, that the release of some of the radioactivity from a reprocessing plant, due to destruction of the plant by a third agent (e.g. an act of war), cannot be completely prevented. In order to prevent this, the dispersible radioactive substances would have to be stored underground (see 5.844). However, the studies found that the plutonium storage facility of a reprocessing plant could be constructed and secured in such a way that a terrorist attack from the outside would be impossible, although the embezzlement of plutonium by plant workers could not be as completely prevented (93).

As mentioned in 5.822.3, highly active waste is not yet being permanently disposed of in any country, but studies have indicated that the problems of permanent disposal can be satisfactorily solved. Penetration of the permanent depot by water could possibly endanger the biosphere, but for this to happen, the water would have to dissolve large amounts of salt and mechanisms leading to disruption of the conditioned waste products would also have to come into play. Finally, the dissolved radionuclides would also have to be transported to the surface. In the geological formations being considered for permanent disposal, however, the chances of such an event are extremely slight; furthermore, there are several barriers which would inhibit the release of radioactive material (see 5.822.3). Studies have also shown that the times needed for radionuclides to be transported from a breeched permanent depot are extremely long, on the order of 10^6 years. It follows that if radioactive materials were to escape from a sealed depot, the danger would be not from these materials, but from their decay products. The doses of radioactivity which could, in the far distant future ($> 5 \cdot 10^5$ years), be expected from the accidental release from a permanent depot would be far below the natural background radiation (see 5.822.3), regardless of whether the material in storage were spent fuel elements or reprocessing wastes.

In summary, thorough studies have shown that nuclear fuel can be disposed of in such fashion that neither the public nor the plant employees are at greater risk of death than they are through other industrial and technological facilities to which the public has become accustomed (93, 94).

It is true that nuclear facilities are very complicated. Some of this is due, however, precisely to the fact that many safety features are built in to reduce the residual risk from a large number of thinkable accidents. Two suggestions to increase safety should still be given serious consideration: to concentrate nuclear facilities in "nuclear parks" in thinly populated areas, or to build them underground.

5.843 External influences

External influences can be either of natural or of technological origin. (The influence of third parties, in sabotage or war, is discussed in 5.844.) The natural external influences are such things as floods, lightning, storms and earthquakes; the technological external influences might be caused by an aircraft crashing into the power plant or chemical explosions in the neighborhood (154). (Fires are also considered external influences.)

The present safety regulations require that nuclear power plants be protected against external influences, i.e. the safety systems must be capable of shutting down the reactor and removing the decay heat. The emergency system (compare Fig. 5-12) must be able to do this even if important operational and safety systems have been destroyed. However, it is not required that the plant be so well protected against external influences that it can be operated after such an accident.

In principle, damage from external influences can be prevented by suitable construction, physical separation of important operational and safety systems, or a suitable combination of both. Nuclear power plants have generally been designed to withstand the influences listed above. The corresponding accident analyses must be based on a specific reference plant or site, however, and their results cannot be directly applied to another plant or site, because the probabilities of occurrence and stresses caused by various types of external influence are highly dependent on location.

Designs for handling accidents due to high water, lightning or fire can be based on conventional architectural experience. Modern technology is well able to seal a building against floods or to prevent fires (e.g. by using non-combustible building materials). The physical separation of redundant electrical cables and supply lines is an important aspect of fire protection.

Storms cause external mechanical stress on the reactor building of the containment vessel, which makes them in some respects similar to the effects of an aircraft crash. If a nuclear power plant is designed to withstand the crash of a fast military craft, as is the case in the Federal Rep. of Germany, it can also withstand storms (see below).

Earthquakes have only been systematically recorded by instruments for the last 80 years or so. A world-wide, closely-spaced net of measurement stations was first built up in the sixties, in order to detect underground nuclear weapon tests; these instruments record extremely weak tremors. Even the historical record of the damage from earthquakes goes back only about a thousand years. The history of earthquakes in Basel, Switzerland, is an interesting example. There have been 118 earthquakes in the area around Basel since the 11th century. There were severe earthquakes in 1021, 1346, 1356, 1428, 1531, 1538 and 1635. These dates demonstrate the irregularity of occurrence of severe quakes. Also, the severity of future earthquakes cannot be predicted from the record of past quakes. Research in this area, particularly in California and Japan, has increased our understanding of the causes and processes in earthquakes, so that there has been some progress toward earthquake prediction. Earthquakes occur when the tension and pressure on rock layers exceed the strength of the material. A sudden breaking of the rock release the tension and pressure, and the two sides of the fault slide past each other to reach a new equilibrium position, which only holds, however, until the tension has again built up to the breaking point of the rock. The energy of deformation released during an earthquake is dispersed in the form of elastic waves. The focus or hypocenter of the earthquake is the point where the original break occured; the epicenter is the point on the earth's surface directly above the hypocenter. An earthquake is characterized by its magnitude M and its intensity I. The magnitude M is a measure of the displacement of the seismograph needle 100 km away from the epicenter. (It is not an immediate indicator of the damage caused by the earthquake.) If no instrumental values are available, the intensity can be estimated from the damage caused, e.g. on the Mercalli or the Richter Scale (154, 191).

To examine the events subsequent to a possible accident which might be caused by an earthquake, a series-of-events analysis is required, which is similar in conception to the series-of-events analysis for the emergency power diagram. In general, nuclear power plants are designed to withstand possible earthquakes of predetermined magnitude. The plant must withstand a "normal operation earthquake" intact, and even in a "safe shut down earthquake", which is expected to damage the plant, the shut-down and shut-down cooling systems must function, and any released activity must be contained (154). The normal operation earthquake is defind as the most severe earthquake which has occured in the past in the same seismotectonic unit, up to about 50 km distance from the plant site. The safe shut down earthquake is defined as the most severe earthquake of the past which occurred within 200 km of the site. According to N. C. Rasmussen, earthquakes therefore contribute little to the total risk of reactor accidents. The probability of a core-melt accident due to an earthquake is between 10^{-6} and 10^{-8} per year and reactor (172).

The risk posed by air traffic is determined by fast military craft, as has been shown by analysis (164). The risks associated with large commercial or small private planes, or helicopters, are low. Large commercial planes have a much lower probability of crashing, and the damage from small planes or helicopters is much less than from a fast military plane. In the Federal Rep. of Germany, for example, the 20 t Phantom II, crashing at a velocity of 215 m/s, defines the stress which the plant must withstand. (The Phantom II is flown by the air forces of nearly all NATO countries.) Also, the risk is calculated on the basis of craft in free flight, not in the process of taking off or landing, or circling an airfield (7). The effects of burning or exploding fuel and wreckage must also be taken into account. In designing a power plant to withstand an aircraft crash, it must be assumed that the craft can hit any point in the plant which is not protected by buildings in front of it. It is assumed that the target is a circular area of 7 m^2. The desired protection against an aircraft crash can be achieved by physical separation of redundant systems or by appropriate construction. Electric cables or pipes, for instance, can be laid far enough apart within the building so that in case of a crash, only one of the redundant cables or pipes can be destroyed. They are laid deep enough so that flying wreckage cannot harm them. For the reactor building itself, however, protection through heavy construction is absolutely necessary (7). The German risk study showed that the probability of a core-melt accident due to an aircraft crash into the reference plant is less than $2 \cdot 10^{-7}$ per year and reactor. If follows that, compared to other accidents, the chance of an aircraft crash against a correspondingly designed reactor does not make a significant contribution to the overall risk (164). It should also be mentioned that a power plant designed to withstand the crash of a fast military craft is also protected to a certain degree against other unforeseeable events, for instance, the influence of third parties (see 5.844).

Experience has taught that chemical explosions can occur both in industrial sites and on transportation facilities (streets, railroads, rivers or pipelines) near a nuclear power plant. This holds, for example, for power plants on navigable rivers plied by liquid gas tankers. An explosion causes pressure waves, and the pressures which arise as a result of explosions are known. In general, nuclear power plants are designed to withstand pressure waves from explosions. (This is true in the Federal Rep. of Germany, for example.) The German risk study showed that the chance of a core-melt accident due to a chemical explosion is extremely small, and does not contribute a significant amount to the risk (164).

Other studies have shown that a reprocessing plant, like any other nuclear installation, can be designed to withstand the above external influences (7, 93). Permanent disposal depots would lie in geological formations and regions in which the tectonic tensions are low, so that the probability of an earthquake is slight. In addition, the depth of the depot can be chosen so that external influences cannot result in inacceptible release of radionuclides (see 5.822.3) (7, 93).

5.844 Influences of third parties

The possible effects of third parties include deliberate human attacks on nuclear installations, either sabotage or acts of war. Sabotage could be either a direct attack by a group, possibly of terrorists who wanted to extort fulfillment of some demand, and to this end occupy a nuclear power plant and threaten to destroy the core and thus release massive amounts of radioactivity; or it could be a subversive act on the part of employees of the plant. It is possible, for instance, that an employee could change or damage components in such a way that damage to the environment would result (154). The possible influences of third parties is analysed in the same way as operational accidents.

The principle of physical separation of redundant systems, which is applied for protection against operational accidents, makes it difficult completely to incapacitate the important safety systems for rapid shut down and removal of decay heat (see 5.842). The presence of the radiation field makes it even more difficult. (The important safety systems could be completely incapacitated only by using highly technical weapons at several positions simultaneously.) Also, unauthorized entry into the "sensitive" areas of the plant cannot be indefinitely prevented by technical means, but it can be delayed sufficiently to allow for the arrival of enough security personnel to protect the reactor. According to D. Smidt, the terrorists would in all probability be immediate victims of their acts, because of the special technical features of a nuclear power plant and because of the time required to prepare the action and to carry out their blackmail. A successful escape after the act had succeeded would be impossible (154).

Subversive actions against the power plant can largely be prevented by administrative and technical supervision of the employees. The smuggling in of saboteurs can be prevented, for instance, by careful selection of employees and security clearances, by supervision of certain areas by television cameras and security personnel, and by systems of keys. If saboteurs should succeed in entering in spite of this, the comments made above with regard to violent action still apply (154, 192).

In the following, we shall discuss the possible dangers presented by nuclear installations in the event of war. As discussed in 5.843, nuclear power plants are now generally designed to withstand a large variety of possible external influences, and these protective measures represent a certain degree of security against the effects of war. Fig. 5-14 shows the degree of protection offered by the reactor building against various intensities of pressure wave released by explosions or earthquakes (192, 193). The superimposed line represents the maximum distance from the reactor building at which explosions of various intensities could damage it. For example, a conventional bomb with the equivalent of 1 ton TNT would have to explode within 20 m of the reactor building in order to damage it. A 100 kt TNT nuclear explosive would have to be detonated within 1 km of the reactor building in order to damage it. (The atom bomb which destroyed Hiroshima had the equivalent explosive power of about 20 kt TNT.)

As mentioned above, a 1000 MW LWR contains about 10^{10} curies of radioactivity after several months of operation (theoretical hazard potential) (3, 10). There are two important differences between this radioactivity and that released by conventional nuclear weapons. First, the isotopes generated in the reactor are in the main longer lived than those released by a conventional nuclear explosion, and second, there is several times as much radioactivity in a 1000 MW reactor as a conventional nuclear bomb (depending of course on the size of the bomb) (194, 195). Studies in the United States have come to the conclusion that a 100 kt TNT bomb would have to be detonated within 70 m of the reactor in order to destroy the reactor pressure vessel and the core (196, 197). In an extreme case, the radioactive material in the reactor core should be sucked up into the "mushroom cloud" and carried to great heights, which would increase the fallout from the nuclear bomb (195). In other words, it is only partly possible to protect above-ground nuclear installations from precision-guided munitions or missiles, which can hit "point targets" (198), by protective construction.

Political efforts to give special protection to nuclear power plants in the event of war have been concluded. The diplomatic conference on reconfirmation and further development of human rights applicable in armed conflicts ended on June 10, 1977. It adopted two protocols supplementary to the Geneva Convention of August 12, 1949, Protocol I referring to international conflicts, and Protocol II referring to non-international conflicts. Because the protocols must be ratified, their adoption by the member states is expected no earlier than the beginning of the eighties. The special provisions for nuclear power plants in case of international conflicts are found in Protocol I, Article 56, 1–7 "Protection of works and installations containing dangerous forces"[1] (199). The content of Article 56 must be taken into consideration in locating sites for nuclear power plants, i.e. they should not be built in the neighborhood of military targets. According to H. Zünd, the risk from reactors can also be reduced if they are shut down in time (193). The radioactivity in the reactor core drops to 1/5 of the original amount in 1 hour after shut down, 1/10 in 1 week, 1/18 in 1 month, and about 1/50 after 7 months. However, not only the Western

---

[1] The text of this article is as follows:

"1. Works or installations containing dangerous forces, namely dams, dykes and nuclear electrical generating stations, shall not be made the object of attack, even where these objects are military objectives, if such attack may cause the release of dangerous forces and consequent severe losses among the civilian population. Other military objectives located at or in the vicinity of these works or installations shall not be made the object of attack if such attack may cause the release of dangerous forces from the works or installations and consequent severe losses among the civilian population.

2. The special protection against attack provided by paragraph 1 shall cease:

(a) for a dam or a dyke only if it is used for other than its normal function and in regular, significant and direct support of military operations and if such attack is the only feasible way to terminate such support;

(b) for a nuclear electrical generating station only if it provides electric power in regular significant and direct support of military operations and if such attack is the only feasible way to terminate such support;

(c) for other military objectives located at or in the vicinity of these works or installations only if they are used in regular, significant and direct support of military operations and if such attack is the only feasible way to terminate such support.

3. In all cases, the civilian population and individual civilians shall remain entitled to all the protection accorded them by international law, including the protection of the precautionary measures provided for in Article 57. If the protection ceases and any of the works, installations or military objectives mentioned in paragraph 1 is attacked, all practical precautions shall be taken to avoid the release of the dangerous forces.

4. It is prohibited to make any of the works, installations or military objectives mentioned in paragraph 1 the object of reprisals.

5. The Parties to the conflict shall endeavour to avoid locating any military objectives in the vicinity of the works or installations mentioned in paragraph 1. Nevertheless, installations erected for the sole purpose of defending the protected works or installations from attack are permissible and shall not themselves be made the object of attack, provided that they are not used in hostilities except for defensive actions necessary to respond to attacks against the protected works or installations and that their armament is limited to weapons capable only of repelling hostile action against the protected works or installations.

6. The High Contracting Parties and the Parties to the conflict are urged to conclude further agreements among themselves to provide additional protection for objects containing dangerous forces.

7. In order to facilitate the identification of the objects protected by this article, the Parties to the conflict may mark them with a special sign consisting of a group of three bright orange circles placed on the same axis, as specified in Article 16 of Annex I to this Protocol. The absence of such marking in no way relieves any Party to the conflict of its obligations under this Article."

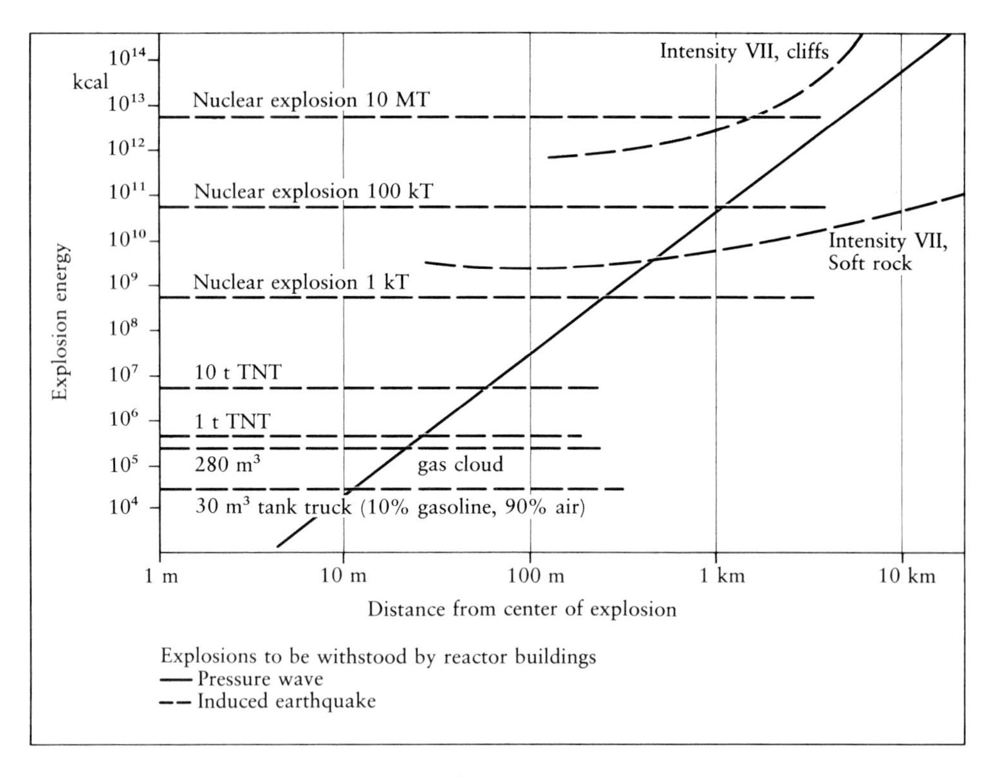

Fig. 5-14: Explosions which can be withstood by the reactor building

Source: E. Münch, The security of nuclear installations and fissionable materials, Atomkernenergie-Kerntechnik, Vol. 33, 229–234, 1979.

countries are building more nuclear power plants, but also the Eastern countries (compare Table 4-2). In the age of precision-guided missile systems, which can hit "point targets", this brings a reciprocity of hazards from nuclear installations in the case of war.

Essentially the same considerations apply to reprocessing plants as to power plants. Thorough studies have shown that it is possible to protect a reprocessing plant against sabotage, but it is not as easy to prevent embezzlement of plutonium by plant employees (93). The hazards from an above-ground intermediate storage depot or a reprocessing plant in the event of war could be eliminated by storing the radioactive material sufficiently far underground (93, 198). A permanent depot for radioactive waste can be laid deep enough underground to prevent sabotage or bomb damage in war.

In summary, it cannot be denied that the peaceful use of nuclear fission energy brings serious problems of environmental protection and safety. However, it is only realistic to say that at the present level of technology, there is no other "new" source of energy (not counting new forms of conventional fuels) which can make a significant contribution to the world's energy economy.

5.9 Environmental and safety problems specific to nuclear fusion

As discussed earlier, the realization of controlled nuclear fusion would provide humanity with a practically unlimited energy supply (see 3.36 and 4.22), and this is the reason that the achievement of a controlled thermonuclear reactor (CTR) is one of the main goals of research and development (200). In addition to the question of fuel supplies and costs, pollution of the environment and safety are increasingly important criteria for judging an energy source. With respect to heat pollution, a CTR power plant is in principle no different from any other thermal power plant (see 5.2); it produces low-temperature waste heat. The relative proportion of the heat that is wasted depends on the efficiency of the plant. In this discussion, we shall first examine the potential hazard from a D-T CTR, then the extent to which this potential can become an actual danger, especially for the people living close to a CTR, and then the problem of radioactive waste will be treated (201–203).

The hazard potential from a CTR can be subdivided into a radiological, a toxicity and an energetic component. The radiological hazard potential, regardless of the reactor concept adopted, comes mainly from the tritium and materials activated by the neutron flux. The toxicity hazard comes both from the inventory of toxic materials and from the elements (e.g. beryllium and lithium) which can form toxic substances by chemical reactions. The energetic hazard potential comes from the energy stored in physical or chemical form, e.g. the plasma and magnetic field energies, the energies of the coolants and working substances in the thermal conversion part of the plant, and from the latent chemical heat of these substances, and from decay heat after the reactor is shut down (204).

In a D-T reactor, tritium is bred in the blanket by two reactions (see equations 11 and 12 in section 4.22, and Fig. 4-6): $^{6}Li(n, \alpha)\,T$ and $^{7}Li(n, n'\alpha)\,T$. The tritium is used in a closed fuel cycle inside the reactor installation. The tritium storage is fed from the tritium extracted from the blanket; the deuterium storage is fed from external supplies. The required amounts of deuterium and tritium are removed from the storages, mixed, and fed into the injector, which shoots it into the reactor in the form of frozen pellets. Only part of the fuel reacts in the plasma; the burn-up rate is only a few percent. The unreacted part of the fuel is sucked off with the helium formed in the reaction and led into a gas-separation facility, where the helium is removed and the D-T mixture is cycled back into the plasma. The materials fed into this fuel cycle from the outside, lithium and deuterium and the product, helium, are not radioactive. (When the reactor is started up, to be sure, some tritium is required until a sufficient amount can be formed in the blanket.) The total tritium inventory depends on system parameters like the choice of blanket-cooling principle and the method of tritium recovery, but lies in the range between 1 and 10 g/MW thermal power, which amounts to a specific activity between 10 and 100 Ci/kW thermal power (203). The biological hazard potential[1)] associated with the tritium lies between 0.05 and 0.5 km^3/kW thermal power (205, 206). This is negligible in comparison to that of the structural materials (compare Fig. 5-16a).

The second source of radiological hazard potential is the materials activated by neutron flux. These are primarily the structural materials, since they are exposed to the highest neutron flux, and to a much lesser degree, the shielding materials. In the last few years, an entire series of structural materials have been studied with regard to their activation behavior in a fusion reactor: niobium and its alloys, stainless steels, vanadium and its alloys, aluminum, and molybdenum alloys (e.g. the TZM alloy, which is 99.4% Mo, 0.5% Ti, and 0.1% Zr) (204, 207, 208). Niobium has good mechanical properties at high temperatures, but its specific activity in Ci/kW thermal power is relatively high. Stainless steels, vanadium, aluminum and molybdenum have more favorable activation properties. For example, after 1 to 2 years of irradiation, aluminum, steel and vanadium have around 1000 Ci/kW thermal power; the values for niobium and molybdenum alloys are 4 to 5 times as high. Vanadium and its alloys have relatively favorable activation properties. In general, the effect of impurities – a common impurity in vanadium is niobium – is important here, so that the long-term activity of the vanadium depends very strongly on the purity of the starting material (compare Fig. 5-15).

Because the specific activity inventory of the materials depends very strongly on the design of the CTR, e.g. on the thickness of the first wall and the volume of structural material in the breeding zone, the amount of specific activity generated during operation is not the only critical factor. (For instance, niobium and molybdenum are relatively strong, so that it would probably be possible to use less of them as structural material.) Another important criterion for the choice of a structural material is its decay behavior after the CTR is shutdown. Fig. 5-15 shows the behavior of a few materials after shutdown of the fusion reactor (200). The activity of steel (316-SS) decays little during the first year after shutdown, but thereafter there is a considerable drop in activity. The activity of the TZM molybdenum alloy decays about 4 orders of magnitude during the first year, but thereafter it would have a relatively high long-lived activity. It would thus be necessary to store either material as radioactive waste for a long period. It should be emphasized that the activity of a vanadium/titanium alloy would be essentially gone after 20 years, but the vanadium must be very pure, as mentioned above. As soon as other elements, for instance

[1)] The biological hazard potential (BHP) of a nuclide is the quotient of the specific radioactivity inventory (Ci/kW thermal power) and the maximum permissible concentration (MPC) in the air or water, given in Ci/km^3. In other words, the biological hazard potential of a nuclide can be thought of as the volume of air or water which would be required to dilute an activity inventory to the maximum permissible concentration.

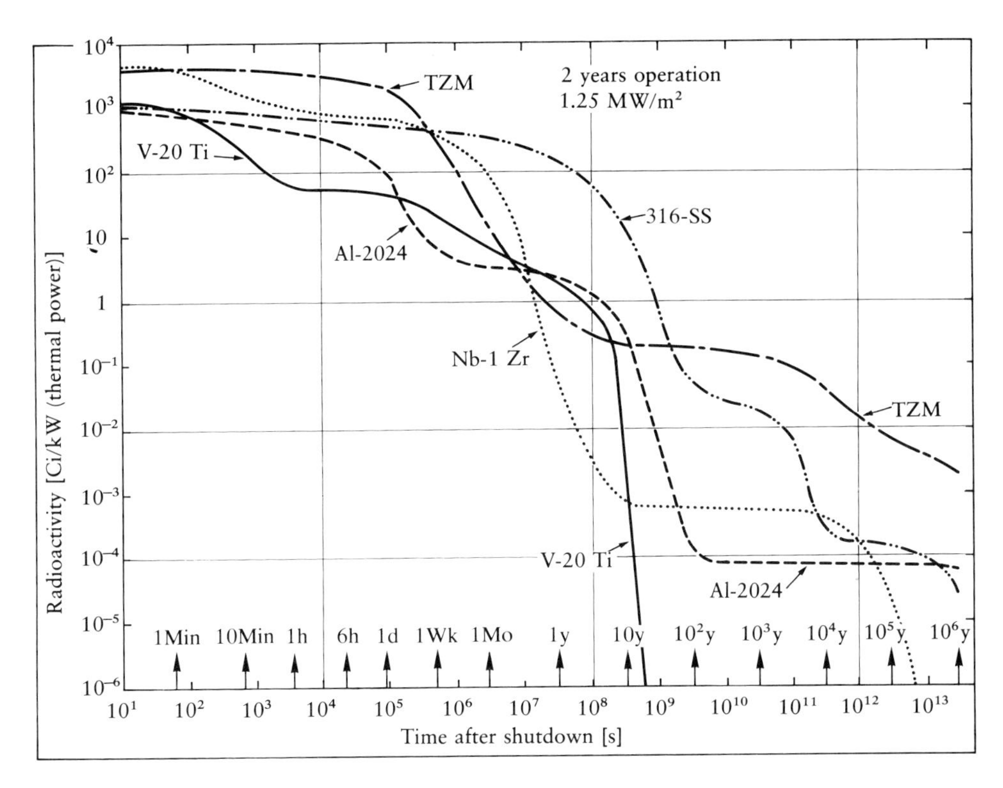

Fig. 5-15: Radioactivity of CTR blankets after shutdown

Source: W. Häfele, J. P. Holdren, G. Kessler, G. L. Kulcinski, Fusion and Fast Breeder Reactor, IIASA, Laxenburg/Vienna, RR-77-8, July 1977.

chromium, are added, this material also has a long-lived activity (due to ^{53}Mn), which is comparable to that of aluminum.

The activation and decay properties (compare Fig. 5-15) are not the only criteria to be considered in the choice of structural materials for maximum safety of a CTR; the biological hazard potential (BHP) has to be considered as well. Fig. 5-16a shows the BHP in air after shutdown of a CTR, and Fig. 5-16b shows the corresponding BHP in water (200). For example, immediately after the shutdown, the BHP in air is greatest with the molybdenum TZM alloy, and a little less with steel (316-SS). They are followed by aluminum (Al-2024) and niobium (Nb-1Zr), and finally by vanadium (V-20 Ti). This order changes several times in the course of the decay; over the long run, steel (316-SS) has the highest values. The most suitable material appears to be vanadium, as is the case with activation as well (compare Fig. 5-15). It is the only one of these materials which has a negligible BHP after 20 years. Steel (316-SS), by contrast, is not well suited, either in terms of activation or of BHP. The use of steel might at first seem preferable, because a mature technology exists, and its

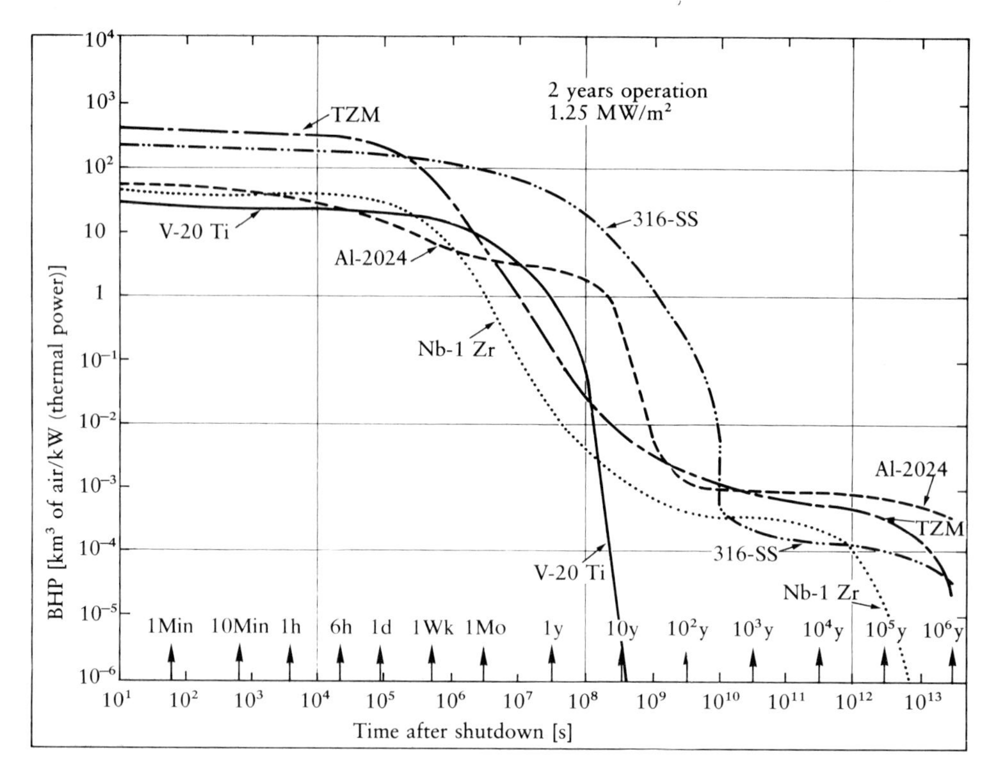

Fig. 5-16a: Biological Hazard Potential (BHP of air) of various CTR structural materials after shutdown

Source: W. Häfele, J. P. Holdren, G. Kessler, G. L. Kulcinski, Fusion and Fast Breeder Reaktor, IIASA, Laxenburg/Vienna, RR-77-8, July 1977.

irradiation behavior is well known, but in the long run other alternatives, like vanadium, are unquestionably more favorable. The comparison of the activity inventories of a CTR – even with the unfavorable values for steel – and a fast breeder (LMFBR) are compared in Fig. 5-17a (200). The decay curve for the LMFBR is determined essentially by the decay behavior of the fission products. It can be seen that the curve for the CTR, except for the short period between years 1 and 10, is always below that of the LMFBR, and the difference ranges up to 2.5 orders of magnitude. If one uses vanadium instead of steel in the CTR, the starting point for the curves is about the same, but after only about 3 hours, the difference is 1.5 to 2 orders of magnitude, and would remain this large for a long time; in about 20 years the activity would be practically gone from the CTR. Fig. 5-17b compares the BHP of a CTR with that of an LMFBR (200). The biological hazard potential of the LMFBR is at first determined by the inventory of actinides, especially plutonium. It is assumed that after one year, 99% of the plutonium will be reprocessed and returned to

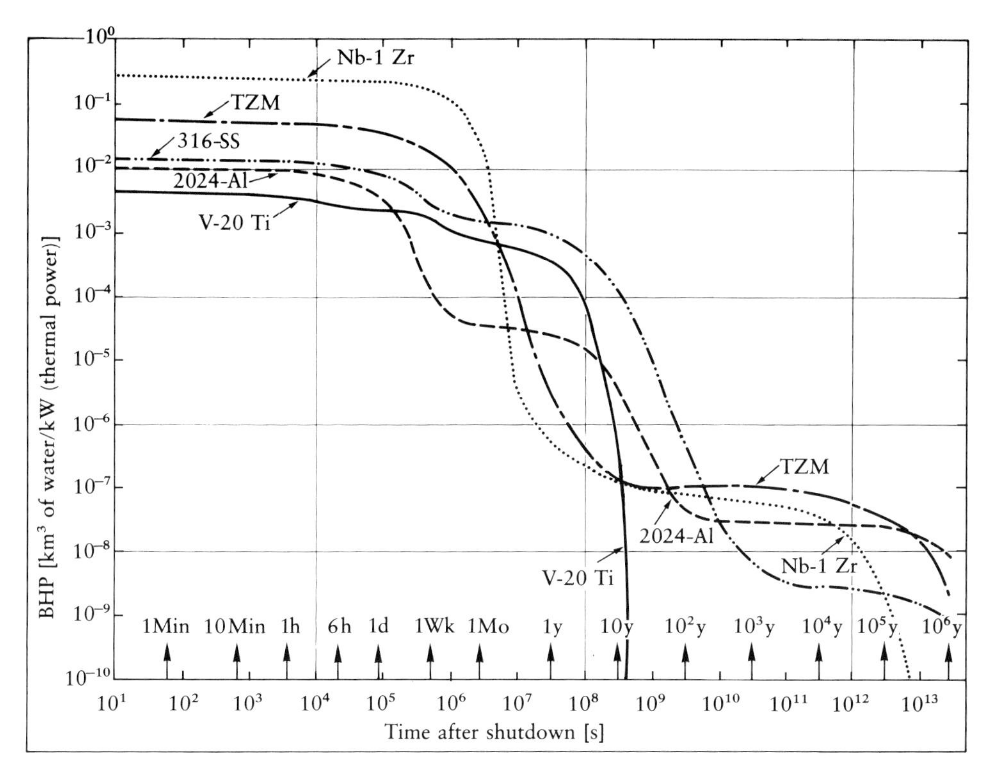

Fig. 5-16b: Biological Hazard Potential (BHP of water) of various CTR structural materials after shutdown

Source: W. Häfele, J. P. Holdren, G. Kessler, G. L. Kulcinski, Fusion and Fast Breeder Reactor, IIASA, Laxenburg/Vienna, RR-77-8, July 1977.

another reactor, and thus will no longer contribute to the BHP of the reactor under consideration. The further decay is determined by the fission products alone. Even on this assumption, the BHP from the CTR with a steel structure is about 1.5 orders of magnitude lower than that of the LMFBR for the first year of the decay period; thereafter the difference rises to 4 orders of magnitude. Thus the BHP of a CTR with a steel structure is considerably less than that of a LMFBR over the entire decay period (200). For a CTR with a vanadium or vanadium/titanium (V-20 Ti) structure, the curve begins about 0.5 order of magnitude lower than the curve for steel, and drops in the course of the first year to a point 1.5 orders of magnitude below steel; after 20 years, there is again a sharp decline (see Fig. 5-16a and b).

Although the radiological hazard potential from nuclear power plants is almost the only source of danger which is considered, because it is a new type of danger, these plants also contain or can accidently produce substances which are potentially dangerous for humans due to their chemical toxicity (203). For example, the beryl-

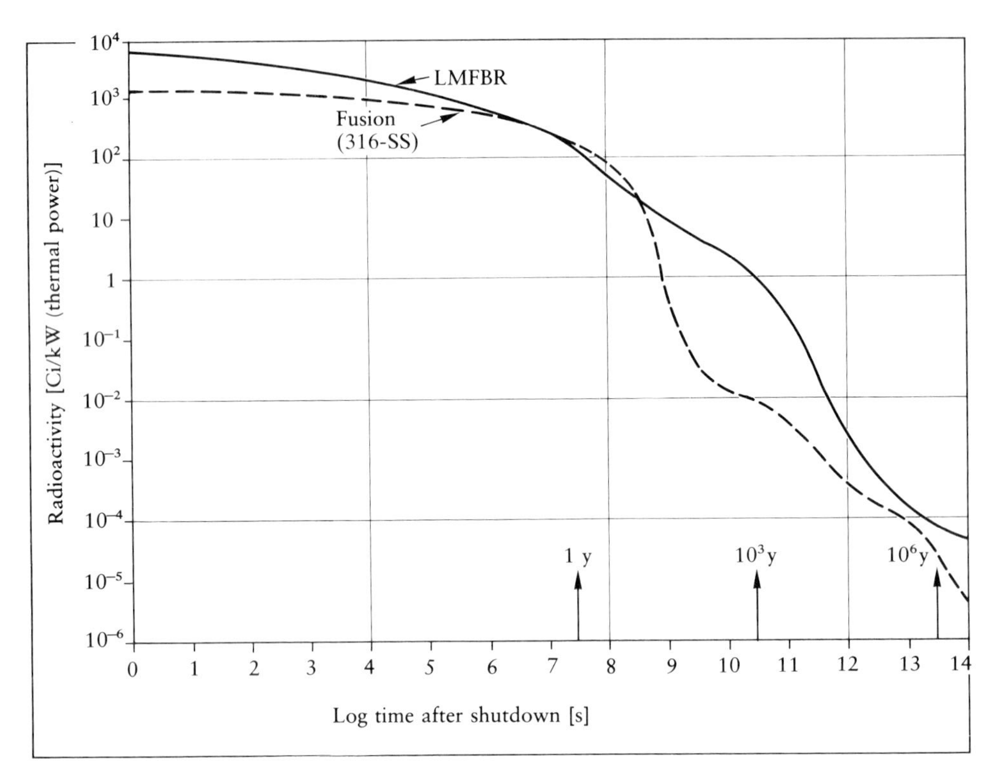

Fig. 5-17 a: Comparsion of radioactivity inventory for fission and fusion reactors with 316-SS structure

Source: W. Häfele, J. P. Holdren, G. Kessler, G. L. Kulcinski, Fusion and Fast Breeder Reactor, IIASA, Laxenburg/Vienna, RR-77-8, July 1977.

lium which is used in metallic or oxidized form in the blanket as a neutron multiplier is poisonous, although the hazard potential it represents is orders of magnitude smaller than the biological hazard potential due to the radioactivity inventory (204). If liquid metallic lithium were accidently released into the air, it would react to form toxic substances like lithium oxide, Li_2O, lithium peroxide, Li_2O_2, lithium nitride, Li_3N and, if the air were moist, lithium hydride LiH and lithium hydroxide, LiOH. The hazard potential from these compounds is of the same order of magnitude as that from beryllium, i.e. it is orders of magnitude less than the BHP from the radioactivity.

The energy hazard potential from a CTR comes from energy stored in physical or chemical form. The plasma energy in a CTR is proportional to the density, temperature and volume of the plasma. For tokamak reactors, typical values for the thermal energy of the plasma range from 0.3 to 0.8 MJ/MW thermal power.

The magnetic field energy is another form of stored energy typical for a CTR. For tokamak reactors, the values range from 40 to 50 MJ/MW thermal power.

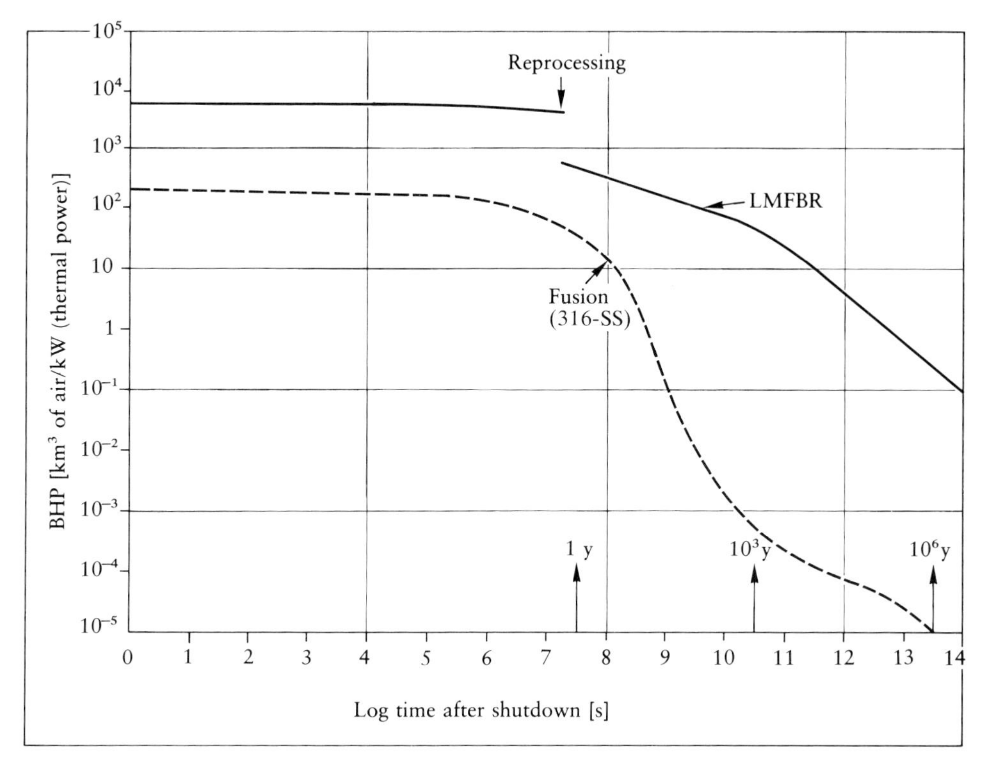

Fig. 5-17b: Comparsion of Biological Hazard Potential (BHP of air) of an LMFBR and a D-T fusion reactor with 316-SS structure

Source: W. Häfele, J. P. Holdren, G. Kessler, G. L. Kulcinski, Fusion and Fast Breeder Reactor, IIASA, Laxenburg/Vienna, RR-77-8, July 1977.

Quasi-steady-state operation of a tokamak reactor requires the maintenance of a strong stationary magnetic field, which would probably be done with superconducting magnets. These are cooled with liquid helium. For example, the UWMAK I proposal (see 4.222) provides for 450000 litres of helium, and the enthalpy difference between operating temperature (4.2 K) and room temperature (about 300 K) for this helium is 84 GJ. The addition of heat would convert at least some of this enthalpy difference into mechanical energy as the helium evaporated and built up pressure.

If liquid lithium is used as breeder material, the thermal energy content of the lithium, which amounts to several GJ/MW thermal power has to be taken into consideration. (In some reactor proposals, lithium would also be used as primary coolant, which would increase the thermal energy content of the reactor.) In addition, lithium is a reactive material, which reacts exothermically with water and atmospheric oxygen and nitrogen. The latent chemical energy of lithium can exceed its thermal energy content by about 60%.

The energy stored in a CTR includes the heat produced by decay of radioactive isotopes after the reactor is shutdown. As in the case of the radioactivity or the biological hazard potential, this can be differentiated according to source: structural materials, breeding materials, etc. Shortly after the reactor is shutdown, the decay heat amounts to 0.3 to 3% of the thermal power, depending on the type and amount of structural materials. Although the decay heat from vanadium structural elements is especially low, it can be shown that for any type of structural materials, the decay heat problem is insignificant after a relatively short time.

Up to this point, we have discussed the hazard potential of a CTR. In the following, we shall discuss the extent to which this potential can become an actual danger in normal operation or in case of accident. In normal operation, the only radioactive material which could be released is tritium, which is a gas. It is an advantage that tritium is circulated in a closed system within the CTR. There are two mechanisms by which it can be lost: permeation and leakage. The walls of nearly every container and pipe in which tritium is stored or transported are surfaces through which tritium can be lost. The permeation rate depends on the material, the thickness, surface area and temperature of the walls, and the partial pressure of tritium on both sides of the permeation surface. Since the partial pressure of tritium also increases with temperature, the losses are especially great where tritium is handled at high temperature, e.g. in the blanket and the tritium recovery system. In reactors in which the breeding material also serves as the primary heat transfer medium, the surface area of the necessary pipes considerably increases the permeation surface. Leaks are especially likely at fittings and welded seams. For safety analysis, there are essentially two important loss pathways; losses to the containment atmosphere, and from there via the ventilation outlet into the atmosphere, and in the energy conversion system, from the heat-transfer system into the steam system, and from there into the atmosphere or hydrosphere. Losses into the containment atmosphere can be kept down by making all tritium-carrying components and pipes double-walled and surrounded with a carrier gas or vacuum; in addition, all tritium-carrying systems can be built in separate spaces with monitored atmospheres (tritium cells). It appears that it would be possible to limit the tritium losses to the containment atmosphere to about 2 Ci/day (203). The losses in the energy exchange system are more of a problem. The tritium can permeate the hot heat-exchange surfaces and get into the steam system. Although it is not difficult to remove the tritium from an intermediate coolant, removing it from steam would require isotope separation, which would be elaborate and difficult. Therefore, if a steam generator is used, the tritium concentration in the steam cannot be allowed to exceed a certain level, so that the steam can be released into the environment. In the published reactor proposals, a total tritium release at the rate of 10–15 Ci/day has been considered acceptable; the radiation exposure under normal operating conditions would then be about 0.30 mrem/year at a distance of 600 m from the point of release. Compared to the regulations which limit the radiation exposure in the neighborhood of nuclear power plants to

30 mrem/year, and the fluctuation range of natural background radiation, the radiation resulting from the controlled release of tritium is slight (203).

Because of the relatively high hazard potential which a CTR represents, the examination of possible accidents is very important. However, it is not yet possible to carry out a thorough accident analysis for a CTR, because the systems and components which would have to be studied in detail have not yet been assembled. Therefore, it is only possible to consider what kinds of accident are in principle possible with a CTR. Basically any accident in which radioactive or toxic substances are released in such a way that they endanger the plant, its personnel or the population of the surrounding area must be considered (203).

Radioactive solids could only be released by the uncontrolled application of energy to the activated structure components. The only sources of such energy, aside from mechanical breaking, are the plasma and the decay heat. Such an accident might conceivably happen when the plasma containment is disturbed in some way, for example, if a disturbance in one of the magnet fields occurred and could not be controlled by the regulating system. If this should happen, there would be a thermal overload. The effects would depend on the area of the affected wall surface and the kinetics of energy dissipation. They could range from an even adiabatic heating of the walls to local melting or evaporation. However, in any case there is an upper limit to the amount of energy which could be released, because the reactor would shutdown automatically. A power excursion is not physically possible (inherent safety). The results of such an accident would be that radioactive substances would get into the reactor vessel, and from there, in the worst case, into the fuel circulation, from which they could be removed by filters. The results would be similar if the first wall or any other part of the blanket should be damaged in some other way, e.g. by mechanical failure or loss of cooling in the operating or shut-down reactor. The power plant would have to be shutdown, possibly for a long time, but there would be no unacceptable damage to the environment.

Accidents in which decay heat plays a role are collectively referred to as loss of coolant accidents, because decay heat only becomes a problem if the cooling fails. The decay heat in the reactor proposals considered so far is low. It amounts to $\leqq 1\%$ of the thermal power of the reactor, and is not a problem. Even with total loss of coolant, the rates of temperature increase would be expected to be in the range of 0.1 to 0.2 K/s. (The rates of increase in a pressurized water reactor are several times as high.) These rates are so low that a separate cooling system could be brought into action. If this should also fail, damage to the blanket could be expected, the results of which would be similar to the ones discussed above.

A sudden release of all the energy stored in the magnetic field is physically impossible, due to the high inductivity of the coils. Some of the energy could be released if the superconducting material should revert to normal conductivity. However, zones of normal conductivity also do not arise suddenly, and can be detected by monitoring systems which have already been tested. The failure of a magnet coil does have consequences, however. For example, it could distort the magnetic field in such a way as to endanger the plasma containment, resulting in the type of accident discussed above, if the reactor were not shutdown.

As far as is now known, more serious safety problems would be caused by the accidental release of tritium from a fusion reactor. There are several possible causes of tritium release, from small leaks to a break in a main coolant pipe. The results would depend on the cause. Even the worst case, the release of large amounts of tritium due to the break of a main coolant pipe, could be controlled if the reactor were surrounded by a pressure-resistant containment, and the containment vessel in turn with a gas-tight safety shell (209). If one assumes that the entire tritium inventory were inside the containment, leakage rates up to 100 kCi/day from the containment would be possible. If the containment were surrounded by an additional gas-tight shell, it would be possible to pump these leakage losses back into the containment vessel. If they were released continuously into the atmosphere through a 100 m chimney, the area around the plant would be exposed to several mrem/day. If one assumes that the containment broke, the fusion power plant could certainly cause considerable damage. For example, if one considers the critical dose to bone marrow in rem, the limiting dose in case of accidents, 25 rem, recommended by the ICRP, would only be exceeded in an area about the size of a power plant site (200, 209). (This area is about 40 times larger for a fast breeder.)

As discussed earlier, the CTR, unlike a fission reactor, does not generate fission products as a direct result of its power reaction. The only highly active waste it generates is in the form of solid structural materials. The radioactive material produced by a CTR can be classified in three categories: 1) Continuously produced radioactive materials, such as the filters of the fuel cycle and erosion and corrosion products from the breeder and coolant processing system. It was estimated that the UWMAK I reactor would produce 2.75 t/year of these materials. 2) Intermittantly produced radioactive materials, such as components of the blanket, which would have to be replaced after a certain length of time. It was estimated that the UWMAK I reactor would produce several hundred tons of these per year. 3) Radioactive material to be disposed of when the plant is retired; in addition to the components mentioned above, there would be activated structures like shielding and magnet coils.

It has been shown that the decay behavior of most of the materials suggested for structural components is such that long-term storage would be required (compare Fig. 5-15), but there are also materials, e.g. vanadium, whose activity would decay completely within a reasonably short time, so that long-term storage would not be necessary. It would also be conceivable that this material could be recycled after an appropriate time, to conserve raw material supplies.

In summary, in addition to offering a nearly unlimited source of energy (see 3.36), a CTR would have considerable advantages over any fission reactor with respect to safety and environmental pollution: 1) There is no recognizable mechanism by which a CTR could release activated structural materials, which represent most of the hazard potential, into the environment. 2) By the choice of suitable structural materials, and the corresponding structural design, the problem of decontaminating a CTR can be made much simpler than with a fission reactor. A CTR produces no fission products in its power reaction, and reprocessing – e.g. to recover fuel – is not necessary. 3) A decisive advantage of the CTR with respect to political safety is that it does not produce any materials which could be used in the production of nuclear weapons.

In all these considerations, however, it must be remembered that if nuclear fusion is realized at all, it will probably be 50 years before commercial fusion power reactors are available; the HTRs and FBRs have a developmental lead of 4 to 5 decades over the CTR (see 4.222).

6. Conclusions

The energy problem is a global problem. Because of the relationship between primary energy consumption and gross national product and population, it can be assumed that the world energy demand will continue to rise, although at a decreasing rate.

In the past, economic growth in the industrial countries has been a prerequisite for raising the standard of living of all social classes. However, prosperity is limited to a few countries, and the vast majority of humanity still lives in poverty. For this reason, most countries have a great need for economic growth which, however, at least in the past, has always required a concomitant increase in the growth rate of primary energy consumption. Presumably this will not change in the near future. Therefore, for the medium-term future, a global growth in the consumption of primary energy can be seen as a prerequisite for equalizing the standards of living of the poor and the rich. This is not only desirable, but in the end, it will be necessary to insure world peace. If energy is not to become the limiting factor of economic growth, the necessary amounts of energy will have to be provided, and in time (1, 2).

The analysis of various ways to meet this energy demand shows that, aside from problems of distribution and production of individual energy carriers, the problem is not so much the limited amounts of energy resources, but the global problems of environmental impact and security related to the processes of energy release. In the present situation, taking into account the problems specific to each energy carrier, the following conclusions may be drawn:

1.) In the long run, there is probably no alternative to utilization of solar energy on a scale which makes a significant contribution to the energy economy, because this is a regenerative source of energy, and its utilization is, within wide limits, not harmful to the environment. Essentially the only differences in opinion are concerned with the question of when the various methods for utilization of solar energy will begin to meet a significant fraction of the world energy demand. From the present point of view, it would seem that the chances of success of the various direct and indirect methods differ greatly. The form in which an economically significant utilization of solar energy is achieved will probably vary from one country to the other, depending on a number of factors such as geographical location, the availability and price of other primary energy carriers, and the technological and economic level of development.

The economically exploitable hydropower potential in the industrial countries has already been mostly developed, but in some developing countries there is a large potential, especially compared to the local energy demand.

The development of means to extract energy from waves, ocean heat and currents is only beginning, and these techniques cannot be expected to make a significant contribution in the near future.

There are proven methods for the utilization of wind energy, but the space requirements for their large-scale application are a problem. Also, in some countries the wind can probably only be expected to make a significant contribution if effective energy storage is available, or if the energy provided can be integrated into the existing energy supply system. However, in view of the situation in the world energy market, some countries with a large potential for wind energy are interested in developing it, especially to supply remote areas (decentralized energy supply).

From the present point of view, the most promising methods for utilization of solar energy, even in temperate zones, are low temperature collectors and heat pumps. The basic technology has already been developed, although the individual components still need improvement, and the overall systems need to be optimized, both technically and in terms of cost. It has been shown that a large fraction of the primary energy consumed by industrial countries is used to generate low temperatures heat, which is precisely what the available solar energy could provide, but which is now often provided by petroleum products or natural gas. In other words, the solar energy could provide a substitute for much of the present demand for petroleum and natural gas.

There are possibilities for utilization of high-temperature collectors with concentrating elements in countries with high values of solar radiation.

Of the photochemical methods for utilization of solar energy, biomass produced by photosynthesis has a certain significance. However, the production of biomass for energy purposes alone has not yet proven to be economically feasible in most cases. The disadvantage of renewable biomass, in comparison to fossil fuels, is the relatively low specific energy content and the large surface areas required for its production, which leads among other things to a large expenditure for collection and transport. In addition, the use of land for biomass production competes with its use for more valuable products, such as food. Partly for these reasons, the basic possibility of utilizating solar energy via photolytically generated hydrogen deserves recognition. Because hydrogen can be transported and stored, this would be an interesting possibility for utilizing solar energy in wide areas of the earth. The use of hydrogen as an alternative fuel for mobile consumers would open further prospects for solar energy.

The energy demand is characterized, aside from the large demand for heat, by a large proportion of electricity. Electricity can be produced either from solar heat or from solar cells; the basic technologies exist for both processes. Solar cells have been used successfully in satellites for years, but the investment required for producing electricity from solar cells is many times higher than that for conventional power plants, and the investments required for producing electricity from solar heat are even higher.

An increasingly important criterion for judging solar energy is the high degree of security of national supply. In addition, as far as we know, solar energy is the only source of energy which can be utilized on a global scale without affecting the climate, at least for a period of time long enough to solve many of humanity's current problems.

In summary, given the serious problems associated with other forms of energy, a greater utilization of solar energy all over the world is desirable.

2.) Petroleum is at present the world's main energy carrier ("petroleum phase"). However, it is also the energy carrier which will first be exhausted – at least, as far as economically recoverable reserves are concerned. The problem is exacerbated by the very uneven geographical distribution of the economically recoverable reserves. In spite of this, it appears that it will be extremely difficult to substitute another energy carrier for petroleum.

Thorough studies have shown that – ignoring geographical distribution and recovery costs – there are still large amounts of petroleum which could be recovered, especially if one considers the oil shales and oil sands, which are almost all located in North and South America. On the basis of predictable progress in technology, it can be assumed that in the next 20 years, the offshore reserves of petroleum and natural gas lying under 200–2000 m water will be tapped, even in areas where the climatic conditions are unfavorable. There are also great hopes for the development of new production methods (secondary and tertiary methods). The world average recovery rate of oil was 32% in 1976, compared to only 26% in 1955. It is thought that the world average recovery rate can be increased to about 36% by 1985. (With the use of secondary methods, it is possible in some cases to extract more than 50% of the oil, and with tertiary methods, up to 95%.) An increase in the recovery rate of 1% of the total is equivalent to more than the present yearly world demand for petroleum. In other words, the present *proved recoverable reserves* of nearly $100 \cdot 10^9$ t oil (32% recovery rate) would be increased to $112 \cdot 10^9$ t oil if the recovery rate were 36%, and to $125 \cdot 10^9$ t oil by a recovery rate of 40%. In addition, it can be expected that with intense worldwide exploration, the *proved recoverable reserves* will continue to increase. If one includes the *estimated additional recoverable resources*, which amount to $212 \cdot 10^9$ t oil, a total of $300 \cdot 10^9$ t oil is still available. (The present world production of oil is about $3 \cdot 10^9$ t.)

The "petroleum problem" may thus be primarily a problem of distribution, due to the uneven geographical distribution of the resource, and a price problem for many oil-importing countries, due to the rising prices on the world oil market. So long as petroleum is the primary source of energy, the countries of the Middle East will have a dominant role, especially Saudi Arabia, which owns about a fourth of the *proved recoverable reserves*. However, it cannot be overlooked that the availability of petroleum to individual countries can depend on politics.

Because natural gas can be used in many of the same applications as petroleum, it will be included in these considerations. The technology of natural gas is also similar to that of petroleum in many respects. The hydrocarbons at present supply almost two thirds of the world's primary energy. In light of the growth of demand, the reserve situation for natural gas is somewhat more favorable than that for petroleum, and exploration for and production of natural gas is only beginning in many countries.

Summary: The recoverable reserves of petroleum and natural gas are probably greater than was assumed a few years ago, due to predictable technological developments. This means that the length of time the hydrocarbons can supply a given fraction of the energy demand will depend, aside from distribution problems, on the price people will pay for the their production.

3.) Coal is the only conventional fuel which will certainly be available in sufficient quantity in the next century and thereafter. It is an advantage, also, that unlike petroleum the coal is present in the regions with the largest primary energy consumption – except for Japan. The coal reserves of the important industrial countries will allow a considerable increase in production for many decades to come. However, being a solid, coal is at a disadvantage compared to liquid and gaseous fuels, both technologically and physical-chemically.

In light of the world's coal reserves, it will be necessary to overcome the disadvantages of coal's solid state. By converting it to gaseous and liquid secondary energy carriers (coal refinement products), coal can be made suitable for more applications. The large-scale underground gasification of coal deserves attention, because it makes available reserves of coal which cannot be economically mined. Furthermore, more efforts should be made to generate electricity from coal with less environmental damage, and at the same time, to use the primary energy more efficiently, for instance by heat/power coupling.

A prerequisite for commercial coal refinement will be the availability of cheap process heat. Because high-temperature process heat can be provided by a high temperature reactor, the development of this reactor technology is very important for the realization of commercial coal refinement. The large-scale realization of coal gasification and liquefaction could be a way to substitute coal products for conventional hydrocarbons. Methanol, for example, which in principle can already be produced on a large scale from coal, is a promising liquid fuel which can also be used as an alternative fuel for motor vehicles.

At this point let it be repeated that possible global environmental problems must be taken into account by plans for meeting the future energy demand. On the basis of present predictions of the world demand for primary energy, it would appear that the temperature increase caused by direct heat stress on the environment would be very small compared to the increase to be expected from an increasing atmospheric CO_2 content (indirect heat stress). The limiting factor for the use of coal and its refinement products may be the climatic effects of increased atmospheric CO_2 levels, so that only a fraction of the large coal reserves will be able to be used to meet a growing energy demand. Since this is a global environmental problem, the question may well be raised whether it should be tolerated that the smaller part of humanity applies many times more stress to the environment than the larger part.

Summary: Given the reserve situation, coal is the only conventional primary energy carrier which could supply a considerably higher demand than at present, and that far into the future; it could even meet this demand alone. However, it may not be the coal reserves but the possible climatic effects of the atmospheric CO_2 concentration which limit the use of coal.

4.) The economic potential for tidal energy is, even on a world scale, very slight. The proportion of the world's energy now supplied by the tides is negligibly small and will probably remain so in future. However, tidal energy may well be of local significance.

5.) The global economic potential for geothermal energy is, from the present standpoint, still very slight, but it is rising in a number of countries. Geothermal energy now supplies only a very small fraction of the world's energy, and this is not likely to change in the near future. However, geothermal energy can be significant on a national scale.

6.) Of the "new" energy sources, nuclear fission can contribute an economically significant amount to the world's energy supply.

The reserves of nuclear fission fuels are something of a problem, considering the sources which are now economically recoverable, and the problem is intensified by the concentration of the reserves in a relatively small number of countries. (The USA has a key position.) To be sure, exploration for uranium or thorium is only beginning in many countries, but the forced development of breeders in a few countries ought to be an indication that they consider the economically recoverable reserves of uranium or thorium to be limited. It is possible to make 60 to 70 times better use of the energy content of uranium with fast breeder reactors. Whether the predicted demand for nuclear fission fuels can be met will depend, in the final analysis, on the price people are willing to pay for extracting them from poorer ores, or, in the case of uranium, from sea water.

Years of generally satisfactory operation of more than 250 nuclear power plants and other nuclear installations show that the quality of nuclear safety technology is very high. Experience with nuclear installations and thorough studies of possible accidents, even those which might be caused by natural or human external influences, show that the existing technical standards are such that the population and the plant personnel are not exposed to higher risk of death from reactors than from other industrial installations to which the public has become accustomed. This holds in part even for the influence of third parties on nuclear installations, such as sabotage and war. Although the measures for controlling accidents and other external influences can protect against most acts of sabotage, and to some extent, even against acts of war, a reactor core could be destroyed by a sufficiently powerful nuclear bomb exploding close to it. In the extreme case, the fallout from the core could be added to that of the bomb. The political efforts to place nuclear power plants under special protection in the event of war have been successful. The Diplomatic Conference for Reaffirmation and Further Development of Human Rights in Armed Conflicts was concluded on June 10, 1977. It adopted two amendments to the Geneva Convention of August 8, 1949 which included regulations for the special protection of nuclear power plants in the case of armed conflicts. However, in principle the existence of precision-guided munitions and missiles, which can hit "point targets" make for a reciprocity of hazard from nuclear installations in the event of war.

Aside from a long-term intermediate storage with the intention of reprocessing them later, there are basically two ways to deal with spent fuel elements: conditioning them for permanent storage or reprocessing them to retrieve the energy raw materials they contain. These are then recycled, and the radioactive waste is converted to suitable form for permanent disposal. The direct disposal of fuel elements amounts to throwing away usable energy raw materials.

Fast breeders can only be used if the fuel elements are reprocessed (uranium-plutonium cycle); thorium high-temperature reactors cannot be optimally utilized without reprocessing (uranium-thorium cylce). The reprocessing of spent fuel also has technical advantages, because the reprocessing wastes decay about 1000 times faster than unrecycled fuel. There is no standard method for conditioning spent fuel elements for permanent storage, nor any experience with it on an industrial scale, in contrast to the situation for reprocessing.

Permanent storage of highly active waste is not yet being undertaken in any country, but at present there are no recognizable technical or scientific reasons why it should not be practical. There can be no doubt, however, that much work in this area is still needed.

Because there are certain connections between the peaceful use of nuclear energy and the ability to produce nuclear explosives, the danger that the expansion of peaceful utilization of nuclear energy around the world will increase the number of nuclear powers or threshold powers must be taken very seriously. None of the presently available fuel cycles is absolutely resistant to misuse for the production of nuclear weapons. The international conference INFCE therefore made it clear that the problem of non-proliferation of nuclear weapons is primarily a political one.

Summary: It cannot be denied that there are difficulties associated with the peaceful use of nuclear energy. However, it is only realistic to say that at the present level of technology, there is no other "new" source of energy (ignoring conventional fuels) which can make a significant contribution to the energy economy of the world.

7.) If controlled nuclear fusion can be achieved, the world's lithium and deuterium reserves are large enough to provide humanity with another practically unlimited energy source besides solar energy. The estimates as to when – if they are possible – the various stages on the way to an economical fusion power reactor will be reached vary, sometimes widely; in general, the fast breeder and the high temperature reactor are thought to have a lead of four to five decades.

Everything known so far suggests that fusion would have considerable advantages over fission. Some examples are the high degree of regional or national security of supply, because lithium and deuterium can be extracted in practically unlimited amounts from sea water; the greater safety of the reactors, because a power excursion is not physically possible; the greater ease of waste disposal, given the appropriate choice of structural materials, because the products of the fusion reaction are not radioactive; and the fact that no materials which might be used in the production of nuclear weapons are produced. On the assumption that it will not be possible to use solar energy in the thickly populated areas of the temperature zones to produce an economically significant amount of secondary energy carriers (e.g. electricity or hydrogen), controlled nuclear fusion could be a primary energy source from which these could be made in the long term.

However, it should be mentioned here that commercial energy production from nuclear reactions, either fission or fusion, necessitates handling of radioactive materials on an industrial scale, including the disposal of wastes from the nuclear reactors. However, as mentioned above, the disposal problems arising from nuclear fusion are likely to be more easily solved than those from fission.

Summary: The realization of controlled nuclear fusion would make it possible to meet any desirable future energy demand, and everything we know suggests that fusion would have considerable advantages over fission. Therefore research on the realization of controlled nuclear fusion deserves intensive support.

8.) Secondary energy carriers and supply systems are very important, especially with regard to possibilities for saving energy, since many countries are now utilizing only about a third of the primary energy they consume. The production, transport, storage, environmental impact and safety considerations are all important factors in the evaluation of secondary energy carriers. The trend toward liquid and gaseous secondary fuels and towards electricity will probably continue around the world. In addition, the development of primary and secondary energy carriers cannot be considered separately, in many cases. For example, commercial coal refinement depends on the availability of cheap process heat.

Even in the future, there will be a number of applications in which no other secondary energy carrier will be able to replace electricity. District heating is a practical method to supply the demand for low temperatures heat in population centers. Petroleum products will undoubtedly continue to dominate in the transportation sector for a long time. It is also likely that gaseous and liquid energy carriers, produced on an industrial scale from coal, will substitute in many applications for conventional hydrocarbons, especially because they could easily fit into the existing supply structures. If large-scale production of hydrogen should become economically feasible it would be another versatile secondary energy carrier.

9.) Because the price of energy can be expected to continue to rise, and because of the possible global problems of environmental impact (e.g. the carbon dioxide problem) which might result from too high a consumption of primary energy, as well as the problems with the use of nuclear fission, it will be absolutely necessary, especially for the main consumer countries, to exhaust all possibilities for using energy economically and rationally. Many of the present technologies were developed at a time in which various forms of energy were cheap and readily available. Comparisons between countries show that a lower primary energy consumption per capita is not necessarily indentical with a lack of goods and services, or with a lower standard of living. A prerequisite for the success of energy-saving measures will be a deep change in awareness, a new energy-consciousness among the public, and a responsible attitude towards "energy". However, it should not be forgotten that the present economic structures, which include some energy-intensive technologies, still need to grow to meet great human needs, and the necessary world-wide structures will need time in which to change.

Realistically, in the process of deciding the amounts and combination of energy carriers to be used, not only the scientifically and technologically possible should be considered, but also other questions relevant to energy supply concepts, such as economic, ecological, political, social, legal and ethical questions. The optimal combination may differ from country to country.

7. Apendices

7.1 *Literature*

1. Introduction

(1) W. Häfele (Ed.): Energy in a finite World, Vol I, II, Cambridge: Ballinger 1981.
(2) World Energy Outlook, Public Affairs Department, Exxon Corporation 1981.

2. Primary energy sources and world economics

(1) D. N. Lapedes (Ed. in Chief): Encyclopedia of ENERGY, New York, St. Louis, San Francisco: McGraw-Hill Book Company 1976.
(2) G. Falk, W. Ruppel: Energie und Entropie, pp. 2–41, Berlin, Heidelberg, New York: Springer 1976.
(3) M. Mesarovic, E. Pestel: Mankind at the Turning Point, the Second Report to the Club of Rome, New York: Dutton 1974.
(4) United Nations: Statistical Yearbook 1977, 1978, New York 1978 and 1979.
(5) H. Kahn et al.: The Next 200 Years, A Scenario for America and the World, New York: William Morrow and Company 1976.
(6) World Development Report 1980, World Bank, 1818 H Street, N.W., Washington, D.C. 20433, 1980.
(7) United Nations: Statistical Yearbook 1959–1978, New York 1960–1979.
(8) BP statistical review of the world oil industry 1979, 1980, London 1980 and 1981.
(9) Energy Program of the Government of the Federal Republic of Germany, Federal Ministry of Economics, Bonn, December 1977 and November 1981.
(10) I. J. Bloodworth, E. Bossanyi, D. S. Bowers, E. A. C. Crouch, R. J. Eden, C. W. Hope, W. S. Humphrey, J. V. Mitchell, D. J. Pullin, J. A. Stanislaw: World Energy Demand, The Full Report to the Conservation Commission of the World Energy Conference, Guildford (UK) and New York: IPC Science and Technology Press 1978.
(11) Bulletin Nr. 30, Sonderausgabe: Grundlinien und Eckwerte für die Fortschreibung des Energieprogramms, Presse- und Informationsabteilung der Bundesregierung, March 1977.
(12) D. R. Price: Energy for Food in the Next Century, Division 1 B, 11th World Energy Conference, Munich, pp. 755–776, Sept. 1980.
(13) R. K. Pauchari: Energy and Economic Development in India, New York: Praeger 1977.
(14) M. Chou: World Food Projects and Agricultural Potential, New York: Praeger 1977.
(15) W. F. Martin, F. J. Pinto: Energy for the Third World, Technology Review, Vol. 80, June/July, 48–56 (1978).
(16) M. Grathwohl: Zukunftsperspektiven der Energieversorgung, Naturwissenschaftliche Rundschau, Vol. 30, p. 2 (1977).

(17) Energiebilanzen der "Arbeitsgemeinschaft Energiebilanzen", Düsseldorf 1981.
(18) Shell Oil Company: Public Affairs and Information Department, Hamburg August 1979, and Shell International, Energy Conservation. The Prospects of Improved Energy Efficiency, London 1979.
(19) Exxon Corporation: Public Affairs and Information Department, Hamburg 1978.
(20) Project Interdependence: U.S. and World Energy Outlook through 1990, A Report printed by the Congressional Research Service, Library of Congress, U.S. Government Printing Office, Washington D.C. 20402, November 1977.
(21) E. T. Hayes: Energy Resources Available to the United States, 1985 to 2000, Science, Vol. 203, 233–239 (1979).
(22) H. Schneider, D. Schmitt, W. Pluge: Die Energiekrise in den USA, München, Wien: Oldenburg 1974.
(23) J. Grawe: Aspekte der amerikanischen Energiepolitik, Zeitschrift für Energiewirtschaft, Vol. 1, 20–32 (1978).
(24) E. Pestel et al.: Das Deutschland-Modell, Stuttgart: Deutsche Verlags-Anstalt 1978.
(25) OECD: International Energy Trends, p. 10, Paris, May 1978.
(26) World Energy: looking ahead to 2000. Report by the Conservation Commission of the World Energy Conference, Guildford (UK) and New York: IPC Science and Technology Press 1978.
(27) W. Ungerer: Die Energieprobleme und ihre Perspektiven, Außenpolitik, Vol. 30, 149–160, 2. Quartal 1979.
(28) E. Penrose: OPEC's Importance in the World Oil industry, INTERNATIONAL AFFAIRS, Vol. 55, January, 18–32 (1979).
(29) M. Blair: The Control of Oil, London: McMillan 1977.
(30) P. R. Odell, L. Vallenilla: The Pressure of Oil: A Strategy for Economic Revival, London: Harper and Row 1978.
(31) F. E. Niering Jr.: A new force in world oil, Petroleum Economist, Vol. 46, No 3, 105–113 (1979).
(32) D. O. Beim: Rescuing the LDCS, Foreign Affairs, Vol. 55, 717– 731 (1977).
(33) H. van B. Cleveland, W. H. Bruce Brittain: Are the LDCS in over their heads? Foreign Affairs, Vol. 55, 732–750 (1977).
(34) J. C. Campbell: Oil Power in the Middle East, Foreign Affairs, Vol. 56, 89–110 (1977).
(35) W. J. Levy: The Years that the Locust Has Eaten: Oil Policy and OPEC Development Prospects, Foreign Affairs, Vol. 57, 287–305 (1978) and New York Times, 5. 1. 1979.
(36) J. Amuzegar: OPEC and the Dollar Dilemma, Foreign Affairs, Vol. 56, 740–750 (1978).
(37) Dennis Meadows, Donella Meadows, E. Zahn, P. Milling: The Limits to Growth, a Report for the Club of Rome's Project on the Predicament of Mankind, New York: Universe Books 1972.
(38) Energy Needs, Uses, and Resources in Developing Countries, Brookhaven National Laboratory Developing Countries Energy Program, Report No. BNL 50784, pp. 79–82, March 1978.
(39) D. Gabor, U. Colombo, A. Kling, R. Galli: Beyond the Age of Waste. Science, technology and the management of natural resources, energy, materials, food. A Report to the Club of Rome, Oxford: Pergamon Press 1978.
(40) B. A. Rahmer: Long-term outlook hopeful, Petroleum Economist, Vol. 46, No 3, 91–92 (1979).
(41) BP statistical review of world oil industry 1980, London 1981.
(42) C. Norman: U.S. Details Energy Plan for Third World, Science, Vol. 212, 21–24 (1981).
(43) Ph. H. Abelson: World Energy in Transition, Science, Vol. 210, p. 1311 (1980).
(44) G. Philip: Mexican Oil and Gas, INTERNATIONAL AFFAIRS, Vol. 56, No 3, 474–483 (1980).
(45) World Bank hinking production loans, THE OIL AND GAS JOURNAL, Vol. 76, Oct. 2, p. 64 (1978).
(46) E. Friedmann, R. Goodman: Oil and gas prospects in developing countries, Finance & Development, Vol. 16, June, p. 7 (1979).
(47) L. Auldridge: World's oil flow gains slightly, reserves dip, THE OIL AND GAS JOURNAL, Vol. 76, Dec. 25, 99–106 (1978).
(48) S. M. Billo: Future petroleum resources seen great in Saudi Arabia, THE OIL AND GAS JOURNAL, Vol. 77, Jan. 1, 98–103 (1979).
(49) T. W. Mermel: Contribution of Dams to the Solution of Energy Problems, 10th world Energy Conference (Division 1), pp. 1–14, Istanbul, Sept 1977.
(50) D. O. Hall: Plants as an energy source, Nature, Vol. 278, 114–117 (1979).

(51) Ph. H. Abelson: Bio-Energy, Science, Vol. 204, p. 1161 (1979).
(52) A. K. N. Reddy: Energy Options for the Third World, Bulletin of the Atomic Scientists, Vol. 34, 28–33 (1978).
(53) N. L. Brown, J. W. Howe: Solar Energy for Village Development, Science, Vol. 119, 651–657 (1978).
(54) Auf dem Wege zu neuen Energiesystemen, Teil I, Federal Ministry of Research and Technology (Ed.), Bonn 1975.
(55) H. Schäfer: Energy Technology Research Institute, Munich 1980.
(56) Energiebilanzen in der Bundesrepublik Deutschland, Arbeitsgemeinschaft Energiebilanzen (Ed.), VWEW, Frankfurt/M 1979.
(57) W.-J. Schmidt-Küster, H.-F. Wagner: New Energy Technologies, 10th World Energy Conference (Division 4, 4.8–5), pp. 1–21, Istanbul, Sept. 1977.
(58) H. Schulte: The Combined Generation of Heat and Electricity as a Means of Saving Primary Energy, 10th World Energy Conference (Division 3, 3.3–3), p. 4, Sept. 1977.
(59) Y. Nagano: Japan's Future Energy Structure and the Effect of Selection of Fuels by the Power Industry, 10th World Energy Conference (Division 2, 2.5–6), pp. 1–20, Istanbul, Sept. 1977.
(60) N. Holmström, M. Höjeberg, U. Norhammar: Energy Conservation in Swedish Industry, 10th World Energy Conference (Division 2, 2.1–3), pp. 1–23, Istanbul, Sept. 1977.
(61) Energy Statistics Yearbook 1973–1977, Statistical Office of the European Communities, Bruxelles 1979.
(62) Energy R & D, OECD, Paris 1975.
(63) Statistiques de base de la Communauté 1977. Office statistique des Communautés Européennes, Bruxelles 1978.
(64) N. N.: Market forces slowing U.S. energy consumption. THE OIL AND GAS JOURNAL, Vol. 76, Nov. 6, 34–35 (1978).
(65) N. N.: Citibank: U.S. must intensify conservation moves, THE OIL AND GAS JOURNAL, Vol. 77, Mar. 19, 54–55 (1979).
(66) W. G. Dupree, J. West: United States Energy Trough the Year 2000, Department of the Interior, Washington, D. C., 1972.
(67) Federal Energy Administration, Project Independence Blueprint, Government Printing Office, Washington, D.C., 1974.
(68) S. D. Freeman: Ford Foundation Policy Project, A Time to Choose: America's Energy Future, Cambridge, Mass.: Ballinger 1974.
(69) A National Plan for Energy Research, Development and Administration: Creating Energy Choices for the Future, Vol. 1, ERDA 76-1, Government Printing Office, Washington, D.C., 1976.
(70) Demand and Conservation Panel of the Committee on Nuclear and Alternative Energy Systems, Science, Vol. 200, p. 142 (1978).
(71) Department of Commerce, Domestic and International Business Administration, Forecast of Likely U.S. Energy Supply/Demand Balances for 1985 and 2000 and Implications for U.S. Energy Policy, NTIS PB 266 240, National Technical Information Service, Springfield, Va., 1977.
(72) Executive Office of the President, Office of Energy Policy and Planning. The National Energy Plan, Government Printing Office, Washington, D.C., 1977.
(73) G. Leach: A future with less energy, New Scientist, Vol. 81, 81–83, 11 January 1979.
(74) C. Lewis: A Low Energy Option for the UK, Energy Policy, Vol. 7, 131–148 (1979).
(75) C. P. L. Zaleski: Energy Choices for the Next 15 Years: A View from Europe, Science, Vol. 203, 849–851 (1979).
(76) C. Henderson: The tragic failure of energy planning, Bulletin of the Atomic Scientists, Vol. 34, 15–19, December 1978.
(77) Ph. H. Abelson: Energy Conservation, Science, Vol. 204, p. 695, 1979.
(78) H. Franssen: Energy – An Uncertain Future: An Analysis of U.S. and World Energy Projections through 1990, Washington, D.C., 1978.
(79) R. Halvorsen: Energy Substitution in U.S. Manufacturing, Review of Economics and Statistics, Vol. 59, 381–388 (1977).
(80) K. K. S. Dadzie: Economic Development, Scientific American, Vol. 243, No 3, 55–61 (1980).
(81) E. A. Hudson, D. W. Jorgenson: Energy Policy and U.S. Economic Growth, The American Economic Review, Papers and Proceeding, Vol. 68, 118–123 (1978).

(82) E. A. Hudson, D. W. Jorgenson: Energy Prices and the U.S. Economy, 1972–1976, Natural Resources Journal, Vol. 18, 877–897 (1978).
(83) W. S. Humphrey, J. Stanislaw: Economic Growth and Energy Consumption in the UK, 1700–1975, Energy Policy, Vol. 7, 29–42 (1979).
(84) C. Norman: Energy Conservation: The Debate Begins, Science, Vol. 212, 424-426 (1981).
(85) M. A. Fuss: The Demand for Energy in Canadian Manufacturing, An Example of the Estimation of Production Structures with Many Inputs, Journal of Econometrics, Vol. 5, 89–116 (1977).
(86) M. Denny, J. D. May, C. Pinto. The Demand for Energy in Canadian Manufacturing: Prologue to an Energy Policy, Canadian Journal of Economics, Vol. 11, 300–313 (1978).
(87) Shell Briefing Service, Public Affairs and Information Department, Hamburg, July 1979.
(88) E. Hirst, B. Hannon: Effects of Energy Conservation in Residential and Commercial Buildings, Science, Vol. 205, 656–661 (1979).
(89) K. M. Meyer-Abich (Ed.): Energie, Energieeinsparung als neue Energiequelle, München, Wien: Hanser 1979.
(90) N. N.: VDI-Nachrichten, No. 35, p. 36, 31. 8. 1979.
(91) W. Müller, B. Stoy: Die Entkopplung von Wirtschaftswachstum und Energiemehrverbrauch – ein lang übersehener, gangbarer Weg, Zeitschrift für Energiewirtschaft, Vol. 1, 220– 223 (1978).
(92) W. Müller, B. Stoy: Entkopplung, Wirtschaftswachstum ohne mehr Energie? Stuttgart: Deutsche Verlags-Anstalt 1978.
(93) S. Özatalay, S. Grubaugh, T. Veach Long II: Energy Substitution and National Energy Policy, The American Economic Review, Papers and Proceedings, Vol. 69, 369–371 (1979).
(94) B. Fritsch, G. Kirchgässner: Zur Interdependenz von Wirtschaftswachstum, Energie- und Ressourcenverbrauch, Arbeitstagung des Vereins für Socialpolitik, Mannheim, 24.–26. 9. 1979.
(95) W. Schulz: Wirtschaftstheoretische und empirische Überlegungen zur These der Entkopplung von Wirtschaftswachstum und Energieverbrauch, Arbeitstagung des Vereins für Socialpolitik, Mannheim, 24.–26. 9. 1979.
(96) E. R. Berndt, D. O. Wood: Engineering and Econometric Interpretations of Energy-Capital Complementary, The American Economic Review, Vol. 69, 342-354 (1979).
(97) F. Roberts: The Scope of Energy Conservation in the EEC, Energy Policy, Vol. 7, 117–130 (1979).
(98) V. H. Oppenheim: Why Oil Prices Go Up The Past: We pushed them, Foreign Policy, No 25, pp. 24–57, Winter 1976/77.
(99) Message from the President of the United States (H. Doc. No. 92–201), in: Congressional Record, Vol. 120, No. 3, H 151-H 156.
(100) Carter's Fireside Chat on Energy, in: United States Wireless Bulletin (USWB), 19. 4. 1977, and Carter's Address to Congress on Energy, in: USWB, 21. 4. 1977.
(101) ERDA: A National Plan for Energy Research, Development and Demonstration: Creating Energy Choices for the Future, ERDA-Report 76-1, Washington, D.C., 1976.
(102) James R. Schlesinger: Business Week, p. 71, 25. 4. 1977.
(103) N. N.: Financial Times, p. 18, 21. 4. 1977.
(104) J. Lesourne: INTERFUTURES: facing the future mastering the probable and managing the unpredictable, OECD, Paris 1979.
(105) Landsberg: Washington Post, p. A 21, 16. 5. 1977.
(106) M. Willrich, M. A. Conant: The International Energy Agency: An Interpretation and Assessment, American Journal of International Law, Vol. 71, No. 2, 213 (1977).
(107) A. D. Sakharov: Nuclear energy and the freedom of the West, Bulletin of the Atomic Scientists, Vol. 34, 13–14, June 1978.
(108) A. B. Lovins: Energy Strategy: The Road Not Taken?, Foreign Affairs, Vol. 55, 65-96 (1976).
(109) D. C. White: Energy Choices for the 1980s, Technology Review, Vol. 82, No 8, 30-40 (1980).
(110) A. B. Lovins: Soft Energy Paths: Toward a Durable Peace, New York: Harper and Row 1979.
(111) G. Neuwirth, P. Weish: From Austria: 1978 referendum, Bulletin of the Atomic Scientists, Vol. 35, p. 49, May 1979.
(112) G. Petitpierre, B. Giovannini: From Switzerland: 1979 initiative, Bulletin of the Atomic Scientists, Vol. 35, p. 49, May 1979.
(113) Nuclear Power Issues and Choices. Report to the Nuclear Energy Policy Study Group (sometimes called the Ford-Mitre-Report), Cambridge: Ballinger 1977.

(114) M. Miller: The Nuclear Dilemma: Power, Proliferation and Development, Technology Review, Vol. 81, 18-29, May 1979.
(115) T. Greenwood, H. A. Feiveson, T. B. Taylor: Nuclear Proliferation, New York, St. Louis, San Francisco: McGraw-Hill Book Company 1977.
(116) Projections of Energy Supply and Demand and Their Impacts, DOE/EIA – 0036/2, Energy Information Administration, Department of Energy, Washington, D.C., 1978.
(117) C. Marchetti: On Strategies and Fate, in: Second Status, Report of the IIASA Project on Energy Systems 1975, RR-76-1 Laxenburg/Vienna, Austria.
(118) R. Strobaugh, D. Yergin: Energy Future-Report of the Energy Project at the Harvard Business School, Random House N.Y. 1979.
(119) W. Metz, A. L. Hammond: Solar Energy in America, American Association for the Advancement of Science 1978.
(120) R. H. Bezdek, A. S. Hirshberg, W. H. Babook: Economic Feasibility of Solar Water and Space Heating, Science, Vol. 203, 1214–1220 (1979).
(121) N. N.: Sun should provide 20% of US energy, says Carter, Nature, Vol. 279, p. 747 (1979).
(122) P. Goldman: Carter's New Energy Plan, Newsweek, April 1979.
(123) B. Ante: Sonnenenergienutzung 1977, Stand, Erfahrungen, Aussichten, in: Heizen mit Sonne, Tagungsbericht der Deutschen Gesellschaft für Sonnenenergie, U. Bossel (Hrsg.), Göttinger Dissertationsdruck 1977.
(124) Japan's Sunshine Project, Ministry of International Trade and Industry, Tokyo, July 1977.
(125) H. Kobayashi: Technical Development of Solar Energy Utilization in Japan (Division 4, 4.2–5), 10th World Energy Conference, Istanbul, pp. 1–10, Sept. 1977.
(126) World Energy Conference 1980: Survey of Energy Resources 1980, Munich, September 1980.
(127) N. N.: Paraho: Put oil shale development on crash basis, THE OIL AND GAS JOURNAL, Vol. 77, Jan. 22, 28-29 (1979).
(128) P. Crow: Canada looks to tar sands, heavy oil to fill conventional-oil deficit, THE OIL AND GAS JOURNAL, Vol. 77, Feb. 26, 81–87 (1979).
(129) N. N.: Oil Sands Projects Move Forward, Petroleum Economists, Vol. 46, No 1, 32 (1979).
(130) N. N.:Carter budget cuts coal, oil outlays, THE OIL AND GAS JOURNAL, Vol. 77, Jan. 29, 94–96 (1979).
(131) W. Sassin: A Global Scenario, in: Second Status Report of the IIASA Project on Energy Systems 1975, RR-76-1, Laxenburg/Vienna, Austria.
(132) W. Häfele, W. Sassin: A Future Energy Scenario, 10th World Energy Conference (Volume 1), pp. 477–506, Istanbul, Sept. 1977.
(133) F. Adler: Steinkohle, in: Das Energiehandbuch, 2. Aufl., G. Bischoff; W. Gocht (Ed.), p. 90, Braunschweig: Vieweg, 1976.
(134) W. Gumpel: Energiepolitik in der Sowjetunion. (Abhandlungen des Bundesinstituts für ostwissenschaftliche und internationale Studien, Bd. XXIV), p. 99, Köln: Wissenschaft und Politik 1970.
(135) Ch. Starr: Energy System Options, Volume 1, 10th World Energy Conference, Istanbul, pp. 437–473, Sept. 1977.
(136) J. P. Hardt, R. A. Bresnick, D. Levine: Soviet Oil and Gas in the Global Perspective, pp. 787–858, in: Project Interdependence: U.S. and World Energy outlook through 1990 (see 20).
(137) Essam El-Hinnawi: Energy, Environment and Development (Division 4, 4.8–1), 10th World Energy Conference, Istanbul, pp. 1–26, Sept. 1977.
(138) R. J. Budnitz, J. P. Holdren: Social and Environmental Costs of Energy Systems, in: J. M. Hollander (Ed.), Annual Review of Energy, Vol. 1, Annual Review Inc., Palo Alto, Calif., 1976.
(139) E. J. Barron, S. L. Thompson, S. H. Schneider: An Ice-Free Cretaceous? Results from Climate Model Simulations, Science, Vol. 212, 501–508 (1981).
(140) H. Siebert: Externalities, Environmental Qualiaty, and Allocation, Review of World Economics, pp. 17-32 (1975).
(141) H. Trenkler: An Attempt to give a Survey of the Environmental Influence being exercised by the Electricity Supply Industry within the Overall Economy System, Division 3, 11th World Energy Conference, Munich, pp. 305–324, Sept. 1980.
(142) A. C. Fischer, J. V. Krutilla: Resource Conservation, Environmental Preservation, and the Rate of Discount, The Quarterly Journal of Economics, Vol. 89, 358–370 (1975).

(143) W. J. Baumol: On Recycling as a Moot Environmental Issue, Journal of Environmental Economics and Management, Vol. 4, 83–87 (1977).
(144) K. Jaeger: Eine ökonomische Theorie des Recycling, Kyclos, Vol. 29, 660-677 (1976).
(145) R. Lusky: A Model of Recycling and Pollution Control, Canadian Journal of Economics, Vol. 9, 91–101 (1976).
(146) R. M. Solow: The Economics of Resources or the Resource of Economics, American Economic Review, Vol. 61, 1–21 (1974).
(147) P. Nounou: La pollution pétrolière des océans, La RECHERCHE, Vol. 10, 147–155, Fevrier 1979.
(148) M. Grathwohl: Die Energieversorgung der Bundesrepublik Deutschland und Westeuropas unter besonderer Berücksichtigung des Öl- und Gaspotentials der Nordsee, Brennstoff-Wärme-Kraft, Vol. 29, 6, 233–238 (1977).
(149) E. Symonds: U.S. Energy Demand and Supply 1975–1990 – Financing Problems, in: Project Interdependence: U.S. and World Energy Outlook through 1990, A Report printed by the Congressional Research Service. U.S. Government Printing Office, Washington, D.C., November 1977.
(150) Financial Analysis of a Group of Petroleum Companies, The Chase Manhattan Bank, New York, N.Y. September 1976.
(151) Energy Financing: A New Look at FEA's National Energy Outlook, Resources for the Future, Washington, D.C., August 1976.
(152) Capital Resources for Energy Through the Year 1990, Bankers Trust Co., New York, N.Y. 1976.
(153) W. W. Rostow: Energy Target for the United States: A Net Export Position by 1990, Orbis, Vol. 24, No 3, 459–489 (1980).
(154) H. W. Kendall, S. J. Nadis (Eds.): Energy Strategies, Cambridge: Ballinger 1980.
(155) W. Müller-Michaelis, K. Harms: Der zukünftige Investitionsbedarf der Mineralöl- und Energiewirtschaft, British Petroleum Company, Public Affairs and Information Department, Hamburg 1974.
(156) G. Schürmeyer: The economic aspects of North Sea oil, Marine Technology, Vol. 7, 37–45, April 1976.
(157) C. F. v. Weizsäcker: Wege in der Gefahr. Eine Studie über Wirtschaft, Gesellschaft und Kriegsverhütung, 4. Aufl., München, Wien: Carl Hanser, 1977.
(158) W. J. Levy: Oil and the Decline of the West, Foreign Affairs, Vol. 58, 999–1015 (1980).
(159) UN Economic Commission for Europe (ECE), Increased Energy Efficiency in the ECE Region, United Nations, E/ECE/883, Rev. 1, New York 1976.
(160) Energy Models for the European Community, Guildford (UK) and New York: IPC Science and Technology Press 1979.

3. The world's energy potential

(1) Angewandte Systemanalyse, Nukleare Primärenergieträger, Teil II, Energie durch Kernfusion, Authors: R. Bünde, W. Dänner, H. Herold, J. Raeder, M. Söll, Max-Planck-Institut für Plasmaphysik, Garching/München 1978.
(2) Die künftige Entwicklung der Energienachfrage und deren Deckung, Abschnitt III, Authors: F. Barthel, P. Kehrer, J. Koch, F. K. Mixius, D. Weigel, Federal Insitute for Geosciences and Natural Resources, Hannover 1976.
(3) G. B. Fettweis: World coal resources: methods of assessments and results, Essen: Glückauf 1979.
(4) L. Bauer, G. B. Fettweis, W. Fiala: Classification Schemes and their Importance for the Assessment of Energy Supplies (Division 1, 1.1–2), 10th World Energy Conference, Istanbul, pp. 1–20, Sept. 1977.
(5) V. E. McKelvey: Mineral Resources Estimates and Public Policy, American Scientist, Vol. 60, 32–40 (1972).
(6) J. J. Schanz: Resource Terminology. An Examination of Concepts and Terms and Recommendations for Improvement. Final Report 1975. Supported by Electric Power Research Institute Palo Alto, Calif. 1975.

(7) Department of Energy, Mines and Resources, Ottawa, Departmental terminology and definitions of resources and reserves. Interim document, January 1975.
(8) G. I. S. Govett, M. H. Govett: The Concept and Measurement of Mineral Reserves and Resources, Resource Policy, Vol. 1, 46-55 (1974).
(9) L. Bauer, G. B. Fettweis, W. Fiala: Classification Schemes and their Importance for the Assessment of Energy Supplies (Division 1, 1.1–2), pp. 1–21, 10th World Energy Conference, Istanbul, Sept. 1977.
(10) M. Grenon: On Fossil Fuel Reserves and Resources, International Institute for Applied Systems Analysis, Laxenburg/Vienna RM-78-35, June 1978.
(11) L. Auldridge: World oil flow gains slightly, reserve dip (worldwide issue), THE OIL AND GAS JOURNAL, Vol. 76, Dec. 25, 99–106 (1978).
(12) World Energy Conference 1980: Survey of Energy Resources 1980, prepared by Federal Institute for Geoscience and Natural Resources Hannover, Federal Republic of Germany for 11th World Energy Conference, Munich, 8.–12. September 1980.
(13) S. B. Alpert, D. F. Spencer, A. Flowers: Coal Utilization for Advanced Power Generating Systems and Pipeline Gas (Division 3, 3.4–1), pp. 1–21, 10th World Energy Conference, Istanbul, Sept. 1977.
(14) J. Gibson, G. F. Kennedy: Coal on the Environment, Division 3, 11th World Energy Conference, Munich, pp. 325–346, Sept. 1980.
(15) World Energy: looking ahead to 2020. Report by the Conservation Commission of the World Energy Conference, Guildford (UK) and New York: IPC Science and Technology Press 1978.
(16) M. F. Duret et al.: The contribution of nuclear power to world energy supply, 1975 to 2020, A Report prepared for the Conservation Commission of the World Energy Conference, Ottawa, July 1977.
(17) B. Badger et al.: UWMAK-I, A Wisconsin Toroidal Fusion Reactor Design. University of Wisconsin, Report UWFDM-68 (Vol. I 1974, Vol. II 1975);UWMAK-II, a Conceptual Tokamak Power Reactor Design. University of Wisconsin, Report UWFDM-112 (1975); UWMAK-III, a High Performance Non Circular Tokamak Power Reactor Design. University of Wisconsin, Report UWFDM-150 (1976).
(18) R. Bünde, W. Dänner, W. Hofer, M. Hüls, R. Pöhlchen, M. Söll, E. Taglauer, H. Weichselgartner: Aspects of Energy Supply by Fusion Reactors, Max-Planck-Institut für Plasmaphysik, Garching/München, IPP V 1/1-Report März 1974.
(19) Nukleare Primärenergieträger (Teil II): Energie durch Kernfusion, Max-Planck-Institut für Plasmaphysik, Garching/München, ASA-ZE/09/78, Authors: R. Bünde, W. Dänner, H. Herold, J. Reader, M. Söll.
(20) Essam El-Hinnawi: Energy, Environment and Development (Division 4, 4.8–1), 10th World Energy Conference, Istanbul, Turkey, Sept. 1977.
(21) American Gas Association, Gas Supply Review, Vol. 5, 6 (1977).
(22) F. M. Peterson, A. C. Fisher: The Exploitation of Extractive Resources, Economic Journal, Vol. 87, 681–721 (1977).
(23) C. L. Wilson (Ed.): Coal-Bridge to the Future, Vol. I, II, Cambridge: Ballinger 1980.
(24) J. P. Holdren: Energy and Prosperity: Some Elements of a Global Perspective, Bulletin of Atomic Scientists, Vol. 31, No 1, 26–28 (1975).
(25) W. J. Levy: New York Times, 4. 1. 1979.
(26) W. Sassin: Energy, Scientific American, Vol. 243, No. 3, 106–110 (1980).
(27) Exxon Corporation, Public Affairs and Information Department, Hamburg 1981.
(28) Energy: Global Prospects 1985–2000, Report of the Workshop on Alternative Energy Strategies (WAES), New York, St. Louis, San Francisco: McGraw-Hill Book Company 1977.
(29) W. Peters: Possibilities and Limitations of a Future Utilization of Coal for Energy Supplies, Division 1 A, 11th World Energy Conference, Munich, pp. 612–631, Sept. 1980.
(30) H. Aoki: International Cooperation in World Coal Development, Division 2, 11th World Energy Conference, Munich, pp. 461-479, Sept. 1980.
(31) Jahrbuch für Bergbau, Energie, Mineralöl und Chemie, Essen: Glückauf 1950–1974.
(32) A. F. Agnew: Coal Reserves, Resources and Production, pp. 208–263, in: Project Interdependence: U.S. and World Energy Outlook through 1990, A Report printed by the Congressional Research Service, Library of Congress, U.S. Government Printing Office Washington, D.C. 20402, November 1977.

(33) M. Grenon: Coal: Resources and Constraints, in Second Status Report of the IIASA Project on Energy Systems PR-76-1, International Institute for Applied Systems Analysis, Laxenburg/Vienna 1976.
(34) E. D. Griffith, A. W. Clarke: World Coal Production, Scientific American, Vol. 240, No 1, 28–37 (1979).
(35) E. T. Hayes: Energy Resources Available to the United States. 1985 to 2000, Science, Vol. 203, 233-239 (1979).
(36) Statistik der Kohlenwirtschaft e.V., Essen und Köln, September 1980.
(37) H.-G. Frank, A. Knop: Kohleveredelung, Berlin, Heidelberg, New York: Springer 1979.
(38) P. F. Chester, A. J. Clarke, R. B. Hyde, F. R. Hunt: The Effect on the Environment of Producing Electricity from Coal, Division 3, 11th World Energy Conference, Munich, pp. 347–366, Sept. 1980.
(39) K. Bund, W. Bellingrodt, F. C. Erasmus, R. Lenhartz: The Working of Coal Deposites in Deep Depth in the Federal Republic of Germany (Division 1, 1.2–7), 10th World Energy Conference, Istanbul, pp. 1–10, Sept. 1977.
(40) M. Weber: Hydraulischer Feststofftransport, VDI-Nachrichten, No. 26, 30. 6. 1978.
(41) J. D. Moody, H. T. Halbouty: 10th World Petroleum Congress, Bukarest 1979.
(42) D. Levine: Petroleum Geology, pp. 821–837, in: Project Interdependence: U.S. and World Energy Outlook through 1990, A Report printed by the Congressional Research Service, Library of Congress, U.S. Government Printing Office, Washington, D.C. 20402, November 1977.
(43) J. D. Moody, R. W. Esser: World Crude Resource May Exceed 1500 Billion Barrels, World Oil, Vol. 181, No 4, 47–50 (1975).
(44) M. K. Hubbert: World Oil and Natural Gas Reserves and Resources, in: Project Interdependence: U.S. and World Energy Outlook through 1990 (see 42).
(45) P. Desprairies: Report on Oil Resources, 1985 to 2020, Executive Summary, 10th World Energy Conference, Conservation Commission, London, 15. August 1977.
(46) J. Barnea, M. Grenon, R. F. Meyer (Eds.): International Conference sponsored by the United Nations Institute for Training and Research (UNITAR) and the IIASA, New York: Pergamon Press 1977.
(47) P. Odell: The Future of Oil, A Rejoinder, Geographical Journal, Vol. 139, No 3, 436–454 (1973).
(48) R. A. Dick, S. P. Wimpfen: Oil Mining, Scientific American, Vol. 243, No 4, 156–161 (1980).
(49) N. N.: How much oil in the world? (a study conducted under the auspices of the World Energy Conference), Petroleum Economist Vol. 45, No 3, 86–87 (1978).
(50) Exploration in Developing Countries, Exxon Corporation, June 1978.
(51) E. Gabriel: Erdölwirtschaft in Afrika, Geographische Rundschau, Vol. 31, No 2, 46–50, (1979).
(52) N. N.: World Bank hiking production loans, THE OIL AND GAS JOURNAL, Vol. 76, Oct. 2, p. 64 (1978).
(53) B. Enright: Saudis add another Mexico, THE OIL AND GAS JOURNAL, Vol. 77, Jan. 1, p. 19 (1979).
(54) F. E. Niering Jr.: A new force in world oil, Petroleum Economist, Vol. 46, No 3, 105–113, (1979).
(55) W. D. Metz: Mexico: The Premier Oil Discovery in the Western Hemisphere, Science, Vol. 202, 1261-1265 (1978).
(56) B. A. Rahmer: Big potential for offshore oil, Petroleum Economist, Vol. 45, No 1, 9–10 (1978).
(57) N. N.: Soviets hit China oil potential reports, THE OIL AND GAS JOURNAL, Vol. 77, Feb. 12, 40–41 (1979).
(58) S. S. Harrison: A study from the Carnegie Endowment for International Peace, New York: Columbia University Press 1977.
(59) A. A. Meyerhoff, J. O. Willums: China's potential still a guessing game, Offshore, Vol. 39, No 1, 54–56 (1979).
(60) N. N.: China pushing expansion of oil and gas, THE OIL AND GAS JOURNAL, Vol. 77, Apr. 23, 26–29 (1979).
(61) F. Müller: Die Situation des Energiesektors in der Sowjetunion mit Blick auf die 80er Jahre, Osteuropa Wirtschaft, Vol. 24, No 1, 24–34 (1979).
(62) H. M. Wilson: How operators view Prudhoe Bay now, THE OIL AND GAS JOURNAL, Vol. 77, Feb. 26, 67–71 (1979).

(63) H. M. Wilson: Alaska explorers still sure big finds coming, THE OIL AND GAS JOURNAL, Vol. 77, Feb. 26, 72–77 (1979).
(64) B. M. Miller et al.: Geological Estimates of Undiscovered Recoverable Oil and Gas Resources in the United States (U.S. Geological Survey Circular 725; available at no charge from the Branch of Distribution, U.S. Geological Survey, 1200 South Eads Street, Arlington, VA. 22202).
(65) P. Crow: Canadian frontier potential still high: but production distant, THE OIL AND GAS JOURNAL, Vol. 77, Feb. 26, 88–92 (1979).
(66) L. Dienes: Soviet Energy Resources and Prospects, Current History, March, 117–135 (1978).
(67) S. M. Billo: Future petroleum resources, THE OIL AND GAS JOURNAL, Vol. 77, Jan. 1, 98–103 (1979).
(68) P. A. Ziegler: Geology and Hydrocarbon Provinces of the North Sea, GeoJournal, No. 1, 7–31 (1977).
(69) Exxon Corporation (Oeldorado 78–80), Public Affairs and Information Department, Hamburg, 1979–1981.
(70) BP statistical review of world oil industry 1980, London 1981.
(71) M. Grathwohl: The significance of North Sea oil for energy supplies to Western Europe and the Federal Republic of Germany, Marine Technology, Vol. 8, No 1, 1–8 (1977).
(72) M. Quinlan: One million barrels daily, Petroleum Economist, Vol. 45, pp. 337–339, August 1978.
(73) M. Bendell: Norway's go-slow goes even slower, Petroleum Economist, Vol. 45, pp. 137–139, April 1978.
(74) R. A. Kerr: How Much Oil? It depends on whom you ask, Science, Vol. 212, 427–429 (1981).
(75) P. A. Abelson: Dependence on Imports of Oil, Science Vol. 203, p. 1297 (1979).
(76) N. N.: World Oil Production, Petroleum Economist, Vol. 46, p. 131, March 1979.
(77) N. N.: World wide crude oil production, THE OIL AND GAS JOURNAL, Vol. 77, Feb. 26, p. 166 (1979).
(78) H. William Menard: Toward a Rational Strategy for Oil Exploration, Scientific American, Vol. 244, No 1, 47–57 (1981).
(79) G. Schürmeyer: The economic aspects of North Sea oil, Marine Technology, Vol. 7, 37–45, April 1976.
(80) N. N.: Iran crude export capacity not expected to exceed 4 million b/d by late 1980's, THE OIL AND GAS JOURNAL, Vol. 77, 22–23, Jan. 8 (1979).
(81) R. Vielvoye, B. Tippee: Iran shutdown pinches global supply of crude, THE OIL AND GAS JOURNAL, Vol. 77, 75–78, Jan. 29 (1979).
(82) B. Tippee: Iran shutdown slashes global 'faest' of crude, THE OIL AND GAS JOURNAL, Vol. 77, 35–39, Feb. 26 (1979).
(83) A. R. Flower: World Oil Production, Scientific American, Vol. 238, 42–49, March 1978.
(84) Shell Briefing Service, Public Affairs and Information Department, Hamburg, October 1979.
(85) R. Vielvoye: OPEC price hike benefits North Sea, THE OIL AND GAS JOURNAL, Vol. 77, pp. 82–87, June 4 (1979).
(86) L. Le Blanc: Platform economics: a costly game, Offshore, the Journal of Ocean Business, pp. 46–56, December 1978.
(87) Offshore progress-technology and costs, Shell Briefing Service, London, September 1975.
(88) H.-G. Goethe: New Technologies for the development of energy resources as demonstrated in the production of crude oil and natural gas from shelf areas, VDI-Berichte 338, Düsseldorf: VDI-Verlag, 1979.
(89) R. A. Kerr: Petroleum Exploration: Discouragement about the Atlantic Outer Continental Shelf Deepens, Science, Vol. 204, 1069–1072 (1979).
(90) "Enhanced Oil Recovery", National Petroleum Council, December 1976. (This report contains over 235 selected references).
(91) Exxon Corporation (Informationsprogramm Nr. 13) Public Affairs and Information Department, Hamburg 1976.
(92) H.-J. Neumann: Erdölforschung, Die Naturwissenschaften, Vol. 63, No 10, 471–476 (1976).
(93) G. Pusch: Tertiärölgewinnungsverfahren, Erdöl und Kohle, Vol. 30, No 1, 13–25 (1977).
(94) R. J. Blackwell: Enhanced Oil Recovery Processes, 10th World Energy Conference (Division 1, 1.2–1), pp. 1–16, Istanbul, Sept. 1977.

(95) Enhanced Oil Recovery Potential in the United States, Washington, D.C., Office of Technology Assessment, US Congress, January 1978.
(96) Exxon Corporation, Public Affairs and Information Department, Hamburg, 1979.
(97) H. Gaensslen: Economic Analysis of Coal – and Oil – Based Chemical Processes, Energy Systems and Policy, Vol. 2, No 4, 369–379 (1978).
(98) D. Mutch: Rising prices stagger UK progress, Offshore, Vol. 39, 70–73, February 1978.
(99) R. Steven: Stormy politics deter UK drilling, Offshore, Vol. 39, 61–64, February 1978.
(100) Ch. J. Holland Jr.: Drilling for gas: It's complicated now, World Oil, Vol. 188, 55–58, March 1979.
(101) P. Crow: Need seen for pipeline grid changes, THE OIL AND GAS JOURNAL, Vol. 76, 120–123, Nov. 13 (1978).
(102) G. Bonfiglioli: Trans-Med pipeline will stretch offshore laying technology, THE OIL AND GAS JOURNAL, Vol. 76, 108–115, Sept. 25 (1978).
(103) D. Ewringmann: The Importance of the Gas Economy with Regard to the Protection of the Environment, Division 3, 11th World Energy Conference, Munich, pp. 491–513, Sept. 1980.
(104) N. N.: Semac lays pipeline in record speed, Offshore, Vol. 39, 70–73, February 1979.
(105) P. Alby, G. Kardaun, K. Liesen: The International Natural Gas Trade – an Example of World-Wide Cooperation, Division 2, 11th World Energy Conference, Munich, pp. 351–373, Sept. 1980.
(106) E. A. Giorgis: Natural Gas in a Changing World – The Importance of Worldwide Cooperation, Division 2, 11th World Energy Conference, Munich, pp. 247–274, Sept. 1980.
(107) N. N.: U.S. potential gas resource hiked 5–10%, THE OIL AND GAS JOURNAL, Vol. 77, 82-83, Apr. 9 (1979).
(108) R. Sumpter: U.S. gas supply/demand seen nearing balance, THE OIL AND GAS JOURNAL, Vol. 76, 57–62, Sept. 25 (1978).
(109) K. Beissner, W. Dreyer, G. Fürer, R. Koch, H. Schleicher: Speicherung von Kohlenwasserstoffen in Kavernen, Erdöl und Kohle, Vol. 29, No 5, 193–198 (1976).
(110) E. Meinen: LNG storage enclosed in prestressed concrete safety walls, THE OIL AND GAS JOURNAL, Vol. 77, 117–120, May 14 (1979).
(111) S. M. Wolf: Liquefied natural gas, The Bulletin of the Atomic Scientists, Vol. 34, 20–25, December 1978.
(112) E. Drake, R. C. Reid: The Importation of Liquefied Natural Gas, Scientific American, Vol. 236, 22–29, April 1977.
(113) R. L. Keeney, R. B. Kulkarni, K. Nair: Assessing the Risk of an LNG Terminal, Technology Review, Vol. 81, 64–72, October 1978.
(114) J. M. Stuchly, G. Walker: LNG long-distance pipelines – a technology assessment, THE OIL AND GAS JOURNAL, Vol. 77, 59–63, Apr. 16 (1979).
(115) A. Lumsden: The LNG Market, Petroleum Economist, Vol. 45, 465–466, November 1978.
(116) G. Bischoff, W. Gocht (Ed.): Das Energiehandbuch, Braunschweig: Vieweg, 1976.
(117) N. N.: Paraho: Put oil shale development on crash basis, THE OIL AND GAS JOURNAL, Vol. 77, 28– 29, Jan. 22 (1979).
(118) Th. Maugh II: Tar Sands: A New Fuels Industry takes Shape, Science, Vol. 199, 756–760 (1978).
(119) E. Marschall: OPEC Prices make heavy oil look profitable, Science, Vol. 204, 1283–1287 (1979).
(120) N. N. Oil Sands Projects Move Forward, Petroleum Economist, Vol. 46, p. 32, January 1979.
(121) C. W. Bowman, G. W. Govier: Status and Challenges in the Recovery of Hydrocarbons from the Oil Sands of Alberta, Canada, 10th World Energy Conference, (Division 1, 1.2–6), pp. 1–21, Istanbul, Sept. 1977.
(122) G. T. Seaborg: Albert Einstein – a reflection, Bulletin of the Atomic Scientists, Vol. 35, No 3, 20–26 (1979).
(123) B. T. Feld: Einstein and the politics of nuclear weapons, Bulletin of the Atomic Scientists, Vol. 35, No. 3, 5–16 (1979).
(124) M. Gowing: Reflections on atomic energy history, Bulletin of the Atomic Scientists, Vol. 35, No. 3, 51–54 (1979).
(125) S. R. Weart, C. W. Szilard (Eds.): Leo Szilard - His Version of the Facts: Selected Recollections and Correspondence, Cambridge and London: MIT Press 1978.

(126) K. Winnacker, K. Wirtz: Das unverstandene Wunder – Kernenergie in Deutschland, Düsseldorf, Wien: Econ 1975.
(127) G. Walpuski: Verteidigung + Entspannung = Sicherheit, Bonn–Bad Godesberg 1975.
(128) "Treaty on the Non-Proliferation of Nuclear Weapons", Europa-Archiv, Folge 14/1968, p. D 321 ff.
(129) W. Sweet: The U.S.–India safeguards dispute, Bulletin of the Atomic Scientists, Vol. 34, No. 6, 50–52 (1978).
(130) N. N.: Nuclear powers must act quickly, Nature, Vol. 280, p. 1, 5 July 1979.
(131) A. Kidwai: Pakistan considers uranium enrichment, Nature, Vol. 280, p. 436, 9 August 1979.
(132) C. F. v. Weizsäcker (Ed.): Kriegsfolgen und Kriegsverhütung. Mit Beiträgen v. H. Afheldt, A. Pfau, U.-P. Reich, P. Sonntag, A. Künkel, K. Rajewski, E. Rahner, H. Roth, München, Wien: Carl Hanser 1971.
(133) C. F. v. Weizsäcker: Wege in der Gefahr. Eine Studie über Wirtschaft, Gesellschaft und Kriegsverhütung, 4. Aufl., München, Wien: Carl Hanser 1977.
(134) "Treaty between the United States of America and the Union of the Soviet Socialist Republics on underground nuclear explosions for peaceful purposes." Joint Committee on Atomic Energy Development, use and control of nuclear energy for the common defence and security and for peaceful purposes, Washington D.C. 1976, pp. 108–125.
(135) "Treaty between the United States of America and the Union of Soviet Socialist Republics on the Limitation of Underground Nuclear Weapon Tests." US Arms Control and Disarmament Agency, Arms Control and Disarmament Agreements, Washington D.C. 1977, pp. 158–161.
(136) "Protocol to the treaty between the United States of America and the Union of Soviet Socialist Republics on underground nuclear explosions for peaceful purposes." The Department of State Bulletin, Washington, Vol. LXXIV, No. 1931, 28. 6. 1976, p. 802 ff.
(137) Th. Ginsburg: Die friedliche Anwendung von nuklearen Explosionen, Thiemig-Taschenbücher, Bd. 21, pp. 3–15, München: Thiemig 1965.
(138) F. M. Kaplan: Enhanced – Radiation Weapons, Scientific American, Vol. 238, 44-51, May 1978.
(139) K. N. Lewis: The prompt and delayed effects of nuclear war, Scientific American, Vol. 241, 27–39, July 1979.
(140) H. Motz: The Physics of Laser Fusion, London, New York, San Francisco: Academic Press 1979.
(141) L. K. Isaacson: Laser-Fusion Program, Summary Report, EPRI-SR-9, Electric Power Research-Institute, Palo Alto, Calif. 1975.
(142) K. A. Brueckner et al.: Assessment of Laser-Driven Fusion, EPRI-ER-203, Electric Power Research Institute, Palo Alto, Calif. 1976.
(143) J. P. Reilly: High-Power Lasers, Technology Review, Vol. 80, p. 57, June/July 1978.
(144) I. Spalding: Moscow revisited-laser fusion, 1979, Nature, Vol. 277, p. 431, 8 February 1979.
(145) Joint Committee on Atomic Energy, Development, use, and control of nuclear energy for the common defense and security and for peaceful purposes, Washington, D.C. 1976.
(146) International Atomic Energy Agency, Treaty on the Non-Proliferation of Nuclear Weapons. Review Conference, May 1975, Bulletin, Vol. 17, No. 2, New York 1975.
(147) V. Schuricht: Kernexplosionen für friedliche Zwecke, Braunschweig: Vieweg 1978.
(148) Peaceful Nuclear Explosions V. Proceedings of a Technical Committee, Vienna, 22–24 November 1976, International Atomic Energy Agency 1978.
(149) Environment and Natural Resources Policy Division, Congressional Research Service, Nuclear Proliferation Factbook, Washington D.C. 1977.
(150) E. Teller et al.: The constructive uses of nuclear explosives, New York: McGraw Hill 1968.
(151) M. Langer: Physikalische Aspekte der nuklearen Sprengtechnik, Physik in unserer Zeit, Vol. 2, No. 3, 66–73 (1971).
(152) W. T. Harvey: Nuclear cratering simulation techniques, The Military Engineer, Vol. 67, No. 6, 204–207 (1975).
(153) Uranium Resources, Production and Demand, A Joint Report by the OECD Nuclear Energy Agency and the International Atomic Energy Agency, Paris, December 1977.
(154) R. J. Sherman: An Industry Perspective of the United States Nuclear Power Situation, Division 4 A, 11th World Energy Conference, Munich, pp. 565-575, Sept. 1980.
(155) World Uranium and Thorium Resources, OECD/ENEA Report, Paris, 1965. Uranium Resources, Revised Estimates, Joint OECD/ENEA – IAEA Report, Paris, 1967. Uranium Produc-

tion and Short Term Demand, Joint OECD/ENEA – IAEA Report, Paris, 1969. Uranium Resources, Production and Demand, Joint OECD/NEA – IAEA Report, Paris, 1970, 1973, 1975, 1977.
(156) Energy Statistics Yearbook 1970–1975 (eurostat), Statistical Office of the European Communities, Brussels 1976.
(157) M. Miller: The Nuclear Dilemma: Power, Proliferation and Development, Technology Review, Vol. 81, 18–29, May 1979.
(158) W. D. Metz: Fusion Research I, II, III, Science, Vol. 192, 1320–1323 (1976), Science, Vol. 193, 38–40, 307–309 (1976).
(159) R. L. Hirsch: Status and Future Directions of the World Program in Fusion Research and Development, Annual Review of Nuclear Sciences, Vol. 25, 79 (1975).
(160) M. Kenward: Fusion research – the temperature rises, New Scientist, Vol. 82, 626–630, 24 May 1979.
(161) J. A. Duffie, W. A. Beckmann: Solar Energy-Thermal Processes, New York: John Wiley & Sons, Inc. 1974.
(162) G. Lehner: Possibilities of utilization of new non-nuclear and non-fossil energy sources, Kerntechnik, Vol. 20, No. 4, 157–168 (1978).
(163) Solar Energy, UK Section of the International Solar Energy Society, The Royal Institution. Published by UK-ISES, London 1976.
(164) W. D. Metz: Energy Storage and Solar Power: An Exaggerated Problem, Science, Vol. 200, 1471–1473 (1978).
(165) J. C. C. Fan: Solar Cells: Plugging into the Sun, Technology Review, Vol. 80, 14–36 (1978).
(166) A. L. Robinson: American Physical Society Panel Gives a Long-Term Yes to Electricity from the Sun, Science, Vol. 203, p. 629 (1979).
(167) S. R. Ovshinsky: A new amorphous silicon-based alloy for electronic applications, Nature, Vol. 276, 482–483 (1979).
(168) J. C. C. Fan, C. O. Bozler, R. L. Chapman: Simplified Fabrication of GaAs Homojunction, Solar Cells with increased Conversion Efficiencies, Appl. Phys. Lett., Vol. 32, p. 390 (1978).
(169) R. Reisfeld: Possibilities of Solar Energy Utilization, Die Naturwissenschaften, Vol. 66, No. 1, 1 –8 (1979).
(170) US Council on Environmental Quality. Solar Energy: Progress and Promise. Washington 1978.
(171) H. Moesta: Possibilities and Restraints in the Use of Solar Energy, Die Naturwissenschaften, Vol. 63, No. 11, 491–498 (1976).
(172) D. Borgese, J. J. Fauré, J. Gretz, A. Strub, H. Treiber, L. Tuardich: Eurelios, The 1 MW(el) Solar Electric Power Plant: A European Community Research Project in the Field of New Energy Sources, Division 2, 11th World Energy Conference, Munich, pp. 374–387, Sept. 1980.
(173) P. Brosche: Gezeiten, Die Naturwissenschaften, Vol. 62, No. 1, 1–9 (1975).
(174) "Final Report on Tidal Power Study for U.S. Energy Research and Development Agency", U.S. Department of Energy contract number EX-76-C-01-2293, Stone and Webster Engineering Corp., Boston, March, 1977.
(175) L. P. Leibowitz: California's Geothermal Resource Potential, Energy Sources, Vol. 3, No. 3/4, 293–311 (1978).
(176) D. Dickson: Energy search comes down to earth, Nature, Vol. 279, 94–95 (1979).

4. Energy supply systems

(1) A. L. Hammond, W. D. Metz, Th. H. Maugh: Energy and the Future, American Association for the Advancement of Science, Washington D.C., 1973.
(2) Chauncey Starr: Energy System Options, 10th World Energy Conference, Vol. 1, pp. 437–476, Istanbul, Sept. 1977.
(3) E. W. Scholski: Atomphysik, Teil II, 5. Aufl., Berlin: VEB Deutscher Verlag der Wissenschaften 1969.
(4) D. Smidt: Reaktortechnik, Bd. 1 (Grundlagen), pp. 4–31, Karlsruhe: G. Braun 1976.

(5) J. A. L. Robertson: The CANDU-Reactor System: An Appropriate Technology, Science, Vol. 199, 675–664 (1978).
(6) E. F. Emley: Principles of Magnesium Technology London: Pergamon Press 1966.
(7) W. Kliefoth, E. Sauter: Kernreaktoren, No. 2, 5. Auflage, Deutsches Atomforum (Hrsg.), Karlsruhe: C. F. Müller 1973.
(8) D. Smidt: Reaktortechnik, Bd. 2 (Anwendungen), Karlsruhe: G. Braun 1976.
(9) W. Oldekop: Einführung in die Kernreaktor- und Kernkraftwerkstechnik, I und II, München: Thiemig 1975.
(10) Uranium Resources, Production and Demand, A Joint Report by the International Atomic Energy Agency, Paris, December 1977.
(11) H. Michaelis: Kernenergie, München: Deutscher Taschenbuch Verlag 1977.
(12) S. Rippon: The Commercial Steam Generating Heavy Water Reactor, Nucl. Engin. Intern., Vol. 19, 659 (1974).
(13) Power Reactors in Member States, 1979 Edition, IAEA, Vienna 1979 and 1980.
(14) N. N.: Atomwirtschaft – Atomtechnik, Vol. 24, p. 526, November 1979.
(15) IAEA – Bulletin, Vienna, No. 1, February 1976.
(16) Nuclear Power in Developing Countries, Vol. 6 of Nuclear Power and its Fuel Cycle, Proceedings of the International Conference, Salzburg, May 2–13, 1977, IAEA, Vienna, 1977.
(17) L. P. Bloomfield: Nuclear Spread and World Order, FOREIGN AFFAIRS, Vol. 53, 743–755, July 1975.
(18) H. H. Fewer, W. Mattick: Economic and Safety Advantages of Standardization (Division 3, 3.2–2), 10th World Energy Conference, Istanbul, pp. 1–20, Sept. 1977.
(19) W. Keller: Energieversorgung durch Leichtwasserreaktoren, Atomwirtschaft – Atomtechnik, Vol. 20, No. 10, 476–482 (1975).
(20) W. Grüner, H. Kumpf: Stand der LWR-Technik in der Bundesrepublik Deutschland, Atomwirtschaft – Atomtechnik, Vol. 22, No. 4, 180–184 (1977).
(21) D. Stünke, F. Bremer, R. Ruf, F. E. Schilling: Stand und Entwicklung der Reaktordruckgefäße, Atomwirtschaft – Atomtechnik, Vol. 19, No. 11, p. 530 (1974).
(22) B. J. Baumgartl: Betriebserfahrungen und Zuverlässigkeit bei Siedewasserreaktoren, Kerntechnik, Vol. 17, No. 9/10, 415–418 (1975).
(23) G. E. Rajakovics: Höhere Kraftwerkswirkungsgrade durch neue Technologien, Atomwirtschaft – Atomtechnik, Vol. 20, No. 1, p. 24 (1975).
(24) D. H. Imhoff: The ESADA Vallecitos Experimental Superhead Reactor Power Reactor Experiments, Vol. 2, SM-21/12 IAEA, Vienna, 1962.
(25) K. Traube, L. Seyfferth: Der Heißdampfreaktor – Konstruktion und Besonderheiten, Atomwirtschaft – Atomtechnik, Vol. 14, p. 539 (1969).
(26) H. M. Carruthers: The Evolution of Magnox Station Design, J. Brit. Nucl. Energy Soc., Vol. 4, No. 3, 171–180 (1965).
(27) G. B. Greenough, J. S. Naim, P. Waine: The AGR Fuel, Nucl. Eng., Vol. 10, p. 373 (1965).
(28) K. H. Dent: The Advanced Gas-Cooled Reactor System, Int. Nuclear Industry Fair (Nuclex) No. 3/8, Basel 1966.
(29) Dungeness B, AGR Nuclear Power Station, Atom, No. 107, 168–174 (1965).
(30) W. Häfele: Schnelle Brutreaktoren, ihr Prinzip, ihre Entwicklung und ihre Rolle in der Kernenergiewirtschaft. Kernforschungszentrum Karlsruhe, KfK 480 (1966).
(31) Lane et al.: Thorium Utilization Systems, 2nd UN Conf., Genf, A/Conf. 28/P/214 (1964).
(32) Nukleare Primärenergieträger (Teil I): Energie durch Kernspaltung, GKSS-Geesthacht, STE-KFA-Jülich, AFAS-KfK-Karlsruhe, ASA-ZE/08/78, Authors: W. Jaek, D. Bünemann, H.-J. Zeck, J. Raeder, Köln 1978.
(33) H. Krämer: The gas-cooled reactor, current development status, Kerntechnik, Vol. 20, 357–361 (1978).
(34) K. Knizia, G. Hirschfelder, D. Schwarz, K. Weinzierl: Improvements at the Energy Conversion Technique (Division 3, 3.1–5), 10th World Energy Conference, Istanbul, pp. 1–22, Sept. 1977.
(35) W. Peters, R. Schulten, P. Speich: Future Availability of liquid and gaseous Hydrocarbons by Coal Gasification and the long-term Prospects for a Hydrogen Technology (Division 3, 3.4–2), 10th World Energy Conference, Istanbul, pp. 1–19, Sept. 1977.

(36) E. Teuchert: Once Through-Cycles in the Pebble Bed HTR, KFA-Jülich, Jül – 1470, December 1977.
(37) H. Krämer: Der gasgekühlte Hochtemperaturreaktor – Entwicklungsaussichten thermischer Reaktoren mit hoher Wärmenutzung, Kerntechnik, Vol. 17, No. 9/10, 399–403 (1975).
(38) D. Klein, H. Prik: Projekt eines Weiterlagers für abgebrannte Brennelemente aus Hochtemperaturreaktoren, Kerntechnik, Vol. 19, No. 4, 180–187 (1977).
(39) A. M. Angelini: The Fuel Supply (U, Pu, Th), in: Nuclear Energy Maturity, Proceedings of the European Nuclear Conference, Paris 1975, Oxford, New York: Pergamon Press 1976.
(40) A. M. Perry, A. M. Weinberg: Thermal Breeder Reactors, Annual Review of Nuclear Science, Vol. 22, 317–354 (1972).
(41) Energy Research and Development Agency, Environmental Impact Statement, Liquid Metal Cooled Fast Breeder Reactor Program, WASH – 1535, part 4.6, 1975.
(42) The Clinch River Breeder Reactor Project: A Briefing for Engineers, Proceeding of the Breeder Reactor Corporation April 1975, Information Session, PMC – 74 – 02, CONF-74 1087, USAEC Technical Information Center, Oak Ridge, Tenn., 1975.
(43) R. N. Smith, W. H. Perry, G. C. Wolz: Operating Experience with Experimental Breeder Reactor II, Joint ASME/ANS International Conference on Advanced Nuclear Energy Systems, Pittsburg, March 14–17, 1976.
(44) G. Kessler, G. L. Kulcinski: Reference Reactor Systems in: Fusion and Fast Breeder Reactors, International Institute for Applied Systems Analysis, Laxenburg, Austria, RR-77-8, pp. 121–162, July 1977.
(45) G. A. Vendreyes: Superphenix, a Full-Scale Breeder Reactor, Scientific American, Vol. 236, p. 26, March 1977.
(46) R. Miki et al.: Brief Description of Planned Prototype FBR MONJU of Japan, Fast Reactor Power Stations, Proceedings of the International Conference, London, March 1–14, 1974, British Nuclear Energy Society, 101–104, 1974.
(47) G. Vendreyes, M. Banal, J. C. Leny: L.M.F.B.R Present Status and Prospects (Division 3, 3.2–7), 10th World Energy Conference, Istanbul, pp. 1–16, Sept. 1977.
(48) F. R. Farmer: How safe is the fast reactor? Nature, Vol. 278, 593–594, 12. April 1979.
(49) D. Smidt et al.: Safety and Cost Analysis of a 1000 MW Sodium Cooled Fast Breeder Reactor. Kernforschungszentrum Karlsruhe, KfK 398 (1965).
(50) R. Krymm: A New Look at Nuclear Power Costs, IAEA Bulletin, Vol. 18, 2 (1976).
(51) J. M. Morelle, K.-W. Stöhr, J. Vogel: The Kalkar Station, Design and Safety Aspects, Nucl. Eng. Int., 43–48, July 1976.
(52) W. M. Jacobi: The Clinch River Breeder Reactor Project Nuclear Steam Supply System, Nucl. Eng. Int., 846–850, October 1974.
(53) M. Banal: Work Starts on Super Phénix at the Creys-Malville Site, Nucl. Eng. Int., 41–45, May 1977.
(54) H. H. Hennies, A. Brandstetter: Stand und Aussichten des Schnellen Brüters in der Bundesrepublik Deutschland, Atomwirtschaft – Atomtechnik, Vol. 22, No. 4, 199–202 (1977).
(55) D. Smidt: Reaktor-Sicherheitstechnik, New York, Berlin, Heidelberg: Springer 1979.
(56) N. Domberg: Can we afford to make the fast reactor safe?, Nature, Vol. 280, 270–272 (1979).
(57) W. Häfele, J. P. Holdren, G. Kessler, G. L. Kulcinski: Fusion and Fast Breeder Reactors, International Institute for Applied Systems Analysis, Laxenburg, Austria, RR-77-8, July 1977.
(58) W. Frisch et al.: System Analysis of a Fast Steam-Cooled Reactor of 1000 MW, Kernforschungszentrum Karlsruhe, KfK-636 (SM 101/10 oder EUR 3680e), 1967.
(59) C. A. Goetzmann, M. Dalle Donne: Design and Safety Studies for a 1000 MW Gas-Cooled Fast Reactor, Proc. Fast Reactor Safety Meeting, Conf. – 740401-P 2, Los Angeles 1974.
(60) G. Melese-d'Hospital, L. Meyer: Status of the US Gas-Cooled Fast Reactor Demonstration Plant Program, NUCLEX 75, International Nuclear Industries Fair, October 7–11, 1975, Basel.
(61) Fusion Technology: Proceedings of the 9th Fusion Technology, Garmisch-Partenkirchen (FGR), June 14–18, (1976), Association Euratom/IPP Garching, Commission of the European Communities, Oxford, New York: Pergamon Press 1976.
(62) W. D. Metz: Fusion Research I: What is the Program Buying the Country? Science, Vol. 192, 1320–1323 (1976); Fusion Research II: Detailed Reactor Studies Identify More Problems, Science, Vol. 193, 38–40 (1976); Fusion Research III: New Interest in Fusion-Assisted Breeders, Science, Vol. 193, 307–309 (1976).

(63) B. Badger et at.: UWMAK-I, a Wisconsin Toridal Fusion Reactor Design. University of Wisconsin, Report UWFDM-68 (Vol. I 1974, Vol. II 1975).
(64) R. G. Mills (Ed.): A fusion power plant, Princeton Plasma Physics Laboratory, MATT-1050, June 1974.
(65) B. Badger et al.: UWMAK-II, a Conceptual Tokamak Power Reactor Design. University of Wisconsin, Report UWFDM-112 (1975).
(66) Nukleare Primärenergieträger (Teil II), Energie durch Kernfusion, Max-Planck-Institut für Plasmaphysik, Garching/München, ASA-ZE/09/78, Authors: R. Bünde, W. Dänner, H. Herold, J. Reader, M. Söll, Köln 1978.
(67) J. Raeder: The status of fusion development, Kerntechnik, Vol. 19, No. 6, 253–262 (1977).
(68) R. Wienecke: Der Fusionsreaktor: Physikalische und technische Probleme, in: Plenarvorträge der 38. Physikertagung (Deutsche Physikalische Gesellschaft), Physik-Verlag: Weinheim 1974.
(69) J. D. Lawson: Some Criteria For a Power Producing Thermonuclear Reactor, Proceedings of the Physical Society, Vol. 70, 6 (1957).
(70) G. Kessler, G. L. Kulcinski: Present Status of Fission and Fusion Reactors, in: Fusion and Fast Breeder Reactors, International Institute for Applied Systems Analysis, Laxenburg, Austria, RR-77-8, pp. 61–120, July 1977.
(71) J. Nuckolls, J. Emmett, L. Wood: Laser-induced thermonuclear fusion, Physics Today, Vol. 26, 46 (1973).
(72) J. L. Emmett, J. H. Nuckolls, L. Wood: Fusion Power by Laser Implosion, Scientific American, Vol. 230, p. 24 (1974).
(73) J. P. Reilly: High-Power Lasers, Technology Review, Vol. 80, p. 57, June/July (1978).
(74) L. K. Isaacson: Laser-Fusion Program, Summary Report, EPRI-SR-9, Electric Power Research Institute, Palo Alto, Ca., 1975.
(75) M. V. Babykin: The investigation on the powerful electron beam application for thermonuclear fusion initiation, 7th Europ. Conf. on Contr. Fus. and Plasma Phys., Lausanne, Vol. 2, 172 (1975).
(76) P. A. Miller et al.: Propagation of pinched electron beams for pellet fusion, Phys. Rev. Let., Vol. 39, 92 (1977).
(77) G. Yonas: Fusion Power with Particle Beams, Scientific American, Vol. 239, 40–51, November (1978).
(78) H. Hora: New developments in laser fusion for future energy, Atomkernenergie – Kerntechnik, Vol. 34, 182–187 (1979).
(79) R. L. Hirsch: Status and Future Directions of the World Program in Fusion Research and Development, Annual Review of Nuclear Sciences, Vol. 25, 79 (1975).
(80) R. S. Pease: Towards a Controlled Nuclear Fusion Reactor, IAEA Bulletin, Vol. 20, No. 6, 9–20 December 1978.
(81) G. Gruber, R. Wilhelm: The belt pinch-a high-β tokamak with non-circular cross-section, Nuclear Fusion, Vol. 16, 243 (1976).
(82) H. P. Furth: Tokamak research, Nuclear Fusion, Vol. 15, 487 (1975).
(83) D. J. Dudziak, R. A. Krakowski: Radioactivity in a Theta Pinch Fusion Reactor, Nuclear Technology, Vol. 25, 33, 1975.
(84) R. Parker et al.: High density discharges in ALCATOR. 6th International Conferences on Plasma Physics and Controlled Nuclear Fusion, Research IAEA, Berchtesgaden 1976, Paper IAEA, CN. 35/A 5.
(85) A. B. Berlizov et at.: First results in the T 10 Tokamak 6th International Conference on Plasma Physics and Controlled Nuclear Fusion IAEA, Berchtesgaden 1976, Paper IAEA, CN. 35/A 1.
(86) M. Nozawa, D. Steiner: An assessment of the power balance in fusion reactors, Oak Ridge National Laboratory, Report ORNL-TM-4421 (1974).
(87) J. F. Clarke: Suggestions for an Updated. Fusion Power Program, ORNL/TM-5280, Oak Ridge National Laboratory, Oak Ridge, Tenn., 1976.
(88) C. C. Baker et al.: Experimental power reactor, Conceptual design study – final report for the period July 1, 1974 through June 30, 1976. General Atomic Company Report GA-1400 (1976).
(89) J. W. Davis, G. L. Kulcinski: Major Features of D-T Tokamak Fusion Reactor Systems, Nuclear Fusion, Vol. 16, 355 (1976).
(90) J. Darvas et al.: Energy balance and efficieny of power stations with a pulsed tokamak reactor, KFA Jülich, Report JÜL-1304 (1976).

(91) J. A. Phillips: Recent Developments in Nuclear Fusion (Division 4, 4.1–1), 10th World Energy Conference, Istanbul, pp. 1–14, Sept. 1977.
(92) M. Roberts et al.: Oak Ridge tokamak experimental power reactor studies – 1976. Oak Ridge National Laboratory, Report ORNL-TM-5572 (1976).
(93) W. M. Stacey et al.: Tokamak experimental power reactor conceptional design. Argonne National Laboratory, Report ANL/CIR-76-3 (1976).
(94) A. Gibson: The JET project, Die Naturwissenschaften, Vol. 66, No. 10, 481–488 (1979).
(95) B. Badger et al.: UWMAK-III, a High Performance Non Circular Tokamak Power Reactor Design. University of Wisconsin, Report UWFDM-150 (1976).
(96) A. P. Fraas: Conceptional design of the blanket and shield region and related systems for a full scale toroidal fusion reactor, Oak Ridge National Laboratory, ORNL-TM-3096 (1973).
(97) R. G. Mills (Ed.): A fusion power plant, Princeton Plasma Physics Laboratory, MATT-1050, June 1974.
(98) K. Ehrlich: First wall materials for fusion reactors, Kerntechnik, Vol. 19, No. 6, 263–267 (1977).
(99) H. Vernickel: Impurities in fusion reactor plasmas, Kerntechnik, Vol. 19, No. 6, 279–284 (1977).
(100) M. Söll: Superconducting magnet systems for fusion reactors. Kerntechnik, Vol. 19, No. 6, 272–278 (1977).
(101) P. Komarek, H. Krauth: The "Large Coil Task", an international contribution to the development of superconducting magnets for nuclear fusion, Kerntechnik, Vol. 20, 274–281 (1978).
(102) H. Brockmann, H. Clermont, J. Darvas, S. Förster, H. F. Niessen, U. Ohlig, P. Quell, B. Sack: Das Potential von Fusionsreaktoren als Prozeßwärmequelle, Brennstoff-Wärme-Kraft, Vol. 31, No. 2, 61–66, 1979.
(103) S. Förster et al.: Conceptional Design of a TOKAMAK-Reactor, Workshop on Fusion Reactor Design, October 10–21, 1977, University of Wisconsin, Madison, USA.
(104) H. Frey: Fusionsreaktor oder Schneller Brüter, VDI-Zeitschrift, Vol. 121, No. 10, 535– 541 (1979).
(105) J. Redfearn: EEC Commissioner opens European nuclear fusion centre, Nature, Vol. 279, 277–278, 24 May 1979.
(106) L. M. Lidsky: Fission-Fusion Systems: Hybrid, Symbiotic, and Augean, Nuclear Fusion, Vol. 15, 151–173 (1975).
(107) C. E. Taylor et al.: Proceedings of the US-USSR Symposium on Fusion-Fission Reactors, CONF-76-0733, Technical Information Center, Oak Ridge, Tenn., 1976.
(108) S. I. Abdel-Khalik, P. Jansen et al.: Impact of fusion-fission hybrides on world nuclear future, Atomkernenergie – Kerntechnik, Vol. 36, No. 1, 23–25 (1980).
(109) Solar Energy, UK Section of the International Solar Energy Society, The Royal Institution, Published by UK-ISES, London 1976.
(110) J. A. Duffie, W. A. Beckmann: Solar Energy-Thermal Process, New York: John Wiley & Sons, Inc. 1974.
(111) J. Fricke, W. L. Borst: Nutzung der Sonnenenergie, Physik in unserer Zeit, Vol. 10, No. 6, 183–192 (1979).
(112) D. Orth: Heat Pump Operation to utilize regenerative Energy Resources, 1st German Solar Energy Forum, Proceedings, Vol. I, Göttingen, 1977.
(113) R. Winston: Principles of Solar Concentrators of a Novel Design, Solar Energy, Vol. 16, 89- 95 (1974).
(114) World Energy: looking ahead to 2020. Report by the Conservation Commission of the World Energy Conference, Guildford (UK) and New York: IPC Science and Technology Press 1978.
(115) G. Lehner: Possibilities of utilization of new non-nuclear and non-fossil energy sources, Kerntechnik, Vol. 20, No. 4, 157–168 (1978).
(116) E. Rummich: Nichtkonventionelle Energienutzung, New York, Wien: Springer 1978.
(117) J. Schröder: Thermal Energy Storage Using Fluorides of Alkali and Alkaline Earth Metals, 148th Meeting of the Electrochemical Society Dallas, Texas, 1975.
(118) B. Anderson: Solar Energy: Fundamentals in Building Design, New York, St. Louis, San Francisco: McGraw-Hill Book Company 1977.
(119) A. B. Meinel, M. P. Meinel: Applied Solar Energy, London: Addison-Wesley Publishing Company 1977.
(120) R. H. Bezdek, A. S. Hirshberg, W. H. Babcock: Economic Feasibility of Solar Water and Space Heating, Science, Vol. 203, 1214–1220, 23 March, 1979.

(121) B. Stoy, J. E. Feustel: Solar Energy and Wind Energy-Contribution to Covering Future Energy Demand, Division 1 A, 11th World Energy Conference, Munich, pp. 587– 611, Sept. 1980.
(122) M. Clemot, B. Dessus, C. Etievant, F. Pharabod: Solar Power Plants: French Realisations and Projects (Division 4, 4.2–4), 10th World Energy Conference, Istanbul, p. 1–22, Sept. 1977.
(123) W. Grasse: Primärenergie Solarstrahlung für Kraftwerke, VDI-Nachrichten, No. 3, p. 4, 18. Januar 1980.
(124) M. Selders, D. Bonnet: Solarzellen, Physik in unserer Zeit, Vol. 10, No. 1, 3–16 (1979).
(125) R. Reisfeld: Possibilities of Solar Energy Utilization, Die Naturwissenschaften, Vol. 66, No. 1, 1–8 (1979).
(126) G. H. Hewig: Solar cells, Atomenergie – Kerntechnik, Vol. 34, 165–171 (1979).
(127) J. C. C. Fan: Solar Cells: Plugging into the Sun, Technology Review, Vol. 80, 14–36, August/September 1978.
(128) P. E. Glaser: Satellite Solar Power Station, Solar Energy, Vol. 12, 353–361 (1969).
(129) R. A. Herendeen, T. Kary, J. Rebitzer: Energy Analysis of the Solar Power Satellite, Science, Vol. 205, 451–454 (1979).
(130) A. L. Hammond: An International Partnership for Solar Power, Science, Vol. 197, p. 623 (1977).
(131) H. Kelly: Photovoltaic Power Systems: A Tour Through the Alternatives, Science, Vol. 199, 634–643 (1978).
(132) H. K. Köthe: Solargeneratoren für terrestrische Energieversorgung, Elektrotechnische Zeitschrift, Vol. 28, 396–400 (1976).
(133) US Council on Environmental Quality. Solar Energy: Progress and Promise. Washington, D.C., 1978.
(134) Application of Solar Technology to Today's Energy Needs, Office of Technology Assessment, (OTA), Washington, D.C., June 1977.
(135) T. W. Mermel: Contributions of Dams to the Solution of Energy Problems (Division 1), 10th World Energy Conference, Istanbul, pp. 1–14, Sept. 1977.
(136) H. Christaller: Wasserkraft, in: Das Energiehandbuch, 2. Aufl., G. Bischoff, W. Gocht (Ed.), Braunschweig: Vieweg 1976.
(137) BP statistical review of the world oil industry 1978, London 1979.
(138) Statistical Office of the European Communities 1980, Office for Official Publications of the European Communities, Bruxelles 1980.
(139) S. H. Salter: Energy from waves of the sea, Marine Technology, Vol. 6, 131–134, August (1975).
(140) F. Séguier: L'énergie au creux de la vague, la Recherche, Vol. 9, 502–503, Mai 1978.
(141) N. Ambli, K. Budal, J. Falnes, A. Sorenssen: Wave Power Conversion (Division 4, 4.5–2), 10th World Energy Conference, Istanbul, pp. 1–17, Sept. 1977.
(142) Zur friedlichen Nutzung der Kernenergie, Eine Dokumentation der Bundesregierung, Federal Ministry of Research and Technology (Ed.), Bonn, 1978.
(143) J. Fricke, W. L. Borst: Energie aus dem Meer, Physik in unserer Zeit, Vol. 10, 85–93 (1979).
(144) J. R. Justus: Renewable Sources of Energy from the Ocean, in: Project Interdependence: U.S. and World Energy Outlook through 1990, A Report printed by the Congressional Research Service, Library of Congress, U.S. Government Printing Office, Washington, D.C. 20402, pp. 405– 468, November 1977.
(145) M. Melíß, D. Oesterwind, A. Voß: Non-nuclear and non-fossil energy resources and their possibilities for the future power generation, Kerntechnik, Vol. 17, No. 7, 301–306 (1975).
(146) C. Zener: Solar Sea Power, Physics today, Vol. 26, 10 (1973).
(147) W. F. Whitmore: OTEC: Electricity from the Ocean, Technology Review, Vol. 81, 58–63, October 1978.
(148) L. Jarrass, L. Hoffmann, G. Obermair: Windenergie, Berlin, Heidelberg, New York: Springer 1980.
(149) Energiequellen für morgen? Federal Ministry of Research and Technology (Ed.), Frankfurt/M.: Umschau Verlag 1976.
(150) M. R. Gustavson: Limits to Wind Power Utilization, Science, Vol. 204, 13–17 (1979).
(151) A. Fritsche: The Chance of Wind Power, Die Naturwissenschaften, Vol. 68, No. 4, 157–162 (1981).
(152) M. Michel, J. Pottier, J. Jaéglé: Compression and Absorption Heat Pumps, Fields of Use and Future (Division 4, 4.6–1), 10th World Energy Conference, Istanbul, pp. 1–22, Sept. 1977.

(153) L. R. Glicksman: Heat Pumps: Off and Running ... Again, Technology Review, Vol. 80, 64–70 (1978).
(154) W. Borst: Vom sparsamen Umgang mit Energie, Physik in unserer Zeit, Vol. 8, No. 5, 131–144 (1977).
(155) D. Oesterwind: Aspects of the soft-hard discussion – Centralized and dezentralized energy systems as a common option, Atomkernenergie – Kerntechnik, Vol. 34, No. 3, 172–176 (1979).
(156) Rheinisch-Westfälisches Elektrizitätswerk, Essen, Information No. 157 (1979).
(157) J. R. Gosz, R. T. Hohnes, G. E. Likens, F. H. Bormann: The Flow of Energy in a Forest Ecosystem, Scientific American, Vol. 238, No. 3, 92–102 (1978).
(158) J. R. Bolton: Solar Fuels, Science, Vol. 202, 705–711 (1978).
(159) D. O. Hall: Plants as an energy source, Nature, Vol. 278, 114–117 (1979).
(160) E. S. Lipinsky: Fuels from Biomass: Integration with Food and Material Systems, Science, Vol. 199, 644–651 (1979).
(161) P. Böger: Utilization of solar energy by biological production of hydrogen – an approach from fundamental research, Atomkernenergie – Kerntechnik, Vol. 33, No. 1, 13–18 (1979).
(162) P. Böger: Photosynthese in globaler Sicht, Naturwissenschaftliche Rundschau, Vol. 28, No. 12, 429–435 (1975).
(163) S. C. Trindale: The Brasilian Alcohol Program, Division 4 A, 11th World Energy Conference, Munich, pp. 241–260, Sept. 1980.
(164) K.-J. Euler: Reflections regarding a future hydrogen economy, Atomkernenergie – Kerntechnik, Vol. 37, No. 1, 3–12 (1981).
(165) R. Buvet et al. (Eds.): Living Systems as Energy Converters, Amsterdam, New York, Oxford: North Holland 1977.
(166) M. Calvin: Solar Energy by Photosynthesis, Science, Vol. 184, 375–381 (1973).
(167) Ph. Abelson: Bio-Energy, Science, Vol. 204, p. 1161 (1979).
(168) V. Smil: China Claims Lead in Biogas Energy Supply, Energy International, Vol. 14, 25–27 (1977).
(169) M. R. Ladisch, K. Dyck: Dehydration of Ethanol: New Approach Gives Positive Energy Balance, Science, Vol. 205, pp. 898-900 (1979).
(170) V. Balzani, L. Moggi, M. F. Manfrin, F. Bolletta; M. Gleria: Solar Energy Conversion by Water Photodissociation, Science, Vol. 189, 852–856 (1975).
(171) N. Getoff, et al.: Wasserstoff als Energieträger, Herstellung, Lagerung, Transport, Wien, New York: Springer 1977.
(172) N. Getoff, S. Solar, M. Gohn: Solar Energy Utilization. Studies on Photoelectrochemical Systems, Die Naturwissenschaften, Vol. 67, No. 1, 7–13 (1980).
(173) J. O'M. Bockris, W. E. Justi: Wasserstoff, Energie für alle Zeiten. Konzept einer Sonnen-Wasserstoff-Wirtschaft, München: Pfriemer 1980.
(174) Ch. Garret: Tidal Resonance in the Bay of Fundy and Gulf of Maine, Nature, Vol. 238, 441–443 (1972).
(175) G. D. Duff: Tidal Power in the Bay of Fundy, Technology Review, Vol. 81, 34–42 (1978).
(176) Ch. Garret: Predicting Changes in Tidal Regime: The Open Boundary Problem, Journal of Physical Oceanography, Vol. 7, 171–181 (1977).
(177) J. S. Rinehart: Geysers and Geothermal Power Production, Die Naturwissenschaften, Vol. 63, 218–223 (1976).
(178) K. Beck: Rückgewinnung eingesetzter Energie, Brennstoff-Wärme-Kraft, Vol. 27, No. 3, pp. 100, 101 (1975).
(179) J. E. Tillmann: Eastern Geothermal Resources: Should we pursue them?, Science, Vol. 210, 595–600 (1980).
(180) J. Buter: Heat and electricity generating methods, in: K.-J. Euler, A. Scharmann (Ed.), Energy Supply of the Future, Munich: Thiemig 1977.
(181) R. Davis, J. R. Knight: Operation of GaAs Solar Cells at High Solar Flux Density, Solar Energy, Vol. 17, 145 (1975).
(182) S. Polgar: Use of Solar Generators in Africa for Broadcasting Applicanes, ISES Conf., Los Angeles, Paper 13/1, 44–45, July 1975.
(183) F. de Hoffmann: Energy storage and methods of energy transport, Die Naturwissenschaften, Vol. 64, 166–173 (1977).

(184) W. Häfele, W. Sassin: A Future Energy Scenario (Volume 1), 10th World Energy Conference, Istanbul, pp. 477–506, Sept. 1977.
(185) A. E. Sheindlin, W. D. Jackson, W. S. Brzozowski, L. H. Th. Rietjens: Magnetohydrodynamic Power Generation (Divison 3, 3.6–1), 10th World Energy Conference, Istanbul, pp. 1–21, Sept. 1977.
(186) W. Siefritz: Soviet Union paces MHD research, Energy International, Vol. 12, No. 8, 25–27 (1975).
(187) V. Rich: No crisis, but save energy, say Soviets, Nature, Vol. 277, 258 –259 (1979).
(188) C. Gary: Le transport de l'énergie électrique, La Recherche, Vol. 10, 222–231, Mars 1979.
(189) E. Jeff: Binational UHV Research Project Moves Forward, Energy International, Vol. 13, 28–29 (1976).
(190) S. Richter, E. Böhm: Power cables – an international comparison, Kerntechnik, Vol. 20, No. 11, 494–503 (1978).
(192) O. Weber: Huntorf air storage gas turbine power station, BBC-Nachrichten, Vol. 57, 401–406 (1975).
(193) E. Pestel et al.: Das Deutschland-Modell, Stuttgart: Deutsche Verlags-Anstalt 1978.
(194) H. Paschen: Consequences of Large-scale Implementation of Nuclear Energy in the Federal Republic of Germany, Part III (D. Brune, R. Coenen, F. Conrad, S. Klein, H. Paschen, H. Scheer, Economic Problems of the Large-scale Implementation of Nuclear Energy in the FR Germany), Kernforschungszentrum Karlsruhe, 1978.
(195) H. Weible: Power generation costs in coal-fired and in nuclear power plants, Kerntechnik, Vol. 20, No. 11, 504–511 (1978).
(196) C. J. Kim: Review of Nuclear Power Program in Korea (Division 3, 3.2–11), 10th World Energy Conference, Istanbul, pp. 1–20, Sept. 1977.
(197) H. Schulte: The Combined Generation of Heat and Electricity as a Means of Saving Primary Energy (Division 3, 3.3–3), 10th World Energy Conference, Istanbul, pp. 1–20, Sept. 1977.
(198) H. Gallenberger, K. Grabenstätter, T. Olles, K. Korn: Karlsruhe Nuclear Research Centre supplied with nuclear heat, Atomkernenergie – Kerntechnik, Vol. 33, No. 3, 166–169 (1979).
(199) K. Möglich: Exxon Magazin, Vol. 30, No. 2 (1978).
(200) K. Künstle, Ch. Koch, K. Reiter, H.-J. Thelen: Possibilities for the application of process steam from nuclear steam supply systems for production and treatment of liquid and gaseous prime movers, Atomkernenergie – Kerntechnik, Vol. 33, No. 3, 170–173 (1979).
(201) A. Ziegler, R. Holighaus: Technical possibilities and economic prospects for coal refining ENDEAVOUR, Vol. 3, No. 4, 150–157 (1979).
(202) F. Benthaus et al.: Rohstoff Kohle, Weinheim, New York: Verlag Chemie 1978.
(203) C. C. Ingram, O. C. Davis, B. Schlesinger: Coal Gasification and the Natural and Economic Environment, Division 3, 11th World Energy Conference, Munich, pp. 184–198, Sept. 1980.
(204) G. J. Pitt, G. R. Millward (Ed.): Coal and Modern Coal Processing, London, New York, San Francisco: Academic Press 1979.
(205) H. K. Völkel, M. J. van der Burgt: Gas from Coal, ENERGY DEVELOPMENTS, pp. 3–5, September (1978).
(206) T. Wett: U.S. synthetic fuels: Big potential, slow progress, THE OIL AND GAS JOURNAL, Vol. 76, Nov. 6, 19–23 (1978).
(207) G. Hirschfelder, K. Knizia, D. Schwarz, K. Weinzierl: Improvements at the Energy Conversion Technique (Division 3, 3.1–5), 10th World Energy Conference, Istanbul, pp. 1–22, Sept. 1977.
(208) J. Rondest: Quand le charbon est trop profond, LA RECHERCHE, Vol. 9, 498-500, Mai 1978.
(209) J. Ribesse: Hydrogasification "in situ" of Coal, a Potential Source of Synthetical Natural Gas for the Future (Division 3, 3.4–3), 10th World Energy Conference, Istanbul, pp. 1–19, Sept. 1977.
(210) N. Jenkins: Underground Gasification Offers Clean Safe Route to Coal Energy, Energy International, Vol. 14, 28–30 (1977).
(211) J. W. Hand: Picking coal deposits for in situ gasification, THE OIL AND GAS JOURNAL, Vol. 77, May, 119–121 (1979).
(212) B. A. Rahmer: Fair prospects for coal liquids, Petroleum Economist, Vol. 46, 67–68, February 1979.
(213) B. Tippee: Coal's problems to keep pressure on oil and gas, THE OIL AND GAS JOURNAL, Vol. 77, Mar 26, 47–51 (1979).

(214) Statistik der Kohlenwirtschaft e.V., Essen und Köln, September 1978 und Sept. 1980.
(215) W. L. Nelson: Coal-conversion plants cost 5 times refineries, THE OIL AND GAS JOURNAL, Vol. 77, April 16, p. 91 (1979).
(216) W. L. Nelson: What are prices of gas and oil from coal? THE OIL AND GAS JOURNAL, Vol. 77, April 23, p. 67 (1979).
(217) L. E. Swabb: Liquid Fuels from Coal: From R & D to an Industry, Science, Vol. 199, 619–622 (1978).
(218) R. S. Wishart: Industrial Energy in Transition: A Petrochemical Perspective, Science, Vol. 199, 614–618 (1978).
(219) R. Schulten, C. B. von der Decken, K. Kugeler, H. Barnet: Chemical Latent Heat for Transport of Nuclear Energy over Long Distances, The High Temperature Reactor and Process Applications, BNES Conference London, November 1974, British Nuclear Energy-Society, London, paper No. 38, 1976.
(220) Th. Bohn: Energy requirements and problems of its covering, in: K. J. Euler, A. Scharmann (Eds.), Energy Supply of the Future, Munich: Thiemig 1977.
(221) D. P. Gregory: The Hydrogen Economy, Scientific American, Vol. 234, 13–21 (1973).
(222) E. Supp, H. Jockel: Verfahren zur Herstellung von Wasserstoff, Erdöl und Kohle, Vol. 29, No. 3, 117–122 (1976).
(223) R. W. Coughlin, M. Farooque: Hydrogen production from coal, water and electrons, Nature, Vol. 279, 301–303, 24 May (1979).
(224) Auf dem Wege zu neuen Energiesystemen, Teil III, Wasserstoff und andere nicht-fossile Energieträger, Federal Ministry of Research and Technology (Ed.), Bonn, 1975.
(225) C. Keller: Wasserstoff: Energieträger mit Zukunft, Bild der Wissenschaft, Vol. 13, No. 10, 76–82 (1976).
(226) S. Kakaç, T. N. Veziroglu: Production of Hydrogen as a Means of Storing Energy (Division 3, 3.5–4), 10th World Energy Conference, Istanbul, pp. 1–20, Sept. 1977.
(227) B. Sweetmann: Hydrogen stands by for take-off, New Scientist, Vol. 82, 818–820 (1979).
(228) Shell Briefing Service, Public Affairs and Information Department, Hamburg, Juli 1979.
(229) Exxon Corporation, Public Affairs and Information Department, Hamburg 1978.
(230) D. Gwinner: Alternativ-Kraftstoffe für Straßenfahrzeuge der Zukunft, VDI-Zeitschrift, Vol. 118, No. 22, 1053–1060 (1976).
(231) R. Strobaugh, D. Yergin: Energy Future-Report of the Energy Project at the Harvard Business School, Random House, N.Y. 1979.
(232) A. B. Hart: Storing energy as acid, Nature, Vol. 277, 15–16 (1979).
(233) D. W. Murphy, P. A. Christian: Solid State Electrodes for High Energy Batteries, Science, Vol. 205, 651–656 (1979).
(234) W. Fischer, W. Haar: Die Natrium-Schwefel-Batterie, Physik in unserer Zeit, Vol. 9, No. 6, 184–191 (1978).
(235) E. S. Tucker: The future for electric vehicles, Petroleum Economist, Vol. 46, 59–60 (1979).

5. Environmental impact and safety problems

(1) P. Harnik: The ethics of energy-production and use: debate within the National Council of Churches, The Bulletin of the Atomic Scientists, Vol. 35, No. 2, pp. 5–9 (1979).
(2) H. Alfvén: Science, progress and destruction, The Bulletin of the Atomic Scientists, Vol. 35, No. 3, 68–71 (1979).
(3) H. Michaelis: Kernenergie, München: Deutscher Taschenbuch Verlag 1977.
(4) B. Bolin, E. T. Degens, S. Kempe, P. Ketner (Eds.): The global carbon cycle, SCOPE Report 13, Chichester, New York: John Wiley 1979.
(5) K. M. Meyer-Abich: Die ökologische Grenze des herkömmlichen Wirtschaftswachstums, in: H. v. Nußbaum (Ed.), Die Zukunft des Wachstums, Bertelsmann Universitätsverlag 1973.
(6) H. Flohn: Energie und Klima im 21. Jahrhundert, Bild der Wissenschaft, Vol. 12, No. 11, 83–86 (1975).

(7) Zur friedlichen Nutzung der Kernenergie; Federal Ministry of Research and Technology (Ed.), Bonn 1978.
(8) L. W. Chaney: Carbon Monoxide Automobile Emissions Measured from the Interior of a Traveling Automobile, Science, Vol. 199, 1203–1204 (1978).
(9) K. Bullrich: Atmosphärische Spurenstoffe, Die Naturwissenschaften, Vol. 63, 171–179 (1976).
(10) Air Quality and Stationary Source Emission Controls, Natural Academy of Sciences (NAS), March 1975. Prepared for the Senate Committee on Public Works, U.S. Government Printing Office, Washington, D.C.
(11) N. N.: Strahlenbelastung durch Kraftwerke, Naturwissenschaftliche Rundschau, Vol. 31, No. 10, 433–434 (1978).
(12) J. Mehl: Natural and man-made radiation exposure, Kerntechnik Vol. 21, No. 5, 221–228 (1978).
(13) P. Davids, G. Günthner, N. Hang, M. Lange: Luftreinhaltung bei Kraftwerks- und Industriefeuerung, Brennstoff-Wärme-Kraft, Vol. 31, No. 4, 158–167 (1979).
(14) F. Niehaus: The Problem of Carbon Dioxide, IAEA-Bulletin, Vol. 21, No. 1, 2–10, February 1979.
(15) C. Marchetti: Constructive solutions of the CO_2 problem, International Institute for Applied Systems Analysis IIASA, Laxenburg/Vienna, 1979.
(16) P. Böger: Ist der Sauerstoff der Luft in Gefahr? Naturwissenschaftliche Rundschau, Vol. 29, 221–223 (1976).
(17) N. N.: Costs and benefits of carbon dioxide, Nature, Vol. 279, p. 1 (1979).
(18) M. Glantz: A political view of CO_2, Nature, Vol. 280, 189–190 (1979).
(19) W. Bach, W. W. Kellogg, J. Pankrath (Eds.): Man's impact on climate, Amsterdam, Oxford, New York: Elsevier 1979.
(20) W. Bach: The Potential Consequences of Increasing CO_2 Levels in the Atmosphere, IIASA Workshop on Carbon Dioxide, Climate and Society, February 1978, International Institute for Applied Systems Analysis, Laxenburg/Vienna, 1978.
(21) D. H. Meadows et al.: The Limits to Growth, a Report for the Club of Rome's Project on the Predicament of Mankind, New York: Universe Books 1972.
(22) D. F. Baes, H. E. Groeller, J. S. Olson, R. M. Rotty: The global carbon dioxide problem, ORNL-5194, Oak Ridge National Laboratory 1976.
(23) H. W. Bernard Jr.: The Greenhouse Effect, Cambridge: Ballinger 1980.
(24) G. M. Woodwell: The Carbon Dioxide Question, Scientific American, Vol. 238, No. 1, 34–43 (1978).
(25) G. Skirrow: A surplus of carbon dioxide, Nature, Vol. 278, 121–122 (1979).
(26) E. T. Degens: Carbon in the sea, Nature, Vol. 279, 191–192 (1979).
(27) U. Hampicke: Man's impact on the earth's vegetation cover and its effects on carbon cycle and climate, in: W. Bach, J. Pankrath, W. W. Kellogg (Eds.), Man's Impact on Climate, Amsterdam, Oxford, New York: Elsevier 1979.
(28) W. Häfele: Possible Impacts of Waste Heat on Global Climate Patterns, International Institute for Applied Systems Analysis IIASA, Research Report RR-76-1, Laxenburg/Vienna.
(29) G. M. Woodwell, E. V. Peacan (Eds.): Carbon and the Biosphere, 24th Brookhaven Sympos. 16–18 May 1972, US Atomic Energy Comm. Conf. – 720501 (1973).
(30) P. M. Gootes: Carbon – 14 Time Scale Extended: Comparison of Chronologies, Science, Vol. 200, 11–15 (1978).
(31) J. Gribbin: Making a date with radio carbon, New Scientist, Vol. 82, 17 May 1979.
(32) M. Stuiver: Atmospheric Carbon Dioxide and Carbon Reservoir Changes, Vol. 199, 253–258 (1978).
(33) E. T. Sundquist, L. N. Plummer, T. M. L. Wigeley: Carbon Dioxide in the Ocean Surface: The Homogeneous Buffer Factor, Science, Vol. 204, 1203–1205 (1979).
(34) Ghen-Tung Chen, F. J. Millero: Gradual increase of oceanic CO_2, Nature, Vol. 277, 205–206 (1979).
(35) H. Oeschger, U. Siegenthaler, U. Schotterer, A. Gugelmann: A Box diffusion Model to study the Carbon Dioxide Exchange in Nature. Tellus XXVII, pp. 168–192 (1975).
(36) C. S. Wong: Atmospheric Input of Carbon Dioxide from Burning Wood, Science, Vol. 200, 197–200 (1978).
(37) J. Gribbin: Woodmann, spare that tree, New Scientist, Vol. 82, 1016–1018 (1979).

(38) S. L. Thompson, S. H. Schneider: Carbon dioxide and Climate, Nature, Vol. 290, 9–10 (1981).
(39) U. Hampicke: Das CO_2-Risiko, Umschau, Vol. 77, No. 18, 599–606 (1977).
(40) F. Niehaus: Carbon Dioxide as a Constraint for Global Energy Scenarios, in: W. Bach, J. Pankrath, W. W. Kellogg (Eds.), Man's Impact on Climate, Amsterdam, Oxford, New York: Elsevier 1979.
(41) H. Flohn: Ice Age or Warm Age?, Die Naturwissenschaften, Vol. 66, 325–330 (1979).
(42) H. Oeschger, B. Messerli, M. Svilar (Eds.): Das Klima, Berlin, Heidelberg, New York: Springer 1980.
(43) O. Haxel: Beitrag der Physik zur Klimageschichte, Die Naturwissenschaften, Vol. 63, 1, 16–22 (1976).
(44) M. Stuiver, C. J. Heusser, In Che Yang: North American Glacial History Extended to 75 000 Years ago, Science, Vol. 200, 16–21 (1978).
(45) W. F. Ruddiman, A. McIntyre: Oceanic Mechanisms for Amplification of the 23 000-Year Ice-Volume Cycle, Science, Vol. 212, pp. 617–627 (1981).
(46) W. Dansgaard: Ice core studies: dating the past to find the future, Nature, Vol. 290, p. 360 (1981).
(47) C. D. Schönwiese: Klimaschwankungen, Berlin, Heidelberg, New York: Springer Verlag, 1979.
(48) H. Flohn: Can Climate History repeat itself? Possible Climatic Warning and the Case of Paleo climatic Warm Phases, in: W. Bach, J. Pankrath, W. W. Kellogg (Eds.), Man's Impact on Climate, Amsterdam, Oxford, New York: Elsevier 1979.
(49) H. E. Landsberg (Ed. in Chief): World Survey of Climatology, Vol. 1 – Vol. 15, Amsterdam, Oxford, New York: Elsevier Publ. Co, 1977, (Vol. 1 bis Vol. 3, H. Flohn (Ed.), General Climatology).
(50) H. H. Lamb: Climate-Present, Past and Future, Vol. I (1972); Vol. II (1977) London: Methuen.
(51) Understanding Climate Change: A program for action. US Committee for GARP, National Academy of Sciences (NAS), Washington D. C. 1977.
(52) J. A. Laurmann: Market Penetration Characteristics for Energy Production and Atmospheric Carbon Dioxide Growth, Science, Vol. 205, 896–898 (1979).
(53) H. Flohn: Some Aspects of man-made climate modification and desertification, Applied Sciences and Development, Vol. 10, 44–58 (1977).
(54) K. E. Zimen: The Carbon Cycle, the Missing Sink, and Future CO_2 Levels in the Atmosphere, in: W. Bach, J. Pankrath, W. W. Kellogg (Eds.), Man's Impact on Climate, Amsterdam, Oxford, New York: Elsevier 1979.
(55) U. Siegenthaler, H. Oeschger: Predicting Future Atmospheric CO_2 Levels, Science, Vol. 199, 388–395 (1978).
(56) F. Niehaus: A non Linear eight level tandem model to calculate the future CO_2 and C-14 burden to the atmosphere, International Institute for Applied Systems Analysis-Research, Memorandum-77-54, Laxenburg, Austria, 1977.
(57) H. Lieth, J. Seeliger, G. Zimmermeyer: The CO_2-Question from Geological and Energy Economical Viewpoint, 11th World Energy Conference (Division 3), pp. 514–533, Munich, Sept. 1980.
(58) J. Williams (Ed.): Carbon Dioxide, Climate and Society, Oxford: Pergamon Press 1978.
(59) T. Augustsson, V. Rmanathan: A radiative-study of the CO_2 climate problem, Journal of Atmospheric Science, Vol. 34, 448–451 (1977).
(60) S. Manabe, R. T. Wetherald: The effects of doubling the CO_2-concentration on the climate of a general circulation model, Journal of Atmospheric Science, Vol. 32, 3–15 (1975).
(61) S. Manabe, R. J. Stouffer: A CO_2-climate sensitivity Study with a Mathematical Model of the Global Climate Nature, Vol. 282, 491–493 (1979).
(62) G. M. Woodwell et al.: The Biota and the World Carbon Budget, Science, Vol. 199, 141–146 (1978).
(63) Environmental and Societal Consequences of a Possible CO_2 – induced Climate Change, Annapolis, April 1979, Washington D.C., US Department of Energy 1980.
(64) P. Collins: World climate conference turns to the weather, Nature, Vol. 278, 3–4 (1979).
(65) J. H. Mercer: West Antarctic Ice Sheet and CO_2 Greenhouse Effect: A Threat of Disaster, Nature, Vol. 271, p. 321 (1978).
(66) R. H. Thomas, T. J. O. Sanderson, K. E. Rose: Effect of climatic warming on the West Antarctic ice sheet, Nature, Vol. 277, 355–358 (1979).
(67) J. Hahn: Man-Made Perturbation of the Nitrogen Cycle and its possible Impact on Climate, in: W. Bach, J. Pankrath, W. W. Kellogg (Eds.), Man's Impact on Climate, Amsterdam, Oxford, New York: Elsevier 1979.

(68) D. Perner: The Consequences of Increasing CFM Concentrations for Chemical Reactions in the Stratosphere and their Impact on Climate, in: W. Bach, J. Pankrath, W. W. Kellogg (Eds.), Man's Impact on Climate, Amsterdam, Oxford, New York: Elsevier 1979.
(69) W. C. Wang, Y. L. Yung, A. A. Lacis, T. Mo, J. E. Hansen: Greenhouse effects due to man-made perturbations of trace gases, Science, Vol. 194, 685–690 (1976).
(70) R. Eiden: The Influence of Trace Substances on the Atmospheric Energy Budget, in: W. Bach, J. Pankrath, W. W. Kellogg (Eds.), Man's Impact on Climate, Amsterdam, Oxford, New York: Elsevier 1979.
(71) P. Fabian: The Current State of the Art of the Ozone Problem, Die Naturwissenschaften, Vol. 67, 109–121 (1980).
(72) Umwelt, No. 76, 25. 4. 1980, Bundesminister des Innern, Federal Republic of Germany, Bonn.
(73) H. Grassel: Possible Changes of Planetary Albedo due to Aerosol Particles, in: W. Bach, J. Pankrath, W. W. Kellogg (Eds.), Man's Impact on Climate, Amsterdam, Oxford, New York: Elsevier 1979.
(74) J. Egger: The Impact of Waste Heat on the Atmospheric Circulation, in: W. Bach, J. Pankrath, W. W. Kellogg (Eds.), Man's Impact on Climate, Amsterdam, Oxford, New York: Elsevier 1979.
(75) H. Flohn: Stehen wir vor einer Klima-Katastrophe?, Umschau in Wissenschaft und Technik, No. 17, 561–569 (1977).
(76) W. Bach: Das CO_2-Problem: Lösungsmöglichkeiten durch technische Gegensteuerung, Umschau in Wissenschaft und Technik, No. 14, 117–118 (1978).
(77) A. B. Meinel, M. P. Meinel: Physics looks at solar energy, Physics Today, Vol. 25, 2, 44–50 (1972).
(78) Energiequellen für morgen? Federal Ministry of Research and Technology (Ed.), Frankfurt am Main: Umschau Verlag 1976.
(79) Essam El-Hinnawi: Energy Environment and Development (Division 4, 4.8–1), pp. 1–26, 10th World Energy Conference, Istanbul, September 1977.
(80) Proceedings of the 2nd US-Symposium on the Development and Use of Geothermal Resources, San Francisco, May 1975.
(81) A. L. Hammond, W. D. Metz, Th. H. Maugh: Energy and the future, American Association for the Advancement of Science, Washington, D.C., 1973.
(82) C. F. v. Weizsäcker: Wege in der Gefahr. Eine Studie über Wirtschaft, Gesellschaft und Kriegsverhütung, 4. Aufl., pp. 21–32, München, Wien: Carl Hanser 1977.
(83) M. Kranzberg: Technology and Human Values, DIALOGUE, Vol. 11, 21–29 (1978).
(84) K. M. Meyer-Abich (Ed.): Frieden mit der Natur, Freiburg: Herder 1979.
(85) Project Interdependence: US and World Energy outlook through 1990, A Report printed by the Congressional Research Service, U.S. Government Printing Office, Washington, D.C., November 1977.
(86) S. Villani (Ed.): Uranium Enrichment, Berlin, Heidelberg, New York: Springer 1979.
(87) A. Chamberlain: International Conference on Nuclear Power and its Fuel Cycle, IAEA, Salzburg, 2.–13. 5. 1977.
(88) D. R. Olander: The Gas Centrifuge, Scientific American, Vol. 239, No. 2, 27–33 (1978).
(89) S. Rippon: Eurodif starts production, Nuclear News, Vol. 22, 62–65 (1979).
(90) V. S. Letokhov: Laser isotope separation, Nature, Vol. 277, 605–610 (1979).
(91) Federal Ministry of Research and Technology (Ed.): Zur friedlichen Nutzung der Kernenergie, Bonn 1978.
(92) W. Stoll: Plutonium – Ziel und Problem des Kernbrennstoffkreislaufs, Chemiker Zeitung, Vol. 101, 1–8 (1977).
(93) Niedersächsische Landesregierung (Federal Republic of Germany), Dokumentation des Gorleben-Symposiums, Hannover, v. 28. 3. bis 3. 4. 1979; E. Albrecht, Regierungserklärung zum geplanten Nuklearen Entsorgungszentrum in Gorleben, Hannover, 16. 5. 1979.
(94) Comparison of Different Back End Fuel Cycle, Concepts and Evaluation of their Feasibility, Kernforschungszentrum Karlsruhe, September 1980.
(95) Int. Conf. Nucl. Power and its Fuel Cycle, IAEA-CN-36/304, Salzburg 1977.
(96) Final Generic Environmental Statement on the Use of Recycled Plutonium in Mixed Oxide Fuel in Light Water Cooled Reactors, NUREG-0002, Vol. 3, August 1976.
(97) International Nuclear Fuel Cycle Evaluation "Reprocessing, Plutonium Handling, Recycle". Report of Working Group 4. Published by the IAEA, Vienna 1980.

(98) Verordnung über den Schutz vor Schäden durch ionisierende Strahlen (Strahlenschutzverordnung – StrSchV) v. 13. 10. 1976, Bundesgesetzblatt Nr. 125, Bonn 1976.
(99) E. Merz: Endlagerformen für hochradioaktive Spaltproduktabfälle, Atomwirtschaft – Atomtechnik, Vol. 24, 409–414 (1979).
(100) T. E. Rummery, D. R. McLean: The Canadian Nuclear Fuel Waste Management Program. Paper presented to the Atomic Industrial Forum Workshop on "The Management of Spent Fuel and Radioactive Wastes", Washington D.C., September 1979.
(101) Report to the President by the Interagency Review Groups on Nuclear Waste Management, TID-29442, Washington D.C., March 1979.
(102) International Nuclear Fuel Cycle Evaluation: "Waste Management and Disposal". Report of INFCE Working Group 7. Published by the IAEA, Vienna 1980.
(103) J. Mischke: Stand der technischen Projektierung des deutschen Entsorgungszentrums, Atomwirtschaft – Atomtechnik, Vol. 23, 342–347 (1978).
(104) Nukleare Primärenergieträger (Teil I): Energie durch Kernspaltung, GKSS-Geesthacht, STE-KFA-Jülich, AFAS-KfK-Karlsruhe, ASA-ZE/08/78, Federführung: W. Jaek, Köln 1978.
(105) Consequences of Large-scale Implementation of Nuclear Energy in the Federal Republic of Germany, Part II, Kernforschungszentrum Karlsruhe, ASA/ZE-11/78.
(106) R. A. Kerr: Nuclear Waste Disposal: Alternatives to Solidification in Glass Proposed, Science, Vol. 204, 289–291 (1979).
(107) A. E. Ringwood, S. E. Kesson, N. G. Ware, W. Hibberson, A. Major: Immobilisation of high level nuclear reactor wastes im SYNROC, Nature, Vol. 278, 219–223 (1979).
(108) G. de Marsily: High level nuclear waste isolation: borosilicate glass versus crystals, Nature, Vol. 278, 210–212 (1979).
(109) W. Hunzinger: Sicherheitsaspekte der Meeresversenkung radioaktiver Abfälle, Marine Technology, Vol. 6, No. 1, 23–27 (1975).
(110) W. L. Lennemann, H. E. Parker, P. I. West: Management of Radioactive Wastes, Annals of Nuclear Energy, Vol. 3, No. 5/6, pp. 285–295 (1976).
(111) R. Lipschutz: Radioactive Waste, Politics, Technology and Risk, Cambridge: Ballinger 1980.
(112) H.-Ch. Breest, H. Holtzem: Strahlenschutz und Sicherheit bei der Entsorgung der Kerntechnik, in: Jahrbuch der Atomwirtschaft, 7. Jahrg., W. D. Müller, R. Hossner (Eds.), Düsseldorf, Frankfurt: Verlag für Wirtschaftsinformation 1976.
(113) R. W. Ramsey: Alternativlösungen in der Beseitigung radioaktiver Abfälle, in: Entsorgung der Kerntechnik, Berichte des Symposiums am 19. u. 20. 1. 1976, Mainz. Deutsches Atomforum (Ed.), Bonn.
(114) M. H. Ross, R. H. Williams: Our Energy: Regaining Control, New York: McGraw-Hill Book Co. 1981.
(115) A. A. Ribicoff: A Market-Sharing Approach of the Nuclear Sales Problem, Foreign Affairs, Vol. 54, No. 4, 763–787 (1976).
(116) A. B. Lovins, L. Hunter Lovins, L. Ross: Nuclear Power and Nuclear Bombs, Foreign Affairs, Vol. 58, No. 5, pp. 1137–1177 (1980).
(117) W. Epstein: The Proliferation of Nuclear Weapons, Scientific American, Vol. 232, No. 4, 18–33 (1975).
(118) W. Epstein: Nuclear Proliferation in the Third World, INTERNATIONAL AFFAIRS, Vol. 29, No. 2, 185–202 (1975).
(119) "Treaty on the Non-Proliferation of Nuclear Weapons", Europa-Archiv, Folge 14/1968, p.D 321 ff.
(120) International Atomic Energy Agency, Bulletin, Vol. 22, No. 3/4, August 1980.
(121) The Present Status of IAEA Safeguards on Nuclear Fuel Cycle Facilities, IAEA, Bulletin, Vol. 22, No. 3/4, 1980 and Vol. 23, No. 2, 1981.
(122) L. A. Dunn: Half Past India's Bang, Foreign Policy, No. 36, 71–89 (1979).
(123) J. R. Redick: Regional Restraint: U.S. Nuclear Policy and Latin American, Orbis, Vol. 22, No. 1, 161–200 (1978).
(124) D. J. Rose, R. K. Lester: Nuclear Power, Nuclear Weapons and International Stability, Scientific American, Vol. 238, No. 4, 45–57 (1978).
(125) P. Lellouche: International Nuclear Politics, Foreign Affairs, Vol. 58, No. 2, 336–350 (1979/80).

(126) Z. Khalilzad: Pakistan and the Bomb, Survival, Vol. 21, 244–250 (1979).
(127) J. Leite Lopes: Atoms in the developing nations, The Bulletin of the Atomic Scientists, Vol. 34, No. 4, 31–34 (1978).
(128) A. S. Krass: Laser Enrichment of Uranium: The Proliferation Connection, Science, Vol. 196, 721–731 (1977).
(129) K. H. Beckurts, A. Carnesale, R. Darendorf, B. Flowers, H. Gruhl, H. Matthöfer, D. Schmitt, B. Svenson: Unter der Wolke des Atoms, Die Zeit, Sonderdruck aus No. 3, 4, 5, Hamburg vom 21. 1. 1977.
(130) C. F. v. Weizsäcker: Wege in der Gefahr. Eine Studie über Wirtschaft, Gesellschaft und Kriegsverhütung, 4. Aufl., pp. 21–32, München, Wien: Carl Hanser 1977.
(131) T. B. Taylor: Nuclear Safeguards, Annual Review of Nuclear Science, Vol. 25, pp. 407–421 (1975).
(132) L. Weiss: Nuclear Safeguards: A congressional perspective, The Bulletin of the Atomic Scientists, Vol. 34, No. 3, 27–33 (1978).
(133) E. F. Wonder: Nuclear Commerce and Nuclear Proliferation: Germany and Brazil, Orbis, Vol. 21, No. 2, pp. 277–306 (1977).
(134) W. W. Lowrance: Nuclear Futures for Sale. To Brazil from West Germany, International Security, No. 2, pp. 147–166 (1976).
(135) J. S. Nye (Jr.): Balancing Nonproliferation and Energy Security, Technology Review, Vol. 81, 48–57 (1978).
(136) S. Keeny (Ed.): Nuclear Power Issues and Choices: Report of the Nuclear Energy Policy Study Group, Cambridge: Ballinger 1977.
(137) H. A. Feiveson: Proliferation Resistent Fuel Cycles, Annual Review of Energy, Vol. 3, 357–394 (1978).
(138) W. Bennet Lewis: New Prospects for Low-Cost Thorium Cycles, Annals of Nuclear Energy, Vol. 5, p. 297 (1978).
(139) T. Greenwood, H. A. Feiveson, T. B. Taylor: Nuclear Proliferation, New York: McGraw-Hill Book Company, 1977.
(140) F. von Hippel, R. H. Williams: On thorium cycles and proliferation, The Bulletin of the Atomic Scientists, Vol. 35, No. 5, 50–52 (1979).
(141) The Thorium Fuel Cycle, International Conference of Nuclear Power and its Fuel Cycle, Salzburg, 2.–13. 5. 1977 (IAEA – CN – 36/96).
(142) A. B. Lovins: Thorium cycles and proliferation. The Bulletin of the Atomic Scientists, Vol. 35, No. 2, 16–22 (1979).
(143) P. Fortescue, R. C. Dahlberg: Assessing the thorium cycle, ENDEAVOUR, Vol. 4, No. 1, 14–19 (1980).
(144) E. Merz: Wiederaufarbeitung thoriumhaltiger Kernbrennstoffe im Lichte proliferationssicherer Brennstoffkreisläufe, Die Naturwissenschaften, Vol. 65, 8, 424–431 (1978).
(145) H. A. Feiveson, F. von Hippel, R. H. Williams: Fission Power: An Evolutionary Strategy, Science, Vol. 203, 330–337 (1979).
(146) P. Jansen, G. Keßler: Versorgungs- und proliferationspolitische Aspekte des Schnellen Brüters, Kernforschungszentrum Karlsruhe, KfK-Nachrichten, Jahrg. 10, No. 3–4, p. 126 (1978).
(147) Nuclear Proliferation and Safeguards. Congress of the United States, Office of Technology Assessment, 1977.
(148) N. N.: Die Richtlinien des London Club, Atomwirtschaft – Atomtechnik, Vol. 23, No. 2, p. 65 (1978).
(149) J. S. Nye: Nonproliferation: A Long Term Strategy, Foreign Affairs, Vol. 56, No. 3, 601–623 (1978).
(150) International Nuclear Fuel Cycle Evaluation: "INFCE Summary Volume", Published by the IAEA, Vienna 1980.
(151) E. E. Lewis: Nuclear Power Reactor Safety, Chichester: John Wiley 1977.
(152) L. D. Hamilton, A. S. Manne: Health and Economic Costs of Alternative Energy Sources, IAEA Bulletin, Vol. 20, No. 4, 44 (1979).
(153) G. Hensener, H. Hübel, W. Roßbach: The safety of fast breeder reactors with regards to accidents involving destruction of the core, Atomkernenergie – Kerntechnik, Vol. 36, No. 4, 282–287 (1980).

(154) D. Smidt: Reaktor-Sicherheitstechnik: Sicherheitssysteme und Störfallanalyse für Leichtwasserreaktoren und Schnelle Brüter, Berlin, Heidelberg, New York: Springer Verlag 1979.
(155) E. Münch (Ed.): Tatsachen über Kernenergie, Essen: Energiewirtschaft- und -technik Verlagsgesellschaft, 1980.
(156) K. H. Beckurts, T. J. Connolly, U. Hansen, W. Jack: Nuclear Energy Future, Division 4 B, 11th World Energy Conference, Munich, pp. 239–258, Sept. 1980.
(157) Verordnung über den Schutz vor Schäden durch ionisierende Strahlen (Strahlenschutzverordnung – StrlSchV) vom 13. Oktober 1976.
(158) ICRP Publication 1: Recommendations of the Commission (1958), Oxford: Pergamon Press, 1959; ICRP Publication 6: Recommendations of the Commission (1962), Oxford: Pergamon Press, 1964; ICRP Publication 9: Recommendations of the Commission (1965), Oxford, 1966; ICRP Publication 26: Recommendations of the Commission (1977), Oxford: Pergamon Press, 1977.
(159) BEIR-Report: The Effects on Populations of Exposure to Low Levels of Ionizing Radiation. Report of the advisory committee on the biological effects of ionizing radiations. Division of medical sciences. Nat. Research Council Washington, D.C. 20006, November 1972.
(160) IAEA, Late Biological Effects of Ionizing Radiation, Vol. Ia, II, Vienna 1978.
(161) ICRP, International Commission on Radiation Protection. Heft 8: Abschätzung der Strahlenrisiken, Stuttgart, New York: Gustav Fischer 1977.
(162) K. Aurand: Kernenergie und Umwelt, Berlin: Erich Schmidt 1976.
(163) H. Schüttelkopf: Untersuchungen zur Radioökologie des J 129, Projekt Nukleare Sicherheit, 1978/2, KFK 2750, Okt. 1979.
(164) Federal Ministry of Research and Technology (Ed.), Deutsche Risikostudie – Kernkraftwerke, Bonn 1979.
(165) D. Smidt: Reaktor-Sicherheit und menschliche Unzulänglichkeit, Die Naturwissenschaften, Vol. 66, 593–600 (1979).
(166) Reactor Safety Study, Appendix II, Fault Trees. USNRC, PB-248 203, Oct. 1975.
(167) E. Marshall, L. J. Carter: The Crisis at Three Mile Island: Nuclear Risks are reconsidered, Science, Vol. 204, 152–155 (1979).
(168) E. Marshall: A preliminary Report on Three Mile Island, Science, Vol. 204, 280–281 (1979).
(169) B. T. Feld: Three Mile Island, The Bulletin of the Atomic Scientists, Vol. 35, No. 5, p. 6 (1979).
(170) R. Kirchhoff, H.-J. Linde (Eds.): Reaktorunfälle und nukleare Katastrophen – Ärztliche Versorgung Strahlengeschädigter, Erlangen: Perimed 1979.
(171) Recommendation of the International Commission of Radiological Protection, ICRP Publication No. 26, 1977.
(172) N. C. Rasmussen: Reactor Safety Study: An Assessment of Accident Risks in U.S. Commercial Nuclear Power Plants. U.S. Nuclear Regulatory Commission (Ed.), WASH-1400 (NUREG-751014), Oct. 1975.
(173) H. W. Lewis: The Safety of Fission Reactors, Scientific American, Vol. 240, 33–45, March 1980.
(174) H. W. Lewis: NUREG - CR – 0400, Risk Assessment Review Group Report to the U.S. Nuclear Regulatory Commission, Sept. 1978.
(175) W. Hanle: Kernreaktoren anstatt OPEC-Öl, Atomkernenergie – Kerntechnik, Vol. 34, 241 –242 (1979).
(176) F. R. Farmer: Risk Quantification and Acceptability, Nuclear Safety, Vol. 17, 418–421 (1976).
(177) International Meeting on Fast-Reactor Safety and Related Physics, Chicago, CONF – 761001, Oct. 5–6, 1976.
(178) J. F. Meyer et al.: An Analysis and Evaluation of the Clinch River Breeder Reactor Core Disruptive Accident Energetics, NUREG – 0122 (1977).
(179) R. Fröhlich et al.: Analyse schwerer hypothetischer Störfälle für den SNR-300, KfK 2310, Kernforschungszentrum Karlsruhe 1976.
(180) W. Häfele, J. P. Holdren, G. Kessler, G. L. Kulcinski: Fusion and Fast Breeder Reactors, International Institute for Applied Systems Analysis, Laxenburg, Austria, RR-77-8, July 1977.
(181) N. Dombey: Can we afford to make the fast reactor safe?, Nature, Vol. 280, 270–272 (1979).
(182) B. Flowers: Nuclear power, The Bulletin of the Atomic Scientists, Vol. 34, 21–57 (1978).
(183) F. R. Farmer: How safe is the fast reactor?, Nature, Vol. 278, 593–594 (1979).

(184) Nukleare Primärenergieträger (Teil I): Energie durch Kernspaltung, GKSS-Geesthacht, STE-KFA-Jülich, AFAS-KfK-Karlsruhe, ASA-ZE/08/78, Authors: W. Jaek, D. Bünemann, H.-J. Zeck, J. Raeder, Köln 1978.
(185) G. H. Lohnert, B. Craemer, A. Diekmann, H. G. Spillekothen: Ein Abschaltkonzept für große HTR, Atomwirtschaft – Atomtechnik, Vol. 24, 372–375 (1979).
(186) W. Schenk: Untersuchungen zum Verhalten von beschichteten Brennstoffteilchen und Kugelbrennelementen bei Störfalltemperaturen, Jül-1490, Mai 1980.
(187) HTGR Accident Initiation and Progression Analysis Status Report, GA-Report 13617, UC-77, January 1976.
(188) W. Kröger et al.: Sicherheitsstudie für Hochtemperaturreaktoren unter deutschen Standortbedingungen, Jül-Spez-19, August 1978.
(189) A. F. Abdul-Fattah, A. A. Husseiny: Failure analysis of loss of HTGR core auxiliary cooling system, Atomkernenergie – Kerntechnik, Vol. 34, 195–198 (1979).
(190) E. Pestel: Energie im Blick der Gesellschaft – Erwartungen und Möglichkeiten, Deutscher Ingenieurtag, Nürnberg 1979.
(191) G. Schneider: Wie überraschend treten Erdbeben auf?, Die Naturwissenschaften, Vol. 66, 65–72 (1979).
(192) E. Münch: The security of nuclear installations and fissionable materials, Atomkernenergie – Kerntechnik, Vol. 33, 229–234 (1979).
(193) H. Zünd: SVA-Fagung Sicherheit von Kernkraftwerken, Nov. 1974, Schweizerische Vereinigung für Atomenergie (SVA): Zürich 1974.
(194) E. Teller et al.: The constructive uses of nuclear explosives, New York: McGraw-Hill Book Comp. 1968.
(195) C. F. v. Weizsäcker: Wege in der Gefahr. Eine Studie über Wirtschaft, Gesellschaft und Kriegsverhütung, 4. Aufl., München, Wien: Carl Hanser 1977.
(196) C. V. Chester, R. O. Chester: Civil Defense Implications of a Pressurized Water Reactor in a Thermonuclear Target Area, Nucl. Appl. Technol., Vol. 9, 786–795 (1970).
(197) C. V. Chester, R. O. Chester: Civil Defense Implications of an LMFBR in a Thermonuclear Target Area, Nucl. Appl. Technol., Vol. 21, 190–200 (1974).
(198) C. F. v. Weizsäcker: Zwölf Thesen zur Kernwaffen-Rüstung, Die Zeit, No. 47, v. 16. Nov. 1979.
(199) Auswärtiges Amt, Federal Republic of Germany, Bonn, 13. Oct. 1977.
(200) W. Häfele, J. P. Holdren, G. Kessler, G. L. Kulcinski: Fusion and Fast Breeder Reactors, IIASA, Laxenburg/Vienna, RR-77-8, July 1977.
(201) W. Dänner: Safety and environmental impact of fusion power plants, Kerntechnik, Vol. 19, 6, 268–271 (1977).
(202) J. Darvas et al.: Energy balance and efficiency of power stations with a pulsed tokamak reactor, KFA Jülich, Report Jül-1304 (1976).
(203) R. Bünch, W. Dänner, W. Hofer, M. Hüls, P. Pöhlchen, M. Söll, E. Taglauer, H. Weichselgartner: Aspects of Energy Supply by Fusion Reactors, Max-Planck-Institut für Plasmaphysik, Garching/München, IPP V 1/1-Report März 1974.
(204) Nukleare Primärenergieträger (Teil II): Energie durch Kernfusion, Max-Planck-Institut für Plasmaphysik, Garching/München, ASA-ZE/09/78, Authors: R. Bünde, W. Dänner, H. Herold, J. Reader, M. Söll, Köln 1978.
(205) ICRP, International Commission on Radiation Protection, Publication 11: A Review of the Radiosensitivity of the Tissues in Bone, Oxford: Pergamon Press 1968.
(206) P. S. Rohwer, W. H. Wilcox: Radiological Aspects of Environmental Tritium, Nuclear Safety, Vol. 17, 216–223 (1976).
(207) J. W. Davis, G. L. Kulcinski: Assessment of Titanium for Use in the First Wall/Blanket Structure of Fusion Power Reactors, EPRI-ER-386, Electric Power Research Institute, Palo Alto, Calif., 1977.
(208) D. Steiner: The Nuclear Performance of Vanadium as a Structural Material in Fusion Reactor Blankets, ORNL-TM-4353, Oak Ridge National Laboratory, 1973.
(209) W. Dänner: Sicherheits- und Umweltaspekte bei Fusionskraftwerken, Bericht No. 16, Max-Planck-Institut für Plasmaphysik, Garching/München, Mai 1978.

6. Conclusions

(1) K.-E. Schulz (Ed.): Streitkräfte im gesellschaftlichen Wandel, Bonn: Osang 1980.
(2) C. Holden: Energy, Security, and War, Science, Vol. 211, p. 683 (1981).

7.2 Treaty on the non-proliferation of nuclear weapons (text)

The States concluding this Treaty, hereinafter referred to as the "Parties to the Treaty",

Considering the devastation that would be visited upon all mankind by a nuclear war and the consequent need to make every effort to avert the danger of such a war and to take measures to safeguard the security of peoples,

Believing that the proliferation of nuclear weapons would seriously enhance the danger of nuclear war,

In conformity with resolutions of the United Nations General Assembly calling for the conclusion of an agreement on the prevention of wider dissemination of nuclear weapons,

Undertaking to cooperate in facilitating the application of International Atomic Energy Agency safeguards on peaceful nuclear activities,

Expressing their support for research, development and other efforts to further the application, within the framework of the International Atomic Energy Agency safeguards system, of the principle of safeguarding effectively the flow of source and special fissionable materials by use of instruments and other techniques at certain strategic points,

Affirming the principle that the benefits of peaceful applications of nuclear technology, including any technological by products which may be derived by nuclear-weapon States from the development of nuclear explosive devices, should be available for peaceful purposes to all Parties to the Treaty, whether nuclear-weapon or non-nuclear-weapon States,

Convinced that in furtherance of this principle, all Parties to this Treaty are entitled to participate in the fullest possible exchange of scientific information for, and to contribute alone or in cooperation with other States to, the further development of the applications of atomic energy for peaceful purposes,

Declaring their intention to achieve at the earliest possible date the cessation of the nuclear arms race and to undertake effective measures in the direction of nuclear disarmament,

Urging the cooperation of all States in the attainment of this objective,

Recalling the determination expressed by the Parties to the 1963 Treaty banning nuclear weapon tests in the atmosphere, in outer space and under water in its Preamble to seek to achieve the discontinuance of all test explosions of nuclear weapons for all time and to continue negotiations to this end,

Desiring to further the easing of international tension and the strengthening of trust between States in order to facilitate the cessation of the manufacture of nuclear weapons, the liquidation of all their existing stockpiles, and the elimination from national arsenals of nuclear weapons and the means of their delivery pursuant to a Treaty on general and complete disarmament under strict and effective international control,

Recalling that, in accordance with the Charter of the United Nations, States must refrain in their international relations from the threat or use of force against the territorial integrity or political independence of any State, or in any other manner inconsistent with the purposes of the United Nations, and that the establishment and maintenance of international peace and security are to be promoted with the least diversion for armaments of the world's human and economic resources:

Article I

Each nuclear-weapon State Party to this Treaty undertakes not to transfer to any recipient whatsoever nuclear weapons or other nuclear explosive devices or control over such weapons or explosive devices directly, or indirectly, or indirectly; and not in any way to assist, encourage, or induce any non-nuclear-weapon State to manufacture or otherwise acquire nuclear weapons or other nuclear explosive devices, or control over such weapons or explosive devices.

Article II

Each non-nuclear-weapon State Party to this Treaty undertakes not to receive the transfer from any transferor whatsoever of nuclear weapons or other nuclear explosive devices or of control over such weapons or explosive devices directly, or indirectly; not to manufacture or otherwise acquire nuclear weapons or other nuclear explosive devices; and not to seek or receive any assistance in the manufacture of nuclear weapons or other nuclear explosive devices.

Article III

1. Each non-nuclear-weapon State Party to the Treaty undertakes to accept safeguards, as set forth in an agreement to be negotiated and concluded with the International Atomic Energy Agency in accordance with the Statute of the International Atomic Energy Agency and the Agency's safeguards system, for the exclusive purpose of verification of the fulfillment of its obligations assumed under this Treaty with a view to preventing diversion of nuclear energy from peaceful uses to nuclear weapons or other nuclear explosive devices. Procedures for the safeguards required by this Article shall be followed with respect to source or special fissionable material whether it is being produced, processed or used in any principal nuclear facility or is outside any such facility. The safeguards required by this Article shall be applied on all source or special fissionable material in all peaceful nuclear activities within the territory of such State, under its jurisdiction, or carried out under its control anywhere.

2. Each State Party to the Treaty undertakes not to provide: (a) source or special fissionable material, or (b) equipment or material especially designed or prepared for the processing, use or production of special fissionable material, to any non-nuclear-weapon State for peaceful purposes, unless the source or special fissionable material shall be subject to the safeguards required by this Article.

3. The safeguards required by this Article shall be implemented in a manner designed to comply with Article IV of this Treaty, and to avoid hampering the economic or technological development of the Parties or international cooperation in the field of peaceful nuclear activities, including the international exchange of nuclear material and equipment for the processing, use or production of nuclear material for peaceful purposes in accordance with the provisions of this Article and the principle of safeguarding set forth in the Preamble.

4. Non-nuclear-weapon States Party to the Treaty shall conclude agreements with the International Atomic Energy Agency to meet the requirements of this Article either individually or together with other States in accordance with the Statute of the International Atomic Energy Agency. Negotiation of such agreements shall commence within 180 days from the original entry into force of this Treaty. For States depositing their instruments of ratification or accession after the 180-day period, negotiation of such agreements shall commence not later than the date of such deposit. Such agreements shall enter into force not later than eighteen months after the date of initiation of negotiations.

Article IV

1. Nothing in this Treaty shall be interpreted as affecting the inalienable right of all the Parties to the Treaty to develop research, production and use of nuclear energy for peaceful purposes without discrimination and in conformity with Articles I and II of this Treaty.

2. All the Parties to the Treaty undertake to facilitate, and have the right to participate in the fullest possible exchange of equipment, materials and scientific and technological information for the peaceful uses of nuclear energy. Parties to the Treaty in a position to do so shall also cooperate in contributing alone or together with other States or international organizations to the further development of the applications of nuclear energy for peaceful purposes, especially in the territories of non-nuclear-weapon States Party to the Treaty, with due consideration for the needs of the developing areas of the world.

Article V

Each Party to this Treaty undertakes to take appropriate measures to ensure that, in accordance with this Treaty, under appropriate international observation and through appropriate international procedures, potential benefits from any peaceful applications of nuclear explosions will be made available to non-nuclear-weapon States Party to this Treaty on a non-discriminatory basis and that the charge to such Parties for the explosive devices used will be as low as possible and exclude any charge for research and development. Non-nuclear-weapon States Party to this Treaty shall be able to obtain such benefits, pursuant to a special international agreement or agreements, through an appropriate international body with adequate representation of non-nuclear-weapon States. Negotiations on this subject shall commence as soon as possible after the Treaty enters into force. Non-nuclear-weapon States Party to this Treaty so desiring may also obtain such benefits pursuant to bilateral agreements.

Article VI

Each of the Parties to this Treaty undertakes to pursue negotiations in good faith on effective measures relating to cessation of the nuclear arms race at an early date and to nuclear disarmament, and on a Treaty on general and complete disarmament under strict and effective international control.

Article VII

Nothing in this Treaty affects the right of any group of States to conclude regional treaties in order to assure the total absence of nuclear weapons in their respective tertitories.

Article VIII

1. Any Party to this Treaty may propose amendments to this Treaty. The text of any proposed amendment shall be submitted to the Depositary Governments which shall circulate it to all Parties to the Treaty. Thereupon, if requested to do so by one-third or more of the Parties to the Treaty, the Depositary Governments shall convene a conference, to which they shall invite all the Parties to the Treaty, to consider such an amendment.

2. Any amendment to this Treaty must be approved by a majority of the votes of all the Parties to the Treaty, including the votes of all nuclear-weapon States Party to this Treaty and all other Parties which, on the date the amendment is circulated, are members of the Board of Governors of the International Atomic Energy Agency. The amendment shall enter into force for each Party that deposits its instrument of ratification of the amendment upon the deposit of instruments of ratification by a majority of all the Parties, including the instruments of ratification of all nuclear-weapon States Party to this Treaty and all other Parties which, on the date the amendment is circulated, are members of the Board of Governors of the International Atomic Energy Agency. Thereafter, it shall enter into force for any other Party upon the deposit of its instrument of ratification of the amendment.

3. Five years after the entry into force of this Treaty, a conference of Parties to the Treaty shall be held in Geneva, Switzerland, in order to review the operating of this Treaty with a view to assuring that the purposes of the preamble and the provisions of the Treaty are being realized. At intervals of five years thereafter, a majority of the Parties to the Treaty may obtain, by submitting a proposal to this effect to the Depositary Governments, the convening of further conferences with the same objective of reviewing the operation of the Treaty.

Article IX

1. This Treaty shall be open to all States for signature. Any State which does not sign the Treaty before its entry into force in accordance with paragraph 3 of this Article may accede to it at any time.

2. This Treaty shall be subject to ratification by signatory States. Instruments of ratification and instruments of accession shall be deposited with the Governments of the Union of Soviet Socialist Republics, the United Kingdom of Great Britain and Northern Ireland, and the United States of America, which are hereby designated the Depositary Governments.

3. This Treaty shall enter into force after its ratification by the Depositary Governments, and 40 other States signatory to this Treaty and the deposit of their instruments of ratification. For the purposes of this

Treaty, a nuclear-weapon State is one which has manufactured and exploded a nuclear weapon or other nuclear explosive device prior to January 1, 1967.

4. For the States whose instruments of ratification or accession are deposited subsequent to the entry into force of this Treaty, it shall enter into force on the date of the deposit of their instruments of ratification or accession.

5. The Depositary Governments shall promptly inform all signatory and acceding States of the date of each signature, the date of deposit of each instrument of ratification or of accession, the date of the entry into force of this Treaty, and the date of receipt of any requests for convening a conference or other notices.

6. This Treaty shall be registered by the Depositary Governments pursuant to Article 102 of the Charter of the United Nations.

Article X

1. Each Party shall in exercising its national sovereignty have the right to withdraw from the Treaty if it decides that extraordinary events, related to the subject matter of this Treaty, have jeopardized the supreme interests of its country. It shall give notice of such withdrawal to all other Parties ot the Treaty and to the United Nations Security Council three months in advance. Such notice shall include a statement of the extraordinary events it regards as having jeopardized its supreme interests.

2. Twenty-five years after the entry into force of the Treaty, a Conference shall be convened to decide whether the Treaty shall continue in force indefinitely, or shall be extended for an additional fixed period or periods. This decision shall be taken by a majority of the Parties to the Treaty.

Article XI

This Treaty, the English, Russian, French, Spanish and Chinese texts of which are equally authentic, shall be deposited in the archives of the Depositary Governments. Duly certified copies of this Treaty shall be transmitted by the Depositary Governments to the Governments of the signatory and acceding States.

In witness whereof the undersigned, duly authorized, have signed this Treaty.

Done in ... at this ... of ...

Source: United Nations, General Assembly, A/RES. 2373 (XXII), 12. Juni 1968.

7.3 Abbreviations

Ac	Actinium
Ag	Silver (Argentum)
AGR	Advanced Gas Reactor
Al	Aluminium
ALCATOR	Experimental TOKAMAK Reactor (USA)
Am	Americium
ANL	Argonne National Laboratory (USA)
Ar	Argon
As	Arsenic
At	Astatine
Au	Gold (Aurum)
AVR	Arbeitsgemeinschaft Versuchsreaktor GmbH, Düsseldorf, Federal Republic of Germany
B	Boron
Ba	Barium

Be	Beryllium
BHP	Biological Hazard Potential
Bi	Bismuth
Bk	Berkelium
BN-350	Fast breeder prototype reactor (USSR)
BN-600	Advanced fast breeder prototype reactor (USSR)
BNL	Brookhaven National Laboratory (USA)
BOR-60	A Soviet experimental fast test reactor
Br	Bromine
BR-1, BR-2, BR-5	Soviet experimental fast reactors
BWR	Boiling Water Reactor
C	Carbon
Ca	Calcium
CANDU	Canadian Deuterium Uranium
CCMS	Committee on the Challenges of Modern Society
Cd	Cadmium
Ce	Cerium
CEA	Commisseriat à l'Energie Atomique
Cf	Californium
CFR-1	Commercial Fast Reactor One (UK)
Cl	Chlorine
CLEMENTINE	First Pu-fueled fast research reactor (USA)
Cm	Curium
Co	Cobalt
COMECON	Council for Mutual Economic Aid
Cr	Chromium
CRBR	Clinch River Breeder Reactor, US fast breeder demonstration reactor
Cs	Cesium
CTR	Controlled Thermonuclear Reactor
Cu	Copper (Cuprum)
D	Deuterium
DFR	Dounreay Fast Reactor, fast breeder prototype reactor (UK)
DOE	Department of Energy (USA)
Dy	Dysprosium
EBR-I, II	Experimental Breeder Reactors (USA)
EC	European Community (Common Market)
EPA	Environmental Protection Agency (USA)
EPR	Experimental Power Reactor (USA)
ERDA	Energy Research and Development Adminstration (USA)
Er	Erbium
Es	Einsteinium
Eu	Europium
EURATOM	European Atomic Energy Community
F	Fluorine
FAO	Food and Agriculture Organization
FBR	Fast Breeder Reactor
Fe	Iron (Ferrum)
FFTF	Fast Flux Test Facility
FLIBE	Fluorine-Lithium-Beryllium (a liquid salt used in fusion reactors)
Fm	Fermium
FR	Fusion Reactor
FRG	Federal Republic of Germany
Fr	Francium

Ga	Gallium
Gd	Gadolinium
Ge	Germanium
GFBR	Gas-cooled Fast Breeder Reactor
GGR	Gas-cooled Graphite Reactor
GNP	Gross National Product
GWP	Gross World Product
H	Hydrogen
HAW	High Active Waste
He	Helium
Hf	Hafnium
Hg	Mercury (Hydrargyrum)
Ho	Holmium
HTGR (or HTR)	High Temperature Gas Reactor
HWR	Heavy Water Reactor
IAEA (or IAEO)	International Atomic Energy Agency
IAEO	International Atomic Engery Organization
ICPR	International Commission on Radiological Protection
ICRU	International Commission on Radiological Units (and Measurements)
IEA	International Energy Agency
IIASA	International Institute for Applied Systems Analysis
INFCE	International Nuclear Fuel Cycle Evaluation
In	Indium
I	Iodine
Ir	Iridium
JET	Joint European Torus
JET-60	Experimental TOKAMAK test facility (Japan)
JOYO	Fast experimental test reactor (Japan)
K	Potassium (Kalium)
KEWA	Kernbrennstoff-Wiederaufbereitungsgesellschaft
KNK II	Sodium-cooled fast test reactor with thermal driver zone (Federal Rep. of Germany)
Kr	Krypton
La	Lanthanum
LASL	Los Alamos Scientific Laboratory (USA)
LAW	Low Active Waste
LH_2	Liquid Hydrogen
Li	Lithium
LLL	Lawrence Livermore Laboratory (USA)
LMFBR	Liquid Metal Fast Breeder Reactor
LNG	Liquefied Natural Gas
LOCA	Loss of Coolant Accident
LWR	Light Water Reactor
Lu	Luthenium
MAW	Medium Active Waste
Md	Mendelevium
Mg	Magnesium
MHD	Magneto Hydro Dynamic
Mirror	Type of magnetic confinement for high temperature plasma in fusion reactors
Mn	Manganese
Mo	Molybdenum

MONJU	Fast breeder prototype reactor (USA)
MPC	Maximum Permissible Concentration
N	Nitrogen
n,n'	Neutron
Na	Sodium (Natrium)
NATO	North Atlantic Treaty Organization
Nb	Niobium
Nd	Neodymium
Ne	Neon
NEA	Nuclear Energy Agency
Ni	Nickel
No	Nobelium
Np	Neptunium
NPT (or TNP)	Non-Proliferation Treaty
O	Oxygen
OAPEC	Organization of Arab Petroleum Exporting Countries
OECD	Organization for Economic Cooperation and Development
OPEC	Organization for Petroleum Exporting Countries
Os	Osmium
ORMAC	Experimental TOKAMAK test reactor (USA)
ORNL	Oak Ridge National Laboratory (USA)
p	Proton
P	Phosphorus
Pa	Protactinium
Pb	Lead (Plumbum)
PEC	Primary Energy Consumption
PEC	Prova Elementi Combustibile, fast experimental test reactor (Italy)
Pd	Palladium
PFR	Prototype Fast Reactor (UK)
PHENIX	Fast breeder demonstration reactor (France)
PLT	Princeton Large Torus, experimental TOKAMAK test facility (USA)
Pm	Promethium
PNE	Peaceful Nuclear Explosion
Po	Polonium
PPPL	Princeton Plasma Physics Laboratory (USA)
Pr	Praseodymium
Pt	Platinum
PTR	Pressure Tube Reactor
Pu	Plutonium
PUREX	Plutonium and Uranium Recovery by Extraction
PWR	Pressurized Water Reactor
q	quad (1 q = 10^{15} Btu)
Q	Quality factor
Ra	Radium
RAPSODIE	Experimental fast test reactor (France)
Rb	Rubidium
Re	Rhenium
Rh	Rhodium
Rn	Radon
Ru	Ruthenium
S	Sulfur
Sb	Antimony (Stibium)

Sc	Scandium
SCYLLAC	Toroidal theta pinch experiment (USA)
Se	Selenium
SEFOR	Southwest experimental fast oxide reactor (USA)
SFBR	Sodium Fast Breeder Reactor
SGHWR	Steam Generating Heavy Water Reactor
Si	Silicon
Sm	Samarium
Sn	Tin (Stannum)
SNG	Substitute Natural Gas
SNR-300	Schneller Natriumgekühlter Reaktor, Federal Rep. of Germany
Sr	Strontium
SS	Stainless Steel
SUPER-PHENIX	Fast breeder reactor power plant following upon PHENIX in the French breeder reactor program
T	Half-life
T	Tritium
Ta	Tantalum
Tb	Terbium
Tc	Technetium
Te	Tellurium
TETR	TOKAMAK engineering test reactor (USA)
TFR	Toroidal Fusion Reactor (USA)
Th	Thorium
THETA PINCH	Type of magnetic confinement for fusion plasmas
THOREX	Thorium Recovery by Extraction
THTR (or THTGR)	Thorium High Temperature Reactor
Ti	Titanium
Tl	Thallium
Tm	Thulium
TNP (or NPT)	Treaty on the Non-Proliferation
TOKAMAK	Russian: toroidal chamber machine, fusion reactor concept with a toroidal plasma configuration
U	Uranium
UK	United Kingdom
UKAEA	United Kingdom Atomic Energy Authority
UN	United Nations
USA	United States of America
USAEC	United States Atomic Energy Commission
USSR	Union of Soviet Socialist Republics
UWMAK I, II, III	University of Wisconsin conceptual designs of commercial TOKAMAK reactors
V	Vanadium
W	Tungsten (Wolfram)
WAK	Wiederaufarbeitungsanlage Karlsruhe (Fed. Rep. of Germany)
Xe	Xenon
Y	Yttrium
Ye	Ytterbium
Zn	Zinc
Zr	Zirconium

α	Alpha
β	Beta
γ	Gamma
δ, Δ	Delta
ε	Epsilon
η	Eta
λ	Lambda (Wave length)
μ	My
ν	Ny
σ, Σ	Sigma
τ	Tau

7.4 Units

a	Year
atm	Atmosphere
bar	1 bar = $10^5 N/m^2$
barn	1 barn $\triangleq 10^{-24}$ cm^2
bbl	Barrel
Btu	British thermal unit
c	Centi $\triangleq 10^{-2}$
cal	Calorie
°C	Degree Celsius
Ci	Curie
cm	Centimeter
cm^2	Square Centimeter
cm^3	Cubic Centimeter
cts	Cents
d	Day
$	US-Dollar
E	Exa $\triangleq 10^{18}$
EJ	Exajoule (1 EJ = 10^{18} J)
eV	Electron Volt
°F	Degree Fahrenheit
g	Gram
G	Giga $\triangleq 10^9$
G	Gauss, magnetic induction (1 G $\triangleq 10^4$ Vs/m^2)
GJ	Gigajoule (1 GJ = 10^9 J)
Gt	Gigaton (1 Gt = 10^9 t)
Gtce	Gigatons of coal equivalent
GW	Gigawatt (1 GW = 10^9 W)
h	Hour
J	Joule (1 J = 1 Ws)

K	Kelvin
k	Kilo ≙ 10^3
kcal	Kilocalorie (1 kcal = 10^3 cal)
keV	Kiloelectronvolt (1 keV = 10^3 eV)
kg	Kilogram (1 kg = 10^3 g)
kgce	Kilograms of coal equivalent
kJ	Kilojoule (1 kJ = 10^3 J)
km	Kilometer
km^2	Square Kilometer
km^3	Cubic Kilometer
kp	Kilopond (1 kp = 10^3 p)
kt	Kiloton (1 kt = 10^3 t)
kW	Kilowatt (1 kW = 10^3 W)
kWh	Kilowatthour
l	Litre
lb	Pound (1 lb ≙ 454 g)
lm	Lumen
m	Milli ≙ 10^{-3}
m	Meter
m^2	Square Meter
m^3	Cubic Meter
M	Mega ≙ 10^9
MeV	Megaelectronvolt
Min	Minute
mm	Millimeter
mol	Quantity of a substance whose weight is equal to the formula mass (in g)
mrem	Millirem (see rem)
Mt	Megaton (1 Mt = 10^6 t)
MW	Megawatt (1 MW = 10^6 W)
μ	Micro ≙ 10^{-6}
μg	Microgram (1 μg = 10^{-6} g)
N	Newton
n	Nano ≙ 10^{-9}
ns	Nanosecond (1 ns = 10^{-9} s)
Nm^3	Norm cubic meter (O°C; 1.01325 bar)
p	Pond
ppm	Part per million
%	Per cent
q	quad (1 q = 10^{15} Btu)
rad	Radiation absorbed dose
rem	Radiation equivalent man
s	Second
sh. t.	Short ton (1 sh. t. = 907.185 kg)
T	Temperature
T	Tesla, magnetic flux density (1 T ≙ 1 Vs/m^2)
T	Tera ≙ 10^{12}
t	Ton
tce	Tons of coal equivalent

TNT	Trinitrotoluol
toe	Tons of oil equivalent
TW	Terawatt (1 TW = 10^{12} W)
V	Volt
W	Watt
y	Year

7.5 Conversion table

Energy

1 tce = $29.3 \cdot 10^9$ J
1 toe = $42.2 \cdot 10^9$ J
1 Btu = $1.06 \cdot 10^3$ J
1 kcal = $4.19 \cdot 10^3$ J
1 q = 10^{15} Btu
1 q = $3.62 \cdot 10^7$ tce
1 J = 1 Nm = 1 Ws
1 J = $2.7778 \cdot 10^{-7}$ kWh
1 eV = $1.6022 \cdot 10^{-19}$ J
1 toe = 1.44 tce
1 m³ "average" natural gas ≙ $39.4 \cdot 10^6$ J
1000 m³ "average" natural gas ≙ 1.34 tce
1 t brown coal ≙ 0.3 – 0.78 tce
1 t TNT ≙ $4.2 \cdot 10^9$ J

Length

1 mile = 1609 m
1 sm = 1852 m
1 ft = 0.3048 m
1 inch = 0.0254 m

Area

1 acre = 4046 m^2

Volume

1 l = 0.001 m^3
1 bbl = 0.1588 m^3
1 gal = 0.00378 m^3
1 cft = 0.0283 m^3

Mass

1 t	= 1 000 kg	
1 lb	= 0.454 kg	
1 sh.t.	= 907.185 kg	
1000 kg U_3O_8		≙ 848 kg U
1 lb U_3O_8		≙ 0.38 kg U
1 t crude oil (North Sea)		≙ 7.5 bbl crude oil

Pressure

1 bar = 10^5 N/m^3 = 10^5 Pa
1 atm = 1.013 bar = 101 325 Pa

Activity (of radionuclides)

1 Ci ≙ $3.7 \cdot 10^{10}$ decays/s = $3.7 \cdot 10^{10}$ Bq

Energy Dose (Radiation absorbed Dose) D

1 Gy = 1 J/kg = 100 rad

Equivalent Dose H

H [rem] = Q · D [rad] (Q: Quality factor)

Temperature

$$t\,[^\circ C] = T\,[K] - 273.16$$

$$t\,[^\circ C] = \frac{t\,[^\circ F] - 32^\circ}{1.8}$$

Metric multipliers

Symbol	Prefix	Power
E	Exa	10^{18}
P	Peta	10^{15}
T	Tera	10^{12}
G	Giga	10^9
M	Mega	10^6
k	Kilo	10^3
h	Hecto	10^2
da	Deca	$10^1 = 10$
d	Deci	10^{-1}
c	Centi	10^{-2}
m	Milli	10^{-3}
μ	Micro	10^{-6}

n	Nano	10^{-9}
p	Pico	10^{-12}
f	Femto	10^{-15}
a	Atto	10^{-18}

7.6 Author index

7.7 Index

Walter de Gruyter Berlin · New York

Alexandersson Klevebring

World Resources

Energy Metals Minerals
Studies in Economic and Political Geography

By Gunnar Alexandersson, Professor of International Economic Geography at the Stockholm School of Economics, and Dr. Björn-Ivar Klevebring, Technologist and Metallurgist.

Editor: Wolf Tietze

1978. 17 cm x 24 cm. VII? 248 pages.
With 43 figures and 54 tables. Bound DM 36,–;
approx. US $18.00
ISBN 3 11 006577 0 (GeoSpectrum)

A series of dramatic events in recent years have focused public attention as never before on our supply of energy and minerals. Raw materials more than ever have become pawns in the political game between nations and blocs of nations. The authors provide an economic, geographic, political and technical background to the new raw-material-centered international situation. Maps, tables and texts provide an easily accessible geographic analysis and diagrams give the long historic perspective on production in the world and in some major producing countries.